New Perspectives: Toxicology and the Environment

Target Organ Toxicity in Marine and Freshwater Teleosts

Volume 1 – Organs

T0203903

Edited by

Daniel Schlenk
Department of Environmental Sciences
University of California
Riverside, California
USA

William H. Benson
Gulf Ecology Division, National Health and Environmental Effects
Research Laboratory
US Environmental Protection Agency
Gulf Breeze, Florida
USA

CRC Press
Taylor & Francis Group
Boca Raton London New York

CRC Press is an imprint of the
Taylor & Francis Group, an **informa** business
A TAYLOR & FRANCIS BOOK

First published 2001 by Taylor & Francis

Published 2019 by CRC Press
Taylor & Francis Group
6000 Broken Sound Parkway NW, Suite 300
Boca Raton, FL 33487-2742

First issued in paperback 2019

No claim to original U.S. Government works

ISBN 13: 978-0-367-45530-9 (pbk)
ISBN 13: 978-0-415-24838-9 (hbk)

Visit the Taylor & Francis Web site at
http://www.taylorandfrancis.com

and the CRC Press Web site at
http://www.crcpress.com

Typeset in Times by
Prepress Projects Ltd, Perth, Scotland

British Library Cataloguing in Publication Data
A catalogue record for this book is available
from the British Library

Library of Congress Cataloging in Publication Data

Target Organ Toxicity in Marine and Freshwater Teleosts

Contents

List of contributors   vii
Foreword   viii
Copyright acknowledgments   ix

1   **Toxic responses of the gill**   1
    CHRIS M. WOOD

    Introduction 1
    Functional anatomy 1
    Basic physiology 13
    General responses to toxicants 22
    Specific responses to selected toxicants 31
    Concluding remarks 64

2   **Target organ toxicity in the kidney**   90
    BODIL K. LARSEN AND EVERETT J. PERKINS JR

    Introduction 90
    Functional morphology of the teleost kidney 91
    Overall kidney function 98
    Urine formation 100
    Control of kidney function 117
    Concluding remarks on teleost kidney physiology 124
    Nephrotoxicity 126
    Concluding remarks on nephrotoxicity 139

3   **Toxic responses of the skin**   151
    JAMES M. MCKIM AND GREGORY J. LIEN

    Introduction 151
    Structure–function of fish skin 153
    Skin absorption of xenobiotics 166

Toxic responses of the skin 182
Research needs 191

4 Toxic responses of the liver   224
DAVID E. HINTON, HELMUT SEGNER, AND THOMAS BRAUNBECK

Introduction 224
Important aspects of fish liver anatomy related to interpretation
   of toxicity 225
Important aspects of fish liver physiology related to interpretation
   of toxicity 231
Studies with reference hepatotoxicants in fish 236
Hepatocellular adaptation 241
Hepatobiliary system 243
In vitro *model systems 247*
Chronic toxicity – liver carcinogenesis studies 251
Information gaps and future directions 254

5 Response of the teleost gastrointestinal system to xenobiotics   269
KEVIN M. KLEINOW AND MARGARET O. JAMES

Introduction 269
Structure of the gastrointestinal tract 270
Gastrointestinal tract physiology 281
Xenobiotic absorption and disposition 290
Elimination 315
Gastrointestinal tract toxicity 317
Summary 339

Index   363

Contributors

Thomas Braunbeck is at the Department of Zoology I, University of Heidelberg, Germany

David E. Hinton is at the Nicholas School of the Environment, Duke University, Durham, NC, USA

Margaret O. James is at the Department of Medicinal Chemistry, University of Florida, FL, USA

Kevin M. Kleinow is at the Department of Comparative Biomedical Sciences, School of Veterinary Medicine, Louisiana State University, Baton Rouge, LA, USA

Bodil K. Larsen is at the Department of Pharmacology, University of Mississippi, MS, USA

Gregory J. Lien is at the Mid-continent Ecology Division, National Health and Environmental Effects Laboratory, US Environmental Protection Agency, Duluth, MN, USA

James M. McKim is at the Mid-continent Ecology Division, National Health and Environmental Effects Laboratory, US Environmental Protection Agency, Duluth, MN, USA

Everett J. Perkins Jr is at Eli Lilly and Company, Drug Disposition, Indianapolis, IN, USA

Helmut Segner is at the Sektion Chemische Okotoxikologie, UFZ-Umweltforschungszentrum Leipzig-Halle GmbH, Leipzig, Germany

Chris M. Wood is at the Department of Biology, McMaster University, West Hamilton, ON, Canada

Foreword

This book provides a timely reminder of how far the field of aquatic toxicology, particularly with respect to fishes, has progressed in recent years. This field emerged only two to three decades ago, largely in response to the need for scientifically valid data upon which to base water quality regulations or criteria. To a considerable degree, therefore, the field became synonymous with acute and chronic toxicity testing that focused upon whole-organism effects. And while these remain important activities, many scientists perceived the importance of elucidating mechanisms underlying interactions of chemicals with aquatic organisms, including fishes. Such studies contribute to our basic understanding of toxicological phenomena in these organisms, but also improve our ability to establish regulations with confidence and provide useful tools for monitoring ('biomarkers'). Thus, elegant research is now being performed that is identifying mechanisms of chemical metabolism, toxicity and adaptation in fish models. These studies require an understanding of specific organ systems that chemicals interact with, and cellular and subcellular components of target organs in these models.

Thus, this book provides an immensely useful reference text for scientists pursuing mechanistic aspects of toxicology in teleost models. Each chapter, written by eminent scientists in their field, provides a detailed synopsis of the state of the science regarding our understanding of chemical interactions in a particular organ system. This information constitutes a solid foundation and springboard for students and researchers to extend our understanding of mechanistic toxicology in marine and freshwater teleosts. Such efforts will foster both the maturation of this field and the wise management of chemical releases into aquatic ecosystems.

Richard T. Di Giulio
Duke University

Copyright acknowledgments

The authors and publishers would like to thank the following for granting permission to reproduce material in this work:

John Wiley & Sons, Inc., New York, for permission to reproduce Figure 1.4, originally published as Figure 16 in Pisam, M. *et al.* 1987. Two types of chloride cells in the gill epithelium of a freshwater-adapted euryhaline fish: *Lebistes reticulatus*; their modifications during adaptation to saltwater. *American Journal of Anatomy* 179: 40–50.

NRC Research Press, Ottawa, for permission to reproduce Figure 1.8, originally published as Figure 8 in Mallatt, J. 1985. Fish gill structural changes induced by toxicants and other irritants: a statistical review. *Canadian Journal of Fisheries and Aquatic Sciences* 42: 630–648.

Springer-Verlag, Heidelberg, for permission to reproduce Figure 1.9, originally published as Figures 1, 9 and 10 in Wendelaar Bonga, S.E. *et al.* 1990. The ultrastructure of chloride cells in the gills of the teleost *Oreochromis mossambicus* during exposure to acidified water. *Cell and Tissue Research* 259: 575–585.

Springer-Verlag, Heidelberg, for permission to reproduce Figure 1.10, originally published as Figure 4a and b in Greco A.M. *et al.* 1996. The effects of soft-water acclimation on gill structure in the rainbow trout *Oncorhynchus mykiss*. *Cell and Tissue Research* 285: 75–82.

Springer-Verlag, Heidelberg, for permission to reproduce Figure 1.14, originally published as Figure 7 in Playle, R.C. and Wood, C.M. 1989. Water pH and aluminum chemistry in the gill micro-environment of rainbow trout during acid and aluminum exposures. *Journal of Comparative Physiology B* 159: 539–550.

Academic Press for permission to reproduce Figures 3.1, 3.2 and 3.11, originally published as Figures 2, 3 and 7 in McKim *et al.* 1996. Dermal absorption of three waterborne chloroethanes in rainbow trout (*Oncorhynchus mykiss*) and channel catfish (*Ictalurus punctatus*). *Fundamental and Applied Toxicology* 31: 218–228.

Elsevier Science, Oxford, for permission to reproduce Figure 3.6, originally published as Figure 1A in *Comparative Biochemistry and Physiology* (1993) 105A: 625–641.

Elsevier Science, Oxford, for permission to reproduce Figure 3.8, originally published as Figure 3 in *Respiration Physiology* (1978) 35: 111–118.

Academic Press for permission to reproduce Figure 3.9, originally published as Figure 1 in *General and Comparative Endocrinology* 66: 415–424.

Elsevier Science, Oxford, for permission to reproduce Figures 3.10 and 3.13, originally published as Figures 3 and 4a in *Aquatic Toxicology* (1993) 27: 15–32.

The Society of Environmental Toxicology and Chemistry for permission to reproduce Figure 3.12, originally published as Figure 1 in *Environmental Toxicology and Chemistry* (1994) 13: 1195–1205.

Academic Press for permission to reproduce Figure 3.14, originally published as Figure 1 in *Fundamental and Applied Toxicology* 31: 229–242.

University of Wisconsin Press, Madison, WI, for permission to reproduce Figure 3.15, originally published as Figure 34.1 in Ribelin, W.E. and Migaki, G. 1975. *The Pathology of Fishes*.

Elsevier Science, Oxford, for permission to reproduce Figure 3.16, originally published as Figure 14 in *Aquatic Toxicology* (1988) 11: 241–257.

Battele Press, Columbus, OH, for permission to reproduce Figures 3.17 and 3.18, originally published as Figures 1A and B and 2A and B in Black, J.J. 1982. *Polynuclear Aromatic Hydrocarbons: Formation, Metabolism and Measurement.* Cooke, M.W. and Dennis, A.J. (eds), pp. 99–111.

1 Toxic responses of the gill

Chris M. Wood

Introduction

The gills of a fish constitute a multifunctional organ (respiration, ionoregulation, acid–base regulation, nitrogenous waste excretion) accounting for well over 50 percent of the total surface area of the animal. The branchial epithelium is made up of multiple cell types and is both delicate and geometrically complex. Because of the low oxygen capacitance of water, the gills are irrigated externally with a massive ventilatory flow, in the order of 20 L kg^{-1} h^{-1}, whereas internally they are perfused with the entire cardiac output, approximately 2 L kg^{-1} h^{-1}. The diffusion distance between water and blood is extremely thin (0.5–10 µm), and the effectiveness of diffusive exchange is maximized by the countercurrent flow of water versus blood at the exchange sites. Given these facts, it is not surprising that that the gills are not only the major site of uptake for most waterborne toxicants but also the first, and most important, site of toxic impact for many of them. All four of the major physiologic processes performed by the gills are essential for life and are sensitive to both structural and biochemical disturbance of the branchial epithelium. Interruption of any of them will result in death, whereas sublethal disturbances will chronically depress the fitness of the fish.

Functional anatomy

External anatomy

The basic anatomy of the gills is well described in the older literature (e.g. van Dam, 1938; Hughes, 1966, 1984; Hughes and Morgan, 1973; Wood, 1974; Laurent and Dunel, 1980; Laurent, 1984; Satchell, 1984). Typically, teleost fish have four gill arches (Figure 1.1A), each bearing two hemibranchs comprising up to several hundred gill filaments (Figure 1.1B; primary lamellae in the older literature). The gill filaments are long finger-like processes (e.g. 5–15 mm) which may be seen with the naked eye. The internal cartilaginous skeleton of the filaments and the filament abductor and adductor muscles ensure that, during normal ventilation, the tips of adjacent filaments are held in close apposition, forming a more or less complete curtain of water channels (see below) through which the water flow is pumped. When ventilatory volume increases (exercise, hypoxia, stress), the curtain

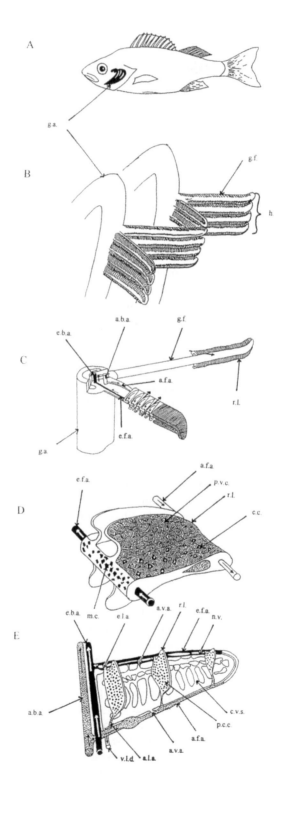

A

B

g.a.

g.f.

h

C

a.b.a.

e.b.a.

g.f.

a.f.a.

r.l.

e.f.a.

g.a.

D

e.f.a.

a.f.a.

p.v.c.

r.l.

c.c.

e.b.a. m.c. e.l.a.

a.v.a. r.l.

e.f.a. n.v.

E

a.b.a.

c.v.s.

p.c.c.

a.f.a.

v.l.d. a.l.a. a.v.a.

may be pulled apart, so that some of the water flow bypasses the channels and respiratory gas exchange per unit water flow becomes less efficient. Because of the dense, viscous nature of water, the cost of ventilation is high even at rest (most estimates are in the range of 10 percent of resting metabolic rate) and increases greatly during hyperventilation (Perry and McDonald, 1993).

On each surface of a filament, there are rows of regularly spaced, leaf-like structures, the respiratory lamellae (or secondary lamellae; Figure 1.1C) which are the actual sites of gas exchange, analogous to the alveoli of the human lung. The respiratory lamellae are generally quite thin (10–25 μm), but vary greatly among species in both spacing (20–100 μm) and shape (oblong to triangular). Typical lamellar heights and lengths are 100–500 μm and 500–1500 μm respectively; the leading edge is generally the highest point of the structure (Figure 1.1D). The spaces between the respiratory lamellae constitute the water channels through which the ventilatory flow occurs in a direction opposite to that of the internal blood flow. In general, more active, pelagic-type species have more filaments, more closely spaced lamellae, more numerous but smaller individual lamellae, and greater total lamellar surface area. In a 1-kg fish, the total number of secondary lamellae might range from 0.5 million in the relatively inactive toadfish to about 6 million in highly active tuna, with corresponding total lamellar surface areas of 1300–13 000 cm². Lamellar surface area increases and the mean blood-to-water diffusion distance decreases in euryhaline fish after transfer to seawater (Laurent and Hebibi, 1989). Total lamellar area is normally taken as the area available for gas exchange, although it should be realized that water-to-blood diffusion distance will be quite heterogeneous within different parts of individual lamellae because of their complex internal structure.

Internal anatomy

Internally, the respiratory lamellae are composed of a web-like anastomosis of blood channels (pillar cell channels; Figure 1.1E) separated and indeed formed

Figure 1.1 Gross and fine anatomy of the gills of a typical teleost fish. Drawing courtesy of S. Wood. (A) Location of four gill arches (*g.a.*). (B) Detail of portions of two gill arches each bearing two hemibranchs (*h.*) composed of many gill filaments (*g.f.*) in close apposition. (C) Detail of a portion of one gill arch with two attached gill filaments, each bearing two rows of respiratory lamellae (*r.l.*), illustrating organization of afferent branchial artery (*a.b.a.*), efferent branchial artery (*e.b.a.*), afferent filamental artery (*a.f.a.*), and efferent filamental artery (*e.f.a.*); arrows indicate the directions of blood and water flow. (D) External detail of epithelium of respiratory lamellae and gill filament illustrating predominant locations of mucous cells (*m.c.*), chloride cells (*c.c.*), and pavement cells (*p.v.c.*), the last marked by a typical 'fingerprint' surface. (E) Internal detail of respiratory lamellae, gill filament, and gill arch illustrating organization of blood vessels: afferent lamellar arteriole (*a.l.a.*), efferent lamellar arteriole (*e.l.a.*), pillar cell channels (*p.c.c.*), nutritive vessels (*n.v.*), arteriovenous anastomoses (*a.v.a.*; which are much more frequent on the post-lamellar blood flow side), central venous sinus (*c.v.s.*), and venolymphatic drainage (*v.l.d.*).

by the bodies of individual pillar cells; these cells run laterally across the width of the lamella, thereby supporting the two opposite epithelial layers in a brace-like fashion. Bundles of collagen traverse the lamellae and are anchored to the basement membrane underlying the epithelium on each side, creating circular depressions in this membrane (Penrice and Eddy, 1993). In parallel to the collagen columns, the pillar cells contain fibrils of an actomyosin-like contractile protein, whose function remains conjectural (autoregulation?) in light of the general belief that the pillar cells are not innervated. The flanges of individual pillar cells abut, forming a pseudoendothelium which contains the blood space. Pillar cell channels are so narrow (e.g. 4 μm diameter) that erythrocytes must deform as they transit. Possible pathways for shunting blood flow within the lamellae include a wider marginal channel around the upper border of each lamellae, through which the erythrocytes appear to move more quickly, and pillar cell channels at the base of the lamellae, which are largely buried in the body of the filament and are therefore unavailable for gas exchange (Tuurala *et al.*, 1984).

Deoxygenated 'venous' blood travels along the ventral aorta under arterial pressure from the heart, then via individual afferent branchial arteries in each arch, individual afferent filamental arteries running down the trailing margin of each filament, and finally short afferent lamellar arterioles leading into each respiratory lamella (Figure 1.1C–E). Oxygenated 'arterial' blood leaves the lamellae along short efferent lamellar arterioles into efferent filamental arteries which course along the leading margin of each filament and then drain into the efferent branchial artery of each gill arch. The branchial arteries coalesce to form the dorsal aorta. Thus, the millions of respiratory lamellae are perfused in parallel with one another at virtually the highest blood pressures (30–60 mmHg) present in the fish. Together with their associated arteries and arterioles, the lamellae constitute the branchial vascular resistance which is arranged in series with the vasculature of the rest of the fish, the systemic resistance. Considering its anatomic complexity, the overall branchial resistance is amazingly low under normal circumstances, such that 60–80 percent of the arterial pressure 'survives' post-gill in the dorsal aorta to allow effective perfusion of the much higher systemic resistance with oxygenated blood.

The preceding paragraph describes the traditional 'arterioarterial pathway'. However, much debate has centered on the functional importance of a recurrent 'arteriovenous' pathway (also variously called 'nutritive', 'venolymphatic', 'secondary', or 'Fromm's arteries') which leads post-lamellar blood flow back through a network of vessels or sinuses ('central venous sinus') in the body of the filament and eventually drains into filamental veins and then branchial veins (Figure 1.1E; Fromm, 1974; Randall, 1985; Vogel, 1985; Satchell, 1991). The origin of the arteriovenous pathway is usually ascribed to short arteriovenous anastomoses (AVAs) leaving the efferent filamental arteries and to longer thin arteries leaving both the efferent filamental and efferent branchial arteries. However, in a few species, additional AVAs have been reported on the afferental filamental arteries, which would allow blood flow to bypass the respiratory lamellae all together. Although early work with perfused gill and head preparations suggested that arteriovenous blood flow could be very high, these findings were likely biased by abnormal outflow resistances in the *in vitro* preparations. *In vivo*, the arteriovenous flow appears to be less than 10 percent of the arterioarterial flow, and the hematocrit of blood collected from the branchial vein is only about 15 percent of that in the dorsal aorta, which indicates considerable 'plasma skimming' at the point of entry into the arteriovenous circulation (Ishimatsu *et al.*, 1988). These measurements indicate that the ability of this pathway to deliver O_2 to the body of the filament is quite limited. Instead, the overlying filamental epithelium probably gets most of its O_2 directly from the water, and the more important function of the arteriovenous flow is supplying and removing ions, acid–base equivalents, nutrients, and wastes to and from this epithelium which appears to be heavily involved in active transport processes (see below).

The possibilities for shunting blood flow within the gills are numerous (Wood, 1974; Booth, 1979a,b; Farrell *et al.*, 1979a,b; Soivio and Tuurala, 1981; Pettersson, 1983; Nilsson, 1984; Bailly *et al.*, 1989). Under resting conditions, only the more proximally located lamellae on the filaments appear to be completely perfused, with the distal lamellae providing reserve capacity which may be opened up during exercise, hypoxia, or stress. This may be achieved by increases in inflow (ventral aortic) pressure, outflow pressure (dorsal aortic) and pressure pulsatility (via greater cardiac stroke volume) which provide transmural pressures which surpass the critical opening pressures of afferent and efferent lamellar arterioles or pillar channels themselves in more distal lamellae. At the same time, higher transmural pressure will increase the thickness of the blood sheet in individual lamellae, may stretch and thereby thin the epithelium, and may favor flow through central pillar cell channels and marginal channels where overall blood-to-water diffusion distances are lower and countercurrent exchange is more efficient. Catecholamines circulating in the blood plasma will tend to augment these effects by β-adrenergically dilating the afferent arterioles and α-adrenergically constricting the efferent arterioles, although the overall effect is vasodilatory. On the outflow side, the α-constrictory response may also be reinforced by catecholamines released from sympathetic neural activity. Serotonin, released from indoleaminergic neurons innervating the efferent filamental sphincters, similarly constricts the outflow of the arterioarterial pathway and may favor filling of more distal lamellae by the back-pressure effect. Additionally, catecholamines from either neural or hormonal sources tend to constrict AVAs so as to favor arterioarterial flow over arteriovenous flow. In contrast, cholinergic stimulation, probably arising *in vivo* only from parasympathetic nerves, tends to reduce the number of lamellae perfused by an unknown mechanism, causes muscarinic vasoconstriction of the sphincters on efferent filamental arteries, and has an overall vasoconstrictory effect on the arterioarterial pathway.

The branchial epithelium

The branchial epithelium (Figure 1.1D) covers the lamellar surfaces, the filamental surface including the region between each respiratory lamella, and extends over the surface of the gill arches themselves, and even to the neighboring inner opercular surface in a few species. The epithelium lies on top of a prominent basement membrane (basal lamina) which is thought to provide important structural support to both filaments and lamellae (Penrice and Eddy, 1993). The branchial epithelium is made up of four or five cell types, only the first of which is abundant on the lamellar surfaces (Laurent and Dunel, 1980; Laurent, 1984).

Pavement cells (respiratory or squamous epithelial cells in the older literature) are generally large, polygonal in shape and thin, and cover the majority of the gills, including virtually all of the lamellar surface in healthy, non-stressed fish (Figures 1.1D and 1.2A). Traditionally, the pavement cells were considered the sites of diffusive gas exchange and passive ion and water movements, but more recently an additional role in active acid–base and ion transport has been proposed, specifically in H^+ excretion and coupled Na^+ uptake (Goss *et al.*, 1992, 1994, 1995; Laurent *et al.*, 1994a; Perry, 1997). The apical membrane of the pavement cell is overlaid with glycocalyx and decorated with characteristic 'fingerprint' ridges, whorls or microvilli, structures which may have some role in increasing the surface area available for gas exchange, trapping mucus, and/or in creating an unstirred boundary layer (Figure 1.2A and B). Pavement cells are closely joined to one another and to neighboring chloride cells by numerous desmosomes and deep tight junctions at the apical surface (Figure 1.2B). Based on their impermeability to lanthanum and their multistranded nature, these junctions are thought to have a high electrical resistance and low ionic conductance. Internally, the cells have relatively sparse mitochondria but are rich in other organelles, including well-developed Golgi, abundant rough endoplasmic reticulum, and numerous vesicles. These last structures are especially prominent under conditions in which the pavement cells are thought to become 'active' in ion and acid–base regulation (Figure 1.2C). Often, several layers of pavement cells overlap (generally two in the respiratory lamellae), with underlying ones being more columnar in shape and less well differentiated (Figure 1.3A). Lacunae between pavement cell layers appear to communicate with the central venous sinus.

Figure 1.2 Electron micrographs of the branchial epithelium on the trailing edge of the gill filament in a freshwater rainbow trout. Plate courtesy of P. Laurent. (A) Representative scanning electron micrograph of the epithelium, illustrating typical surface morphology of pavement cells (*pvc*), chloride cells (*cc*), and mucous cells (*mc*). Note the characteristic 'fingerprint' whorls on pavement cells compared with the smoother surface of the chloride cells decorated only with microvilli, and the similar but smaller surface exposure of the mucous cell (1500×). (B) Representative transmission electron micrograph of a freshwater-type chloride cell and its relationship to neighboring pavement cells, marked by long, presumably very tight, junctions (small arrows) and close contact below (broken line). Note also the density of mitochondria (*m*) closely associated with the internal network of tubules (*t*) and the microvilli (*mv*) in contact with the external water (*w*). The long arrow indicates the faint glycocalyx coat on the pavement cells (18 000×). (C) TEM of an 'active' pavement cell in freshwater completely overlying a chloride cell (on the left). In this case, the pavement cell contains larger mitochondria and a rich complement of Golgi apparatus (*g*), the latter giving rise to vesicles (*v*) (25 000×).

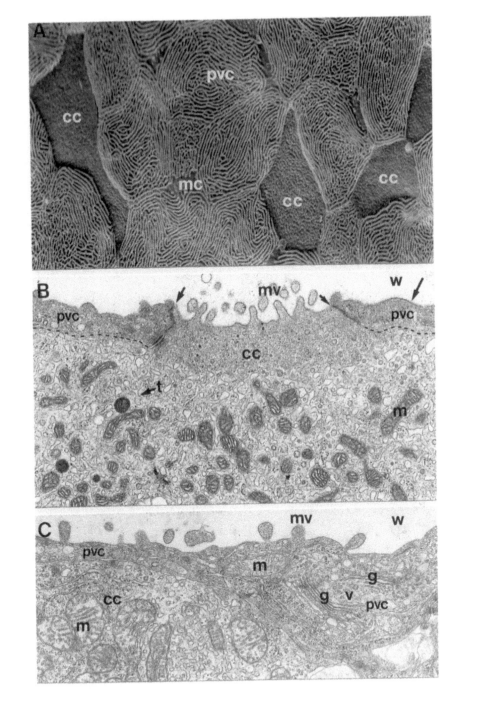

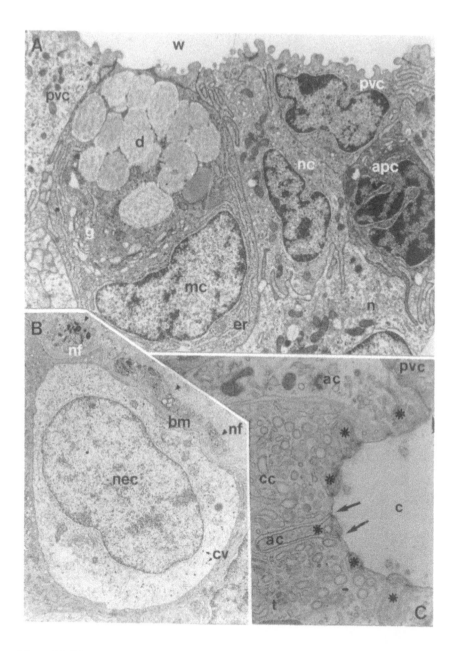

Figure 1.3 Electron micrographs of the branchial epithelium. Plate courtesy of P. Laurent. (A) Representative TEM of the trailing edge region of the gill filament in a freshwater rainbow trout, illustrating typical internal structure of mucous cells (*mc*), surface pavement cells (*pvc*), less-differentiated subsurface pavement cells (*nc*), and an electron-dense apoptotic cell (*apc*). In the mucous cell, note the abundant droplets (*d*) of mucus ready to be released, the small smooth apical contact with the water (*w*), and the rich endoplasmic reticulum (*er*) and Golgi apparatus (*g*) (5000×). (B) Representative TEM of a neuroepithelial cell (*nec*) resting on the basement membrane (*bm*), very close to endings (arrow heads) of nerve fibers (*nf*) within the filament epithelium of a freshwater perch. Note the characteristic

Chloride cells (also called mitochondrial-rich cells or ionocytes) are largely restricted to the interlamellar region (Figure 1.1D) and trailing edge of the filamental epithelium and the base of the respiratory lamellae, but in freshwater fish may proliferate up onto the surfaces of the respiratory lamellae under a variety of stressful conditions (e.g. Figure 1.10B). In seawater fish, their occurrence on the lamellae is rare. The chloride cells generally make up less than 10 percent of the total branchial epithelial cells, but are nevertheless the second most numerous type after the pavement cell. Under the light microscope (LM), chloride cells are recognized by their rounded morphology and their ability to take up zinc and silver stains, and under phase contrast by their granular appearance. The accumulation of mitochondrial-specific fluorochromes, such as DASPEI (dimethylaminostyrylethylpyridinium iodide) or DASPMI (dimethylamino-styrylmethylpyridinium iodide), as seen under epifluorescence, is a standard diagnostic marker for chloride cells. Fluorescent derivatives of ouabain (e.g. anthroylouabain; McCormick, 1990), radioactive ouabain (e.g. Karnaky *et al.*, 1976) or antibodies against Na^+,K^+-ATPase (e.g. Witters *et al.*, 1996a) have also been used to identify chloride cells successfully. Under the scanning electron microscope (SEM), chloride cells manifest as indented patches or protuberant bumps on the filamental surface, especially in freshwater fish, and their exposed apical surfaces with distinctive microvilli appear smoother than those of pavement cells (Figure 1.2A). Under the transmission electron microscope (TEM), they are characterized by electron-dense staining, abundant mitochondria, numerous vesicles and an extensive internal tubular system which is actually part of the basolateral membrane (Figure 1.2B). All these features are accentuated in seawater fish (Figure 1.3C). Na^+,K^+-ATPase activity in chloride cells is up to 30-fold higher in chloride cells than in pavement cells (Sargent *et al.*, 1975) and has been localized exclusively to this basolateral membrane system (e.g. Karnaky *et al.*, 1976).

Pisam and co-workers have presented convincing evidence for the presence of two morphologic types of chloride cells in freshwater fish (Figure 1.4; see Pisam *et al.*, 1987, 1993; Pisam and Rambourg, 1991), but it remains possible that these are different developmental or degenerative stages of the same cell (see Wendelaar Bonga and van der Meij, 1989; Wendelaar Bonga *et al.*, 1990). The more electron-opaque α-cells are elongated in shape, are located at the base of lamellae in intimate contact with the basal lamina overlying lower pillar cell channels (i.e. arterioarterial circulation), and possess apical membranes which are fairly smooth. The more electron-dense β-cells are ovoid in shape, are located in the interlamellar region

dense core vesicles (*cv*) (9000×). (C) TEM illustrating the seawater-type organization of a chloride cell (*cc*) relative to a neighboring accessory cell (*ac*) and a prominent apical crypt (*c*) in a tilapia living in a hyperosmotic medium. The edge of a pavement cell (*pvc*), which curves around (out of field) to cover partially the apical crypt, can also be seen. Note the cytoplasmic processes (asterisks) of the accessory cell which invade the chloride cell apex with multiple very short, presumably leaky, junctions (long arrows), in contrast to the long, presumably tight, junction (arrow head) between the accessory cell and the pavement cell. Note the very different organization from that seen in the freshwater situation (see Figure 1.2B) (23 000×).

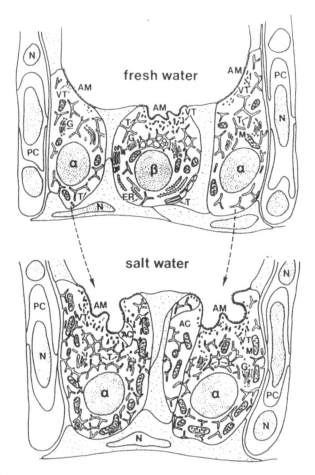

Figure 1.4 Diagrammatic representation of chloride cell types and morphology in the interlamellar region of the filamental epithelium in teleosts adapted to freshwater or seawater (from Pisam *et al.*, 1987, by permission). In freshwater (top), the α-type cells are located beside the basal pillar cell channels (*PC*) of the adjoining respiratory lamellae, whereas the β-type cells lie immediately over the central venous sinus of the filament. Note the different shape of α-type (elongated) and β-type chloride cells (round), the latter having a more corrugated apical membrane (*AM*), a more prominent endoplasmic reticulum (*ER*), more prominent vesicular traffic (*VT*) between the Golgi apparatus (*G*) and the apical membrane, and a less regular system of tubules (*T*). In saltwater (bottom), β-type cells disappear and accessory cells (*AC*) develop, sending apical processes into the α-type cells. The α-type cells become larger, develop contact with the central venous sinus as well as the pillar cell channels, and exhibit more prominent mitochondria (*M*), vesicles, tubule system, and Golgi, and invagination of the apical membrane to form a crypt.

in intimate contact with the basal lamina overlying the central venous sinus (i.e. arteriovenous circulation), and exhibit much more corrugated apical membranes. In addition, the pattern of subapical vesicles and the internal tubular system differs between the two cells. Both cell types have a more elaborate glycocalyx than pavement cells, with a greater depth in β-cells than in α-cells (Powell *et al.*, 1994). The α-cells are thought to become larger and to invade the interlamellar region in seawater fish, whereas the β-cells are thought to disappear (Figures 1.3C and 1.4). However, there is no information on possible functional differences between the two morphotypes, so the significance of these observations remains unclear.

Chloride cells may occur singly or in groups, with the latter arrangement more prevalent in seawater fish in which the groups are often clustered around a shared invagination, the apical crypt, and flanked by accessory cells (Figures 1.3C and 1.4; see below). Individual chloride cells (in freshwater fish; Figure 1.2B) or the whole crypt (in seawater fish; Figure 1.3C) are usually partially covered by pavement cells, indeed sometimes totally covered (Figure 1.2C), and make very close contact with pavement cells. Recent studies show that the apical chloride cell-surface area exposed to the external water can be adjusted, within a few hours, by dynamic changes in pavement cell coverage in response to acid–base and ionic disturbances (Goss *et al.*, 1995; Laurent *et al.*, 1995; Perry, 1997). The deep multistranded tight junctions (Figure 1.2B) and numerous desmosomes between chloride cells and pavement cells are similar to those between adjacent pavement cells themselves, whereas very shallow, presumably 'leaky' (lanthanum permeable) junctions are made with neighboring accessory cells (Figure 1.3C). In freshwater fish, chloride cells appear to be the sites of active uptake of Cl^-, Ca^{2+}, and probably other divalent metals, whereas it remains controversial whether active Na^+ uptake occurs here, via pavement cells, or via both pathways (Wicklund-Glynn *et al.*, 1994; Jurss and Bastrop, 1995; Hogstrand *et al.*, 1996a; Perry, 1997). In seawater fish, chloride cells are principally responsible for the active transcellular excretion of Cl^- and the coupled passive excretion of Na^+ through a paracellular pathway (Wood and Marshall, 1994; McCormick, 1995; Jurss and Bastrop, 1995).

Accessory cells are a characteristic feature of the branchial epithelium only in seawater fish, in which they are closely associated with chloride cells (Figures 1.3C and 1.4). However, their presence has also been reported in some euryhaline species while still living in freshwater, in association with the α-type chloride cell; in salmonids, their frequency increases greatly after smoltification, presumably a preadaptation to life in seawater (Pisam and Rambourg, 1991). Some authorities consider accessory cells to be either immature chloride cells or a modified type of chloride cell (e.g. Hootman and Philpott, 1980; Wendelaar Bonga *et al.*, 1990), but this view is now generally discounted (e.g. Laurent and Dunel, 1980; Laurent, 1964; Pisam and Rambourg, 1991; Jurss and Bastrop, 1995). Like chloride cells, accessory cells are rich in mitochondria, but their organelles are smaller and their cytoplasm is more electron opaque. In marine fish, a small, pear-shaped accessory cell generally lies next to each chloride cell, sandwiched between it and the pavement cells so as to achieve apical exposure to the crypt (Figures 1.3C and 1.4). Characteristically, accessory cells give rise to a number of apical processes which invade the apical membrane of the adjacent chloride cell (Figure 1.3C). On these invaginations are numerous short, electrically 'leaky' junctions between the two cells, which provide the anatomic basis for the paracellular extrusion of Na^+ which is coupled to the active excretion of Cl^- in seawater teleosts.

Mucous cells (also called goblet cells) are variable in distribution among species (Laurent and Dunel, 1980). They are usually most concentrated on the leading edge of the filament epithelium (Figure 1.1D), especially in distal regions, but are sometimes more abundant on the trailing edge, with a generally sparse occurrence in the interlamellar regions and on the respiratory lamellae themselves, at least in disease-free fish (see Ferguson *et al.*, 1992). The glycoprotein mucin, released by these cells, is considered generally to be a multifunctional substance (and perhaps a multicomponent substance) involved in mechanical, toxicant, and pathogen defense and in osmotic insulation, ionic regulation, and respiratory gas exchange (for reviews, see Satchell, 1984; Shephard, 1994). Particularly noteworthy is the ability of mucus to increase Na^+, K^+, Cl^-, and Ca^{2+} concentrations close to the gill surface (Handy, 1989), its ability to complex metals and depurate them by sloughing (Handy *et al.*, 1989; Handy and Eddy, 1991; Playle and Wood, 1991), its ability to slow down the diffusion rate of metals (Part and Lock, 1983), and its content of the respiratory enzyme carbonic anhydrase (Wright *et al.*, 1986). Production of mucus appears to increase under a variety of stressful conditions. Although this may serve to shield the gill surface physically from noxious agents, under severe stress conditions the production may become excessive and contribute to pathology. Specifically, a thickened mucous layer will block the water flow channels between the respiratory lamellae and magnify the diffusion distance between water and blood, in addition to slowing physically the diffusion of O_2. Ultsch and Gros (1979) concluded that the diffusivity of O_2 in mucus was only 70 percent of that in water.

Mucous cells are easily recognized by LM when stained for mucin with alcian blue (for acidic glycoproteins with sulfated or sialic acid residues) and/or periodic acid–Schiff reagent (for neutral glycoproteins). However, this approach will not detect mucous cells which have recently discharged their glycoprotein content. Under TEM, mucous cells are generally smaller than chloride cells, lack a glycocalyx, and have a distinctive tear-drop shape in which most of the intracellular contents comprise large mucin-containing vesicles (Figure 1.3A). Like chloride cells, mucous cells are usually partially covered, and sometimes completely covered, by pavement cells. By SEM, their apical exposure is usually small but can easily be confused with that of a chloride cell (e.g. Figure 1.2A), unless specifically marked by the presence of a mucous globule or strand (van der Heijden and Morgan, 1997). The actual tissue fixation conditions, as well as the degree of 'stress' involved in sampling, probably greatly affect the distribution and appearance of mucus on the branchial surface (Shephard, 1994). For example, glutaraldehyde fixation strips off mucus or causes it to form globules. For these reasons, the actual location of mucus *in vivo* has proven to be a controversial topic. The traditional view holds that the whole branchial epithelium is normally covered by a thin sheet of mucus in healthy fish, with the microridges on pavement cells serving to anchor the mucous layer (Powell *et al.*, 1992). An alternate view is that in truly unstressed fish mucus is absent from the secondary lamellae so that only the filamental surface is coated (Handy and Eddy, 1991). Unfortunately, both viewpoints are supported by observations using the best possible preservation technique – cryofixation – so the truth remains elusive.

Neuroepithelial cells (NECs; Figure 1.3B) are perhaps the least well investigated of the epithelial cells, with detailed anatomic studies only by Laurent and co-workers (e.g. Dunel-Erb *et al.*, 1982; Laurent, 1984; Bailly *et al.*, 1989, 1992). These small innervated cells are located exclusively in the body of the filament, mainly on the leading edge and with a greater concentration at the distal ends. They occur either singly or in clusters, and can be recognized by their clear cytoplasm and their abundance of small dense core vesicles (80–100 nm). The biogenic amine serotonin has been localized to these vesicles, which explains the intense fluorescence of NECs when treated with formaldehyde vapor. These cells lie directly on the basement membrane in close proximity to nerve endings (Figure 1.3B), in intimate contact with both the central venous sinus and the efferent filamental artery (both of which contain arterialized post-lamellar blood). From this location, they also face the incoming external water flow, but without direct exposure to it. Environmental hypoxia causes degranulation of the dense core vesicles. Recent physiologic interest has focussed on the role of NECs as putative chemoreceptors, important in providing afferent information to the central nervous system for the control of ventilation (Burleson and Milsom, 1995). Surprisingly, however, serotonin does not appear to be the neurotransmitter which activates the afferent nerve fibers during hypoxia.

Basic physiology

Respiratory gas exchange and nitrogenous waste excretion

Two of the respiratory gases, O_2 and CO_2, are exchanged in almost equimolar amounts at the gills, whereas the third, ammonia, is excreted at a rate typically 10-fold lower. Ammonia-N is also the principal nitrogenous waste in most teleost fish (> 70 percent), with urea-N making up most of the balance. The exact ratios depend on the fuels being burned (see Lauff and Wood, 1996); under steady-state aerobic conditions, the respiratory quotient (RQ = CO_2 production/O_2 consumption) may vary from 0.7 (100 percent lipid) to 1.0 (100 percent carbohydrate) and the nitrogen quotient (NQ = N waste production/O_2 consumption) is always less than 0.27 (100 percent protein). However, under non-steady-state conditions (e.g. exhaustive exercise, hypoxia, severe stress), CO_2 and ammonia production generally increase more than O_2 consumption because of metabolic acidosis (see next section) and adenylate breakdown respectively (Wood, 1991a). Thus, RQ and NQ may rise well above their theoretical aerobic limits of 1.0 and 0.27. O_2 moves between blood and water entirely by diffusion, and at least 90 percent of the CO_2 excretion moves in the opposite direction by diffusion also. The majority of ammonia efflux also probably moves by diffusion, although this remains controversial (Wood, 1993; Korsgaard *et al.*, 1995; Wilkie, 1997). Urea also moves by diffusion in most fish, but a facilitated diffusion carrier appears to be present in the gills of a few unusual ureotelic teleosts (Wood *et al.*, 1998).

Figure 1.5 summarizes our current understanding of the pathways by which O_2, CO_2, and ammonia move across fish gills. For each of the three respiratory

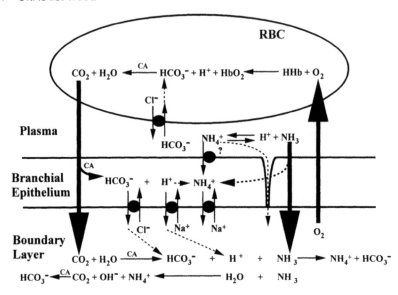

Figure 1.5 Diagrammatic model of the mechanisms of oxygen, carbon dioxide, and ammonia exchange at the fish gill, and their functional interrelationships. CA, carbonic anhydrase; RBC, red blood cell. See text for details.

gases, the driving force for diffusion is provided by the partial pressure gradients. Because O_2 is so much less soluble in water than are CO_2 or NH_3, the ventilatory flow required to meet O_2 uptake requirements effectively hyperventilates the gills with respect to CO_2 and NH_3 excretion. As a result, partial pressures of CO_2 (e.g. $P_{CO_2} = 1–4$ Torr) and NH_3 (e.g. $P_{NH_3} = 20–200$ μTorr) are extremely low in fish arterial blood relative to partial pressures of O_2 (e.g. $P_{O_2} = 30–120$ Torr). Ventilatory control is primarily geared to O_2 such that fish hyperventilate during environmental hypoxia and hypoventilate during hyperoxia (Randall, 1982; Shelton *et al.*, 1986). However, there does appear to be a secondary sensitivity to CO_2 and acid–base status which probably explains hyperventilation in high P_{CO_2} environments and during the acidotic state that accompanies exhaustive exercise and some stress conditions (Wood and Munger, 1994; Perry and Gilmour, 1996).

Blood flow (perfusion = cardiac output) and water flow (ventilation) through the gills are closely matched under normal circumstances, so O_2 uptake is probably diffusion limited (Wood and Perry, 1985; Perry and Wood, 1989; Perry and McDonald, 1993). Thus, negative effects on the capacity for O_2 uptake would be expected from decreases in functional branchial surface area, increases in mean blood-to-water diffusion distance (e.g. Figure 1.6), or decreases in mean water-to-blood P_{O_2} gradient. The Bohr and Root effects, whereby the addition of CO_2 and/or H^+ to the blood decreases the O_2 affinity and capacity of the hemoglobin, falls into this latter category by effectively increasing mean blood P_{O_2} at the gas exchange surface in the respiratory lamellae. During hypoxia, fish compensate for the reduced external P_{O_2} by hyperventilating, a response which not only helps

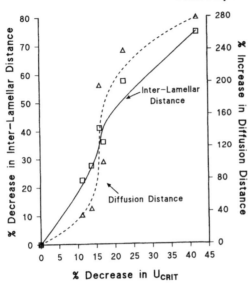

Figure 1.6 An example of the effect on aerobic swimming performance (U_{CRIT} = critical swimming speed) of increases in blood-to-water diffusion distance and associated decreases in interlamellar distance caused by sublethal exposure to a toxicant. Data from Nikl and Farrell (1993) on responses of juvenile chinook salmon to 2-(thiocyanomethylthio)-benzothiazole in freshwater. The predominant lesions were lifting and hypertrophy of pavement cells.

raise the mean water Po_2 at the gas exchange surface but also washes out CO_2 from the blood (Bohr and Root shifts in the opposite direction), thereby lowering the effective mean internal Po_2 at the same surface. Simultaneously, the fish increases functional branchial surface area and decreases diffusion distance in the gills by mechanisms outlined earlier. During exercise, the capacity for increasing area and decreasing diffusion distance by these same mechanisms may reach a plateau in the face of increased blood perfusion, and further compensation is achieved by increasing the diffusion gradient through disproportionate increases in ventilatory flow. Thus, toxicants which reduce branchial surface area, increase diffusion distance, or cause Bohr- and Root-type shifts in the blood O_2 dissociation curve may have relatively benign influence in resting fish but have severe impact on O_2 uptake, ventilatory costs, and aerobic swimming ability during exercise or hypoxia (e.g. Figure 1.6). Additionally, the respiratory adaptations to exercise and hypoxia (hyperventilation, increased blood flow through the respiratory lamellae) will likely increase the effective bioavailability of the toxicant to the gills.

Overall, the factors affecting CO_2 excretion at the gills are more complicated and less well understood than those affecting O_2 consumption (Perry, 1986; Perry and Wood, 1989; Wood and Munger, 1994). Clearly, as with O_2 consumption, functional surface area, diffusion distance, and partial pressure gradient will also affect CO_2 excretion, but additional influences come into play. Despite the fact that the gill is hyperventilated with respect to CO_2 excretion, changes in both ventilatory water flow and perfusive blood flow have been shown to have effects, as have changes in water pH and buffer capacity. The role of enzyme- and transporter-mediated events in both the red blood cell and the water boundary layer at the lamellar surface are probably very important. As blood passes through the respiratory lamellae, the bulk of CO_2 which is excreted originates from HCO_3^- dissolved in the blood plasma (Figure 1.5). This HCO_3^- enters the red blood cells in electroneutral exchange for Cl^- mediated by the 'band 3' anion exchange protein. Within the erythrocyte, the HCO_3^- is then dehydrated to CO_2 and H_2O by reaction with H^+ under the influence of the enzyme carbonic anhydrase; these H^+ ions are pulled off the hemoglobin, thereby increasing its O_2 affinity and capacity (i.e. Bohr and Root effects). Simultaneous oxygenation of the hemoglobin promotes this discharge of H^+ ions by the Haldane effect, which is the mirror image of the Bohr effect and thereby facilitates CO_2 excretion. The newly liberated CO_2 then diffuses from the red blood cell through the lamellar epithelium into the external water. The overall rate-limiting step is thought to be the band 3 Cl^-/HCO_3^- exchanger, so factors which affect its abundance (e.g. hematocrit) or activity may have marked effects. With respect to the latter, there is evidence that catecholamine mobilization into the blood plasma inhibits the rate of erythrocytic Cl^-/HCO_3^- exchange (Wood and Perry, 1991), so stressed fish, especially if anemic, will be expected to retain CO_2.

The boundary layer water may be an additional control point. The higher the water pH and buffer capacity, the faster and more effective will be removal of the excreted CO_2 by conversion back to HCO_3^- and H^+ (hydration reaction), and therefore the more effective will be the maintenance of the blood-to-water P_{CO_2} gradient. Additionally, this hydration of excreted CO_2 in the boundary layer is thought to be accelerated by the presence of carbonic anhydrase on the surface of the pavement cells, perhaps in the mucous sheet (Figure 1.5). Although the histochemical and immunochemical evidence for this external carbonic anhydrase appears strong (Dimberg *et al.*, 1981; Rahim *et al.*, 1988), both the physiologic data (Heming, 1986; Wright *et al.*, 1986; Playle and Wood, 1989a; Henry, 1996; Perry *et al.*, 1999) and the pattern of distribution of mucus on the lamellae (see above) remain controversial. There is also abundant intracellular carbonic anhydrase inside the branchial epithelium (especially within mucous, chloride, and pavement cells), but this is thought to have no role in catalyzing the dehydration of plasma HCO_3^- because of the impermeability of the cell membranes to HCO_3^- (Perry, 1986). However, it is probably involved in accelerating the diffusion rate of CO_2 along its P_{CO_2} gradient, an action of carbonic anhydrase which is most effective at low P_{CO_2} levels typical of fish (Henry, 1996). In addition, up to 10 percent of the CO_2 diffusing through the epithelium may be reconverted to HCO_3^- and H^+ by the intracellular enzyme, with these subsequently being excreted by apical Cl^-/HCO_3^- and Na^+/H^+ (or Na^+/NH_4^+, or Na^+ channel/H^+-ATPase) exchange processes (see next section).

The mechanism of ammonia excretion remains unsettled (Wood, 1993; Korsgaard *et al.*, 1995; Wilkie, 1997; Salama *et al.*, 1999). In freshwater fish,

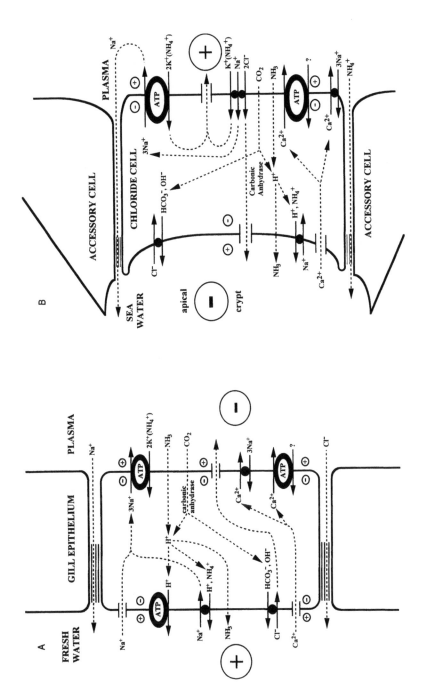

Figure 1.7 Diagrammatic models of the mechanisms of ionic and acid–base transport, and their functional interrelationships, at the gills of teleost fish in (A) freshwater and (B) seawater. In seawater, the processes have been localized to the chloride cells and the paracellular pathway between chloride cells and neighboring accessory cells, but the localization of processes to individual cell types remains unresolved in freshwater. The '+' and '–' symbols represent the directionality of the transepithelial potential difference (TEP). See text for details.

early belief in the predominance of Na^+/NH_4^+ exchange at the apical surface somewhere in the gill epithelium (Krogh, 1939) was later reinforced by some studies (e.g. Maetz and Garcia-Romeu, 1964; Payan, 1978; McDonald and Prior, 1988) but opposed by others which demonstrated a predominance of simple diffusion of NH_3 along a P_{NH_3} gradient from blood to water (e.g. Cameron and Heisler, 1983; Wilson *et al.*, 1994). Still others have provided evidence for a flexible combination of the two mechanisms (e.g. Maetz, 1973; Wright and Wood, 1985). In light of this uncertainty, Figures 1.5 and 1.7 show both mechanisms, although the balance of recent evidence seems to be shifting against the occurrence of apical Na^+/NH_4^+ exchange in freshwater fish (Wilkie, 1997). The initial entry of ammonia into the branchial epithelial cells from the blood plasma may be by diffusion of NH_3, or by substitution of NH_4^+ on a basolateral carrier such as the H^+ site of basolateral Na^+/H^+ exchange, the K^+ site on Na^+,K^+-ATPase, or, in seawater fish only, the K^+ site on the $Na^+,K^+,2Cl^-$ co-transporter (see next section). An attractive model linking CO_2 and ammonia excretion has been presented and partially validated (Randall and Wright, 1989; Wright *et al.*, 1989; Wilson *et al.*, 1994). The catalyzed hydration of excreted CO_2 by external carbonic anhydrase acidifies the water boundary layer of the lamellae, thereby providing H^+ ions which instantaneously convert NH_3 to NH_4^+, and so maintains the blood-to-water P_{NH_3} gradient by keeping the water P_{NH_3} close to zero (Figure 1.5). Even if this catalyzed CO_2 hydration remains controversial, it is probable that H^+ ions excreted by the Na^+/H^+ and/or by the Na^+ channel/H^+-ATPase exchange system (see next section) will provide similar diffusion trapping of NH_3 in the boundary layer. Indeed, as pointed out by Wilson *et al.* (1994), such a coupling would masquerade as Na^+/NH_4^+ exchange. The reverse co-operation also probably occurs, whereby NH_3 excretion raises boundary layer pH (NH_4^+ forms instantaneously without catalysis at low pH, removing H^+ and liberating OH^-), thereby facilitating CO_2 excretion, especially when water pH is low.

In seawater fish, the same mechanisms may occur, and indeed the much higher levels of water $[Na^+]$ render it much more likely that Na^+/NH_4^+ exchange makes a significant contribution to overall ammonia excretion (Evans, 1977; Wilkie, 1997). An additional pathway, the direct paracellular diffusion of NH_4^+, probably also occurs in seawater fish (and only in seawater fish; Figures 1.5 and 1.7B), made possible by the short 'leaky' junctions between accessory cells and chloride cells (e.g. Evans *et al.*, 1989). As in freshwater, the quantitative partitioning among the various pathways remains unclear.

Ionic exchange and acid–base regulation

Acid–base regulation is here considered with respect to its linkage with ionic exchange, but it is also intimately linked with respiratory gas exchange because blood pH is set by the ratio of dissolved HCO_3^- to dissolved CO_2 in the blood plasma as approximated by the Henderson–Hasselbalch equation:

$$\text{pH} = pK^1 + \frac{\log\left[HCO_3^-\right]}{\left[\alpha_{CO_2} \cdot P_{CO_2}\right]}$$

(1.1)

αCO_2 (CO_2 solubility in plasma) and pK^1 (the apparent dissociation constant) are constants; all other buffers fall into line by the isohydric principle (see Albers, 1970; Boutilier *et al.*, 1984). Fish typically maintain arterial pH about 0.6–0.8 units above the pH of neutrality, which varies with temperature (Cameron, 1984). Thus, at 15°C, typical values would be 7.8–8.0, with $P\text{co}_2$ levels of 1–3 Torr and [HCO_3^-] of 3–8 mM. $P\text{co}_2$ build-up ('CO_2 retention', caused by, for example, hypoventilation, high environmental $P\text{co}_2$, or agents which interfere with any of the key steps in CO_2 excretion) will cause an almost instantaneous fall in blood pH, i.e. 'respiratory' acidosis. Conversely, excessive CO_2 excretion (caused by, for example, hyperventilation or high environmental pH) will lower $P\text{co}_2$ and thereby raise pH, i.e. 'respiratory alkalosis'. 'Metabolic' acid–base disturbances and corrections occur more slowly, principally through manipulation of ion exchange processes at the gills, many of which are simultaneously involved in ionoregulation. These processes raise or lower [HCO_3^-], with the addition of 'acid' (e.g. H^+, NH_4^+) being equivalent to removal of 'base' (HCO_3^-, OH^-) and vice versa. The two 'counterions' most involved in the movement of 'acid' and 'base' across the gills are Na^+ and Cl^-; the difference between their net fluxes, no matter how achieved, constrains a matching net acid flux in the opposite direction (McDonald and Prior, 1988; Wood, 1991b; Goss *et al.*, 1992), in accord with the strong ion difference theory (Stewart, 1978).

Freshwater teleost fish maintain internal osmolalities around 280–350 mosmol relative to external values < 10 mosmol, which corresponds to 100- to 500-fold concentration ratios of Na^+ and Cl^- between blood plasma and water. This is achieved by the active uptake of Na^+, Cl^-, and other ions across the gills, in combination with voluminous production of dilute urine by the kidney to bail out the water which continually enters by osmosis. Diffusive efflux probably occurs via the paracellular channels between branchial epithelial cells (Figure 1.7A). The transepithelial potential (blood side negative) is viewed as a diffusion potential reflecting the differential permeability of these channels to Na^+ over Cl^-. Differential passive losses of Na^+ versus Cl^- across the gills constrain net acid or base flux, and some recent studies suggest that this diffusive efflux mechanism can be dynamically regulated by fish for the purpose of acid–base regulation (Goss *et al.*, 1992). However, most attention has focussed on the Na^+ and Cl^- uptake processes which appear to be sensitively adjusted for acid–base homeostasis. Both processes display standard saturation kinetics (e.g. Figure 1.16) which are well described by the Michaelis–Menten equation:

$$J_{in} = \frac{J_{max} \cdot K_m}{[X] + K_m} \tag{1.2}$$

where J_{in} is the unidirectional influx rate, [X] is the concentration of substrate (i.e. Na^+ or Cl^-), J_{max} is the maximum transport rate when the system is saturated with substrate, and K_m is the concentration of substrate which provides a J_{in} of half the J_{max} value. K_m is therefore a measure of the affinity of the transport system for the ion, which is normally close to the environmental concentration. An agent which directly competes for the same transport system will increase K_m (competitive inhibition), whereas an agent which interferes with transport by any other mechanism will decrease J_{max}, i.e. a functional loss of transport sites.

Despite intensive study since the time of Krogh (1939), our detailed understanding of these uptake mechanisms remains fragmentary (Wood, 1991b; Goss *et al.*, 1992; Evans, 1993; Kirschner, 1997). The traditional view is that Na^+ and Cl^- uptake are active, independent processes mediated by electroneutral exchange carriers on the chloride cells, with Na^+ exchanged on a one-for-one basis with H^+ or NH_4^+ ('acid') and Cl^- for HCO_3^- or OH^- ('base') (Figure 1.7A). The acidic and basic counterions are provided by intracellular carbonic anhydrase through the hydration of CO_2, as outlined earlier (Figure 1.5). This view also holds that the energy input for transport occurs entirely via the activity of Na^+,K^+-ATPase on the basolateral tubular membrane system.

At a 'blackbox' level, this general scheme is supported by literally thousands of studies, but it has fared poorly in recent years in the face of detailed mechanistic investigations (Jurss and Bastrop, 1995; Lin and Randall, 1995; Goss *et al.*, 1995; van der Heijden and Morgan, 1997; Kirschner, 1997; Perry, 1997). In particular, measurements of intracellular ions and acid–base status indicate that intracellular Na^+ and Cl^- levels are probably too high and H^+ and HCO_3^- gradients too low to support the apical entry of Na^+ and Cl^-, suggesting that these processes must be additionally energized. Instead, for Na^+ uptake, some evidence has been presented for a $Na^+,NH_4^+/H^+$-dependent ATPase in freshwater fish gills (Balm *et al.*, 1988). However, the bulk of recent evidence points to an active electrogenic extrusion of H^+ ions across the apical membrane by H^+-ATPase which creates an electrochemical gradient for the entry of Na^+ through Na^+-selective channels (Figure 1.7A). Part (the Na^+ channels alone) or all (the Na^+ channels plus the H^+-ATPases) of this system may in fact occur on the pavement cells, based on recent TEM (Laurent *et al.*, 1994a; Goss *et al.*, 1995), antibody (Sullivan *et al.*, 1995), and molecular evidence (Sullivan *et al.*, 1996). In functional terms, this Na^+ channel/H^+-ATPase exchange system would achieve the same one-for-one exchange of Na^+ for 'acid' as an electroneutral Na^+/H^+ exchanger, but by a very different mechanism; also, if NH_3 diffusion was coupled to the resulting $[H^+]$ gradient, it could also mimic a Na^+/NH_4^+ exchanger (Wilson *et al.*, 1994). For Cl^- uptake, most evidence points to the chloride cells as the actual sites and agrees that Cl^- is exchanged against 'base', but it is now impossible to see how this exchange can be energized by the basolateral Na^+,K^+-ATPase. Early evidence for the presence of a Cl^-,HCO_3^--ATPase in freshwater gills (Kerstetter and Kirschner, 1974; De Renzis and Bornancin, 1977) was largely overlooked by most authorities, but clearly deserves consideration now. A Cl^--selective channel has been proposed as the basolateral exit mechanism for Cl^- (Marshall, 1995).

Ca^{2+} uptake from freshwater is the best understood and least controversial pathway (Flik and Verbost, 1993; Flik *et al.*, 1993, 1995; Perry, 1997). Transbranchial Ca^{2+} transport appears to be active under most circumstances and exhibits standard Michaelis–Menten kinetics (e.g. Figure 1.17B). It is exclusive to the chloride cells and occurs via apical Ca^{2+}-selective channels (lanthanum sensitive) and a basolateral high-affinity Ca^{2+}-ATPase, perhaps with some assistance from a basolateral Na^+/Ca^{2+} exchanger (Figure 1.7A). The latter would be driven by the Na^+ gradient, and therefore indirectly energized by the basolateral Na^+,K^+-ATPase. Intracellular Ca^{2+} activities are extremely low (submicromolar) because of the presence of Ca^{2+} binding proteins. Thus, the basolateral exit is active, whereas the apical entry is passive, along the electrochemical gradient. Other divalent metals such as Zn^{2+}, Cd^{2+}, and Co^{2+} also likely enter through these apical Ca^{2+} channels in chloride cells, but the mechanism(s) of their basolateral transport remains unknown (Verbost *et al.*, 1989; Hogstrand *et al.*, 1996a; Comhaire *et al.*, 1998). Unidirectional Ca^{2+} flux rates are low, generally less than 10 percent of Na^+ or Cl^- flux rates.

Marine teleost fish maintain internal osmolalities only slightly higher than those of freshwater fish (320–400 mosmol) despite high external osmolality (1050 mosmol in 100 percent seawater) which continually extracts water from the fish, mainly across the gills. External Na^+ and Cl^- levels are four- to fivefold higher than plasma levels, but it is the osmotic dehydration which appears to be the most critical problem. To counteract this, plus a small additional water loss in obligatory urine production, marine fish continually drink the external medium. To absorb the ingested water, they transport Na^+ and Cl^- inwards across the intestinal surface. Indeed, as much as 80 percent of the total NaCl load may come from drinking rather than from branchial and skin uptake. All of this NaCl load must be pumped out at the gills, with net flux rates many fold higher than those in the opposite direction in freshwater fish. In addition, marine fish have the problem of excluding toxic divalent ions such as Mg^{2+}, SO_4^{2-}, and Ca^{2+}, all of which are much more concentrated in seawater than in teleost plasma.

The details of this Na^+ and Cl^- transport are much better understood in marine than in freshwater fish (Figure 1.7B). This difference is due to the availability of flat skin preparations rich in branchial epithelial cells, for example opercular (killifish, tilapia) and jaw skin (goby) preparations, which serve as convenient physiologic models for the gill itself (Zadunaisky, 1984; Wood and Marshall, 1994; Marshall, 1995).Transport is electrogenic, explaining essentially all of the transepithelial potential (blood side positive), and has been localized exclusively to chloride and accessory cells. Unlike the freshwater situation, most of the Na^+ and Cl^- fluxes are coupled on a one-to-one basis and therefore play no role in acid–base balance. The abundant Na^+,K^+-ATPase on the basolateral tubular membrane system and the 'leaky' short junctions with the apical processes of the adjacent accessory cells are two key points in this transport system. Na^+,K^+-ATPase is the overall transport engine, creating and maintaining a Na^+ electrochemical gradient that strongly favors the entry of Na^+ on the basal side (Figure 1.7B). A $Na^+,K^+,2Cl^-$ co-transporter facilitates this passive entry, with the movement of Na^+ down its gradient dragging Cl^- and K^+ into the cell against their gradients. Thus, Na^+ is continually entering on the co-transporter and exiting on the Na^+,K^+-ATPase (i.e. Na^+ recycling). Cl^- and K^+ are accumulated above electrochemical equilibrium inside the chloride cell; Cl^- is extruded through apical channels into the seawater crypt, whereas K^+ exits basally back to the blood plasma (i.e. K^+ recycling). The transepithelial potential created by this Cl^- efflux, together with the cation-selective conductance of the 'leaky' junctions, drives the diffusive efflux of Na^+ through the paracellular channels in an amount almost equimolar to that of Cl^-.

Curiously, despite the abundance of Ca^{2+} in seawater, marine fish still appear to take up this cation actively at the gills (Figure 1.7B) using identical mechanisms to those of freshwater fish, although perhaps with greater reliance on basolateral Na^+/Ca^{2+} exchange (Flik *et al.*, 1995; Marshall, 1998). Acid–base regulation via the gills occurs more rapidly in marine fish than in freshwater teleosts, but the details have not yet been determined (Evans, 1993; Wood and Marshall, 1994). The chloride cells are rich in carbonic anhydrase. Most probably, electroneutral, independently regulated $Na^+/H^+, NH_4^+$ and $Cl^-/HCO_3^-, OH^-$ exchangers occur on the apical membranes, exclusively for the purposes of acid–base regulation (Figure 1.7B). In contrast to the freshwater situation, the high levels of Na^+ and Cl^- in seawater relative to the intracellular fluid are certainly sufficient to drive these processes. The resultant apical entry of Na^+ and Cl^- will oppose the net direction of transport outlined above. However, these fluxes are likely so small as to be a tolerable cost for the benefits of acid–base regulation.

General responses to toxicants

General structural responses

Mallatt (1985) has provided a comprehensive and quantitative synthesis of more than 100 toxicologic studies in which structural changes in the gills were examined by light or electron microscopy. This heroic review produced a number of important conclusions, which in general were supported by both earlier (Eller, 1975) and later (Evans, 1987; Lauren, 1991; Laurent and Perry, 1991; Wendelaar Bonga and Lock, 1992; McDonald and Wood, 1993; Heath, 1995; Wendelaar Bonga, 1997) reviews, that remain valid today. Most importantly, the majority of lesions reported in the literature, although concentration dependent (i.e. more severe in acute lethal exposures than in chronic sublethal exposures), are *non-specific* in nature and are not related to any one type of toxicant, exposure level (acute versus chronic), exposure medium (freshwater versus seawater), or fish species. However, for a few, which involve cell proliferation, there is an association with chronic exposures, and for one, which involves fluid movement, an association with salinity. There is usually considerable variability among different individuals exposed to the same toxicant in the same experiment. This could because many lesions result from stereotyped physiologic responses to stress, and can logically be considered defense responses; the more stressed the fish, the greater the lesions. Many of these same lesions are also seen in fish which are injected with toxicants rather than exposed to waterborne agents. They are also frequently seen in fish subjected to non-toxicant stresses (e.g. handling, crowding, hypoxia, low social rank). This latter conclusion emphasizes the extreme caution that must be exercised when holding and killing fish for gill morphology in toxicologic studies and the lack of value in sampling fish close to or at death for gill structural indices. Most lesions are independent of temperature, and, in a few cases, physiologic perturbation and death may occur in the absence of obvious gill pathology. Thus, structural changes in the gills, although obviously indicative of damage, are not a good diagnostic tool for the toxicant which is the causative agent of that damage, for its concentration or exposure duration, or for its key toxic mechanism of action at a physiologic level.

Mallatt (1985) identified fourteen different structural lesions of the gills which have been commonly recorded in toxicologic studies and tabulated the frequency with which they have been reported. However, Mallatt cautioned against assuming that this represents frequency of occurrence because different investigators may overlook different lesions to different degrees. All of them relate to the structure of the branchial epithelium or the apparent distribution of blood. The actual cellular structure of the branchial blood vessels (i.e. pillar cells, endothelium and smooth muscle of afferent and efferent vessels) appears very resistant. The fourteen different lesion types are summarized in Mallatt's very useful diagram of a freshwater trout gill (reproduced as Figure 1.8 here), numbered from 1 (most often reported) to 14 (least often reported). Unfortunately, both at the time of Mallatt's review (in which only 15 percent of the studies were in seawater or brackish water) and subsequently, there have been few investigations on toxic responses in marine fish.

Lifting of the pavement cells (1) away from the basement membrane at both the lamellar and filamental surfaces because of an infiltration of fluid is the most common lesion, an effect which would be expected to increase passive fluxes of ions and water by disrupting tight junctions, while inhibiting respiratory gas exchange by increasing diffusion distance and decreasing interlamellar distance (e.g. Figure 1.6). Possibly, this may also retard toxicant uptake. This is the only lesion with a greater reported occurrence in freshwater than in seawater, suggesting that osmotic water influx through the impaired epithelium may be a contributory factor to the fluid infiltration in freshwater and/or that osmotic dehydration may be a protective factor in seawater. *Fusion of neighboring lamellae* (3) and *epithelial rupture* (6) are probably the direct consequences of pavement cell lifting and indicate severe gill damage.

Necrosis (2) of both chloride and pavement cells is the second most frequently reported response overall, but is more common for metals than for organics or other toxicants, perhaps because metals directly interact with and inhibit ion transport proteins. Necrosis would be expected to inhibit ion and acid–base transport capacity and again probably increase diffusive permeability to ions and water. However, it should be realized that, until very recently, most investigators did not have the molecular and microscopic tools needed to distinguish easily between necrosis (accidental or externally induced cell death) and apoptosis (programmed cell death, resulting from natural or accelerated senescence), so both may have been classified as necrosis in many studies. The lesion described as *early chloride cell damage* (11) could also reflect early stages of either phenomenon. In true necrosis, TEM indicates that organelles and cytoplasmic

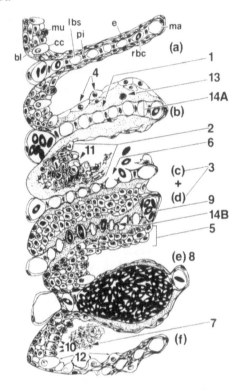

Figure 1.8 Composite diagram of common morphologic lesions in the gills caused by exposure to toxicants (from Mallatt, 1985, by permission). Six respiratory lamellae of a freshwater rainbow trout (*a–f*) are shown, with the top one (*a*) representing the control situation. The individual types of lesions, numbered 1–14, are described in the text. The abbreviations are: *bl*, basal lamina; *cc*, chloride cell; *e*, normal lamellar pavement cells; *lbs*, lamellar blood sinus (pillar cell channel); *ma*, marginal channel; *mu*, mucous cell; *pi*, pillar cell; *rbc*, red blood cell.

volume swell and become more electron opaque; cell membranes eventually rupture, with the contents probably lost by sloughing to the external water (Figure 1.9A and B). However, in apoptosis, the chromatin condenses, and the cytoplasmic volume and organelles generally shrink and become more electron dense (Figure 1.9A and C; see also Figure 1.3A), although the basolateral membrane tubular system of chloride cells may swell in some cases. Eventually, these condensed cells become apoptotic bodies which are engulfed by macrophages. *Leukocyte infiltration* (13) is therefore likely associated with apoptosis and should be considered an adaptive response. Very recently, a technique using confocal laser scanning microscopy and specific fluorescent markers for increased cell membrane permeability (ethidium bromide for necrotic cells) and for DNA breakdown [Tdt-mediated dUTP nick end label (TUNEL) for apoptosis] has been developed and applied to fish gills (Bury *et al.*, 1998). The results suggest that the stress hormone cortisol, in itself, promotes apoptosis but protects against necrosis of chloride

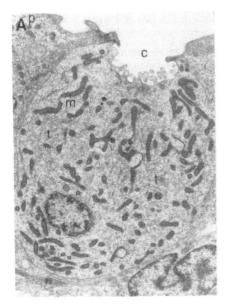
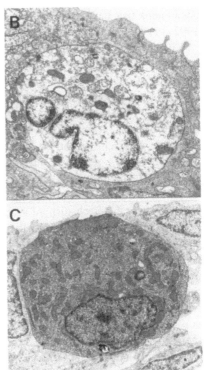

Figure 1.9 Transmission electron micrographs of chloride cells in the branchial epithelium of tilapia in freshwater illustrating chloride cell degeneration by necrosis versus apoptosis, associated with sublethal exposure to low environmental pH (from Wendelaar Bonga *et al.*, 1990, by permission). (A) Typical normal structure of a mature chloride cell on the filamental epithelium, probably of the β-type according to the criteria of Pisam *et al.* (1987); the apical crypt (*c*), mitochondria (*m*), and internal tubules (*t*) of the chloride cell, together with a neighboring pavement cell (*p*), are labeled (5500×). (B) Typical structure during necrosis; note that the cell has become more electron opaque, and has lost its apical crypt and most mitochondria, whereas tubules, nucleus, and remaining mitochondria have swollen (9500×). (C) Typical structure during early stages of apoptosis; note that the cell has become more electron dense as the chromatin and most cytoplasmic organelles are starting to shrink, and has lost its apical crypt (7200×).

cells, whereas copper as a toxicant promotes necrosis and not apoptosis in chloride cells.

Hypertrophy of the pavement cells (4) is probably a phenomenon related to necrosis as volume regulatory processes break down and the cells swell; again, this lesion is more frequently associated with metals. However, it is also possible that this hypertrophy sometimes reflects the rounding of pavement cells which occurs when they pull back to uncover increased chloride cell-surface area in response to acid–base and ionic disturbances or reflects the thickening and increased apical ornamentation that occurs specifically in response to acidosis, phenomena that have only been recognized in the last few years (Goss et al., 1992, 1995; Perry and Laurent, 1993).

Excess mucus secretion (7) may in fact occur more commonly than is reported because of the problems of preservation of mucus in histology discussed earlier. It is often associated with *mucous cell proliferation* (10) and seems to occur more commonly in response to metals than to organic toxicants. Its action is considered protective in limiting toxicant access to the branchial surface and perhaps also in limiting ion and water fluxes, but hypersecretion may block respiratory gas exchange, causing suffocation.

Proliferation of pavement cells (5), *mucous cells* (10), and *chloride cells* (12), with the last two often invading the surface of the respiratory lamellae, are viewed as protective or adaptive responses and are central to the 'damage-repair' hypothesis for the mechanism of acclimation (McDonald and Wood, 1993). This interpretation is in accord with the fact that these lesions are reported more commonly during chronic exposures than during acutely lethal exposures. Chloride cell proliferation is probably stimulated by the stress hormone cortisol (Perry and Wood, 1985; McCormick, 1990; Laurent et al., 1994b) and mucous cell proliferation by the ionoregulatory hormone prolactin, which is mobilized in freshwater fish in response to low plasma electrolyte levels. Indeed, there is considerable circumstantial evidence (summarized by Wendelaar Bonga, 1997) that the whole cell cycle of mitosis, maturation, senescence, and death is accelerated during sublethal toxicant exposure. Obviously, proliferation of ion-transporting cells should help increase active electrolyte transport, and the general thickening of the epithelia and increased mucous cell activity should help reduce diffusive ion permeability. However, ion transport rates or activities of key transport enzymes may not be as high as indicated by the number of chloride cells because a greater percentage of them may be immature, dying, or otherwise impacted by the toxicant (Wendelaar Bonga and Lock, 1992).

An important 'penalty' to the proliferation response has been highlighted in studies on fish adapting to ion-poor, very soft, often acidic waters. In this treatment, both mucous cells and chloride cells proliferate, and the latter (probably α-type cells) almost cover the respiratory lamellae as they migrate up from stem cells in the body of the filament (Figure 1.10; Laurent et al., 1985; Spry and Wood, 1988; Laurent and Hebibi, 1989; Greco et al., 1996). This greatly increases the blood-to-water diffusion distance, decreases interlamellar distance, and causes an overall reduction in the diffusive conductance of the gills to respiratory gases, thereby decreasing arterial Po_2, increasing ventilatory costs, and decreasing hypoxia tolerance (Thomas et al., 1988; Greco et al., 1995; Perry et al., 1996). Likely aerobic swimming performance would also be reduced, as with lifting and hypertrophy of pavement cells (see Figure 1.6), but this has not yet been tested.

The final three of Mallatt's lesions, *lamellar aneurysms* (8), *vascular congestion of the marginal channels* (9), and *lamellar blood sinus constriction or dilation* (14), together with *leukocyte infiltration* (13) could be considered part of an inflammatory response, although Mallatt argued against this interpretation. Alternatively or additionally, these blood flow-related lesions may be associated

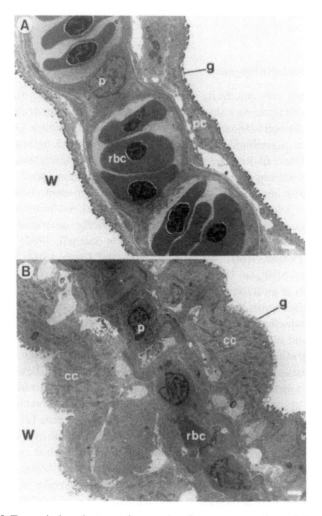

Figure 1.10 Transmission electron micrographs of transverse sections of the respiratory lamellae of freshwater rainbow trout, illustrating chloride cell proliferation, and its effect on the blood-to-water diffusion distance (from Greco *et al.*, 1996, by permission). (A) Typical morphology under control conditions. (B) Typical morphology after 2 weeks of exposure to ion-poor water. Chloride cells (*cc*), glycocalyx (*g*), pillar cells (*p*), pavement cells (*pc*), red blood cells (*rbc*), and the external water (*w*) in the interlamellar channel are labeled (5000×; bars, 1 μm).

with blood shunting in the gills that is driven by mobilization of the stress hormones adrenaline (epinephrine) and noradrenaline (norepinephrine) as well as by autonomic nervous discharge, as discussed earlier. In addition, the marked rise in blood pressure associated with some chronic toxicant stresses (e.g. Milligan and Wood, 1982; Wilson and Taylor, 1993a), again driven by plasma catecholamine elevation and autonomic nervous activity, would be expected to dilate physically the afferent and efferent lamellar arterioles and the pillar cell channels in the gills.

General physiologic responses

In light of the non-specificity of most structural responses, Mallatt (1985) concluded that future work should more valuably concentrate on the physiologic responses of the fish to aquatic toxicants. Indeed, interest has turned more strongly to mechanistic physiologic research on gill effects of specific toxicants in the past 15 years, and these findings are the focus of the remainder of this review. However, as with structural responses, there is a standard suite of non-specific physiologic responses which are seen at 'industrial', acutely toxic concentrations of virtually all toxicants (i.e. levels well above the 96-h LC_{50} value). As these responses are of limited mechanistic relevance, they will be only briefly summarized here, based on a synthesis of several sources (Hughes, 1976; Satchell, 1984; Evans, 1987; Barton and Iwama, 1991; Wendelaar Bonga and Lock, 1992; Heath, 1995; McDonald and Milligan, 1997; Wendelaar Bonga, 1997).

High levels of most toxicants generally cause an initial increase in flux rates of respiratory gases across the gills. These are associated with increased consumption of O_2 and production of CO_2 and ammonia at the systemic tissues as a result of excitement and escape responses. Increased gas exchange is facilitated by the rapid mobilization (within seconds) of catecholamines into the plasma from chromaffin tissue and increased sympathetic nervous activity. These agents dilate the gills, open up more respiratory lamellae, favor the arterioarterial pathway and reduce flow in the arteriovenous pathway, increase blood pressure and cardiac output, and generally shunt blood flow through pathways more favorable for gas exchange in lamellae. The catecholamine surge may stimulate hyperventilation at this time. Irritant effects on the branchial neuroepithelial cells (putative chemoreceptors) may also contribute to immediate hyperventilation. The increases in gill blood flow and gill water flow favor increased uptake of the toxicant, especially in the case of lipophilic contaminants that readily diffuse across the gills. Cortisol mobilization, which is slower (within minutes) from inter-renal tissue, heightens catecholamine responsiveness, and additionally promotes protein breakdown, thereby augmenting ammonia excretion. Later, as the host of severe structural changes discussed above develop (particularly epithelial lifting and hypertrophy, fusion of lamellae, and hypersecretion of mucus) and the catecholamine response is exhausted, gas exchange falls progressively until the fish dies. This is due to progressively increased diffusion distance, decreased surface area for gas exchange, and blockade of water flow channels by mucus and swollen fused lamellae. The accompanying drop in Po_2, rise in Pco_2, and decrease in pH in arterial blood reinforce the costly and futile hyperventilation. The Bohr and Root effects caused by CO_2 retention and acidosis further inhibit O_2 uptake, and severe lactacidosis develops as tissues switch to anaerobic metabolism.

A common ventilatory response to many, but not all, toxicants is an increase in the rate of coughing. For some toxicants (e.g. copper, mercury, methylmercury), increased coughing is an extremely sensitive indicator of sublethal levels of the pollutant around the threshold level for chronic toxicity (Drummond *et al.*, 1973, 1974). Indeed, the recording of cough rate has occasionally been advocated as a pollution-monitoring tool (e.g. Schaumburg *et al.*, 1967). Coughs may occur in several forms and have been described in detail by Satchell (1984). All involve a rapid expulsion of water from the buccal and opercular cavities, and usually include a reversal of water flow over the gills. They seem to be a reflex response to stimulation of mechanoreceptors or nociceptors in the gills and buccal cavity, and probably normally serve to clear particles from the gill mesh and eject noxious materials that are accidentally inhaled.

Diffusive fluxes of electrolytes (principally Na^+ and Cl^-) and water (occurring in opposite directions to each other, and opposite directions in freshwater versus marine fish) are initially elevated by the branchial surface area and blood-shunting effects of catecholamines and sympathetic activity. Increased blood pressure may play an important role here, stretching the lamellar tissue and distorting tight junctions. Later, the net losses or gains of ions and water persist, albeit at a reduced rate, as the catecholamine response attenuates and the gradients tend to run down. However, damage and death of transporting cells decreases active ion transport against the gradients. At the same time, passive permeability is elevated by the non-specific branchial damage, especially lifting of pavement cells and epithelial rupture, whereas lamellar fusion and excessive mucus secretion may tend to attenuate it, although at the cost of severe respiratory maladaptation. Loss of 35–40 percent of the body NaCl content of a freshwater fish can occur within the first several hours of high toxicant exposure, sufficient to precipitate death. At this stage, the moribund fish is usually severely acidotic, which results in an increased efflux of acidity across the gills.

General protective effects of water hardness

A general paradigm that has stood the test of time is that water hardness, the sum of the concentrations of the divalent cations Ca^{2+} and Mg^{2+}, has a direct protective effect against the toxicity of a wide range of aquatic toxicants in freshwater, even those that exert their toxic effects internally rather than at the gills (Brown, 1968; Hunn and Allen, 1974; Alabaster and Lloyd, 1980). The harder the water, the lower the toxic response, with an approximate logarithmic–linear or logarithmic–logarithmic relationship between hardness and toxicity reduction. For most toxicants, $[Ca^{2+}]$ exerts a greater protective effect than does $[Mg^{2+}]$ on a molar basis; additionally, $[Ca^{2+}]$ is normally two- to tenfold higher than $[Mg^{2+}]$ in most natural freshwaters. In many jurisdictions, water hardness has been the sole feature of water geochemistry which is incorporated into environmental regulations as a modifying factor for allowable toxicant level (e.g. Alabaster and Lloyd, 1980; USEPA, 1986; CCME, 1995; Bergman and Dorward-King, 1997; Meyer, 1999). All of the protective effects of hardness can be ascribed to actions at the gills, but may involve several different mechanisms, as follows:

1   The gill surface, and the mucus covering the gill, comprises a variety of functional anionic groups (e.g. sulfhydryl groups, carboxylic acids) to which positively charged water toxicants (e.g. free metal ions) can bind. Some of these anionic groups may be relatively non-specific (i.e. low-affinity binding sites) whereas others may be highly Ca^{2+} specific (e.g. high-affinity binding sites such as the apical Ca^{2+} channels in the chloride cells; Figure 1.7). In either case, Ca^{2+} and Mg^{2+} can directly compete with the toxicant for these same sites, thereby preventing the binding or entry of the toxicant into the gill (Zitko and Carson, 1976; Pagenkopf, 1983; Bergman and Dorward-King, 1997; Playle, 1998).

2   The divalent nature of the hardness cations makes them particularly effective at stabilizing the junctional complexes between gill cells, thereby limiting permeability to ions and water (Oduleye, 1975; Hunn, 1985). In harder water, fish have lower overall gill permeability and osmoregulatory costs. Hardness will retard toxicant uptake and minimize non-specific or specific permeability-increasing effects. This protection should apply to toxicants in general, regardless of their charge. However, some toxicants such as low environmental pH directly titrate Ca^{2+} away from the junctions, so the protective effect of elevated water $[Ca^{2+}]$ is obvious here (McDonald, 1983; McWilliams, 1983). Furthermore, differential paracellular permeability to Na^+ and Cl^- is responsible for the diffusion potential across freshwater fish gills (Potts, 1984). Through their stabilizing effect on this permeability, elevations in $[Ca^{2+}]$ reduce the (inside) negative transepithelial potential across the gills, thereby decreasing the electrochemical gradient for uptake of cationic toxicants such as metals.

3   When fish are acclimated to very soft water, the proliferation of chloride cells and mucous cells which occurs (see above; e.g. Figure 1.10) may result in increased uptake or binding of a toxicant, especially if it is transported via the chloride cells (Heath, 1995). The thickening of the epithelium may also exacerbate respiratory toxicity. These problems are attenuated by increasing hardness.

4   Hardness is often a surrogate for other protective aspects of water chemistry, such as greater pH, alkalinity, $[Na^+]$, $[Cl^-]$, total ionic strength, or organic ligands, all of which can protect in a variety of different ways. These include effects on diffusive gradients for passive ion loss, on substrate concentrations for active ion uptake, on the speciation of the toxicant, or on speciation of the charged groups on the branchial surface (Pagenkopf, 1983; Bergman and Dorward-King, 1997; Playle, 1998).

Surface active toxicity and gill binding models

With respect to the concepts of the preceding section, the proposal of the Gill Surface Interaction Model by Pagenkopf (1983) represented a fundamental breakthrough in the understanding of the toxic responses of the gill. In brief, drawing on earlier work by Zitko and Carson (1976), Pagenkopf introduced the now widely accepted idea that the gill surface presents a family of anionic sites to which cationic metals can bind, thereby inducing toxic physiologic effects, i.e.

'surface-active' toxicity. These sites are finite in number, and interactions with them are rapid relative to the time necessary for physiologic toxicity to develop, so that as a first approximation equilibrium modeling theory can be applied. Furthermore, Pagenkopf postulated that both H^+ ions and the hardness cations Ca^{2+} and Mg^{2+} will compete with metals for these same sites, whereas anionic ligands in the water column such as hydroxide, bicarbonate, and dissolved organic matter would tend to complex the metals, thereby reducing their availability to bind to the gills. All of these reactions, including those with anionic ligands on the gill, could be characterized by conditional equilibrium stability constants (commonly termed log K_D values). The higher the log K_D value, the stronger the binding. Thus, the final equilibrium partitioning of the metal could be described by a series of geochemical equations, with the important output being the amount of metal bound to the gill. Pagenkopf postulated that the latter could be related to toxicity. The later advent of computer-based geochemical modeling programs such as MINTEQA2 (Allison *et al.*, 1991) and MINEQL+ (Schecher and McAvoy, 1992) in which most common K_D values are available, and into which estimates for metal–gill K_D values can easily be entered, greatly simplified computations.

Subsequent refinement of these ideas included the realization that there may be different gill anionic sites for metals of different chemistry, that these sites may in fact represent important apical or basolateral membrane transport proteins, that there may be different populations of sites with different log K_D values – i.e. high-affinity and low-affinity sites or specific and non-specific binding sites – and that a high-affinity gill site could strip metal away from a low-affinity ligand in the water column, and vice versa (McDonald *et al.*, 1989; Reid and McDonald, 1991; Reid *et al.*, 1991; Wood, 1992; Allen and Hansen, 1995; Meyer, 1999). However, a second major breakthrough came from the work of Playle and co-workers (Playle *et al.*, 1992, 1993a,b; Janes and Playle, 1995; Playle, 1998; Richards and Playle, 1998). This group has articulated simple methods based on a combination of Langmuir adsorption isotherm analysis and competition experiments with organic acids of known K_D values and concentration to estimate metal–gill K_D values. The basic approach is to measure metal deposition on fish gills after relatively short-term exposure (e.g. 3 h) of the fish to the metal in a synthetic softwater system of known composition in the presence of known concentrations of various competing (e.g. Na^+, Ca^{2+}, H^+) and complexing (e.g. Cl^-, HCO_3^-, OH^-, organic acids, dissolved organic matter) agents. This approach has now been applied to model successfully gill loading of copper, cadmium, silver, and cobalt, with some extrapolation to resultant pathophysiologic effects and mortality; potentially it could be applied to any cationic toxicant. Figure 1.11 shows a recent model for cobalt (Richards and Playle, 1998). The gill binding model has now been advocated as a regulatory approach for deriving site-specific water quality criteria based on known water chemistry and gill binding constants (Bergman and Dorward-King, 1997; Renner, 1997).

Specific responses to selected toxicants

Our focus here will be on those toxicants for which physiologic, mechanistic studies have been performed, and for which there is a reasonably good

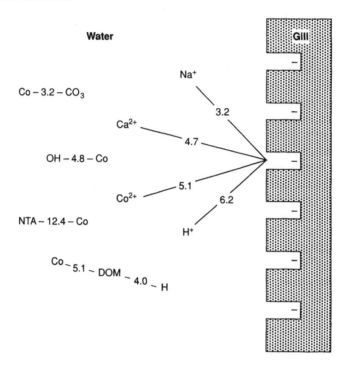

Figure 1.11 An example of an equilibrium gill binding model for cobalt at the gills of freshwater rainbow trout (data from Richards and Playle, 1998). Net negative charge on anionic sites on the gills is illustrated, together with experimentally determined metal–gill K_D values for the binding of Co^{2+} and other cations to these sites, and the comparable K_D values for various ligands (NTA, nitrilotriacetic acid; DOM, dissolved organic matter) which may complex Co^{2+} in the water column. The higher the K_D value, the higher the binding strength.

understanding of their toxic actions at the gills. Ideally, such studies have followed the responses of fish over time to environmentally realistic concentrations of toxicant, i.e. concentrations at or below the 96-h LC_{50} level. Most useful here are studies which used optimal methodology. For exposures, open flow-through systems are best for avoiding speciation changes and/or depletion of the toxicant. Measurements of the movements of ions, acid–base equivalents, and respiratory gases across the gills are most sensitively determined by monitoring their fluxes with the water, with unidirectional fluxes of electrolytes separated by means of radiotracers (e.g. ^{22}Na, ^{36}Cl, ^{45}Ca), as reviewed by Wood (1992). The system should be closed for the minimum time necessary to make the measurements. Blood samples for blood gases, acid–base status, electrolytes, metabolites, and catecholamines should be drawn via chronic indwelling catheters, and for cortisol by rapid anesthetic overdose followed by caudal puncture (Barton and Iwama, 1991; also acceptable for electrolytes other than K^+). Tissue metabolites should be taken by anesthetic overdose followed by freeze-clamping.

Low pH

Environmental acidity is one of the most thoroughly studied toxicants, reflecting the acid rain problem which drove research in the 1970s and 1980s. This phenomenon is relevant only in freshwater, most usually poorly buffered softwaters, although acid mine drainage may contaminate harder waters. In nature (and probably unwittingly in some laboratory exposures), responses are often complicated by the simultaneous presence of metals, particularly aluminum, mobilized at low pH. At environmentally realistic pH values > 4.0, in the absence of aluminum, respiratory effects are negligible, although ventilation does tend to increase. Low pH is a classic ionoregulatory toxicant, with its key toxic mechanism of action occurring at the gills; these branchial effects have been reviewed extensively (Wood and McDonald, 1982; McDonald, 1983; Wood, 1989; Heath, 1995; Reid, 1995).

Low pH inhibits the net uptake of both Na^+ and Cl^- at the gills by the dual action of inhibiting the active uptake components (unidirectional influxes) and stimulating their passive diffusive effluxes (also K^+ efflux). Ca^{2+} influx is also reduced and Ca^{2+} efflux stimulated, but as these effects are transient they are considered inconsequential relative to the effects on Na^+ and Cl^- balance. Influxes appear to be more sensitive than effluxes. In rainbow trout, the threshold for Na^+ and Cl^- efflux stimulation is usually below pH 5.0, whereas an acute drop of only 1.0 pH unit below the circumneutral acclimation level causes a significant 50 percent inhibition of the influxes. By pH 4.0, influx blockade is almost 100 percent, although certain strains may be more resistant (Balm and Pottinger, 1993). Certainly, other species such as tilapia are far more resistant (Wendelaar Bonga *et al.*, 1984). The progressive loss of Na^+ and Cl^- from the blood volume (Figure 1.12A and B) entrains secondary fluid volume disturbances and cardiovascular failure which may eventually kill the fish (see Milligan and Wood, 1982; Wood, 1989). At more moderate pH values (4.8–5.2), trout achieve a new steady state by reducing the effluxes to match the inhibited influxes. Chronic cortisol elevation promotes chloride cell proliferation (e.g. Leino and McCormick, 1984; Tietge *et al.*, 1988; Wendelaar Bonga *et al.*, 1990) which may facilitate partial restoration of influxes. However, plasma Na^+ and Cl^- levels generally do not recover fully in trout (Audet *et al.*, 1988), although they may in tilapia (Wendelaar Bonga *et al.*, 1984).

In the older model of gill Na^+ transport, the mechanism of Na^+ uptake blockade is readily explained by direct competitive blockade of the apical $Na^+/H^+,NH_4^+$ exchangers by external H^+s, or in the newer model by low external pH creating a $[H^+]$ gradient against which the H^+-ATPase cannot function (Figure 1.7A). Ammonia excretion is modestly inhibited (Wright and Wood, 1985), in accord with either model. However, the marked inhibition of Cl^- uptake has never been satisfactorily explained. The effect could result from a direct effect of external H^+ on the Cl^- pumping mechanism, perhaps via depletion of necessary basic counterions (HCO_3^-, OH^-) from the chloride cells. The activity of basolateral Na^+,K^+-ATPase is reduced to a variable degree, probably dependent on the extent of chloride cell necrosis and apoptosis (Wendelaar Bonga *et al.*, 1990), and there

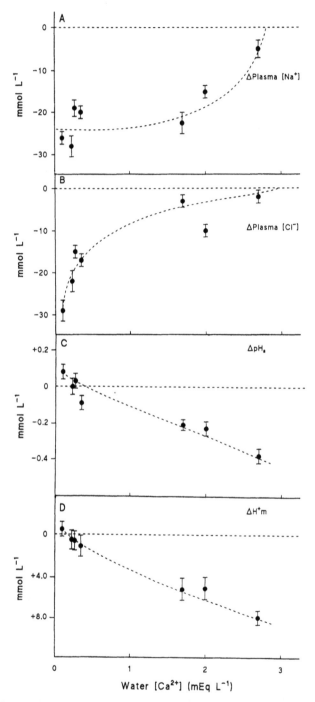

Figure 1.12 The relationship between water calcium concentration ($[Ca^{2+}]$) and the extent of various ionic and acid–base disturbances in the arterial blood of freshwater rainbow trout exposed to a pH of approximately 4.3 for 3 days (data from Wood, 1989). Changes, relative to pre-exposure control value at circumneutral pH, in (A) plasma sodium, (B) plasma chloride, (C) arterial plasma pH, and (D) metabolic acid load in the arterial blood. Note that higher water hardness (i.e. $[Ca^{2+}]$) protects against Na^+ versus Cl^- loss differentially, but exacerbates acid–base disturbance.

are marked changes in both the phospholipid and fatty acid composition of the gill cells (Bolis *et al.*, 1984). The increases in diffusive efflux of Na^+, Cl^-, and K^+ appear to be due to an opening of tight junctions, probably as a result of Ca^{2+} displacement by H^+ ions (McWilliams, 1983; Reid, 1995). The protective action of water hardness is against this diffusive efflux, not against the inhibition of the influx. Interestingly, not only does hardness ameliorate the Na^+, K^+, and Cl^- diffusive effluxes but it also changes their ratio, presumably by changing the cation versus anion selectivity of the paracellular channels (Figure 1.12A and B; Wood, 1989). Thus, in trout at pH ~ 4.3, where active influxes are strongly inhibited, Na^+ plus K^+ loss greatly exceeds Cl^- loss, constraining a large drop in blood pH and sizeable metabolic acidosis (Figure 1.12C and D) when the 3-day exposure is conducted in hardwater (e.g. $[Ca^{2+}] = 1.0–3.0$ mequiv L^{-1}). However, when the same exposure is performed in softwater (e.g. $[Ca^{2+}]$ ~ 0.5 mequiv L^{-1}) typical of acidified lakes in Northern Europe and North America, Na^+ and Cl^- losses are much larger but are approximately equal (and K^+ loss is low), so there is negligible internal acidosis. This modifying effect of water hardness is also seen in complex alterations of the transepithelial potential at low pH (McWilliams and Potts, 1978), the interpretation of which remains controversial (Wood, 1989).

High pH

Alkaline exposure may occur as a result of liming operations, industrial discharges, photosynthetic pH surges, or the introduction of fish into naturally alkaline lakes, and has received detailed study in recent years (Wright and Wood, 1985; Lin and Randall, 1990; Wilkie and Wood, 1991, 1994; Yesaki and Iwama, 1992; Wilkie *et al.*, 1993, 1994, 1999; Wright *et al.*, 1993; Iwama *et al.*, 1997; McGeer and Eddy, 1998). The problem is largely confined to freshwater, although some alkaline lakes are saline. The branchial responses, which are quite complex, have been recently reviewed (Wilkie and Wood, 1996).

The principal problems at high pH levels (e.g. 9.5–10.5) appear to be a blockade of ammonia excretion and an acceleration of CO_2 excretion, both explicable by the altered conditions in the boundary layer of the gills. Because the pH is close to or above the pK (9.1–9.5) of the $NH_4^+ \leftrightarrow NH_3$ conversion, a significant proportion of excreted ammonia will remain as NH_3 in the water, thereby raising P_{NH_3} in the boundary layer and inhibiting diffusive NH_3 flux (see Figure 1.5). Additionally, any NH_4^+ excretion coupled directly or indirectly to Na^+ influx is reduced (Figure 1.7A) because high pH also inhibits this process. In trout at high pH, ammonia excretion drops close to zero, and may even reverse to net uptake if any ammonia is present in the water. Ammonia accumulates internally and blood P_{NH_3} (Figure 1.13D), $[NH_4^+]$ (Figure 1.13E), and total ammonia concentration increase dramatically. If the fish survives, ammonia excretion is restored to normal levels over 48–72 h as the gradients are reset, but blood ammonia levels remain greatly elevated. A transient increase in urea excretion, perhaps a compensation to sustain N waste excretion, may occur during the period of inhibited ammonia excretion. Interestingly, elevated water hardness helps to protect against the

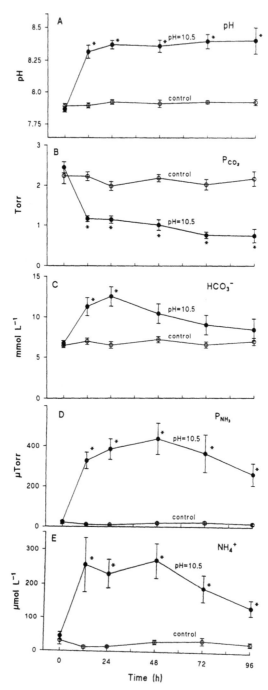

Figure 1.13 Responses of acid–base and ammonia status in the arterial blood plasma of rainbow trout exposed to high environmental pH (approximately 10.5) over a 4-day period (data from McGeer and Eddy, 1998). (A) Arterial plasma pH, (B) partial pressure of CO_2, (C) bicarbonate ion concentration, (D) partial pressure of NH_3, and (E) ammonium ion concentration. Control fish were held at circumneutral pH.

inhibition of ammonia excretion by an unknown mechanism (Yesaki and Iwama, 1992; Iwama *et al.*, 1997).

Simultaneously, high boundary layer pH accelerates the hydration of excreted CO_2 ('CO_2 vacuum effect'), thereby keeping boundary layer P_{CO_2} close to zero (Figure 1.5). Blood P_{CO_2} levels fall by up to 80 percent, which causes a substantial rise in blood pH ('respiratory alkalosis'; Figure 1.13A). As a result of this alkalosis, a greater proportion of this elevated internal ammonia exists in the highly toxic NH_3 form in the blood and body fluids, which may precipitate mortality by neural and metabolic effects. In some studies (e.g. McGeer and Eddy, 1998), this is further compounded by HCO_3^- accumulation ('metabolic alkalosis'; Figure 1.13B), probably associated with inhibition of apical $Cl^-/HCO_3^-,OH^-$ exchange (Figures 1.5 and 1.7A), whereas in other investigations a partially compensating HCO_3^- reduction by lactacidosis comes into play (e.g. Wilkie and Wood, 1991).

Additionally, ionoregulatory problems occur at high water pH (Wilkie and Wood, 1994; McGeer and Eddy, 1998; Wilkie *et al.*, 1996, 1999), although these appear to be less severe than during similar low pH exposure (i.e. at same pH deviation from neutrality). They result from inhibition of active Na^+ and Cl^- uptake processes, with negligible effects on unidirectional effluxes, suggesting that the basic permeability of the junctional pathways is not altered. The inhibition of Cl^- uptake is easily explained by direct HCO_3^- or OH^- competition at the apical $Cl^-/$ HCO_3^-,OH^- exchange pathway ; the decrease in Na^+ uptake may be a consequence of internal alkalosis, i.e. decreased availability of H^+ for exchange against Na^+ (Figures 1.5 and 1.7A). During continued exposure, ion balance is re-established, with a complete recovery of Cl^- uptake, partial recovery of Na^+ uptake, and a reduction in unidirectional Na^+ efflux. The former, at least, appears to be associated with a marked hyperplasia of chloride cells.

Aluminum

The physiologic toxicology of aluminum has been studied intensively in relation to the acid rain problem (reviewed by Neville, 1985; Wood and McDonald, 1987; McDonald *et al.*, 1989; Playle and Wood, 1989b, 1991; Wood, 1989; Spry and Wiener, 1991; Heath, 1995; Poleo, 1995; Brown and Waring, 1996; Sparling and Lowe, 1996; Wilson, 1996). It is the third most abundant element in the earth's crust and is highly soluble at low pH, so most naturally acidified freshwaters contain substantial amounts of aluminum, which may pose a larger threat to the fish than the acidity itself. Aluminum is a specific surface-active toxicant at the gills; indeed, its ability to penetrate into the bloodstream is negligible. At neutral pH, aluminum is relatively innocuous, but becomes highly toxic at low pH, with effects superimposed on those of acidity itself. However, because of its complex aqueous speciation and solubility relationships, both the extent and the mechanism of aluminum toxicity are critically dependent on water pH and hardness ($[Ca^{2+}]$); for a fixed concentration of aluminum, toxicity is actually greater to salmonids at bulk water pH 5.0 than pH 4.4, and at pH 4.0 aluminum may even be protective. These toxic effects are attenuated, but not completely eliminated, when aluminum is complexed by strong anionic ligands such as dissolved organic carbon or fluoride (Wilkinson *et al.*, 1990; Roy and Campbell, 1997).

Aluminum is unique in being the only metal which is a potent respiratory toxicant at environmentally realistic concentrations (20–500 µg L^{-1}). In general, the toxic mechanism switches from mainly respiratory toxicity at the higher acidic pH values (> 5.2) to mainly ionoregulatory toxicity at lower acidic pH values (< 4.8), with synergistic actions in the region of overlap. Hardness either exacerbates or has no effect on this respiratory toxicity, and either protects against or has no effect on ionoregulatory toxicity. At temperatures close to 0°C, the respiratory effect may predominate at lower pH values as well (Laitinen and Valtonen, 1995). Ionoregulatory toxicity is principally manifested by large losses of Na$^+$ and Cl$^-$ from the fish, but K$^+$ and Ca^{2+} are also lost in much smaller amounts. Interestingly, there is some evidence that aluminum may inhibit Ca^{2+} influx in the pH range of respiratory toxicity, although the mechanism is unknown (e.g. Sayer *et al.*, 1991; Verbost *et al.*, 1992). The protective action of aluminum at very low pH (where it exists as Al^{3+}) may reflect its ability to exert Ca^{2+}-like stabilizing actions at the gills. Wilkinson *et al.* (1990) estimated a log K_D value for Al^{3+} binding to gills of about 6.5.

The solubility of aluminum is at a minimum close to pH 6.0, and increases exponentially both above and below this point (Figure 1.14). However at acidic pH values, the water pH at the gill surface (as estimated by expired water pH measurements; Playle and Wood, 1989a; Lin and Randall, 1990) is raised considerably above that in the bulk medium because of the alkalinizing effect of excreted ammonia (see Figure 1.5). As illustrated in the model in Figure 1.14 from Playle and Wood (1989b), the effective solubility of aluminum at the gill surface is therefore greatly reduced, so dissolved aluminum in inhaled water will fall out of solution onto the lamellae, especially if the complex polymerization and precipitation processes have already been nucleated. In addition, the speciation will shift towards Al(OH)$_3$ and Al(OH)$_2{}^+$, and away from Al(OH)$^{2+}$ and Al^{3+} which predominate at lower pH values. The respiratory problems are therefore thought to result from precipitation of the neutral and monovalent hydroxides, with resultant irritation, excessive mucus production, and swelling of the lamellar epithelium to the point of lamellar fusion and rupture, resulting in low blood P_{O_2}, high blood P_{CO_2}, lactic acid production, respiratory and metabolic acidoses, catecholamine mobilization, coughing, and hyperventilation (Playle *et al.*, 1989; Witters *et al.*, 1991). Most of the precipitated aluminum appears to be bound to mucus and is continually being sloughed off (Wilkinson and Campbell, 1993) because only 10–15 percent of the aluminum which is extracted from the ventilatory water flow actually builds up on the gills (Playle and Wood, 1991). Ionoregulatory problems (losses of Na$^+$, K$^+$, Cl$^-$, and Ca^{2+}) at lower pH would be caused by the divalent and trivalent species acting like H$^+$ ions alone to increase paracellular permeability and inhibit apical Na$^+$ and Cl$^-$ uptake processes. In addition, a significant portion of the influx reduction would occur through direct inhibition of basolateral Na$^+$,K$^+$-ATPase activity of the ion-transporting cells (Staurnes *et al.*, 1984; Rosseland *et al.*, 1992). The time taken for these aluminum species to

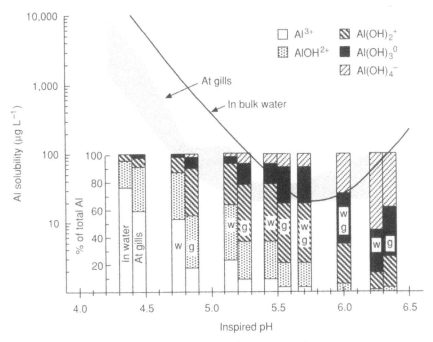

Figure 1.14 A model describing changes in solubility and speciation of aluminum which contribute to toxic precipitation and polymerization in the microenvironment of the gills of rainbow trout in soft freshwater (from Playle and Wood, 1989b, by permission). Trout were exposed to a total aluminum concentration of 93 μg L^{-1} at various acidic pH values in the inspired (bulk) water. The solid line shows the theoretically predicted solubility limit of aluminum at the various inspired pH values ; the shaded area indicates the solubility limit (with 95 percent confidence limits) actually predicted to occur in the gill microenvironment based on measured pH changes at the gills. The bars show the predicted speciation of aluminum (as a percent of the total 93 μg L^{-1}) in the gill microenvironment (*g*) in comparison with that in the inspired bulk water (*w*) at various inspired pH values. Note that the solubility scale is logarithmic, whereas the speciation scale is linear.

penetrate to the basolateral tubule system would explain why inhibition of Na$^+$ influx develops slowly whereas stimulation of ionic efflux is almost immediate.

If the fish survives the first few days of exposure, at least a partial recovery of resting respiratory and ionoregulatory homeostasis occurs. This is associated with a reduction in gill aluminum burden and mucus content (McDonald *et al.*, 1991) and with an apparent compartmentalization of the remaining aluminum into inert membrane-bound precipitates within the gill cells (Youson and Neville, 1987). There is also change in the binding affinity of the gill ligands for aluminum and a decreased ability of aluminum to displace calcium (Reid *et al.*, 1991). Mucous cell and chloride cell hyperplasia and a general repair of lamellar structure occurs, thickening the body of the filament. There is a chronic reduction in respiratory surface area associated with shorter respiratory lamellae and an increase in water-to-blood diffusion distance; exercise performance and maximal O$_2$ uptake rate are correspondingly reduced (Tietge *et al.*, 1988; Mueller *et al.*, 1991; Wilson *et al.*, 1994; Wilson, 1996). No studies appear to have checked for metallothionein induction.

Aluminum sulfate is often used to clarify drinking water. Sewage treatment plants and acidic, aluminum-rich rivers often discharge into well-buffered estuaries, and acidified lakes are often limed. Under these neutral and mildly alkaline pH conditions, the toxicity of environmentally relevant levels of aluminum appears to be negligible, as long as the water chemistry is at equilibrium (Freeman and Everhart, 1971; Heming and Blumhagen, 1988). However, under non-steady-state conditions in mixing zones, acute lethality may result from aluminum precipitation and polymerization on the gills (Rosseland *et al.*, 1992; Poleo, 1995; Witters *et al.*, 1996b).

Silver

The physiologic toxicology of silver (reviewed by Hogstrand and Wood, 1998; Wood *et al.*, 1999) has received recent attention because of its discharge in photographic processing and mining effluents. In freshwater, only the free ion Ag^+ appears to exert acute toxicity, and, indeed, in this form it appears to be the most toxic of the metals, with LC_{50} values in the range of a few $\mu g\ L^{-1}$. Hardness exerts a weak beneficial influence, whereas pH and alkalinity are of negligible influence. However, natural anionic ligands with a strong affinity for Ag^+, such as Cl^-, dissolved organic carbon, sulfide, and thiosulfate (the last in photographic effluent), provide almost complete protection when present in excess. Therefore, in seawater, all free Ag^+ is bound up by Cl^-, and silver is at least three orders of magnitude less toxic than in freshwater. Even at very high lethal levels ($\sim 1000\ \mu g\ L^{-1}$), there is no clear evidence of gill toxicity in marine fish, although drinking and intestinal water absorption appear to be inhibited (Wood *et al.*, 1999).

In freshwater, silver is a highly specific ionoregulatory toxicant to the gill with resulting internal effects in the fish reminiscent of those of low pH, i.e. progressive decline in blood Na^+ and Cl^- levels, fluid volume disturbance, and circulatory failure (Wood *et al.*, 1996). However, the mechanism of action appears to be far more specific than that of low pH: an extremely potent, non-competitive inhibition of the active uptake of both Na^+ and Cl^-, with negligible stimulation of the diffusive efflux components (Figure 1.15). Indeed, effluxes generally decline as the internal gradients run down. The blockade of active uptake is explained by a potent inhibition of Na^+,K^+-ATPase activity in the ion-transporting cells, related to the strong affinity of Ag^+ for sulfhydryl groups (Morgan *et al.*, 1997); carbonic anhydrase activity is also mildly inhibited (see Figure 1.7A). The log K_D value for Ag–gill binding is very high, about 10.0 (Janes and Playle, 1995). The effect on influxes takes about 8 h to develop fully (Figure 1.15), presumably reflecting the time taken to penetrate to the basolateral membranes. Internal metabolic acidosis develops associated with a uptake of acidic equivalents across the gills, and a marked decline in plasma HCO_3^-. The mechanism for this is unclear, but as net Na^+ and Cl^- losses are approximately equimolar it is likely related to the large loss of K^+ across the gills which occurs in the face of unchanged plasma $[K^+]$

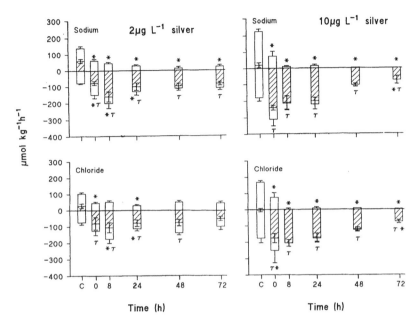

Figure 1.15 Responses of branchial fluxes of sodium and chloride in rainbow trout exposed to two concentrations of total silver (2 or 10 µg L⁻¹) in freshwater over a 3-day period (data from Morgan *et al.*, 1997). Upward open bars represent unidirectional influx rates; downward open bars represent unidirectional efflux rates; and hatched bars represent net flux rates. Significant differences relative to the pre-exposure values (*C*) are indicated by asterisks for influx, and by *T* for net fluxes. Note that net fluxes quickly became negative because of potent and persistent inhibition of the influx components with only a small initial stimulation and later attenuation of the efflux components.

(Webb and Wood, 1998). There is no inhibition of ammonia excretion, but internal ammonia levels and ammonia excretion rates rise as the result of an increased metabolic production rate, probably caused by mobilization of the stress hormone cortisol which drives proteolysis. Ca^{2+} uptake is not affected (Wood *et al.*, 1996). Interestingly, silver also enters the fish at a substantial rate through the gills, but the mechanism is unknown. There are no data on histologic effects or on recovery mechanisms during continued exposure, although there is a clear induction of metallothionein in the gills within 7 days (Hogstrand *et al.*, 1996b).

Copper

The physiology and toxicology of copper in fish has been studied intensively because of its use as a prophylactic agent in fish culture, its diffuse release from industrial and domestic sources (e.g. copper plumbing), high solubility at low pH, and its status as both an essential micronutrient and potent toxicant (second or third to silver among the metals in potency, approximately equal to cadmium).

Normal nutritional uptake is via the diet, but the possibility of supplementary nutritional uptake via the gills does not appear to have been investigated. Rather, all attention has been on damaging effects. As with silver, copper readily enters across the gills and builds up in internal organs (Lauren and McDonald, 1987a; Grosell *et al.*, 1997), but the entry mechanism remains unknown. Taylor *et al.* (1996) have recently reviewed the physiology of copper toxicity, whereas Evans (1987), McDonald *et al.* (1989), Lauren (1991), McDonald and Wood (1993), and Heath (1995) provide useful syntheses of branchial effects.

In freshwater, copper is primarily an ionoregulatory toxicant, exerting substantial pathophysiologic effects at total concentrations well below 100 μg L^{-1}. Copper exposure induces large net losses of Na^+ and Cl^- across the gills, with internal sequelae (declining osmolality, fluid volume disturbance, cardiovascular collapse) similar to those caused by low pH and silver. There is a strong correlation between gill total copper burden and acute toxicity (Playle *et al.*, 1993b). Most but not all studies agree that both hardness and alkalinity are strongly protective against copper toxicity, whereas the reported influence of pH is variable, ranging from antagonism to synergism at very low pH (< 5.0) and from no effect to protection by pH elevations in the circumneutral and alkaline pH range (e.g. Chakoumakos *et al.*, 1979; Miller and MacKay, 1980; Pagenkopf, 1983; Lauren and McDonald, 1986; Erickson *et al.*, 1996, 1997). This uncertainty stems from two reasons. First, there is some indication that not only the free Cu^{2+} ion but also various copper hydroxide complexes may be able to bind to the gill and exert toxicity. Second, for Cu^{2+} itself, the log K_D value for Cu–gill binding is about 7.4 (Playle *et al.*, 1993a,b), much weaker than for Ag–gill binding (10.0), so competing and complexing agents exert a much greater influence. Thus, dissolved organic carbon, with a log K_D value for binding Cu^{2+} about two units higher than that of the gill for Cu^{2+}, is very strongly protective (Playle *et al.*, 1993a,b; Hollis *et al.*, 1997).

There are clearly two mechanisms of action by which copper induces ionoregulatory dysfunction. One is a mixed competitive (i.e. increased K_m) and non-competitive (reduced J_{max}) inhibition of Na^+ and Cl^- influx (Figure 1.16; Lauren and McDonald, 1987b), which occurs at a much lower threshold than the other which is a stimulation of passive effluxes of these ions, e.g. 12.5 versus 100 μg L^{-1} in rainbow trout (Lauren and McDonald, 1985) and at much higher levels in the more resistant tilapia (Pelgrom *et al.*, 1995). The inhibition of influx is insensitive to $[H^+]$ and develops progressively over the first 24 h, reminiscent of the actions of silver and aluminum. Like silver, copper has a strong affinity for sulfhydryl groups which explains its well-documented ability to inhibit branchial Na^+,K^+-ATPase activity both *in vitro* (Li *et al.*, 1996) and *in vivo* (Lorz and McPherson, 1976; Lauren *et al.*, 1987b; Beckman and Zaugg, 1988; Pelgrom *et al.*, 1995). Indeed, there was a linear negative correlation between gill total copper burden and branchial Na^+,K^+-ATPase activity in chronically exposed tilapia (Li *et al.*, 1998). This, rather than inhibition of the apical entry processes, probably accounts for the influx blockade (Figure 1.16; see Figure 1.7A). The efflux stimulation at higher levels is more rapid and occurs because copper is quite potent

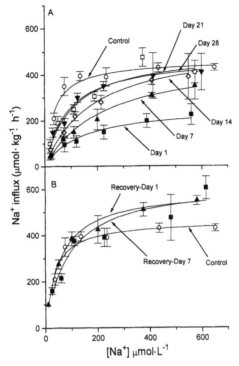

Figure 1.16 The influence of long-term sublethal exposure to copper on the concentration kinetics of unidirectional sodium influx across the gills of freshwater rainbow trout (data from Lauren and McDonald, 1987b). (A) Measurements were made before (control), on days 1, 7, 14, 21, and 28 of exposure (55 μg L^{-1} of total copper), and (B) on days 1 and 7 of the post-exposure period (recovery), and curves were fitted using the Michaelis–Menten equation (equation 1.2, see text). In (A), note the immediate downward shift (decrease in J_{max}, indicative of non-competitive inhibition) and rightward shift (increase in K_m, indicative of competitive inhibition) of the kinetic curves. Over 28 days of continuous exposure, J_{max} recovered fully, whereas K_m recovered partially. In (B), note that during the post-exposure period J_{max} overshot the original control level, whereas K_m returned to normal.

in weakening the junctions between the branchial epithelial cells, thereby increasing paracellular permeability. This may involve mechanisms in addition to Ca^{2+} displacement; Lauren and McDonald (1986) reported that increasing hardness did not protect against this action in the short term, whereas increasing alkalinity was ameliorative. Furthermore, Lauren and McDonald (1985) postulated that copper may also increase transcellular permeability because K^+ losses were disproportionately high relative to Na^+ and Cl^- losses. Regardless, the majority of studies indicate that, at least over the longer term, higher levels of $[Ca^{2+}]$ do reduce net ion losses. Effects of copper on Ca^{2+} uptake processes appear to be relatively minor or non-existent (Reid and McDonald, 1988; Sayer *et al.*, 1991; Pelgrom *et al.*, 1995).

Coughing frequency is an extremely sensitive indicator of exposure to very low levels of copper (e.g. 6–15 μg L^{-1}; Drummond *et al.*, 1973), but there is no clear agreement on whether respiratory dysfunction occurs (Sellers et al., 1975; Heath, 1984, 1991; Pilgaard et al., 1994; De Boeck et al., 1995; Beaumont et al., 1995a,b; Wang et al., 1998); likely this is not a primary mechanism of toxicity at realistic levels of copper exposure, although, as with other metals, it becomes important at acutely toxic levels associated with frank lamellar damage and gill mucification (Wilson and Taylor, 1993a). During sublethal exposure, blood pH tends to rise as a result of metabolic alkalosis (Sellers et al., 1975; Pilgaard et al., 1994; Wang et al., 1998). While this may be related to differential strong electrolyte loss, it more likely reflects an internal build-up of ammonia in the plasma and tissues (Lauren and McDonald, 1985; Taylor et al., 1996; Wang et al., 1998), a phenomenon which has been implicated in reduced swimming performance (Beaumont et al., 1995b). This effect may in some way be linked to an inhibition of ammonia excretion associated with the blockade of Na^+ uptake, but, as extensively discussed by Taylor et al. (1996), the exact mechanism remains unclear. Increased metabolic production of ammonia, driven by the stress hormone cortisol, may also be involved. Detailed measurements of ammonia excretion are needed. Copper also greatly slowed the rate of acid–base compensation during experimental hypercapnia (Wang et al., 1998), suggesting an effect on either or both of the Na^+/ 'acid' and Cl^-/'base' exchange processes at the gills (see Figure 1.7A).

The physical damage to the gills induced by sublethal copper is similar to that caused by aluminum, including epithelial lifting, lamellar fusion, necrosis, and apoptosis (Schreck and Lorz, 1978; Daoust et al., 1984; Khangarot and Tripathi, 1991; Kirk and Lewis, 1993; Li et al., 1998); as discussed earlier, apoptosis appears to be caused by cortisol, in contrast to necrosis, which is caused by copper directly (Bury et al., 1998). Repair during continued exposure involves proliferation of chloride and mucous cells, an accompanying increase in the total Na^+,K^+-ATPase content of the gills, a partial or complete recovery of the ion transport capacity of the gill (Figure 1.16), a reduction in diffusive ion losses, and a restoration of internal ionic levels (McKim et al., 1970; Lauren and McDonald, 1987a,b; Benedetti et al., 1989; Pelgrom et al., 1995; Li et al., 1998). In contrast to silver, there is no evidence for metallothionein induction in the gills (Lauren and McDonald, 1987a; Grosell et al., 1997). In contrast to aluminum, no marked changes in gill copper uptake or binding occur, and the gill copper burden remains stable over time (Lauren and McDonald, 1987a; Grosell et al., 1996, 1997). The changes in gill morphology during chronic exposure may cause respiratory limitation, reflected in decreased aerobic swimming performance (Waiwood and Beamish, 1978a; Beaumont et al., 1995a)

In seawater, copper is also an ionoregulatory toxicant (Cardeilhac and Hall, 1977; Cardeilhac et al., 1979), but with an effect opposite to that in freshwater – the fish gains Na^+ and Cl^- from the hypertonic seawater when the active excretion mechanism on the gills (Figure 1.7B) is impacted. The threshold is also much higher, e.g. negligible effects up to 400 $\mu g\ L^{-1}$ on plasma ions in seawater-acclimated trout at 24 h (Wilson and Taylor, 1993b) or up to 170 $\mu g\ L^{-1}$ in seawater flounder at 21 days (Stagg and Shuttleworth, 1982a, b). Wilson and Taylor (1993b) attributed this much higher threshold not to any marked difference in copper speciation between seawater and freshwater (indeed pH and alkalinity levels are broadly similar) but rather to the 26-fold higher $[Ca^{2+}]$ in seawater, i.e. protection by hardness. Dissolved organic carbon (DOC) may be another factor because cod tested in synthetic seawater (i.e. DOC free) exhibited a large ionoregulatory disturbance within a few hours of exposure to 400 µg L^{-1} (Larsen *et al.*, 1997). Regardless, at least two mechanisms of action appear to be similar to those in freshwater. First, there is an inhibition of basolateral membrane Na^+,K^+-ATPase activity, and an associated decrease in the active Na^+ and Cl^- excretion and transepithelial potential which it energizes (Stagg and Shuttleworth, 1982a; Crespo and Karnaky, 1983; see Figure 1.7B). Second, the rate of acid–base compensation during experimental hypercapnia is severely slowed in seawater cod (Larsen *et al.*, 1997), similar to the observations of Wang *et al.* (1998) on freshwater trout. Several other potential toxic mechanisms deserve further attention in seawater fish. Wilson and Taylor (1993b) reported a marked increase in plasma ammonia concentration in seawater trout (but no acid–base disturbance) before any marked ionoregulatory effect and they speculated that apical Na^+/NH_4^+ exchange was inhibited. Larsen *et al.* (1997) also observed a marked metabolic acidosis in seawater cod exposed to copper (ammonia not measured) which conceivably might reflect a similar inhibition of apical Na^+/H^+ exchange. Curiously, hypercapnia (high P_{CO_2} causing respiratory acidosis) ameliorated the osmoregulatory disturbance caused by copper in this same study, an effect which is not easily explained. Dodoo *et al.* (1992) reported that Ca^{2+} uptake was inhibited in juvenile flounder at a copper concentration only a small fraction of that expected to affect Na^+ and Cl^- regulation. Structural effects of copper on the gills appear similar to those in freshwater, except that chloride cell proliferation is accompanied by an apparent regression of mucous cells (Baker, 1969).

Zinc

Hogstrand and Wood (1996) have recently reviewed the physiology and toxicology of zinc in fish; McDonald *et al.* (1989), Lauren (1991), Wood (1992), McDonald and Wood (1993), and Heath (1995) provide additional summaries. Like copper, zinc is both a micronutrient and toxicant which is released diffusely from industrial and domestic sources (e.g. galvanized metals). However, zinc is about fivefold less toxic than copper, and zinc uptake at the gills appears to be a normal branchial function, occurring via the chloride cells. Chloride cell proliferation accompanying soft water acclimation greatly increases the rate of zinc uptake (Spry and Wood, 1988). Zinc is taken up, probably in the form of Zn^{2+}, by an active saturable transport system which displays standard Michaelis–Menten kinetics (Figure 1.17). When freshwater trout were fed zinc-deficient diets, uptake from the water increased to meet nutritional requirements (Spry *et al.*, 1988).

Elevated hardness and decreased pH are both protective against zinc toxicity, whereas carbonate alkalinity has only a very small ameliorative influence (Bradley and Sprague, 1985a,b). These effects are likely due to Ca^{2+} and H^+ competition for Zn^{2+}-binding sites on the gill. The influence of pH is nominally biphasic because

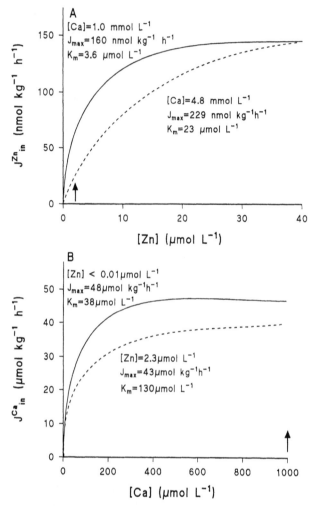

Figure 1.17 The influence of (A) waterborne calcium on the concentration kinetics of unidirectional zinc influx (data from Spry and Wood, 1989a) and of (B) waterborne zinc on the concentration kinetics of unidirectional calcium uptake (data from Hogstrand *et al.*, 1994) across the gills of freshwater rainbow trout. Note the large rightward shifts (increases in K_m, indicative of competitive inhibition) with only small effects on J_{max}. The arrows indicate the respective concentrations of zinc and calcium in the chronic exposure experiments described by Hogstrand *et al.* (1994, 1995, 1998) and Hogstrand and Wood (1996). The calcium concentration is on the 'saturated' portion of the curve, whereas the zinc concentration is far below the K_m, so an increase in K_m for a common Ca^{2+}/Zn^{2+} carrier occurring during chronic exposure to 150 μg L^{-1} of total zinc markedly reduces zinc influx but has only a small influence on the influx of calcium.

at high pH (e.g. 9.0) apparent zinc toxicity falls as $Zn(OH)_x$ and $ZnCO_3$ precipitate out of solution; however, in terms of dissolved zinc, toxicity is not reduced. The $\log K_D$ for Zn–gill binding has only been measured by radiotracer methods because the high endogenous levels of this essential mineral mask the detection of the small increment in 'cold' Zn-binding assays. The value is about 5.0, much lower than for most other metals (Galvez *et al.*, 1998), which probably explains the lower toxicity of zinc.

Early work at very high zinc concentrations (e.g. 1250–40 000 µg L^{-1}) identified rapidly developing hypoxemia (decreased arterial P_{O_2} and blood O_2 content, declining O_2 uptake despite increased ventilation, internal lactacidosis) due to gross morphologic damage of the gills as the mechanism of lethality (Skidmore, 1970; Skidmore and Tovell, 1972; Matthiessen and Brafield, 1973, 1977; Sellers *et al.*, 1975; Brafield and Matthiessen, 1976; Hodson, 1976; Tuurala and Soivio, 1982; Spry and Wood, 1984). However, Spry and Wood (1984, 1985) showed that these symptoms do not occur at more environmentally realistic concentrations at or below the 96-h LC$_{50}$ (e.g. 600–800 µg L^{-1} in softwater), although a modest increase in arterial P_{CO_2} is seen. The latter may reflect a reduction of carbonic anhydrase activity somewhere in the CO_2 excretion pathway (Figure 1.5) as zinc is a potent inhibitor of this enzyme *in vitro* (Christensen and Tucker, 1976). However, the important action of zinc at these lower concentrations, which is masked by the rapid suffocation response at higher levels, is a highly specific inhibition of active Ca^{2+} uptake across the branchial epithelium, resulting in hypocalcemia which may prove fatal. Net Na^+ and Cl^- uptake are not inhibited and plasma levels of these two major electrolytes are not usually affected (although see Heath, 1987). Unidirectional Na^+ and Cl^- influx and efflux rates both increase after several days' exposure (Spry and Wood, 1985). The latter has been attributed to displacement of Ca^{2+} by Zn^{2+} from the paracellular pathway and/or from entry sites on the apical membranes of the ionocytes (Wood, 1992).

Detailed analysis of the hypocalcemic effect of zinc has shown that Ca^{2+} and Zn^{2+} share, at least partially, a common uptake pathway via the chloride cells, such that zinc is a potent inhibitor of calcium uptake and vice versa (Spry and Wood, 1989a; Bentley, 1992; Hogstrand *et al.*, 1994, 1995, 1996a, 1998). The interactions are largely competitive, with large effects on respective K_m values and small effects on J_{max} values (Figure 1.17). Competitive interaction occurs because Zn^{2+} enters through the apical Ca^{2+}-selective channels in the chloride cells (Figure 1.7A), the step which appears to be rate limiting, and because Zn^{2+} inhibits the basolateral high-affinity Ca^{2+}-ATPase enzyme which largely powers transbranchial calcium transport. The latter effect is mainly competitive, with a small non-competitive component. However, Zn^{2+} itself is not transported by this enzyme, nor by Na^+/Ca^{2+} exchange, so the basolateral exit mechanism remains unknown.

Zinc concentrations are usually very low relative to calcium concentrations in freshwater. As illustrated in Figure 1.17, the 'designs' of the Michaelis–Menten kinetics of the transport systems for calcium (high K_m, high J_{max}) and zinc (low K_m, low J_{max}) are such as to allow normally a low level of Zn^{2+} uptake (e.g. a few nmol kg^{-1} h^{-1}) simultaneous to a much higher level of Ca^{2+} uptake (e.g. 45 µmol kg^{-1} h^{-1}). When trout acclimate to chronic sublethal zinc exposure (150 µg L^{-1} = 2.3 µmol L^{-1}), they progressively increase the K_m values of both systems by an unknown mechanism (Hogstrand *et al.*, 1998). Because Zn^{2+}

transport is occurring on the lower portion of its Michaelis–Menten curve (Figure 1.17A), while Ca^{2+} transport is occurring on the upper portion of its curve (Figure 1.17B), the former is reduced without greatly impairing the latter. There is no evidence that this involves any change in gill surface-binding K_D or in chloride cell numbers or apical surface areas (Galvez *et al.*, 1998). However, in other acclimation studies at higher chronic zinc levels (e.g. Matthiessen and Brafield, 1973), chloride cell proliferation onto the respiratory lamellae has been reported, which probably explains long-term increases in branchial Na^+,K^+-ATPase activity (Watson and Beamish, 1980). Repair of structural damage in the gills appears to proceed in much the same way as described for other metals, and there are several reports of the induction of metallothioneins and other low molecular weight zinc-binding proteins in branchial tissue (Bradley *et al.*, 1985; Thomas *et al.*, 1985; Spry and Wood, 1989b; Hogstrand *et al.*, 1995).

In marine fish, uptake of zinc from the water is normally of minor importance relative to uptake from the diet (Hoss, 1964; Pentreath, 1973). Waterborne zinc is less toxic in seawater than in freshwater, probably because of the much higher environmental calcium concentration, but little is known of its mechanism of action. As the calcium uptake pathway is similar in marine and freshwater fish, presumably Zn^{2+} could compete with Ca^{2+} in the same manner, but there are no experimental data. Very high levels of zinc applied to the serosal (blood) side of a killifish opercular epithelium preparation inhibited active Na^+ and Cl^- excretion and transepithelial potential (Crespo and Karnaky, 1983). The effect was only partially explained by inhibition of Na^+,K^+-ATPase activity, and an inhibition of the $Na^+,K^+,2Cl^-$ co-transporter in the chloride cells probably also occurred. Prolonged exposure (≥ 20 days) of an elasmobranch (dogfish) to very high levels (10 000–15 000 µg L^{-1}) caused a modest increase in the zinc content of the gills, chloride cell proliferation, and indications of sublethal internal toxicity (Flos *et al.*, 1979; Crespo *et al.*, 1979, 1981).

Cadmium

Cadmium is a non-nutrient metal released by a variety of industrial processes, often those associated with the refining of other metals. Cadmium toxicity to freshwater fish is high, similar to that of copper, with substantial pathophysiology at total concentrations well below 100 µg L^{-1}. However, in terms of toxic mechanism, cadmium appears more similar to zinc than to copper (reviewed by Sprague, 1987; Lauren, 1991; Spry and Wiener, 1991; Wendelaar Bonga and Lock, 1992; Wood, 1992; McDonald and Wood, 1993). The gills, rather than the diet, are the major route of cadmium uptake (Williams and Giesy, 1978). The affinity of the gills for cadmium is relatively high (log $K_D = 8.6$, which explains its much greater toxicity than that of zinc) and there is a strong relationship between gill cadmium burden and toxicity during acute exposures (Playle *et al.*, 1993b). There is abundant evidence that cadmium is far more toxic in soft water than in hard water, but most studies have failed to separate the effects of alkalinity and pH from those of the two hardness cations (reviewed by Davies *et al.*, 1993). A notable exception is the investigation of Carrol *et al.* (1979), which demonstrated that calcium, and not magnesium or carbonate alkalinity, was the dominant protective factor against cadmium toxicity to brook trout. In accord, geochemical modeling indicates that Ca^{2+} competition rather than carbonate complexation is the major factor keeping Cd^{2+} off the gills in hard water (Hollis *et al.*, 1997, 1999). DOC, which binds Cd^{2+} more weakly than Cu^{2+}, also offers some protection against cadmium toxicity (Playle *et al.*, 1993b).

As for zinc, there is strong evidence that cadmium uptake and toxicity both occur via the chloride cells, the most convincing being the autoradiographical localization of ^{109}Cd to chloride cells reported by Wicklund-Glynn *et al.* (1994). Pavement cells are likely less important as routes of cadmium uptake (Block and Part, 1992). Low concentrations of cadmium effectively block calcium uptake (Verbost *et al.*, 1987, 1989; Reid and McDonald, 1988) and higher concentrations of calcium effectively block cadmium uptake (Part *et al.*, 1985; Bentley, 1991; Wicklund Glynn *et al.*, 1994). Plasma and/or whole body hypocalcemia are the classic symptoms of cadmium exposure in freshwater fish and likely the direct cause of death (Roch and Maly, 1979; Muramoto, 1981; Giles, 1984; Fu *et al.*, 1989; Pratap *et al.*, 1989). Low levels of cadmium have either no effect on Na^+ uptake (Verbost *et al.*, 1987, 1989; Reid and McDonald, 1988; Figure 1.16A) or stimulate it (Reader and Morris, 1988). Negative effects on the uptake rates and plasma levels of Na^+ and Cl^- may occur at higher threshold levels but are much less marked than the effects on calcium regulation (McCarty and Houston, 1976; Giles, 1984; Fu *et al.*, 1989). These are likely attributable to damage or death of chloride cells by both apoptosis and necrosis, and associated loss of Na^+,K^+-ATPase activity (Wendelaar Bonga and Lock, 1992).

Although the kinetics of cadmium versus calcium interactions have not been characterized at a whole animal level, the details have been thoroughly described at the level of chloride cell function in a landmark series of studies (Verbost *et al.*, 1987, 1988, 1989). During exposure to low waterborne levels of cadmium, the inhibition of Ca^{2+} influx is not immediate, but rather develops gradually over time and occurs in the absence of any effect on Na^+ influx (Figure 1.18A). Cadmium inhibits its own influx over a similar time-course (Figure 1.18B). This reflects the fact that cadmium, probably in the form of Cd^{2+}, enters through the Ca^{2+}-selective channels in the apical membrane without appreciably blocking simultaneous Ca^{2+} influx. Both ions are initially sequestered by calmodulin and other intracellular binding proteins. However, as intracellular free $[Cd^{2+}]$ gradually builds up, it potently and competitively inhibits the basolateral Ca^{2+} pump, the high-affinity Ca^{2+}-ATPase. The resulting blockade of Ca^{2+} export causes a rise in intracellular free $[Ca^{2+}]$, which acts as a signal to close the apical Ca^{2+}-selective channels, thereby greatly decreasing the entry of both ions (see Wicklund Glynn, 1996, for additional/alternate hypotheses). Cd^{2+} itself is not pumped by the high-affinity Ca^{2+}-ATPase, and the blockade appears to be more or less irreversible (Reid and McDonald, 1988). The mechanism by which Cd^{2+} exits across the basolateral membrane is unknown, although a facilitated diffusion mechanism has been suggested (Verbost *et al.*, 1989).

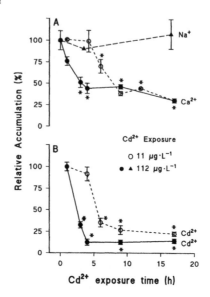

Figure 1.18 The influence of the duration of exposure to waterborne cadmium (at total concentrations of either 11 or 112 µg L^{-1}) on the relative accumulation rates of (A) sodium and calcium and (B) cadmium itself in the gill soft tissue of freshwater rainbow trout (data from Verbost *et al.*, 1989). Accumulation rates, measured by radiotracers, were recorded over 1-h exposure periods, and are thought largely to reflect the entry rate across the gill epithelial cells. Note the progressive and parallel inhibitions of calcium and cadmium uptake without effect on sodium influx.

Respiratory effects of acute cadmium exposure at environmentally realistic concentrations have not been reported, but during chronic sublethal exposure there is a decrease in the O_2-diffusing capacity of the gill, increased ventilatory rate, and associated changes in hematology (Majewski and Giles, 1981). These effects are likely caused by increased blood-to-water diffusion distance in the gills as a result of the cellular proliferation which occurs during damage repair; however, aerobic exercise performance is not impaired (Hollis *et al.*, 1999). Chloride cell hyperplasia is particularly marked. Indeed, this appears to be the most commonly reported morphologic response during chronic sublethal exposure (e.g. Oronsaye and Brafield, 1984; Wendelaar Bonga and Lock, 1992), and is likely responsible for the eventual recovery of plasma calcium concentration that usually occurs during extended exposures (Giles, 1984; Fu *et al.*, 1989; Pratap *et al.*, 1989). Subtle ionic disturbances may persist indefinitely (McCarty and Houston, 1976; Giles, 1984). Although cadmium is often considered a nephrotoxicant because of its selective accumulation in kidney tissue during chronic sublethal exposure (e.g. Sangalang and Freeman, 1979), the work of Giles (1984) demonstrated that chronic disturbances in electrolyte balance do not result from impairment of renal function.

During chronic sublethal exposures, the cadmium-binding characteristics of the gills appear to change; most notably, total gill cadmium burden may increase to up to 40-fold higher than levels known to kill naive fish, without evidence of toxicity (Giles, 1988; Farag *et al.*, 1994; Hollis *et al.*, 1999). This 'non-toxic' metal burden is probably immobilized by various proteins because cadmium appears to be particularly effective in binding to pre-existing proteins (Olsson and Hogstrand, 1987; Wicklund-Glynn and Olsson, 1991) as well as in inducing the synthesis of new metal-binding proteins in the gills. These include both metallothionein (Benson and Birge, 1985; Klaverkamp and Duncan, 1987; Fu *et al.*, 1990) and non-metallothionein low molecular weight (< 3000 Da) polypeptides (Thomas *et al.*, 1983, 1985; Stone and Overnell, 1985; Fu *et al.*, 1990).

As with other metals, the branchial effects of cadmium in seawater are poorly characterized. Morphologic damage occurs at unrealistically high levels (50 000 µg L^{-1}; Gardner and Yevich, 1970). However, cadmium appears to be much less toxic than in freshwater, probably because of the protective effects of high environmental calcium (by competition) and chloride (by complexation). The induction of metal-binding protein in the gills during cadmium exposure has been reported in some studies (Olsson *et al.*, 1989; George *et al.*, 1996) but not others (Noel-Lambot *et al.*, 1978). In full-strength seawater, acute ionoregulatory disturbance occurs only at levels well above environmental relevance (4800 µg L^{-1}; Thurberg and Dawson, 1974). In brackish water slightly hypo-osmotic relative to the blood of exposed flounder, more realistic levels of cadmium (5–500 µg L^{-1} over 4–9 weeks; 100–10 000 µg L^{-1} over 2 weeks) caused concentration-dependent decreases in plasma calcium and potassium concentrations and increases in plasma magnesium and inorganic phosphate levels, with negligible effects on sodium and chloride regulation (Larsson *et al.*, 1976, 1981). Although the site of action was not identified, the calcium uptake pathway in the gills is a likely target.

Cobalt

Cobalt is a micronutrient whose potential toxic properties to fish have been scrutinized recently because of cobalt's abundance in certain fertilizers as well as in wastewater discharges from cobalt mines and nuclear reactors (in various radioisotopic forms). Cobalt uptake from the water appears to predominate normally over cobalt uptake from the diet in freshwater fish (Baudin and Fritsch, 1989). Like zinc, cobalt is not a very potent toxicant (48-h LC$_{50}$ > 5000 µg L^{-1}; Diamond *et al.*, 1992), a fact which probably reflects the low log K_D value of fish gills for Co–gill binding (5.1; Richards and Playle, 1998). This may be an important design feature of the gills for regulation of nutrient metals. Richards and Playle (1998) give a detailed gill binding model for cobalt, which suggests that a number of naturally occurring substances [dissolved organic matter (DOM), CO_3^{2-}, OH$^-$, H$^+$, Ca^{2+}] should all exert significant protective effects if present in high enough concentration because their respective log K_D values for competition and complexation are similar to the log K_D value for Co–gill binding (Figure 1.11). Although this has yet to be systematically tested, increasing levels of CaCO$_3$ were strongly protective against cobalt toxicity (Diamond *et al.*, 1992).

In pharmacology and electrophysiology, Co^{2+} is routinely used as a Ca^{2+}-channel blocker. Therefore, it is not surprising that cobalt, like cadmium and zinc, enters

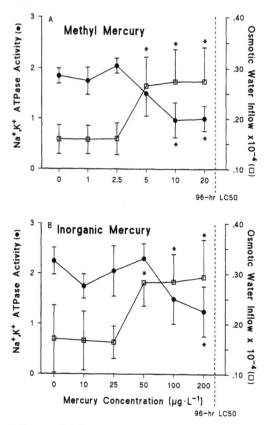

Figure 1.19 The influence of 7 days' exposure *in vivo* to various concentrations of (A) methylmercury and (B) inorganic mercury on branchial Na+,K+-ATPase activity and branchial osmotic water inflow (both determined *in vitro*) in the gills of freshwater rainbow trout. Concentrations representative of 96-h LC_{50} values are indicated. Note the greater potency and toxicity of methylmercury over this time period, and the lower thresholds for stimulation of osmotic water permeability than for inhibition of Na+,K+-ATPase by both forms of mercury (data from Lock *et al.*, 1981).

fish gills at least partially via the Ca^{2+}-uptake pathway of the chloride cells, and probably exerts its toxic effects on this pathway. Cobalt uptake is reduced by increasing levels of calcium in the water as well as by agents which reduce calcium uptake, such as competition by high environmental magnesium, acclimation to high environmental calcium (which reduces chloride cell numbers and apical exposure), and treatment with lanthanum and systemic $CaCl_2$ injections (Comhaire *et al.*, 1994, 1998). Both of the latter treatments tend to block the apical Ca^{2+}-selective channels in the chloride cells. In general, reductions in cobalt uptake are less pronounced than those in calcium uptake caused by these treatments, suggesting that cobalt may penetrate the gills by a second pathway as well. Nothing is presently known about basolateral transfer or inhibition mechanisms, or about cobalt toxicity in seawater.

Mercury

In natural waters, mercury may exist as either inorganic mercury or methylmercury, the latter formed by microbial action. Mercury toxicity has been studied extensively because of its bioaccumulation and biomagnification in food chains (Spry and Wiener, 1991), especially in acidified waters and those with a high input of organic matter (e.g. from logging, or from rotting vegetation in flooded reservoirs). Experimentally, $HgCl_2$ is generally used for inorganic mercury (yielding Hg^{2+}) and CH_3HgCl for methylmercury (yielding CH_3Hg^+). Although inorganic mercury exerts acute toxicity at quite low concentrations (e.g. 96-h LC_{50} values in the range 30–700 µg L^{-1}; Spry and Wiener, 1991), methylmercury is much more toxic, with lethal effects sometimes occurring at levels 1–2 orders of magnitude lower! Nevertheless, it is doubtful that 'natural' waterborne levels of either form (generally in the ng L^{-1} range or below) are ever great enough to damage fish gills, except in cases where the mercury contamination results from mining or industrial effluent (Lauren, 1991).

In contrast to many other metals, the exact mechanism(s) of mercury uptake at the gills and the influence of speciation are still not well understood. For example, the log K_D values for gill–Hg binding, and for competing and complexing agents, have not been determined for either Hg^{2+} or CH_3Hg^+. It is known, however, that DOC reduces both gill binding and whole body uptake of inorganic mercury in rainbow trout (Ramamoorthy and Blumhagen, 1984; Playle, 1998) and Sacramento blackfish (Choi *et al.*, 1998). The binding of inorganic mercury to the gills is inhibited by DOC but not by Ca^{2+} (Playle, 1998). Some methylation of inorganic mercury may occur as a result of the presence of bacteria on the gill surface, but this is not a prerequisite for inorganic mercury uptake (Olson *et al.*, 1973). Both electron microprobe (Olson and Fromm, 1973) and autoradiographic studies (Wicklund-Glynn *et al.*, 1994) indicate that inorganic mercury is distributed diffusely in the gill epithelium, and is not specifically localized to chloride cells, unlike cadmium. Although this suggests that inorganic mercury is taken up non-selectively through all gill cells, this conclusion is confounded by the observation that inorganic mercury uptake was strongly inhibited by lanthanum, which is thought to block the Ca^{2+} channels of the chloride cells (Wicklund-Glynn *et al.*, 1994). These approaches have not been applied to methylmercury uptake.

As for many organic toxicants (e.g. Randall *et al.*, 1996), the uptake rate of methylmercury appears to be directly proportional to both metabolic rate and environmental concentration (Rodgers and Beamish, 1981). The uptake is also highly efficient; Rodgers and Beamish (1981) estimated that the gills of freshwater rainbow trout extracted about 8 percent of a fairly low waterborne concentration of methylmercury (< 8 µg L^{-1}) passing over them, similar to the 10 percent estimated by Phillips and Buhler (1978) at an even lower level (< 1.4 µg L^{-1}). Recent direct measurements by Choi *et al.* (1998) in Sacramento blackfish exposed to 1.4 ng L^{-1} indicate similar values in the presence of DOC, and as high as 33 percent in the absence of DOC. Furthermore, the Q_{10} values for both short-term and long-term uptake by trout were rather low (e.g.1.5–2.0 at 5–20°C; Reinert *et al.*, 1974; Rodgers and Beamish, 1981). All these observations suggest that a lipophilic complex (CH_3HgCl or CH_3HgOH) directly penetrates the gill cell membranes by simple diffusion. However, water chemistry does play a modulatory role. In 30-min uptake experiments with zebrafish, designed to probe the mechanism of apical entry, uptake was stimulated by large (log unit) increases in water chloride and was progressively decreased by reductions in water pH from 7.0 to 3.5, suggesting that CH_3HgCl is more permeant than CH_3HgOH (Block *et al.*, 1997). In this same study, large variations in water [Ca^{2+}] had no clear effect. However, in 60-min uptake experiments with rainbow trout, a 13-fold reduction in water hardness and alkalinity caused a threefold stimulation of methylmercury uptake in rainbow trout (Rodgers and Beamish, 1983). Furthermore, a reduction of water pH from 9.0 to 6.0 approximately doubled the uptake of methylmercury, but did not affect the uptake of inorganic mercury over a 5-day exposure (Drummond *et al.*, 1974). Increasing water DOC from undetectable background levels to 5 mg C L^{-1} reduced methylmercury uptake by about 80 percent in Sacramento blackfish (Choi *et al.*, 1998).

It is well known that mercury accumulation in natural fish populations is generally greater in softwaters of low pH and alkalinity (Wren and MacCrimmon, 1983), but it remains unclear how much of this difference reflects this greater bioavailability at the gills and how much results from greater availability in the environment (both water *and* food chain; Spry and Wiener, 1991; Post *et al.*, 1996). Although uptake from the water does not appear to involve the digestive tract (Olson *et al.*, 1973), uptake from the water and uptake from the diet are independent and additive (Phillips and Buhler, 1978). Inorganic mercury is usually present in greater concentrations than methylmercury in contaminated natural waters (Kudo *et al.*, 1982), but it is usually assumed that most of the uptake occurs in the form of the latter, based on single contaminant uptake experiments (Olson *et al.*, 1973; Wobeser, 1975). However, the simultaneous presence of a higher concentration of inorganic mercury greatly accelerates the uptake of methylmercury by an unknown mechanism (Rodgers and Beamish, 1983).

Both forms of mercury appear to be ionoregulatory toxicants. Unlike acute toxicity, levels of inorganic mercury and methylmercury causing chronic sublethal effects during long-term exposures are similar (Spry and Wiener, 1991). Furthermore, inorganic mercury appears to be actually more potent than methylmercury in causing effects *in vitro* (e.g. permeability increases, Na$^+$,K$^+$-ATPase inhibition; Renfro *et al.*, 1974; Bouquegneau, 1977; Lock *et al.*, 1981). However, when presented at identical concentrations in the water, methylmercury is taken up much more rapidly than inorganic mercury by freshwater trout (Olson *et al.*, 1973; Drummond *et al.*, 1974; Wobeser, 1975). All these observations suggest that waterborne methylmercury is more toxic on an acute basis because it is much more bioavailable rather than because it is intrinsically more potent. This difference reflects not only the greater lipophilicity of methylmercury but also the much greater potency of inorganic mercury to stimulate both the proliferation of mucous cells on the gills and the secretion of mucus into the water (Olson *et al.*, 1973; Lock and van Overbeeke, 1981). Inorganic mercury avidly binds to the mucus, resulting in an apparent change of its molecular structure (Varanisi *et al.*, 1975).

As a result, most of the bound mercury is rendered unavailable for entry into branchial cells, and its removal from the gill surface by mucous sloughing is facilitated. McKone *et al.* (1971) reported that in goldfish exposed to 1000 µg L^{-1} of inorganic mercury for 3 h, 79 percent of the uptake was sloughed in this manner.

Once they enter the gills, the internal fate and metabolism of methylmercury and inorganic mercury appear to differ. Renfro *et al.* (1974) demonstrated that the clearance rate of radiolabeled ^{203}Hg from gill tissue to internal organs in intact freshwater killifish was much faster when the metal was presented as CH$_3$^{203}HgCl than when presented as ^{203}HgCl$_2$. This difference probably explained the faster restoration of positive Na$^+$ balance in the methylmercury-exposed fish. A further complication is the observation that demethylation probably occurs after uptake of methylmercury. Renfro *et al.* (1974) demonstrated that the ^{14}C label on ^{14}CH$_3$HgCl was cleared much more rapidly from both the gills and carcass of these killifish than was the ^{203}Hg label on CH$_3$^{203}HgCl. In similar experiments on rainbow trout (Olson *et al.*, 1978), the differences were smaller but the overall pattern was similar. Notably, the fraction bound to metallothionein in the gills appeared to be demethylated.

Mercury is one of the few toxicants whose physiologic mechanisms of action have been studied to a similar extent in seawater and freshwater fish. In common with other metals, the acute toxicity of mercury appears to be lower in seawater than in freshwater, and in both media, for both forms of mercury, osmoregulatory failure appears to be the proximate cause of acute toxicity (Heath, 1995). Coughing frequency increases in a concentration-dependent manner in response to fairly low levels (3–12 µg L^{-1}) of both forms of mercury (Drummond *et al.*, 1974), but there is no evidence of respiratory toxicity. Concentration-dependent *decreases* in plasma [Na$^+$], [Cl$^-$], and osmolality occur in *freshwater* fish (Lock *et al.*, 1981) and *increases* in plasma [Na$^+$] and [Cl$^-$] in *seawater* fish (Bouquegneau, 1977). In freshwater, these negative effects (for both forms of mercury) appear to be due to two separate mechanisms (Figure 1.19A and B): (1) a rapidly developing increase in osmotic water permeability which occurs at levels well below the 96-h LC$_{50}$, and (2) a more slowly developing inhibition of gill Na$^+$,K$^+$-ATPase activity (Lock *et al.*, 1981) which reduces Na$^+$ uptake (Renfro *et al.*, 1974) at concentrations close to the LC$_{50}$. It is unclear whether the water permeability increase is a specific effect (e.g. opening of apical water channels) or a general increase in permeability because fluxes of other molecules were not measured in the study by Lock *et al.* (1981). At methylmercury concentrations which negatively impacted ionoregulation, unidirectional Na$^+$ efflux did not change in freshwater killifish (Renfro *et al.*, 1974), but increased in freshwater lamprey in the absence of any change in branchial Na$^+$,K$^+$-ATPase activity (Stinson and Mallatt, 1989). Like copper and silver, mercury has a strong affinity for sulfhydryl groups, which explains its ability to inhibit branchial Na$^+$,K$^+$-ATPase activity both *in vivo* and *in vitro*, an effect that can be partially reversed by the addition of cysteine (Bouquegneau, 1977; Lock *et al.*, 1981; Miura and Imura, 1987). Relatively high levels of waterborne inorganic mercury also inhibited branchial Na$^+$,K$^+$-ATPase activity in seawater fish (Bouquegneau, 1977), whereas methylmercury, injected systemically, had no effect (Schmidt-Nielsen *et al.*, 1977). Possible actions of mercury on gill permeability in seawater fish have not been studied.

Curiously, during chronic low level exposures, paradoxical 'improved' osmoregulation has been observed. Thus, *increases* in plasma [Na$^+$] and [Cl$^-$] were seen in *freshwater* brook trout exposed to methylmercury (2.9 µg L^{-1}) for 14 days (Christensen *et al.*, 1977). Similarly, a *decrease* in plasma osmolality occurred in *seawater* winter flounder exposed to inorganic mercury (5 µg L^{-1}) for 60 days (Calabrese *et al.*, 1975), and increases in kidney and urinary bladder Na$^+$,K$^+$-ATPase activities when the same species was treated with repetitive methylmercury injections. Fish are known to acclimate to inorganic mercury (Bouquegneau, 1979), and inorganic mercury is a potent inducer of metallothionein in the gills and other tissues (Klaverkamp *et al.*, 1984; Klaverkamp and Duncan, 1987), so these positive responses may represent part of the acclimatory process.

A variety of morphologic responses in the gills have been described in response to toxic levels of both methylmercury and inorganic mercury, in both freshwater and seawater (Wobeser, 1975; Khangarot and Somani, 1980; Lock and van Overbeeke, 1981; Daoust *et al.*, 1984; Pereira, 1988). Many of these are generalized, non-specific responses (swelling and hyperplasia of pavement cells, necrosis, excess secretion of mucus and mucous cell proliferation, epithelial rupture, fusion of lamellae). Notable exceptions are observations of increased frequency of mitotic figures in pavement cells of freshwater trout exposed to methylmercury (Wobeser, 1975), and chloride cell proliferation in seawater windowpane flounder chronically exposed to inorganic mercury. Both may represent damage–repair phenomena which contribute to acclimation.

Detergents

Detergents are discharged from domestic and industrial sources, and are widely used as oil dispersants. Although not particularly toxic (LC$_{50}$ values in the mg L^{-1} range), they have often been implicated in fish kills. Indeed, in oil spill clean-ups, there is evidence that the dispersant itself, rather than the petroleum hydrocarbon, may be the more important cause of acute toxicity. Most experimental work has concentrated on anionic [e.g. sodium lauryl sulfate (SLS), linear alkylate sulfonate (LAS)] and non-ionic detergents [e.g. nonyl phenol ethoxylate (NP8), synthetic alcohol ethoxylate (A7)]. Although these compounds rapidly pass through the branchial epithelium to internal organs, rather than accumulating in the gills (Tovell *et al.*, 1975a), several disruptive effects on normal branchial physiology have been reported which likely contribute to death. Curiously, the normal protective effect of hardness is reversed with anionic detergents, but not with non-ionics, such that uptake and toxicity of SLS are both potentiated by divalent cations, perhaps throughout the formation of more neutral and permeant compounds, e.g. calcium lauryl sulfate (Tovell *et al.*, 1975b).

Increases in ventilation rate occur in response to both anionic and non-ionic detergents, with response thresholds close to the threshold for sublethal toxicity

(Maki, 1979). It is unclear whether this reflects irritant effects on branchial neuroepithelial cells (or some other segment of the ventilatory control system) or a response to respiratory blockade. The rapid reversibility of the hyperventilation when the detergent is removed favors the former explanation, but the effects on gill blood flow and vascular resistance outlined below could reduce the effectiveness of blood oxygenation. Certainly, higher levels of detergents cause severe morphologic damage to the gills (Schmid and Mann, 1961; Abel and Skidmore, 1975; Abel, 1976; McKeown and March, 1978) which would likely interfere with gas exchange, although this has not been documented.

The best-documented effects of detergents are disturbances in gill vascular resistance and blood flow, although these have only been shown in perfused preparations *in vitro* (Jackson and Fromm, 1977; Bolis and Rankin, 1978, 1980; Stagg *et al.*, 1981; Rankin *et al.*, 1982; Stagg and Shuttleworth, 1986, 1987). The threshold concentrations inducing these effects are in the sublethal range, similar to those causing ventilatory stimulation (Maki, 1979); at these levels, the effects are rapidly reversible, although not at higher concentrations (Stagg *et al.*, 1981). Both observations are suggestive that these actions may be involved in the ventilatory stimulation. There are some discrepancies among different studies, perhaps dependent on fish species and the types and concentrations of detergents used. However, the general picture appears to be a predominant vasodilation of the gills via an opening of the arterioarterial pathway (see Figure 1.1E), although constrictions in response to high concentrations of LAS and SLS have also been reported (Jackson and Fromm, 1977; Stagg and Shuttleworth, 1986, 1987). The latter are thought to reflect direct effects on vascular smooth muscle. The vasodilatory effects can be partially blocked by β-adrenergic antagonists, suggesting an interaction with the adrenoreceptor system (Bolis and Rankin, 1978, 1980; Stagg *et al.*, 1981).

Further complicating the situation is the fact that detergents also reduce the responsiveness (to natural agonists) of the β-adrenergic vasodilatory system which opens the arterioarterial pathway in the gills, without affecting the α-adrenergic constrictory system (Stagg *et al.*, 1981; Stagg and Shuttleworth, 1986, 1987). This β-adrenergic antagonism manifests as a decrease in efficacy (maximum response) but with no change in affinity of the system for natural catecholamines, i.e. classic non-competitive blockade (Figure 1.20). The analysis of Stagg and Shuttleworth (1987) indicates that the site of interference is beyond the β-adrenoreceptor but before the second messenger [cyclic adenosine monophosphate (AMP)], perhaps at the adenylate cyclase step in the response pathway. Thus, while promoting gill vasodilation in themselves, detergents tend to block natural vasodilation induced by endogenous catecholamine mobilization, as might occur during exercise or hypoxia. Negative consequences for performance seem likely, but have not been tested *in vivo*.

At concentrations close to the LC_{50}, detergents cause ionoregulatory disturbance, characterized by a general fall in plasma electrolytes in freshwater fish and a rise in seawater fish (McKeown and March, 1978; Baklien *et al.*, 1986). These effects may arise by several mechanisms. First, in perfused freshwater trout gills, SLS

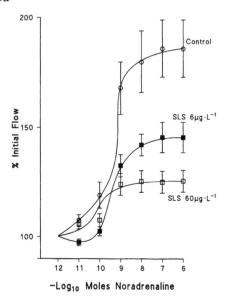

Figure 1.20 The effects of two concentrations of sodium lauryl sulfate (6 and 60 μg L⁻¹) on the concentration–response relationship for noradrenaline-stimulated β-adrenergic vasodilation in isolated-perfused gills of the freshwater European eel (data from Stagg *et al.*, 1981). Note that the maximum response is reduced but that the position of the curves is unchanged, indicating unchanged affinity of the β-adrenoreceptors for noradrenaline, i.e. classic non-competitive inhibition.

greatly increases tritiated water influx (Jackson and Fromm, 1977), whereas anionic, cationic, and neutral detergents all greatly increase [^{14}C]-urea influx (Partearroyo *et al.*, 1992). A non-ionic surfactant increased the efflux of 4-aminoantipyrine from live goldfish (Anello and Levy, 1969). These results all suggest a general increase in gill permeability, perhaps partly due to vasodilatory opening of additional respiratory lamellae and partly due to solubilization of cell membranes and/or disruption of the mucous layer. All these actions could enhance unfavorable passive fluxes of both salt and water. Second, in perfused seawater flounder gills, both non-ionic and ionic detergents inhibit the transepithelial potential (TEP), and reduce the extent of β-adrenergic stimulation of the TEP by catecholamines, indicating an inhibition of active NaCl extrusion (Stagg and Shuttleworth, 1986). It is conceivable that a similar inhibition of active Na$^+$ and Cl$^-$ uptake may occur in freshwater gills.

Ammonia

As well as being the primary nitrogenous waste of teleost fish, ammonia is probably the single most abundant aquatic contaminant because of its ubiquitous discharge from agriculture, aquaculture, and industry. Ammonia toxicity to fish has been thoroughly reviewed (e.g. Alabaster and Lloyd, 1980; Haywood, 1983; Tomasso, 1994). Descriptive studies quantifying ammonia toxicity are numerous, but there have been relatively few mechanistic studies. As a weak base with a pK of about 9.3, ammonia exists largely as NH_4^+ at circumneutral pH, with the NH_3 fraction (which is a dissolved gas) increasing as pH increases. At a fixed total ammonia concentration, toxicity also increases with pH, indicating that NH_3 is the more toxic fraction, and many authors express toxicity solely as a function of NH_3-N. Typical 96-h LC_{50} values are in the range of 0.2–2.0 mg L^{-1} NH_3-N, reflecting widespread variation in the sensitivity of different species, with sublethal thresholds at about 5 percent of these levels. However, this is misleading because there is convincing evidence that NH_4^+ is also toxic. For example, when expressed as a function of NH_3-N, the 96-h LC_{50} for ammonia was markedly lower at water pH 6.5 (where the NH_4^+ fraction was high) than at pH 9.0 (where the NH_4^+ fraction was low; Thurston *et al.*, 1981). Indeed, it is likely that the latter is the major contributor to *branchial* toxicity because of its interference with ionoregulation, as outlined below. Increases in water hardness ($[Ca^{2+}]$) and [NaCl] are both generally protective against ammonia toxicity (Tomasso, 1994), but it is unclear how these protective effects are apportioned between the NH_3 and NH_4^+ components of toxicity. Certainly, both should help resist ionoregulatory disturbance, whereas the former should help resist diffusive NH_3 uptake through the branchial epithelium.

The much greater acute toxicity of NH_3 is undoubtedly due to its very high permeance through the branchial membranes, resulting in internal *systemic* toxicity. This apparently occurs in the absence of serious morphologic damage to the gills (Smart, 1976; Daost and Ferguson, 1984), although this has been disputed (Kirk and Lewis, 1993). Regardless, a variety of systemic symptoms occur rapidly, including increased ventilation, increased heart rate, increased O_2 consumption, moderately decreased arterial Po_2, and elevated plasma concentrations of glucose and cortisol, but none of these responses are considered to be direct causes of toxicity (Smart, 1978; Swift, 1981). Of course at typical fish body fluid pH values (7.0–8.0), most of the ammonia entering as NH_3 becomes protonated again to NH_4^+ in the internal environment. Injection experiments which bypass the gill entirely have pointed to *internal* $[NH_4^+]$ as the more important moiety exerting *systemic* toxicity (Hillaby and Randall, 1979). Likely this reflects an interference with nervous function (by NH_4^+ substitution on cationic sites in neuron channels) and/or energy metabolism in the brain (owing to diversion of α-ketoglutarate away from the Krebs cycle for NH_4^+ detoxification; Tomasso, 1994). This acute toxicity develops rapidly, with symptoms of neurologic disturbance evident less than 10 min after systemic injection of ammonium salts (Hillaby and Randall, 1979). The cardiorespiratory disturbances (Smart, 1978) are likely symptomatic of central neurologic damage. LC_{50} values decrease only slightly after the first day of external ammonia exposure, reflecting this rapid mechanism of internal toxicity (e.g. Buckley, 1978; Person-Le Ruyet *et al.*, 1995). Detoxification mechanisms in various species may involve conversion of ammonia to glutamine, glutamate, or urea so as to reduce ammonia concentrations in the central nervous system (Levi *et al.*, 1974; Arillo *et al.*, 1981a,b; Schenone *et al.*, 1982; Iwata, 1989; Wood *et al.*, 1989; Walsh *et al.*, 1990; Person-Le Ruyet *et al.*, 1997;

Hernandez *et al.*, 1999). Attenuation of ventilatory disturbance may be indicative of the detoxification mechanisms which are induced during prolonged sublethal exposure (Lang *et al.*, 1987) and which lead to acclimation (Haywood, 1983).

The various possible mechanisms by which endogenously produced ammonia may be excreted across the gills, and the current controversies and uncertainties surrounding the relative importance of these mechanisms, have been described earlier (Figures 1.5 and 1.7). Whatever the true situation, it is clear that elevations in waterborne ammonia will reduce the gradients for excretion, leading to internal ammonia accumulation. However, some regulation is clearly possible. Indeed, in many ammonia toxicity studies, fish have been reported to keep arterial blood ammonia levels below those expected from equilibration of the permeant NH_3 moiety across the branchial epithelium (Wilkie, 1997). Some studies have attributed this effect to activation of an ionically coupled active excretion mechanism (e.g. Cameron and Heisler, 1983; Heisler, 1990; Wilson and Taylor, 1992; Rasmussen and Korsgaard, 1998; Salama *et al.*, 1999). Others have attributed the phenomenon to enhanced acidification of the gill boundary layer, thereby reducing the local P_{NH_3} at the gill surface below that in the bulk medium (e.g. Randall and Wright, 1987; Wright *et al.*, 1989; Wilson *et al.*, 1994), an idea first proposed by Lloyd and Herbert (1960). Both mechanisms may be considered as adaptive for avoiding internal ammonia toxicity.

Under conditions (e.g. lower environmental pH values and total ammonia levels) in which rapid internal toxicity due to NH_3 permeance does not occur, direct NH_4^+ toxicity may become important. In marine teleosts which have a higher permeability to monovalent cations, NH_4^+ may penetrate the gills, but this is unlikely in freshwater fish. More probably, direct *branchial* toxicity occurs because NH_4^+ interferes with the mechanism(s) of Na^+ uptake at the gills (Maetz and Garcia-Romeu, 1964; Maetz, 1972, 1973; Wilson *et al.*, 1994), causing large net losses of Na^+ from the body (Paley *et al.*, 1993). This may be caused by NH_4^+ competition with Na^+ for the apical Na^+ channel or $Na^+/H^+,NH_4^+$ antiporter on the apical membrane of the gill transport cells, or by an interference with the H^+-ATPase (Figures 1.5 and 1.7A). A stimulation of unidirectional Na^+ efflux may also occur (Twitchen and Eddy, 1994). In addition, a brisk increase in urine flow observed in the absence of body weight changes during sublethal ammonia exposure suggests that gill water permeability may be increased (Lloyd and Orr, 1969). However, in light of the fact that drinking rate was not measured in this study, and that blood pressure (Smart, 1978) and plasma renin activity (Arillo *et al.*, 1981b) may also increase during ammonia exposure (both responses would increase glomerular filtration rate), this point remains unproven. Imbalances in water entry and excretion rates may explain some observations of increased plasma osmolality or [Na^+] during extended ammonia exposures (Buckley *et al.*, 1979; Oppenborn and Goudie, 1993).

In contrast to the situation during acute exposure (Smart, 1976), a variety of morphologic changes in the gills have been described during chronic sublethal ammonia exposure. Most prominent are an overall swelling of the respiratory lamellae, proliferation of epithelial cells, increased diffusion distance, and an increased prevalence of bacterial gill disease (Burrows, 1964; Haywood, 1983; Lang *et al.*, 1987). These responses would be expected to decrease the respiratory gas exchange capacity of the fish, and thus its swimming performance and tolerance to hypoxia.

Nitrite

The nitrite anion NO_2^- is formed through the reduction of nitrate (NO_3^-) by denitrifying bacteria and through the oxidation of ammonia by nitrifying bacteria. Elevated nitrite levels and associated toxicity to fish is a common problem in eutrophic waters, sites of sewage discharge, aquaculture, and anywhere that excess nitrogen enters the aquatic environment. The relationship between water chemistry and toxicity and the nature of the toxic mechanism have both been studied extensively (reviewed by Lewis and Morris, 1986; Eddy and Williams, 1987; Tomasso, 1994; Jensen, 1995, 1996). Nitrite is not a primary branchial toxicant and does not build up in the gills, but may quickly accumulate in blood plasma and systemic tissues to levels many fold higher than in the external environment (Figure 1.21). This accumulation causes internal toxicity by oxidizing hemoglobin to methemoglobin (which is non-functional in O_2 transport) while increasing the affinity of the remaining hemoglobin for O_2 (thereby impeding O_2 unloading) and increasing red cell fragility (thereby causing anemia). The resultant O_2 starvation of the systemic tissues has multiple internal consequences, most importantly severe liver damage associated with hypoxemia and compounded by direct nitrite attack on mitochondrial cytochromes in hepatocytes (Arillo *et al.*, 1984).

Nitrite is considered here because elucidation of its branchial entry mechanism has proven the key to understanding its widely variant toxicity in different water

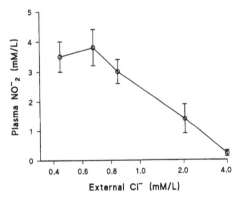

Figure 1.21 The influence of increasing water chloride concentration (by addition of NaCl) on plasma levels of nitrite (NO_2^-) in freshwater rainbow trout exposed to a waterborne nitrite concentration of 0.7 mmol L^{-1} for 24 h (data from Bath and Eddy, 1980). Note that plasma levels may exceed waterborne levels, reflecting active uptake through the Cl^- transport pathway on the gills. Higher chloride levels therefore attenuate nitrite accumulation and toxicity.

qualities and different species; early work indicated 96-h LC_{50} values ranging from the mid-μg L^{-1} to mid-mg L^{-1} range. As first shown by Perrone and Meade (1977) and Russo and Thurston (1977), and later confirmed by many others (e.g. Wedemeyer and Yasutake, 1978; Bath and Eddy, 1980; Russo et al., 1981; Bowser et al., 1989; Mazik et al., 1991), nitrite toxicity is inversely proportional to water Cl^- concentration because elevations in [Cl^-] directly reduce NO_2^- accumulation in the blood plasma (Figure 1.21). Lewis and Morris (1986) estimated that, for rainbow trout, the protective effect of Cl^- could be quantified on a molar equivalent basis as a 0.73 mM elevation in NO_2^- 96-h LC_{50} for every 1 mM elevation in water [Cl^-]. However, Cl^- does not bind NO_2^- in aqueous solution. Extensive studies by Eddy and co-workers (Bath and Eddy, 1980; Eddy et al., 1983; Williams and Eddy, 1986, 1988a,b) have demonstrated that this remarkable protective effect of [Cl^-] is explained by the fact that NO_3^- is actively transported across the freshwater gill via the active Cl^- uptake pathway (Cl^- versus 'base' exchange; see Figure 1.5A) through the chloride cells, with about the same efficiency as Cl^- itself. NO_3^- and Cl^- directly compete for the same carrier system, reciprocally inhibiting each others' uptake, i.e. displacing K_m but not J_{max} values. Not surprisingly, alkalinity has a weak protective influence against NO_2^- toxicity because HCO_3^- is transported by the same system with much lower efficiency. Water hardness also exerts a modest protective effect in some studies (Crawford and Allan, 1977) but not in others (Weirich et al., 1993). In the former, the activity of the Cl^- uptake mechanism is probably reduced in accord with the reduction in branchial permeability by elevated water [Ca^{2+}]. Furthermore, species such as sunfish, bass, carp, and eels with naturally low Cl^- uptake rates at the gills have a high resistance to nitrite (Williams and Eddy, 1986, 1988b; Tomasso, 1986). During chronic sublethal exposures, the only consistent structural change reported in the gills is chloride cell proliferation (Wedemeyer and Yasutake, 1978; Krous et al., 1982; Gaino et al., 1984; Williams and Eddy, 1988b), which could represent a homeostatic mechanism to maintain plasma [Cl^-] levels in the face of NO_2^- competition.

The effects of water pH are on nitrite toxicity are unclear (extensively discussed by Lewis and Morris, 1986, and Eddy and Williams, 1987); the potentiation of toxicity in some studies (but not others) by lower water pH could indicate that nitrite also enters across the gills as free nitrous acid (HNO_2, p$K \sim 3.3$), but the evidence for this is not convincing (Jensen, 1995). Certainly, if the free acid enters at all, it is of minor importance compared with active NO_2^- uptake in freshwater fish. Active Cl^- uptake does not occur in marine fish; here, slow uptake of HNO_2 (and perhaps also NO_2^-) by simple diffusion may explain the much lower toxicity which is observed in seawater (Crawford and Allan, 1977; Eddy et al., 1983; Weirich et al., 1993).

Chlorine

'Chlorine' represents a family of substances (HOCl, OCl$^-$, monochloroamine, dichloroamine, etc., all strong oxidizing agents) which are present as prophylactic agents in domestic drinking water, and as antifouling agents in power plant cooling water effluents. In most studies, the various components have not been separated, although from 1-h pulse exposures carried out at various circumneutral pH values (6.05–8.42) it is known that HOCl is about four times more toxic than OCl⁻ (Mattice *et al.*, 1981). Furthermore, monochloroamine is clearly less toxic than free chlorine (HOCl and OCl⁻), whereas dichloroamine is probably more toxic than free chlorine (Heath, 1977). Mattice and Tsai (1981) developed a model ranking overall toxicity to mosquitofish as follows: OCl⁻ 1.0, monochloroamine 1.7, HOCl 4.8, dichloroamine 6.2. Inasmuch as fish gills continually excrete NH_3, while free chlorine combines instantaneously with NH_3 to form monochloroamine, some detoxification may occur in the gill boundary layer. Toxicity changes might also occur as a result of acidification (more HOCl, greater toxicity) or alkalinization (less HOCl, lower toxicity) of the gill boundary layer water by metabolic CO_2 and NH_3 excretion respectively. Emissions from power plants are generally pulsatile, as are 'chlorine breakthroughs' in dechlorination systems for fish-holding aquaria, so it is difficult to identify a concentration range of 'environmental relevance'. Certainly, the concentrations occurring during intermittent discharge peaks of power plants (up to 0.6 mg L^{-1} total chlorine) as well as concentrations often found in domestic drinking water (up to 2 mg L^{-1}) are sufficient to kill fish in a matter of a few hours (Mattice and Tsai, 1981; Mattice *et al.*, 1981; Mattice, 1985; Tsai *et al.*, 1990; Meyer *et al.*, 1995). For continuous exposures, 96-h LC_{50} values are typically in the 100–300 µg L^{-1} range in both freshwater and seawater. Elevations in water pH (see above) and hardness tend to reduce acute toxicity, but in general protective effects of water chemistry are not marked (Tsai *et al.*, 1990).

Like nitrite, chlorine enters the fish and oxidizes hemoglobin to methemoglobin (Grothe and Eaton, 1975; Buckley *et al.*, 1976; Bass and Heath, 1977; Travis and Heath, 1981) and causes chronic hemolytic anemia (Buckley *et al.*, 1976), although the branchial uptake mechanism(s) are unknown. Monochloroamine appears to be more effective than free chlorine in this regard. However, unlike nitrite, chlorine severely damages branchial structure, with extensive lifting, rupture, and hypertrophy of pavement cells, lamellar fusion, and general gill mucification during acute exposures and general hyperplasia of the lamellar epithelium during chronic exposures (Dandy, 1972; Bass *et al.*, 1977; Middaugh *et al.*, 1980; Mitchell and Cech, 1983). Respiratory blockade ensues. The hypoxemia resulting from this action, rather than methemoglobin production, is the likely mechanism of acute toxicity. This suffocation phenomenon was indicated by a sharp drop in arterial P_{O_2}, coughing, marked hyperventilation, and reflex bradycardia observed during pulse exposures of trout to free chlorine (peaks at 0.4–0.5 mg L^{-1}) lasting only 2 h (Bass and Heath, 1977). Upon repetitive thrice daily exposures (mimicking a power plant discharge), intervening recovery attenuated and responses became more severe until the fish died with extremely low arterial P_{O_2} and severe blood lactacidosis (Figure 1.22). When the same regime was carried out with a similar level of monochloroamine, responses were qualitatively similar but much less severe in magnitude (Travis and Heath, 1981). Exposure of trout to even higher levels of free chlorine caused a severe hemoconcentration response and disturbance of plasma electrolytes (Zeitoun *et al.*, 1977), but at more realistic levels there is little indication of ionoregulatory toxicity (Block, 1977).

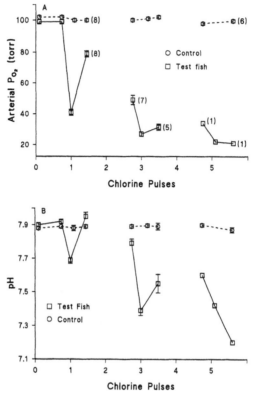

Figure 1.22 The influence of repeated cycles (three per day at 8-h intervals) of 'pulse' exposures to chlorine on the arterial partial pressure of oxygen (A) and on the pH (B) in freshwater rainbow trout (data from Bass and Heath, 1977). The exposure was meant to duplicate the cyclic discharges emitted by power plants. Chlorine was administered as calcium hypochlorite solution, producing peak total chlorine concentrations of about 0.5 mg L^{-1}, and a total pulse duration of about 2 h. Control fish were subjected to similar experimental procedures and blood sampling, but without chlorine exposure. Note the progressive reduction of recovery between pulses, and the declining numbers of fish surviving (numbers in brackets).

Concluding remarks

In this review, only a very few of thousands of toxicants now present in natural waters have been discussed. However, from this it should be clear that, although there are a few important principles that transcend individual toxicants (e.g. the general protective effect of hardness, the critical effect of toxicant speciation and the presence of natural anions and cations in the water), there is an incredible diversity of toxic mechanisms at the gills for contaminants at environmentally realistic concentrations. This diversity reflects the complexity of this multifunctional epithelium, and indicates that it is unwise to propose general mechanisms of toxicant action at gills. The other alarming conclusion is how little we know at present about toxicant action at the gills of seawater fish in contrast to freshwater fish, despite the fact that species abundance, fish productivity, and economic importance are all so much greater in the marine environment.

Acknowledgments

I thank Drs Pierre Laurent, Sjoerd Wendelaar Bonga, and Steve Perry for their generous provision of electron micrographs, Drs Jack Mattice, Joe Meyer, Christer Hogstrand, and Gordon McDonald for much helpful advice, and Erin Fitzgerald and Cheryl Szebedinszky for excellent bibliographic and graphing assistance. Supported by NSERC Research, Strategic, and Industrially Oriented Research Grants to C.M.W.

References

Abel, P.D. 1976. Toxic action of several lethal concentrations of an anionic detergent on the gills of the brown trout (*Salmo trutta* L.). *The Journal of Fish Biology* 9: 441–446.

Abel, P.D. and Skidmore, J.F. 1975. Toxic effects of an anionic detergent on the gills of rainbow trout. *Water Research* 9: 759–765.

Alabaster, J.S. and Lloyd, R. 1980. *Water Quality Criteria for Freshwater Fish*, 2nd edn. Butterworths Scientific, London.

Albers, C. 1970. Acid–base balance. In *Fish Physiology*, Vol. 4. Hoar, W.S. and Randall, D.J. (eds), pp. 173–208. Academic Press, New York.

Allen, H.E., and Hansen, D.J. 1995. The importance of trace metal speciation to water quality criteria. *Water Environment Research* 68: 42–54.

Allison, J.D., Brown, D.S. and Novo-Gradac K.J. 1991. *MINTEQA2/PRODEFA2, a Geochemical Assessment Model for Environmental Systems*. Version 3.0, User's Manual. United States Environmental Protection Agency, Washington, DC.

Anello, J. and Levy, G. 1969. Effect of complex formation on drug absorption. X. Effect of polysorbate 80 on the permeability of biologic membranes. *The Journal of Pharmaceutical Sciences* 58: 721–724.

Arillo, A., Magiocco, C., Melodia, F., Mensi, P. and Schenone, G. 1981a. Ammonia toxicity mechanism in fish: studies on rainbow trout (*Salmo gairdneri* Rich.). *Ecotoxicology and Environmental Safety* 5: 316–328.

Arillo, A., Uva, B. and Vallarino, M. 1981b. Renin activity in rainbow trout (*Salmo gairdneri* Richardson) and effects of environmental ammonia. *Comparative Biochemistry and Physiology* 68A: 307–311.

Arillo, A., Gaino, E., Margiocco, C., Mensi, P. and Schenone, G. 1984. Biochemical and ultrastructural effects of nitrite in rainbow trout: liver hypoxia as the root of the acute toxicity mechanism. *Environmental Research* 34: 135–154.

Audet, C., Munger, R.S. and Wood, C.M. 1988. Long term sublethal acid exposure in rainbow trout (*Salmo gairdneri*) in soft water: effects on ion exchanges and blood chemistry. *The Canadian Journal of Fisheries and Aquatic Sciences* 45: 1387–1398.

Bailly, Y., Dunel-Erb, S., Geffard, M. and Laurent, P. 1989. The vascular and epithelial serotonergic innervation of the actinopterygian gill filament with special reference to the trout, *Salmo gairdneri*. *Cell and Tissue Research* 258: 349–363.

Bailly, Y., Dunel-Erb, S. and Laurent, P. 1992. The neuroepithelial cells of the fish gill filament: indolamine-immunocytochemistry and innervation. *The Anatomical Record* 233: 143–161.

Baker, J.T.P. 1969. Histological and electron microscopical observations on copper poisoning in the winter flounder (*Pseudopleuronectes americanus*). *The Journal Fisheries Research Board of Canada* 26: 2785–2793.

Baklien, A, Lange, R. and Reirsen, L.-O. 1986. A comparison between the physiological effects in fish exposed to lethal and sublethal concentrations of a dispersant and dispersed oil. *Marine Environmental Research* 19: 1–11.

Balm, P., Goosen, N., Vanderij, S. and Wendelaar Bonga, S. 1988. Characterization of transport Na$^+$ATPases in gills of freshwater tilapia. Evidence for branchial Na$^+$/H$^+$(NH$_4^+$) ATPase in fish gills. *Fish Physiology and Biochemistry* 5: 31–38.

Balm, P.H.M. and Pottinger, T.G. 1993. Acclimation of rainbow trout (*Oncorhynchus mykiss*) to low environmental pH does not involve an activation of the pituitary–interrenal axis, but evokes adjustments in branchial ultrastructure. *The Canadian Journal of Fisheries and Aquatic Science* 50: 2532–2541.

Barton, B.A. and Iwama, G.K. 1991. Physiological changes in fish from stress in aquaculture with emphasis on the response and effects of corticosteroids. *Annual Review of Fish Diseases* 1: 3–26.

Bass, M.L. and Heath, A.G. 1977. Cardiovascular and respiratory changes in rainbow trout, *Salmo gairdneri*, exposed intermittently to chlorine. *Water Research* 11: 497–502.

Bass, M.L., Berry, Jr, C.R. and Heath, A.G. 1977. Histopathological effects of intermittent chlorine exposure on bluegill (*Lepomis macrochirus*) and rainbow trout (*Salmo gairdneri*) *Water Research* 11: 731–735.

Bath, R.N. and Eddy, F.B. 1980. Transport of nitrite across fish gills. *The Journal of Experimental Zoology* 214: 119–121.

Baudin, J.P. and Fritsch, J.F. 1989. Relative contributions of food and water in the accumulation of ^{60}Co by a freshwater fish. *Water Research* 23: 817–823.

Beaumont, M.W., Butler, P.J. and Taylor, E.W. 1995a. Exposure of brown trout, *Salmo trutta*, to sub-lethal copper concentrations in soft acidic water and its effect upon sustained swimming performance. *Aquatic Toxicology* 33: 45–63.

Beaumont, M.W., Butler, P.J. and Taylor, E.W. 1995b. Plasma ammonia concentration in brown trout (*Salmo trutta*) exposed to acidic water and sublethal copper concentrations and its relationship to decreased swimming performance. *The Journal of Experimental Biology* 198: 2213–2220.

Beckman, B.R. and Zaugg, W.S. 1988. Copper intoxication in Chinook salmon (*Oncorhynchus tshawytscha*) induced by natural springwater: effects on gill Na$^+$,K$^+$-ATPase, hematocrit, and plasma glucose. *The Canadian Journal of Fisheries and Aquatic Science* 45: 1430–1435.

Benedetti, I., Albano, A.G. and Mola, L. 1989. Histomorphological changes in some organs of the brown bullhead, *Ictalurus nebulosus* LeSueur, following short- and long-term exposure to copper. *The Journal of Fish Biology* 34: 272–280.

Benson, W.H. and Birge, W.J. 1985. Heavy metal tolerance and metallothionein induction in fathead minnows: results from field and laboratory investigations. *Environmental Toxicology and Chemistry* 4: 209–212.

Bentley, P.J. 1991. Accumulation of cadmium by channel catfish (*Ictalurus punctatus*). *Comparative Biochemistry and Physiology* 99C: 527–529.

Bentley, P.J. 1992. Influx of zinc by channel catfish (*Ictalurus punctatus*); uptake from external environment solutions. *Comparative Biochemistry and Physiology* 101C: 215–217.

Bergman, H.L. and Dorward-King, E.J. 1997. Reassessment of metals criteria for aquatic life protection: priorities for research and implementation. In *Proceedings of the SETAC Pellston Workshop on Reassessment of Metals Criteria for Aquatic Life Protection*, 10–14 February 1996. SETAC Press, Pensacola, FL; 114 pp.

Block, M. and Part, P. 1992. Uptake of ^{109}Cd by cultured gill epithelial cells from rainbow trout (*Oncorhynchus mykiss*). *Aquatic Toxicology* 23: 137–151.

Block, M., Part, P. and Glynn, A.W. 1997. Influence of water quality on the accumulation of methyl ^{203}mercury in gill tissue of minnow (*Phoxinus phoxinus*). *Comparative Biochemistry and Physiology* 118C: 191–197.

Block, R.M. 1977. Physiological responses of estuarine organisms to chlorine. *Chesapeake Science* 18: 158–160.

Bolis, C.L., Cambria, A. and Fama, M. 1984. Effects of acid stress on fish gills. In *Toxins, Drugs, and Pollutants in Marine Animals*. Bolis, C.L., Cambria, A. and Fama, M. (eds), pp. 122–129. Springer-Verlag, Berlin.

Bolis., L. and Rankin, J.C. 1978. Vascular effects of acetylcholine, catecholamines, and detergents on isolated perfused gills of pink salmon, *Oncorhynchus gorbuscha*, coho salmon, *O. kisutch* and chum salmon, *O. keta*. *The Journal of Fish Biology* 13: 543–547.

Bolis., L. and Rankin, J.C. 1980. Interactions between vascular actions of detergent and catecholamines in perfused gills of European eel, *Anguilla anguilla* L. and brown trout, *Salmo trutta* L. *The Journal of Fish Biology* 16: 61–73.

Booth, J.H. 1979a. The effects of oxygen supply, epinephrine, and acetylcholine on the distribution of blood flow in trout gills. *The Journal of Experimental Biology* 83: 31–39.

Booth, J.H. 1979b. Circulation in trout gills: the relationship between branchial perfusion and the width of the lamellar blood space. *The Canadian Journal of Zoology* 57: 2183–2185.

Bouquegneau, J.M. 1977. ATPase activity in mercury intoxicated eels. *Experientia* 33: 941–943.

Bouquegneau, J.M. 1979. Evidence for the protective effect of metallothioneins against inorganic mercury injuries to fish. *The Bulletin of Environmental Contamination and Toxicology* 23: 212–213.

Boutilier, R.G., Heming, T.A. and Iwama, G.K. 1984. Appendix: physicochemical parameters for use in fish respiratory physiology. In *Fish Physiology*, Vol. 10A. Hoar, W.S. and Randall, D.J. (eds), pp. 403–430. Academic Press, Orlando.

Bowser, P.K., Wooster, G.A., Aluisio, A.L. and Blue, J.T. 1989. Plasma chemistries of nitrite stressed salmon *Salmo salar*. *Journal of the World Aquaculture Society* 20: 173–180.

Bradley, R.W. and Sprague, J.B. 1985a. The influence of pH, water hardness, and alkalinity on the acute lethality of zinc to rainbow trout (*Salmo gairdneri*). *The Canadian Journal of Fisheries and Aquatic Sciences* 42: 731–736.

Bradley, R.W. and Sprague, J.B. 1985b. Accumulation of zinc by rainbow trout as influenced by pH, water hardness, and fish size. *Environmental Toxicology and Chemistry* 4: 685–694.

Bradley, R.W., DuQuesnay, C. and Sprague, J.B. 1985. Acclimation of rainbow trout, *Salmo gairdneri* Richardson, to zinc: kinetics and mechanism of enhanced tolerance induction. *The Journal of Fish Biology* 27: 367–379.

Brafield, A.E. and Matthiessen, P. 1976. Oxygen consumption by sticklebacks (*Gaserosteus aculeatus* L.) exposed to zinc. *The Journal of Fish Biology* 9: 359–370.

Brown, J.A. and Waring, C.P. 1996. The physiological status of brown trout exposed to aluminium in acidic soft waters. In *Toxicology of Aquatic Pollution – Physiological, Molecular, and Cellular Approaches*. Society for Experimental Biology Seminar Series 57. Taylor, E.W. (ed.), pp. 115–142. Cambridge University Press, Cambridge.

Brown, V.M. 1968. The calculation of the acute toxicity of mixtures of poisons to rainbow trout. *Water Research* 2: 723–733.

Buckley, J.A. 1978. Acute toxicity of un-ionized ammonia to fingerling coho salmon. *The Progressive Fish Culturist* 40: 30–32.

Buckley, J.A., Whitmore, C.W. and Matsuda, R.I. 1976. Changes in blood chemistry and blood cell morphology in coho salmon (*Oncorhynchus kisutch*) following exposure to sublethal levels of total residual chlorine in municipal wastewater. *The Journal of the Fisheries Research Board of Canada* 33: 776–782.

Buckley, J.A., Whitmore, C.M. and Liming, B.D. 1979. Effects of prolonged exposure to ammonia on the blood and liver glycogen of coho salmon (*Oncorhynchus kisutch*). *Comparative Biochemistry and Physiology* 63C: 297–303.

Burleson, M.L. and Milsom, W.K. 1995. Cardio-ventilatory control in rainbow trout. I. Pharmacology of branchial, oxygen-sensitive chemoreceptors. *Respiration Physiology* 100: 231–238.

Burrows, R.E. 1964. Effects of accumulated excretory products on hatchery-reared salmonids. In *Research Report 66, Fish and Wildlife Service, Bureau of Sport Fisheries and Wildlife*. US Government Printing Office, Washington, DC; 12 pp.

Bury, N.R., Li, J., Flik, G., Lock, R.A.C. and Wendelaar Bonga, S.E. 1998. Cortisol protects against copper induced necrosis and promotes apoptosis in fish gill chloride cells in vitro. *Aquatic Toxicology* 40: 193–202.

Calabrese, A., Thurberg, F.P., Dawson, M.A. and Wenzloff, D.R. 1975. Sublethal physiological stress induced by cadmium and mercury in the winter flounder, *Pseudopleuronectes americanus*. In *Sublethal Effects of Toxic Chemicals on Aquatic Fish*. Koeman, J.H. and Strik, J.J.T.W.A. (eds), pp. 15–21. Elsevier, Amsterdam.

Cameron, J.N. 1984. Acid–base status of fish at different temperatures. *The American Journal of Physiology* 246: R452–R459.

Cameron, J.N. and Heisler, N. 1983. Studies of ammonia in rainbow trout: physico-chemical parameters, acid–base behaviour and respiratory clearance. *The Journal of Experimental Biology* 105: 107–125.

Cardeilhac, P.T. and Hall, E.R. 1977. Acute copper poisoning of cultured marine teleosts. *The American Journal of Veterinary Research* 38: 525–527.

Cardeilhac, P.T., Simpson, C.F., Lovelock, R.L., Yosha, S.F., Calderwood, H.W. and Gudat, J.C. 1979. Failure of osmoregulation with apparent potassium intoxication in marine teleosts: a primary toxic effect of copper. *Aquaculture* 17: 231–239.

Carrol, J.J., Ellis, S.J. and Oliver, W.S. 1979. Influences of hardness constituents on the acute toxicity of cadmium to brook trout (*Salvelinus fontinalis*). *The Bulletin of Environmental Contamination and Toxicology* 22: 575–581.

CCME (Canadian Council of Ministers of the Environment). 1995. *Canadian Water Quality Guidelines*. Environment Canada, Ottawa.

Chakoumakos, C., Russo, R.C. and Thurston, R.V. 1979. Toxicity of copper to cutthroat trout (*Salmo clarki*) under different conditions of alkalinity, pH, and hardness. *Environmental Science and Technology* 13: 213–219.

Choi, M.H., Cech, J.J. and Lagunas-Solar, M.C. 1998. Bioavailability of methyl mercury to Sacramento blackfish (*Orthodon microlepidotus*): dissolved organic carbon effects. *Environmental Toxicology and Chemistry* 17: 695–701.

Christensen, G.M. and Tucker, J.H. 1976. Effects of selected water toxicants on the *in vitro* activity of fish carbonic anhydrase. *Chemical and Biological Interactions* 13: 181–192.

Christensen, G., Hunt, E. and Fiandt, J. 1977. The effect of methylmercuric chloride, cadmium chloride, and lead nitrate on six biochemical factors of the brook trout (*Salvelinus fontinalis*). *Toxicology and Applied Pharmacology* 42: 523–530.

Comhaire, S., Blust, R., Van Ginneken, L., Verbost, P.M. and Vanderborght, O.L.J. 1994. Cobalt uptake across the gills of the common carp, *Cyprinus carpio*, as a function of calcium concentration in the water of acclimation and exposure. *Comparative Biochemistry and Physiology* 109C: 63–76.

Comhaire, S., Blust, R., Van Ginneken, L., Verbost, P. M. and Vanderborght, O.L. J. 1998. Branchial cobalt uptake in the carp, *Cyprinus carpio*: effect of calcium channel blockers and calcium injection. *Fish Physiology and Biochemistry* 18: 1–13.

Crawford, R.E. and Allen, G.H. 1977. Seawater inhibition of nitrite toxicity to chinook salmon. *Transactions of the American Fisheries Society* 106: 105–109.

Crespo, S. and Karnaky, Jr, K.J. 1983. Copper and zinc inhibit chloride transport across the opercular epithelium of seawater-adapted killifish *Fundulus heteroclitus*. *The Journal of Experimental Biology* 102: 337–341.

Crespo, S., Flos, R., Balasch, J. and Alonso, G. 1979. Zinc in the gills of the dogfish (*Scyliorhinus canicula* L.) related to experimental aquatic zinc pollution. *Comparative Biochemistry and Physiology* 63C: 261–266.

Crespo, S., Soriano, E., Sampera, C. and Balasch, J. 1981. Zinc and copper distribution in excretory organs of the dogfish, *Scyliorhinus canicula* and chloride cell response following treatment with zinc sulphate. *Marine Biology* 65: 117–123.

van Dam, L. 1938. On the utilisation of oxygen and regulation of breathing in some aquatic animals. Ph.D. Dissertation. University of Groningen, The Netherlands.

Dandy, J.W.T. 1972. Activity response to chlorine in the brook trout *Salvelinus fontinalis* (Mitchill). *The Canadian Journal of Zoology* 50: 405–410.

Daoust, P.Y. and Ferguson, H.W. 1984. The pathology of chronic ammonia toxicity in rainbow trout, *Salmo gairdneri* Richardson. *The Journal of Fish Diseases* 7: 199–205.

Daoust, P.Y., Wobeser, G. and Newstead, J.D. 1984. Acute pathological effects of inorganic mercury and copper in gills of rainbow trout. *Veterinary Pathology* 27: 93–101.

Davies, P.H., Gorman, W.C., Carlson, C.A. and Brinkman, S.F. 1993. Effect of hardness on bioavailability and toxicity of cadmium to rainbow trout. *Chemical Speciation and Bioavailability* 5: 67–76.

De Boeck, G., De Smet, H. and Blust R. 1995. The effect of sublethal levels of copper on oxygen consumption and ammonia excretion in the common carp, *Cyprinus carpio*. *Aquatic Toxicology* 32: 127–141.

De Renzis, G. and Bornancin, M. 1977. A Cl^-/HCO_3^- ATPase in the gills of *Carassius auratus*: its inhibition by thiocyanate. *Biochimica et Biophysica Acta* 467: 192–207.

Diamond, J.M., Winchester, E.L., Mackler, D.G., Rasnake, W.J., Fanelli, J.K. and Gruber, D. 1992. Toxicity of cobalt to freshwater indicator species as a function of water hardness. *Aquatic Toxicology* 22: 163–180.

Dimberg, K., Hoglund, L.B., Knutsson, P.G. and Ridderstrale, Y. 1981. Histochemical localization of carbonic anhydrase in gill lamellae from young salmon (*Salmo salar* L.) adapted to fresh and salt water. *Acta Physiologica Scandinavica* 112: 218–220.

Dodoo, D.K., Engel, D.W. and Sunda, W.G. 1992. Effect of cupric ion activity on calcium accumulation in juvenile flounder (*Paralichthys* spp.). *Marine Environmental Research* 33: 101–109.

Drummond, R.A., Spoor, W.A. and Olson, G.F. 1973. Some short-term indicators of sublethal effects of copper on brook trout, *Salvelinus fontinalis*. *The Journal of the Fisheries Research Board of Canada* 30: 698–701.

Drummond, R.A., Olson, G.F. and Batterman, A.R. 1974. Cough response and uptake of mercury by brook trout, *Salvelinus fontinalis*, exposed to mercuric compounds at different hydrogen-ion concentrations. *Transactions of the American Fisheries Society* 103: 244–252.

Dunel-Erb, S., Bailly, Y. and Laurent, P. 1982. Neuroepithelial cells in fish gill primary lamellae. *The Journal of Applied Physiology* 53: 1342–1353.

Eddy, F.B. and Williams, E.M. 1987. Nitrite and freshwater fish. *Chemistry and Ecology* 3: 1–38.

Eddy, F.B., Kunzlik, P.A. and Bath, R.N. 1983. Uptake and loss of nitrite from the blood of rainbow trout, *Salmo gairdneri* Richardson, and Atlantic salmon, *Salmo salar* L. in fresh water and in dilute sea water. *The Journal of Fish Biology* 23: 105–116.

Eller, L.L. 1975. Gill lesions in freshwater teleosts. In *The Pathology of Fishes*. Ribelin, W.E. and Migaki, G. (eds), pp. 305–330. University of Wisconsin Press, Madison.

Erickson, R.J., Benoit, D.A., Mattso, V.R., Nelson, Jr, H.P. and Leonard, E.N. 1996. The effects of water chemistry on the toxicity of copper to fathead minnows. *Environmental Toxicology and Chemistry* 2: 181–193.

Erickson, R.J., Kleiner, C.F., Fiandt, J.T. and Highland, T.L. 1997. Effect of acclimation period on the relationship of acute copper toxicity to water hardness for fathead minnows. *Environmental Toxicology and Chemistry* 16: 813–815.

Evans, D.H. 1977. Further evidence for Na^+/NH_4^+ exchange in marine teleost fish. *Journal of Experimental Biology* 70: 213–220.

Evans, D.H. 1987. The fish gill: Site of action and model for toxic effects of environmental pollutants. *Environmental Health Perspectives* 71: 47–58.

Evans, D.H. 1993. Osmotic and ionic regulation. In *The Physiology of Fishes*. Evans, D.H. (ed.), pp. 315–341. CRC Press, Boca Raton.

Evans, D.H., More, K.J. and Robbins, S.L. 1989. Modes of ammonia transport across the gill epithelium of the marine teleost fish *Opsanus beta*. *Journal of Experimental Biology* 144: 339–356.

Farag, A.M., Boese, C.J., Woodward, D.F. and Bergman, H.L. 1994. Physiological changes and tissue metal accumulation in rainbow trout exposed to foodborne and waterborne metals. *Environmental Toxicology and Chemistry* 13: 2021–2029.

Farrell, A.P., Daxboeck, C. and Randall, D.J. 1979a. The effect of input pressure and flow on the pattern and resistance to flow in the isolated perfused gill of a teleost fish. *The Journal of Comparative Physiology* 133: 233–240.

Farrell, A.P., Sobin, S.S., Randall, D.J. and Crosby, S. 1979b. Intralamellar blood flow patterns in fish gills. *The American Journal of Physiology* 239: R428–R436.

Ferguson, H.W., Morrison, D., Ostland, V.E., Lumsden, J. and Byrne, P. 1992. Responses of mucus-producing cells in gill disease of rainbow trout (*Oncorhynchus mykiss*). *The Journal of Comparative Pathology* 106: 255–265.

Flik, G. and Verbost, P.M. 1993. Calcium transport in fish gills and intestine. *The Journal of Experimental Biology* 184: 17–29.

Flik, G., Van Der Velden, J.A., Dechering, K.J., Verbost, P.M., Schoenmakers, T.J.M., Kolar, Z.I. and Wendelaar Bonga, S.E. 1993. Ca^{2+} and Mg^{2+} transport in gills and gut of tilapia *Oreochromis mossambicus*: A review. *The Journal of Experimental Zoology* 265: 356–365.

Flik, G., Verbost, P. and Wendelaar Bonga, S.E. 1995. Calcium transport processes in fishes. In *Cellular and Molecular Approaches to Fish Ionic Regulation, Fish Physiology*, Vol. 14. Wood, C.M. and Shuttleworth, T.J. (eds), pp. 317–342. Academic Press, San Diego.

Flos, R., Caritat, A. and Balasch, J. 1979. Zinc content in organs of dogfish (*Scyliorhinus canicula* L.) subject to sublethal experimental aquatic zinc pollution. *Comparative Biochemistry and Physiology* 64C: 77–81.

Freeman, R.A. and Everhart, W.H. 1971. Toxicity of aluminum hydroxide complexes in natural and basic media to rainbow trout. *Transactions of the American Fisheries Society* 100: 644–658.

Fromm, P.O. 1974. Circulation in trout gills: presence of 'blebs' in afferent filamental vessels. *The Journal of the Fisheries Research Board of Canada* 31: 1793–1796.

Fu, H., Lock, R.A.C. and Wendelaar Bonga, S.E. 1989. Effect of cadmium on prolactin cell activity and plasma electrolytes in the freshwater teleost *Oreochromis mossambicus*. *Aquatic Toxicology* 14: 295–306.

Fu, H., Steinebach, O.M., van der Hamer, C.J.A., Balm, P.H.M. and Lock, R.A.C. 1990. Involvement of cortisol and metallothionein-like proteins in the physiological responses of tilapia (*Oreochromis mossambicus*) to sublethal cadmium stress. *Aquatic Toxicology* 16: 257–270.

Gaino, E., Arillo, A. and Mensi, P. 1984. Involvement of the gill chloride cells of trout under acute nitrite intoxication. *Comparative Biochemistry and Physiology* 77A: 611–617.

Galvez, F., Webb, N., Hogstrand, C. and Wood, C.M. 1998. Zinc binding to the gills of rainbow trout: the effect of long-term exposure to sublethal zinc. *The Journal of Fish Biology* 52: 1089–1104.

Gardner, G.R. and Yevich, P.P. 1970. Histological and hematological responses of an estuarine teleost to cadmium. *The Journal of the Fisheries Research Board of Canada* 27: 2185–2196.

George, S.G., Todd, K. and Wright, J. 1996. Regulation of metallothionein in teleosts: induction of MtmRNA and protein by cadmium in hepatic and extrahepatic tissues of a marine flatfish, the turbot (*Scophthalmus maximus*). *Comparative Biochemistry and Physiology* 113C: 109–115.

Giles, M.A. 1984. Electrolyte and water balance in plasma and urine of rainbow trout (*Salmo gairdneri*) during chronic exposure to cadmium. *The Canadian Journal of Fisheries and Aquatic Science* 41: 1678–1685.

Giles, M.A. 1988. Accumulation of cadmium by rainbow trout, *Salmo gairdneri*, during extended exposure. *The Canadian Journal of Fisheries and Aquatic Science* 45: 1045–1053.

Goss, G.G., Laurent, P. and Perry, S.F. 1994. Gill morphology during hypercapnia in brown bullhead (*Ictalurus nebulosus*): role of chloride cells and pavement cells in acid–base regulation. *The Journal of Fish Biology* 45: 705–718.

Goss, G.G., Perry, S.F., Wood, C.M. and Laurent, P.L. 1992. Mechanisms of ion and acid–base regulation at the gills of freshwater fish. *The Journal of Experimental Zoology* 263: 143–159.

Goss, G.G., Perry, S.F. and Laurent, P. 1995. Ultrastructural and morphometric studies on ion and acid–base transport processes in freshwater fish. In *Cellular and Molecular Approaches to Fish Ionic Regulation, Fish Physiology*, Vol. 14. Wood, C.M. and Shuttleworth, T.J. (eds), pp. 257–284. Academic Press, San Diego.

Greco, A.M., Gilmour, K.M., Fenwick, J.C. and Perry, S.F. 1995. The effects of softwater acclimation on respiratory gas transfer in the rainbow trout, *Oncorhynchus mykiss*. *Journal of Experimental Biology* 198: 2557–2567.

Greco, A.M., Fenwick, J.C. and Perry, S.F. 1996. The effects of soft-water acclimation on gill structure in the rainbow trout *Oncorhynchus mykiss*. *Cell and Tissue Research* 285: 75–82.

Grosell, M., Boetius, I., Hansen, H.J.M. and Rosenkilde, P. 1996. Influence of pre-exposure to sublethal levels of copper on ⁶⁴Cu uptake and distribution among tissues of the European eel (*Anguilla anguilla*). *Comparative Biochemistry and Physiology* 114C: 229–235.

Grosell, M.H., Hogstrand, C. and Wood, C.M. 1997. Copper uptake and turnover in both Cu-acclimated and non-acclimated rainbow trout (*Oncorhynchus mykiss*). *Aquatic Toxicology* 38: 257–276.

Grothe, D.R. and Eaton, J.W. 1975. Chlorine-induced mortality in fish. *Transactions of the American Fisheries Society* 4: 800–802.

Handy, R.D. 1989. The ionic composition of rainbow trout body mucus. *Comparative Biochemistry and Physiology* 93A: 571–575.

Handy, R.D., and Eddy, F.B. 1991. The absence of mucus on the secondary lamellae of unstressed rainbow trout, *Oncorhynchus mykiss* (Walbaum). *The Journal of Fish Biology* 38: 153–155.

Handy, R.D., Eddy, F.B. and Romain, G. 1989. *In vitro* evidence for the ionoregulatory role of rainbow trout mucus in acid, acid/aluminium and zinc toxicity. *The Journal of Fish Biology* 35: 737–747.

Haywood, G.P. 1983. Ammonia toxicity in teleost fishes. A review. *Canadian Technical Report of Fisheries and Aquatic Sciences* 1177: 1–35.

Heath, A.G. 1977. Toxicity of intermittent chlorination to freshwater fish: influence of temperature and chlorine form. *Hydrobiologia* 56: 39–47.

Heath, A.G. 1984. Changes in tissue adenylates and water content of bluegill, *Lepomis macrochirus*, exposed to copper. *The Journal of Fish Biology* 24: 299–309.

Heath, A.G. 1987. Effects of waterborne copper or zinc on the osmoregulation response of bluegill to a hypertonic NaCl challenge. *Comparative Biochemistry and Physiology* 88C: 307–311.

Heath, A.G. 1991. Effect of water-borne copper on physiological responses of bluegill (*Lepomis macrochirus*) to acute hypoxic stress and recovery. *Comparative Biochemistry and Physiology* 100C: 559–564.

Heath, A.G. 1995. *Water Pollution and Fish Physiology*, 2nd edn. CRC Press, Boca Raton.

van der Heijden, A.J.H. and Morgan, I.J. 1997. The use of modern microscopical techniques for the study of fish gill. In *Ionic Regulation in Animals: a Tribute to Professor W.T.W. Potts*. Hazon, N., Eddy, F.B. and Flik, G. (eds), pp. 106–124. Springer-Verlag, Berlin.

Heisler, N. 1990. Mechanisms of ammonia elimination in fishes. In *Animal Nutrition and Transport Processes. 2. Transport, Respiration, and Excretion: Comparative and Environmental Aspects.* Truchot, J.-P. and Lahlou, B. (eds), pp. 137–151. Karger, Basel.

Heming, T.A. 1986. CO_2 excretion and ammonia toxicity in fishes: Is there a relationship? In *Proceedings of the USA-USSR Symposium: Problems of Aquatic Toxicology, Biotesting and Water Quality Management.* Borok, Jaroslavl Oblast.

Heming, T.A. and Blumhagen, K.A. 1988. Plasma acid–base and electrolyte status of rainbow trout exposed to alum (aluminum sulphate) in acidic and alkaline environments. *Aquatic Toxicology* 12: 125–140.

Henry, R.P. 1996. Multiple roles of carbonic anhydrase in cellular transport and metabolism. *Annual Review of Physiology* 58: 523–538.

Hernandez, C., Martin, M., Bodega, G., Suarez, I., Perez, J. and Fernandez, B. 1999. Response of carp central nervous system to hyperammonemic conditions: an immunocytochemical study of glutamine synthetase, glial fibrillary acidic protein (GFAP), and 70 kDa heat-shock protein (HSP 70). *Aquatic Toxicology* 45: 195–207.

Hillaby, B.A. and Randall, D.J. 1979. Acute ammonia toxicity and ammonia excretion in rainbow trout (*Salmo gairdneri*). *The Journal of the Fisheries Research Board of Canada* 36: 621–629.

Hodson, P.V. 1976. Temperature effects on lactate–glycogen metabolism in zinc-intoxicated rainbow trout (*Salmo gairdneri*). *Journal of the Fisheries Research Board of Canada* 33: 1393–1397.

Hogstrand, C. and Wood, C.M. 1996. The physiology and toxicology of zinc in fish. In *Toxicology of Aquatic Pollution – Physiological, Molecular, and Cellular Approaches.* Society for Experimental Biology Seminar Series 57. Taylor, E.W. (ed.), pp. 61–84. Cambridge University Press, Cambridge.

Hogstrand, C. and Wood, C.M. 1998. Towards a better understanding of the bioavailability, physiology, and toxicology of silver in fish: implications for water quality criteria. *Environmental Toxicology and Chemistry* 17: 547–561.

Hogstrand, C., Verbost, P.M., Wendelaar Bonga, S.E. and Wood, C.M. 1996a. Mechanisms of zinc uptake in gills of freshwater rainbow trout: interplay with calcium transport. *The American Journal of Physiology* 270: R1141–R1147.

Hogstrand, C., Galvez, F. and Wood, C.M. 1996b. Toxicity, silver accumulation and metallothionein induction in freshwater rainbow trout during exposure to different silver salts. *Environmental Toxicology and Chemistry* 15: 1102–1108.

Hogstrand, C., Webb, N. and Wood, C.M. 1998. Covariation in regulation of affinity for branchial zinc and calcium uptake in freshwater rainbow trout. *The Journal of experimental Biology* 201: 1809–1815.

Hogstrand, C., Wilson, R.W., Polgar, D. and Wood, C.M. 1994. Effects of zinc on the kinetics of branchial calcium uptake in freshwater rainbow trout during adaptation to waterborne zinc. *The Journal of Experimental Biology* 186: 55–73.

Hogstrand, C., Reid, S.D. and Wood, C.M. 1995. Calcium versus zinc transport in the gills of freshwater rainbow trout, and the cost of adaptation to waterborne zinc. *The Journal of Experimental Biology* 198: 337–348.

Hollis, L., Muench, L. and Playle, R.C. 1997. Influence of dissolved organic matter on copper binding, and calcium on cadmium binding, by gills of rainbow trout. *The Journal of Fish Biology* 50: 703–720.

Hollis, L., McGeer, J.C., McDonald, D.G. and Wood, C.M. 1999. Cadmium accumulation, gill Cd binding, acclimation, and physiological effects during long term sublethal Cd exposure in rainbow trout. *Aquatic Toxicology* 46: 101–119.

Hootman, S. and Philpott, C.W. 1980. Accessory cells in teleost branchial epithelium. *The American Journal of Physiology* 238: R199–R206.

Hoss, D.E. 1964. Accumulation of zinc-65 by flounder of the genus *Paralichthys. Transactions of the American Fisheries Society* 93: 364–368.

Hughes, G.M. 1966. The dimensions of fish gills in relation to their function. *The Journal of Experimental Biology* 45: 177–195.

Hughes, G.M. 1976. Polluted fish respiratory physiology. In *Effects of Pollutants on Aquatic Organisms*. Society for Experimental Biology Seminar Series No. 2. Lockwood, A.P.M. (ed.), pp. 163–183. Cambridge University Press, Cambridge.

Hughes, G.M. 1984. General anatomy of the gills. In *Fish Physiology*, Vol. 10A. Hoar, W.S. and Randall, D.J. (eds), pp. 1–72. Academic Press, Orlando.

Hughes, G.M. and Morgan, M. 1973. The structure of fish gills in relation to their respiratory function. *Biological Review* 48: 419–475.

Hughes, G.M. and Perry, S.F. 1979. A morphometric study of effects of nickel, chromium and cadmium on the secondary lamellae of rainbow trout. *Marine Research* 13: 665–679.

Hunn, J.B. 1985. Role of calcium in gill function in freshwater fishes. *Comparative Biochemistry and Physiology* 82A: 543–547.

Hunn, J.B. and Allen, J.L. 1974. Movement of drugs across the gills of fishes. *Annual Review of Pharmacology* 14: 47–55.

Ishimatsu, A., Iwama, G.K. and Heisler, N. 1988. *In vivo* analysis of partitioning of cardiac output between systemic and central venous sinus circuits in rainbow trout: a new approach using chronic cannulation of the branchial vein. *The Journal of Experimental Biology* 137: 75–88.

Iwama, G.K., McGeer, J.C., Wright, P.A., Wilkie, M.P. and Wood, C.M. 1997. Divalent cations enhance ammonia excretion in Lahontan cutthroat trout in highly alkaline waters. *The Journal of Fish Biology* 50: 1061–1073.

Iwata, K. 1989. Nitrogen metabolism in the mudskipper, *Periophthalmus cantonensis*: changes in free amino acids and related compounds of various tissues under conditions of ammonia loading with special reference to high ammonia tolerance. *Comparative Biochemistry and Physiology* 92A: 499–509.

Jackson, W.F. and Fromm, P.O. 1977. Effect of a detergent on flux of tritiated water into isolated perfused gills of rainbow trout. *Comparative Biochemistry and Physiology* 58C: 167–171.

Janes, N. and Playle, R.C. 1995. Modelling silver binding to gills of rainbow trout (*Oncorhynchus mykiss*). *Environmental Toxicology and Chemistry* 14: 1847–1858.

Jensen, F.B. 1995. Uptake and effects of nitrite and nitrate in animals. In *Nitrogen Metabolism and Excretion*. Walsh, P.J. and Wright, P.A. (eds), pp. 289–303. CRC Press, Boca Raton.

Jensen, F.B. 1996. Physiological effects of nitrite in teleosts and crustaceans. In *Toxicology of Aquatic Pollution – Physiological, Molecular, and Cellular Approaches*. Society for Experimental Biology Seminar Series 57. Taylor, E.W. (ed.), pp. 169–186. Cambridge University Press, Cambridge.

Jurss, K., and Bastrop, R. 1995. The function of mitochondria-rich cells (chloride cells) in teleost gills. *Reviews in Fish Biology and Fisheries* 5: 235–255.

Karnaky, Jr., K.J., Kinter, L.B., Kinter, W.B. and Stirling, C.E. 1976. Teleost chloride cell. II. Autoradiographic localization of gill Na^+K^+-ATPase in killifish *Fundulus heteroclitus* adapted to low and high salinity environments. *The Journal of Cell Biology* 70: 157–177.

Kerstetter, T.H. and Kirschner, L.B. 1974. HCO_3^--dependent ATPase activity in the gills of rainbow trout (*Salmo gairdneri*). *Comparative Biochemistry and Physiology* 488: 581–589.

Khangarot, B.S. and Somani, R.C. 1980. Toxic effects of mercury on the gills of a freshwater teleost. *Current Science* 49: 832–834.

Khangarot, B.S. and Tripathi, D.M. 1991. Changes in humoral and cell-mediated immune responses and in skin and respiratory surfaces of catfish, *Saccobranchus fossilis*, following copper exposure. *Ecotoxicology and Environmental Safety* 22: 291–308.

Kirk, R.S. and Lewis, J.W. 1993. An evaluation of pollutant induced changes in the gills of rainbow trout using scanning electron microscopy. *Environmental Technology* 14: 577–585.

74   *Chris M. Wood*

Kirschner, L.B. 1997. Extrarenal mechanisms in hydromineral and acid–base regulation in aquatic vertebrates. In *The Handbook of Physiology*. Dantzler, W.H. (ed.), pp. 577–622. American Physiological Society, Bethesda.

Klaverkamp, J.F. and Duncan, D.A. 1987. Acclimation to cadmium toxicity by white suckers: cadmium binding capacity and metal distribution in gill and liver cytosol. *Environmental Toxicology and Chemistry* 6: 275–289.

Klaverkamp, J.F., McDonald, W.A., Duncan, D.A. and Wagemann, R. 1984. Metallothionein and acclimation to heavy metals in fish: a review. In *Contaminant Effects on Fisheries*. Cairns, V.W., Hodson, P.V. and Nriagu, J.O. (eds), pp. 99–113. John Wiley and Sons, New York.

Korsgaard, B., Mommsen, T.P. and Wright, P.A. 1995. Nitrogen excretion in teleostean fish: adaptive relationships to environment, ontogenesis, and viviparity. In *Nitrogen Metabolism and Excretion*. Walsh, P.J. and Wright, P.A. (eds), pp. 259–287. CRC Press, Boca Raton.

Krogh, A. 1939. *Osmotic Regulation in Aquatic Animals*. Cambridge University Press, Cambridge.

Krous, S.R., Blazer, V.S. and Meade, T.L. 1982. Effect of acclimation time on nitrite movement across the gill epithelia of rainbow trout: the role of 'chloride cells'. *The Progressive Fish Culturist* 44: 126–130.

Kudo, A., Nagase, H. and Ose, Y. 1982. Proportion of methylmercury to total amount of mercury in river waters in Canada and Japan. *Water Research* 16: 1011–1015.

Laitinen, M. and Valtonen, T. 1995. Cardiovascular, ventilatory and haematological responses of brown trout (*Salmo trutta* L.), to the combined effects of acidity and aluminum in humic water at winter temperatures. *Aquatic Toxicology* 31: 99–112.

Lang, T., Peters, G., Hoffmann, R. and Meyer, E. 1987. Experimental investigations on the toxicity of ammonia: effects on ventilation frequency, growth, epidermal mucous cells, and gill structure of rainbow trout *Salmo gairdneri*. *Diseases of Aquatic Organisms* 3: 159–165.

Larsen, B.K., Portner, H.O. and Jensen, F.B. 1997. Extra- and intracellular acid–base balance and ionic regulation in cod (*Gadus morhua*) during combined and isolated exposures to hypercapnia and copper. *Marine Biology* 128: 337–346.

Larsson, A., Bengtsson, B.-E. and Svanberg, O. 1976. Some haematological and biochemical effects of cadmium on fish. In *Effects of Pollutants on Aquatic Organisms*. Society for Experimental Biology Seminar Series No. 2. Lockwood, A.P.M. (ed.), pp. 35–45. Cambridge University Press, Cambridge.

Larsson, A., Bengtsson, B.-E. and Haux, C. 1981. Disturbed ion balance in flounder, *Platichthys flesus* L. exposed to sublethal levels of cadmium. *Aquatic Toxicology* 1: 19–35.

Lauff, R.F. and Wood, C.M. 1996. Respiratory gas exchange, nitrogenous waste excretion, and fuel usage during starvation in juvenile rainbow trout, *Oncorhynchus mykiss*. *The Journal of Comparative Physiology* B165: 542–551.

Lauren, D.J. 1991. The fish gill: a sensitive target for waterborne pollutants. In *Aquatic Toxicology and Risk Assessment: Fourteenth Volume, ASTM STP 1124*. Mayes, M.A. and Barron, M.G. (eds), pp. 223–244. American Society for Testing and Materials, Philadelphia.

Lauren, D.J. and McDonald, D.G. 1985. Effects of copper on branchial ionoregulation in the rainbow trout, *Salmo gairdneri* Richardson. *The Journal of Comparative Physiology* B 155: 635–644.

Lauren, D.J. and McDonald, D.G. 1986. Influence of water hardness, pH, and alkalinity on the mechanisms of copper toxicity in juvenile rainbow trout, *Salmo gairdneri*. *The Canadian Journal of Fisheries and Aquatic Sciences* 43: 1488–1496.

Lauren, D.J. and McDonald, D.G. 1987a. Acclimation to copper by rainbow trout, *Salmo gairdneri*: Biochemistry. *Canadian Journal of Fisheries and Aquatic Sciences* 44: 105–111.

Lauren, D.J. and McDonald, D.G. 1987b. Acclimation to copper by rainbow trout, *Salmo gairdneri*: Physiology. *The Canadian Journal of Fisheries and Aquatic Sciences* 44: 99–104.

Laurent, P. 1984. Gill internal morphology. In *Fish Physiology*, Vol. 10A. Hoar, W.S. and Randall, D.J. (eds), pp. 73–183. Academic Press, Orlando.

Laurent, P. and Dunel, S. 1980. Morphology of gill epithelia in fish. *The American Journal of Physiology* 238: R147–R159.

Laurent, P. and Hebibi, N. 1989. Gill morphometry and fish osmoregulation. *The Canadian Journal of Zoology* 67: 3055–3063.

Laurent, P. and Perry, S.F. 1991. Environmental effects on fish gill morphology. *Physiological Zoology* 64: 4–25.

Laurent, P., Hobe, H. and Dunel-Erb, S. 1985. The role of environmental sodium chloride relative to calcium in gill morphology of freshwater salmonid fish. *Cell and Tissue Research* 240: 675–692.

Laurent, P., Goss, G.G. and Perry, S.F. 1994a. Proton pumps in fish gill pavement cells? *Archives International Physiologica, Biochimica et Biophysica* 102: 77–79.

Laurent, P., Dunel-Erb, S., Chevalier, C. and Lignon, J. 1994b. Gill epithelial cells kinetics in a freshwater teleost, *Oncorhynchus mykiss* during adaptation to ion-poor water and hormonal treatments. *Fish Physiology and Biochemistry* 13: 353–370.

Laurent, P.L., Maina, J.N., Bergman, H.L., Narahara, A., Walsh, P.J. and Wood, C.M. 1995. Gill structure of a fish from an alkaline lake. Effect of exposure to pH 7. *The Canadian Journal of Zoology* 73: 1170–1181.

Leino, R.L. and McCormick, J.H. 1984. Morphological and morphometrical changes in chloride cells of the gills of *Pimephales promelas* after chronic exposure to acid water. *Cell and Tissue Research* 236: 121–128.

Levi, G., Morisi, G., Coletti, A. and Catanzaro, R. 1974. Free amino acids in fish brain: normal levels and changes upon exposure to high ammonia concentrations *in vivo*, and upon incubation of brain slices. *Comparative Biochemistry and Physiology* 49A: 623–636.

Lewis, Jr., W.M. and Morris, D.P. 1986. Toxicity of nitrite to fish: A review. *Transactions of the American Fisheries Society* 115: 183–195.

Li, J., Lock, R.A.C., Klaren, P.H.M., Swarts, H.G.P., Schuurmans Stekhoven, F.M.A.H., Wendelaar Bonga, S.E. and Flik, G. 1996. Kinetics of Cu^{2+} inhibition of Na^+/K^+-ATPase. *Toxicology Letters* 87: 31–38.

Li, J., Quabius, E.S., Wendelaar Bonga, S.E., Flik, G. and Lock, R.A.C. 1998. Effects of water-borne copper on branchial chloride cells and Na^+/K^+ATPase activities in Mozambique tilapia (*Oreochromis mossambicus*). *Aquatic Toxicology* 43: 1–11.

Lin, H. and Randall, D.J. 1990. The effect of varying water pH on the acidification of expired water in rainbow trout. *The Journal of Experimental Biology* 149: 149–160.

Lin, H. and Randall, D.J. 1995. Proton pumps in fish gills. In *Cellular and Molecular Approaches to Fish Ionic Regulation, Fish Physiology*, Vol. 14. Wood, C.M. and Shuttleworth, T.J. (eds), pp. 229–255. Academic Press, San Diego.

Lloyd, R. and Herbert, D.W.M. 1960. The influence of carbon dioxide on the toxicity of un-ionized ammonia to rainbow trout (*Salmo gairdneri* Richardson). *Annals of Applied Biology* 48: 399–404.

Lloyd, R. and Orr, L.D. 1969. The diuretic response by rainbow trout to sublethal concentrations of ammonia. *Water Research* 3: 335–344.

Lock, R.A.C. and van Overbeeke, A.P. 1981. Effects of mercuric chloride and methylmercuric chloride on mucus secretion in rainbow trout, *Salmo gairdneri* Richardson. *Comparative Biochemistry and Physiology* 69C: 67–73.

Lock, R.A.C., Cruijsen, P.M.J.M. and van Oberbeeke, A.P. 1981. Effects of mercuric chloride and methylmercuric chloride on the osmoregulatory function of the gills in rainbow trout, *Salmo gairdneri* Richardson. *Comparative Biochemistry and Physiology* 68C: 151–159.

Lorz, H.W. and McPherson, B.P. 1976. Effects of copper or zinc in fresh water on the adaptation to sea water and ATPase activity, and the effects of copper on migratory disposition of coho salmon (*Oncorhynchus kisutch*). *Journal of the Fisheries Research Board of Canada* 33: 2023–2030.

McCarty, L.S. and Houston, A.H. 1976. Effects of exposure to sublethal levels of cadmium upon water-electrolyte status in the goldfish (*Carassius auratus*). *The Journal of Fish Biology* 9: 11–19.

McCormick, S.D. 1990. Cortisol directly stimulates differentiation of chloride cells in tilapia opercular membrane. *The American Journal of Physiology* 28: R857–R863.

McCormick, S.D. 1995. Hormonal control of gill Na^+,K^+-ATPase and chloride cell function. In *Cellular and Molecular Approaches to Fish Ionic Regulation, Fish Physiology*, Vol. 14. Wood, C.M. and Shuttleworth, T.J. (eds), pp. 285–315. Academic Press, San Diego.

McDonald, D.G. 1983. The effects of H^+ upon the gills of freshwater fish. *The Canadian Journal of Zoology* 61: 691–704.

McDonald, D.G. and Prior, E.T. 1988. Branchial mechanisms of ion and acid–base regulation in the freshwater rainbow trout, *Salmo gairdneri*. *The Canadian Journal of Zoology* 66: 2699–2708.

McDonald, D.G. and Wood, C.M. 1993. Branchial acclimation to metals. In *Fish Ecophysiology*. Rankin, J.C. and Jensen, F.B. (eds), pp. 297–321. Chapman and Hall, London.

McDonald, D.G., Reader, J.P. and Dalziel, T.R.K. 1989. The combined effects of pH and trace metals on fish ionoregulation. In *Acid Toxicity and Aquatic Animals*. Society for Experimental Biology Seminar Series 34. Morris, R., Taylor, E.W., Brown, D.J.A. and Brown, J.A. (eds), pp. 221–242. Cambridge University Press, Cambridge.

McDonald, D.G., Wood, C.M., Rhem, R.G., Mueller, M.E., Mount, D.R. and Bergman, H.L. 1991. Nature and time course of acclimation to aluminum in juvenile brook trout (*Salvelinus fontinalis*). I. Physiology. *The Canadian Journal of Fisheries and Aquatic Sciences* 48: 2006–2015.

McDonald, G. and Milligan, L. 1997. Ionic, osmotic and acid–base regulation in stress. In *Fish Stress and Health in Aquaculture*. Society for Experimental Biology Seminar Series 62. Iwama, G.K., Pickering, A.D., Sumpter, J.P. and Schreck, C.B. (eds), pp. 119–144. Cambridge University Press, Cambridge.

McGeer, J.C. and Eddy, F.B. 1998. Ionic regulation and nitrogenous excretion in rainbow trout exposed to buffered and unbuffered freshwater of pH 10.5. *Physiological Zoology* 71: 179–190.

McKeown, B.A. and March, G.L. 1978. The acute effect of bunker C oil and an oil dispersant on: 1. Serum glucose, serum sodium and gill morphology in both freshwater and seawater acclimated rainbow trout (*Salmo gairdneri*). *Water Research* 12: 157–163.

McKim, J.M., Christensen, G.M. and Hunt, E.P. 1970. Changes in the blood of brook trout (*Salvelinus fontinalis*) after short-term and long-term exposure to copper. *Journal of the Fisheries Research Board of Canada* 27: 1883–1889.

McKone, C.E., Young, R.G., Bache, C.A. and Lisk, D.J. 1971. Rapid uptake of mercuric ion by goldfish. *Environmental Science and Technology* 5: 1138–1139.

McWilliams, P.G. 1983. An investigation of the loss of bound calcium from the gills of the brown trout, *Salmo trutta*, in acid media. *Comparative Biochemistry and Physiology* 74A: 107–116.

McWilliams, P.G. and Potts, W.T.W. 1978. The effect of pH and calcium concentration on gill potentials in the brown trout *Salmo trutta*. *The Journal of Comparative Physiology* 126: 277–286.

Maetz, J. 1972. Branchial sodium exchange and ammonia excretion in the goldfish *Carassius auratus*. Effects of ammonia-loading and temperature changes. *The Journal of Experimental Biology* 56: 601–620.

Maetz, J. 1973. Na^+/NH_4^+, Na^+/H^+ exchanges and NH_3 movement across the gill of *Carassius auratus*. *The Journal of Experimental Biology* 58: 255–275.

Maetz, J. and Garcia-Romeu, F. 1964. The mechanism of sodium and chloride uptake by the gills of a fresh-water fish, *Carassius auratus. The Journal of General Physiology* 47: 1209–1226.

Majewski, H.S. and Giles, M.A. 1981. Cardiovascular–respiratory responses of rainbow trout (*Salmo gairdneri*) during chronic exposure to sublethal concentrations of cadmium. *Water Research* 15: 1211–1217.

Maki, A.W. 1979. Respiratory activity of fish as a predictor of chronic fish toxicity values for surfactants. In *Aquatic Toxicology ASTM ATP 667*. Marking, L.L. and Kimerle, R.A. (eds), pp. 77–95. American Association for Testing and Materials, Philadelphia.

Mallatt, J. 1985. Fish gill structural changes induces by toxicants and other irritants: A statistical review. *The Canadian Journal of Fisheries and Aquatic Science* 42: 630–648.

Marshall, W.S. 1995. Transport processes in isolated teleost epithelia: opercular epithelium and urinary bladder. In *Cellular and Molecular Approaches to Fish Ionic Regulation, Fish Physiology*, Vol. 14. Wood, C.M. and Shuttleworth, T.J. (eds), pp. 1–23. Academic Press, San Diego.

Matthiessen, P. and Brafield, A.E. 1973. The effects of dissolved zinc on the gills of the stickleback *Gasterosteus aculeatus* (L.) *Journal of Fish Biology* 5: 607–613.

Matthiessen, P. and Brafield, A.E. 1977. Uptake and loss of dissolved zinc by the stickleback *Gasterosteus aculeatus* L. *Journal of Fish Biology* 10: 399–410.

Mattice, J.S. 1985. Chlorination of power plant cooling waters. In *Water Chlorination: Chemistry, Environmental Impact, and Health Effects*, Vol. 5. Jolley, R.L., Bull, R.J., Davis, W.P., Katz, S., Roberts, M.H. and Jacobs, V.A. (eds), pp. 479–489. Lewis Publishers, Chelsea, MI.

Mattice, J.S. and Tsai, S.C. 1981. Total residual chlorine as a regulatory tool. In *Water Chlorination: Environmental Impact, and Health Effects*. Vol. 5. Book 2. *Environment, Health, and Risk*. Jolley, R.L., Brungs, W.A., Cotruvo, J.A., Cumming, R.B., Mattice, J.S. and Jacobs, V.A. (eds), pp. 901–912. Ann Arbor Science, Ann Arbor, MI.

Mattice, J.S., Tsai, S.C. and Burch, M.B. 1981. Comparative toxicity of hypochlorous acid and hypochlorite ions to mosquitofish. *Transactions of the American Fisheries Society* 110: 519–525.

Mazik, P.M., Hinman, M.L., Winkelmann, D.A., Klaine, S.J., Simco, B.A. and Parker, N.C. 1991. Influence of nitrite and chloride concentrations on survival and hematological profiles of striped bass. *Transactions of the American Fisheries Society* 120: 247–254.

Meyer, J.S. 1999. A mechanistic explanation for the ln(LC50) *vs.* ln(hardness) adjustment equation for metals. *Environmental Science and Technology* 33: 908–912.

Meyer, J.S., Gulley, D.D., Goodrich, M.S., Szmania, D.C. and Brooks, A.S. 1995. Modeling toxicity due to intermittent exposure of rainbow trout and common shiners to monochloramine. *Environmental Toxicology and Chemistry* 14: 165–175.

Middaugh, D.P., Burnett, L.E. and Couch, J.A. 1980. Toxicological and physiological responses of the fish, *Leistomus xanthurus*, exposed to chlorine-produced oxidants. *Estuaries* 3: 132–141.

Miller, T.G. and MacKay, W.C. 1980. The effects of hardness, alkalinity and pH of test water on the toxicity of copper to rainbow trout (*Salmo gairdneri*). *Water Research* 14: 129–133.

Milligan, C.L. and Wood, C.M. 1982. Disturbances in hematology, fluid volume distribution, and circulatory function associated with low environmental pH in the rainbow trout, *Salmo gairdneri. Journal of Experimental Biology* 99: 397–415.

Mitchell, S.J. and Cech, Jr, J.J. 1983. Ammonia-caused gill damage in channel catfish (*Ictalurus punctatus*): Confounding effects of residual chlorine. *The Canadian Journal of Fisheries and Aquatic Sciences* 40: 242–247.

Miura, K. and Imura, N. 1987. Mechanisms of methylmercury and mercury cytotoxicity. *CRC Critical Reviews in Toxicology* 18: 161–188.

Morgan, I.J., Henry, R.P. and Wood, C.M. 1997. The mechanism of acute silver nitrate toxicity in freshwater rainbow trout (*Oncorhynchus mykiss*) is inhibition of gill Na^+ and Cl^- transport. *Aquatic Toxicology* 38: 145–163.

Mueller, M.E., Sanchez, D.A., Bergman, H.L., McDonald, D.G., Rhem, R.G. and Wood, C.M. 1991. Nature and time course of acclimation to aluminum in juvenile brook trout (*Salvelinus fontinalis*). II. Histology of the gills. *The Canadian Journal of Fisheries and Aquatic Science* 48: 2016–2027.

Muramoto, S. 1981. Vertebral column damage and decrease of calcium concentration in fish exposed experimentally to cadmium. *Environmental Pollution* 24: 125–133.

Neville, C.M. 1985. Physiological response of juvenile rainbow trout, *Salmo gairdneri*, to acid and aluminum-prediction of field responses from laboratory data. *The Canadian Journal of Fisheries and Aquatic Sciences* 42: 2004–2018.

Nikl, D.L. and Farrell, A.P. 1993. Reduced swimming performance and gill structural changes in juvenile salmonids exposed to 2-(thiocyanomethylthio)benzothiazole. *Aquatic Toxicology* 27: 245–264.

Nilsson, S. 1984. Innervation and pharmacology of the gills. In *Fish Physiology*, Vol. 10A. Hoar, W.S. and Randall, D.J. (eds), pp. 185–227. Academic Press, Orlando.

Noel-Lambot, F., Gerday, C.H. and Disteche, A. 1978. Distribution of Cd, Zn, and Cu in liver and gills of eel (*Anguilla anguilla*) with special reference to metallothioneins. *Comparative Biochemistry and Physiology* 61C: 177–187.

Oduleye, S.O. 1975. The effects of calcium on water balance of the brown trout *Salmo trutta*. *The Journal of Experimental Biology* 63: 343–356.

Olson, K.R. and Fromm, P.O. 1973. Mercury uptake and ion distribution in gills of rainbow trout (*Salmo gairdneri*): tissue scans with an electron microprobe. *The Journal of the Fisheries Research Board of Canada* 30: 1575–1578.

Olson, K.R., Bergman, H.L. and Fromm, P.O. 1973. Uptake of methyl mercuric chloride and mercuric chloride by trout: a study of uptake pathways into the whole animal and uptake by erythrocytes in vitro. *The Journal of the Fisheries Research Board of Canada* 30: 1293–1299.

Olson, K.R., Squibb, K.S. and Cosins, R.J. 1978. Tissue uptake, subcellular distribution, and metabolism of $^{14}CH_3HgCl$ and CH_3HgCl by rainbow trout, *Salmo gairdneri*. *The Journal of the Fisheries Research Board of Canada* 35: 381–390.

Olsson, P.-E. and Hogstrand, C. 1987. Subcellular distribution and binding of cadmium to metallothionein in tissues of rainbow trout after exposure to ^{109}Cd in water. *Environmental Toxicology and Chemistry* 6: 867–874.

Olsson, P.-E., Larsson, A., Maage, A., Haux, C., Bonham, K., Zafarullah, M. and Gedamu, L. 1989. Induction of metallothionein in rainbow trout (*Salmo gairdneri*) during long-term exposure to water-borne Cd. *Fish Physiology and Biochemistry* 6: 221–229.

Oppenborn, J.B. and Goudie, C.A. 1993. Acute and sublethal effects of ammonia on striped bass and hybrid striped bass. *The Journal of the World Aquaculture Society* 24: 90–101.

Oronsaye, J.A.O. and Brafield, A.E. 1984. The effect of dissolved cadmium on the chloride cells of the gills of the stickleback, *Gasterosteus aculeatus* L. *The Fisheries Society of the British Isles* 25: 253–258.

Pagenkopf, G.K. 1983. Gill surface interaction model for trace-metal toxicity to fishes: Role of complexation, pH, and water hardness. *The Journal of Environmental Science and Technology* 17: 342–347.

Paley, R.K., Twitchen, I.D. and Eddy, F.B. 1993. Ammonia, Na^+, K^+, and Cl^- levels in rainbow trout yolk sac fry in response to external ammonia. *The Journal of Experimental Biology* 180: 273–284.

Part, P. and Lock, R.A.C. 1983. Diffusion of calcium, cadmium and mercury in a mucous solution from rainbow trout. *Comparative Biochemistry and Physiology* 76C: 259–263.

Part, P., Svanberg, O. and Kiessling, A. 1985. The availability of cadmium to perfused rainbow trout gills in different water qualities. *Water Research* 19: 427–434.

Partearroyo, M.A., Pilling, S.J. and Jones, M.N. 1992. The effects of surfactants on the permeability of isolated perfused fish gills to urea. *Comparative Biochemistry and Physiology* 101A: 653–659.

Payan, P. 1978. A study of the Na^+/NH_4^+ exchange across the gill of the perfused head of the trout (*Salmo gairdneri*). *The Journal of Comparative Physiology* 124: 181–188.

Pelgrom, S.M.G.J., Lock, R.A.C., Balm, P.H.M. and Wendelaar Bonga, S.E. 1995. Integrated physiological response of tilapia, *Oreochromis mossambicus*, to sublethal copper exposure. *Aquatic Toxicology* 32: 303–320.

Penrice, W.S. and Eddy, F.B. 1993. Spatial arrangement of fish gill secondary lamellar cells in intact and dissociated tissues from rainbow trout (*Oncorhynchus mykiss*) and Atlantic salmon (*Salmo salar*). *The Journal of Fish Biology* 42: 845–850.

Pentreath, R.J. 1973. The accumulation and retention of ^{65}Zn and ^{54}Mn by the plaice, *Pleuronectes platessa* L. *The Journal of Experimental Marine Biology and Ecology* 12: 1–18.

Pereira, J.J. 1988. Morphological effects of mercury exposure on windowpane flounder gills as observed by scanning electron microscopy. *The Journal of Fish Biology* 33: 571–580.

Perrone, S.J. and Meade, T.L. 1977. Protective effect of chloride on nitrite toxicity to Coho salmon (*Oncorhynchus kisutch*). *The Journal of Fisheries Research Board of Canada* 34: 486–492.

Perry, S.F. 1986. Carbon dioxide excretion in fishes. *The Canadian Journal of Zoology* 64: 565–572.

Perry, S.F. 1997. The chloride cell: Structure and function in the gills of freshwater fishes. *Annual Review of Physiology* 59: 325–347.

Perry, S.F. and Gilmour, K.M. 1996. Consequences of catecholamine release on ventilation and blood oxygen transport during hypoxia and hypercapnia in an elasmobranch (*Squalus acanthias*) and a teleost (*Oncorhynchus mykiss*). *The Journal of Experimental Biology* 199: 2105–2118.

Perry, S.F. and Laurent, P. 1993. Environmental effects on fish gill structure and function. In *Fish Ecophysiology*. Rankin, J.C. and Jensen, F.B. (eds), pp. 231–264. Chapman and Hall, London.

Perry, S.F. and McDonald, D.G. 1993. Gas exchange. In *The Physiology of Fishes*. Evans, D.H. (ed.), pp. 251- 278. CRC Press, Boca Raton.

Perry, S.F. and Walsh, P.J. 1989. Metabolism of isolated fish gill cells: contribution of epithelial chloride cells. *The Journal of Experimental Biology* 144: 507–520.

Perry, S.F. and Wood, C.M. 1985. Kinetics of branchial calcium uptake in the rainbow trout: effects of acclimation to various external calcium levels. *The Journal of Experimental Biology* 116: 411–433.

Perry, S.F. and Wood, C.M. 1989. Control and co-ordination of gas transfer in fishes. *The Canadian Journal of Zoology* 67: 2961–2970.

Perry, S.F., Reid, S.G., Wankiewicz, E., Iyer, V. and Gilmour, K.M. 1996. Physiological responses of rainbow trout (*Oncorhynchus mykiss*) to prolonged exposure to soft water. *Physiological Zoology* 69: 1419–1441.

Perry, S.F., Gilmour, K.M., Bernier, N.J. and Wood, C.M. 1999. Does gill boundary layer carbonic anhydrase contribute to carbon dioxide excretion: a comparison between dogfish (*Squalus acanthias*) and rainbow trout (*Oncorhynchus mykiss*). *The Journal of Experimental Biology* 202: 749–756.

Person-Le Ruyet, J., Chartois, H. and Quemener, L. 1995. Comparative acute ammonia toxicity in marine fish and plasma ammonia response. *Aquaculture* 136: 181–194.

Person-Le Ruyet, J., Galland, R., Le Roux, A. and Chartois, H. 1997. Chronic ammonia toxicity to juvenile turbot (*Scophthalmus maximus*). *Aquaculture* 154: 155–171.

80   Chris M. Wood

Pettersson, K. 1983. Adrenergic control of oxygen transfer in perfused gills of the cod, *Gadus morhua. The Journal of Experimental Biology* 102: 327–335.
Phillips, G.R. and Buhler, D.R. 1978. The relative contributions of methylmercury from food or water to rainbow trout (*Salmo gairdneri*) in a controlled laboratory environment. *Transactions of the American Fisheries Society* 107: 853–861.
Pilgaard, L., Malte, H. and Jensen, F.B. 1994. Physiological effects and tissue accumulation of copper in freshwater rainbow trout (*Oncorhynchus mykiss*) under normoxic and hypoxic conditions. *Aquatic Toxicology* 29: 197–212.
Pisam, M. and Rambourg, A. 1991. Mitochondria-rich cells in the gill epithelium of teleost fishes: An ultrastructural approach. *International Review of Cytology* 130: 191–232.
Pisam, M., Caroff, A. and Rambourg, A. 1987. Two types of chloride cells in the gill epithelium of a freshwater-adapted euryhaline fish: *Lebistes reticulatus*; their modifications during adaptation to saltwater. *The American Journal of Anatomy* 179: 40–50.
Pisam, M., Auperin, B., Prunet, P., Rentier-Delrue, F., Martial, J. and Rambourg, A. 1993. Effects of prolactin on α and β chloride cells in the gill epithelium of the saltwater adapted tilapia *Oreochromis niloticus. The Anatomical Record* 235: 275–284.
Playle, R.C. 1998. Modelling metal interactions at fish gills. *The Science of the Total Environment* 219: 147–163.
Playle, R.C. and Wood, C.M. 1989a. Water chemistry changes in the gill microenvironment of rainbow trout: experimental observations and theory. *The Journal of Comparative Physiology B* 159: 527–537.
Playle, R.C. and Wood, C.M. 1989b. Water pH and aluminum chemistry in the gill microenvironment of the rainbow trout during acid and aluminum exposures. *The Journal of Comparative Physiology B* 159: 539–550.
Playle, R.C. and Wood, C.M. 1991. Mechanisms of aluminum extraction and accumulation at the gills of rainbow trout (*Salmo gairdneri*) in acidic soft water. *The Journal of Fish Biology* 38: 791–805.
Playle, R.C., Goss, G.G. and Wood, C.M. 1989. Physiological disturbances in rainbow trout (*Salmo gairdneri*) during acid and aluminum exposures in soft water of two calcium concentrations. *The Canadian Journal of Zoology* 67: 314–324.
Playle, R.C., Gensemer, R.W. and Dixon, D.G. 1992. Copper accumulation on gills of fathead minnows: influence of water hardness, complexation and pH of the gill micro-environment. *Environmental Toxicology and Chemistry* 11: 381–391.
Playle, R.C., Dixon, D.G. and Burnison, K. 1993a. Copper and cadmium binding to fish gills: modification by dissolved organic carbon and synthetic ligands. *The Canadian Journal of Fisheries and Aquatic Sciences*: 50: 2667–2677.
Playle, R.C., Dixon, D.G. and Burnison, K. 1993b. Copper and cadmium binding to fish gills: estimates of metal–gill stability constants and modelling of metal accumulation. *The Canadian Journal of Fisheries and Aquatic Sciences*: 50: 2678–2687.
Poleo, A.B.S. 1995. Aluminum polymerization – a mechanism of acute toxicity of aqueous aluminum to fish. *Aquatic Toxicology* 31: 347–356.
Post, J.R, Vandenbos, R. and McQueen, D.J. 1996. Uptake of food-chain and waterborne mercury by fish: field measurements, a mechanistic model, and an assessment of uncertainties. *The Canadian Journal of Fisheries and Aquatic Sciences* 53: 395–407.
Potts, W.T.W. 1984. Transepithelial potentials in fish gills. In *Fish Physiology*, Vol. 10B. Hoar, W.S. and Randall, D.J. (eds), pp. 105–128. Academic Press, Orlando.
Powell, M.D., Speare, D.J. and Burka, J.F. 1992. Fixation of mucus on rainbow trout (*Oncorhynchus mykiss* Walbaum) gills for light and electron microscopy. *The Journal of Fish Biology* 41: 813–824.
Powell, M.D., Speare, D.J. and Wright, G.M. 1994. Comparative ultrastructural morphology of lamellar epithelial, chloride and mucous cell glycocalyx of the rainbow trout (*Oncorhynchus mykiss*) gill. *The Journal of Fish Biology* 44: 725–730.

Pratap, H.B., Fu, H., Lock, R.A.C. and Wendelaar Bonga, S.E. 1989. Effect of waterborne and dietary cadmium on plasma ions of the teleost *Oreochromis mossambicus* in relation to water cadmium levels. *Archives of Environmental Contamination and Toxicology* 18: 568–575.

Rahim, S.M., Delaunoy, J.P. and Laurent P. 1988. Identification and immunocytochemical localization of two different carbonic anhydrase isoenzymes in teleostean fish erythrocytes and gill epithelia. *Histochemistry* 89: 451–459.

Ramamoorthy, S. and Blumhagen, K. 1984. Uptake of Zn, Cd, and Hg by fish in the presence of competing compartments. *The Canadian Journal of Fisheries and Aquatic Sciences* 41: 750–756.

Randall, D.J. 1982. The control of respiration and circulation in fish during hypoxia and exercise. *The Journal of Experimental Biology* 100: 275–288.

Randall, D.J. 1985. Shunts in fish gills. In *Cardiovascular Shunts – A. Benzon Symposium 21*. Johansen, K. and Burggren, W.W. (eds), pp. 71–87. Munksgaard, Copenhagen.

Randall, D.J. and Wright, P.A. 1987. Ammonia distribution and excretion in fish. *Fish Physiology and Biochemistry* 3: 107–120.

Randall, D.J. and Wright, P.A. 1989. The interaction between carbon dioxide and ammonia excretion and water pH in fish. *The Canadian Journal of Zoology* 67: 2936–2942.

Randall, D.J., Brauner, C.J., Thurston, R.V. and Neuman, J.F. 1996. Water chemistry at the gill surfaces of fish and the uptake of xenobiotics. In *Toxicology of Aquatic Pollution – Physiological, Molecular, and Cellular Approaches*. Society for Experimental Biology Seminar Series 57. Taylor, E.W. (ed.), pp. 1–16. Cambridge University Press, Cambridge.

Rankin, J.C., Stagg, R.M. and Bolis, L. 1982. Effects of pollutants on gills. In *Gills*. Houlihan, D.F., Rankin, J.C. and Shuttleworth, T.J. (eds), pp. 207–219. Cambridge University Press, Cambridge.

Rasmussen, R.S. and Korsgaard, B. 1998. Ammonia and urea in plasma of juvenile turbot (*Scophthalmus maximus* L.) in response to external ammonia. *Comparative Biochemistry and Physiology* 120A: 163–168.

Reader, J.P. and Morris, R. 1988. Effects of aluminium and pH on calcium fluxes, and effects of cadmium and manganese on calcium and sodium fluxes in brown trout (*Salmo trutta* L.). *Comparative Biochemistry and Physiology* 91C: 449–457.

Reid, S.D. 1995. Adaptation to and effects of acid water on the fish gill. In *Biochemistry and Molecular Biology of Fishes*, Vol. 5. Hochachka, P.W. and Mommsen, T.P. (eds), pp. 213–227. Elsevier, Amsterdam.

Reid, S.D. and McDonald, D.G. 1988. Effects of cadmium, copper, and low pH on ion fluxes in the rainbow trout, *Salmo gairdneri*. *The Canadian Journal of Fisheries and Aquatic Science* 45: 244–253.

Reid, S.D. and McDonald, D.G. 1991. Metal binding activity of the gills of rainbow trout (*Oncorhynchus mykiss*). *The Canadian Journal of Fisheries and Aquatic Sciences* 48: 1061–1068.

Reid, S.D., McDonald, D.G. and Rhem, R.R. 1991. Acclimation to sublethal aluminum: modifications of metal–gill surface interactions of juvenile rainbow trout (*Oncorhynchus mykiss*). *The Canadian Journal of Fisheries and Aquatic Sciences* 48: 1996–2005.

Reinert, R.E., Stone, L.J. and Willford, W.A. 1974. Effect of temperature on accumulation of methylmercuric chloride and p,p'DDT by rainbow trout (*Salmo gairdneri*). *The Journal of the Fisheries Research Board of Canada* 31: 1649–1652.

Renfro, L.J., Schmidt-Nielsen, B., Miller, D., Benos, D. and Allen, J. 1974. Methyl mercury and inorganic mercury: uptake, distribution, and effect on osmoregulatory mechanisms in fishes. In *Pollution and Physiology of Marine Organisms*. Vernberg, F.J. and Vernberg, W.B. (eds), pp. 101–122. Academic Press, New York.

Renner, R. 1997. Rethinking water quality standards for metals toxicity. *Environmental Science and Technology* 31: 466A–468A.

Richards, J.G. and Playle, R.C. 1998. Cobalt binding to the gills of rainbow trout (*Oncorhynchus mykiss*): An equilibrium model. *Comparative Biochemistry and Physiology* 119C: 185–197.

Roch, M. and Maly, E.J. 1979. Relationship of cadmium-induced hypocalcemia with mortality in rainbow trout (*Salmo gairdneri*) and the influence of temperature on toxicity. *The Journal of the Fisheries Research Board of Canada* 36: 1297–1302.

Rodgers, D.W. and Beamish, F.W.H. 1981. Uptake of waterborne methylmercury by rainbow trout (*Salmo gairdneri*) in relation to oxygen consumption and methylmercury concentration. *The Canadian Journal of Fisheries and Aquatic Sciences* 38: 1309–1315.

Rodgers, D.W. and Beamish, F.W.H. 1983. Water quality modifies uptake of waterborne methylmercury by rainbow trout, *Salmo gairdneri*. *The Canadian Journal of Fisheries and Aquatic Sciences* 40: 824–828.

Rosseland, B.O., Blakar, I.A., Bulger, A., Kroglund, F., Kvellstad, A., Lydersen, E., Oughton, D.H., Salbu, B., Staurnes, M. and Vogt, R. 1992. The mixing zone between limed and acidic river waters: complex aluminium chemistry and extreme toxicity for salmonids. *Environmental Pollution* 78: 3–8.

Roy, R.L. and Campbell, P. 1997. Decreased toxicity of Al to juvenile Atlantic salmon (*Salmo salar*) in acidic soft water containing natural organic matter: a test of the free-ion model. *Environmental Toxicology and Chemistry* 16: 1962–1969.

Russo, R.C. and Thurston, R.V. 1977. The acute toxicity of nitrite to fishes. In *Recent Advances in Fish Toxicology*. Tubb, R.A. (ed.), pp. 118–131. US Environmental Protection Agency, EPA Ecological Research Series EPA-600/3-77-085, Corvallis.

Russo, R.C., Thurston, R.V. and Emerson, K. 1981. Acute toxicity of nitrite to rainbow trout (*Salmo gairdneri*): Effects of pH, nitrite species, and anion species. *The Canadian Journal of Fisheries and Aquatic Sciences* 38: 387–393.

Salama, A., Morgan, I.J. and Wood, C.M. 1999. The linkage between Na^+ uptake and ammonia excretion in rainbow trout: kinetic analysis, the effects of $(NH_4)_2SO_4$ and NH_4HCO_3 infusion, and the influence of gill boundary layer pH. *The Journal of Experimental Biology* 202: 697–709.

Sangalang, G.B. and Freeman, H.C. 1979. Tissue uptake of cadmium in brook trout during chronic sublethal exposure. *Archives of Environmental Contamination and Toxicology* 8: 77–84.

Sargent, J.R., Thomson, A.J. and Bornancin, M. 1975. Activities and localization of succinic dehydrogenase and Na^+/K^+-activated adenosine triphosphatase in the gills of fresh water and sea water eels (*Anguilla anguilla*). *Comparative Biochemistry and Physiology* B51: 75–79.

Satchell, G.H. 1984. Respiratory toxicology of fishes. In *Aquatic Toxicology*, Vol. 2. Weber, L.J. (ed.), pp. 1–50. Raven Press, New York.

Satchell, G.H. 1991. *Physiology and Form of Fish Circulation*. Cambridge University Press, Cambridge.

Sayer, M.D.J., Reader, J.P. and Morris, R. 1991. Effects of six trace metals on calcium fluxes in brown trout (*Salmo trutta* L.) in soft water. *The Journal of Comparative Physiology B* 161: 537–542.

Schaumburg, F.D., Howard, T.E. and Walden, C.C. 1967. A method to evaluate the effects of water pollutants on fish respiration. *Water Research* 1: 731–737.

Schecher, W.D. and McAvoy, D.C. 1992. MINEQL+: a software environment for chemical equilibrium modelling. *Computers, Environment and Urban Systems* 16: 65–76.

Schenone, G, Arillo, A., Margiocco, C., Melodia, F. and Mensi, P. 1982. Biochemical bases for environmental adaptation in goldfish (*Carassius auratus* L.). *Ecotoxicology and Environmental Safety* 6: 479–488.

Schmid, O.J. and Mann, H. 1961. Action of a detergent (dodecylbenzenesulphonate) on the gills of the trout. *Nature* 4803: 673.

Schmidt-Nielsen, B., Sheline, J., Miller, D.S. and Deldonno, M. 1977. Effect of methylmercury upon osmoregulation, cellular volume, and ion regulation in winter flounder, *Pseudopleuronectes americanus*. In *Physiological Responses of Marine Biota to Pollutants*. Vernberg, F.J., Calabrese, A., Thurberg, F.P. and Vernberg, W.B. (eds), pp. 105–117. Academic Press, New York.

Schreck, C.B. and Lorz, H.W. 1978. Stress response of coho salmon (*Oncorhynchus kisutch*) elicited by cadmium and copper and potential use of cortisol as an indicator of stress. *The Journal of the Fisheries Research Board of Canada* 35: 1124–1129.

Sellers, Jr., C.M., Heath, A.G. and Bass, M.L. 1975. The effect of sublethal concentrations of copper and zinc on ventilatory activity, blood oxygen and pH in rainbow trout (*Salmo gairdneri*). *Water Research* 9: 401–408.

Shelton, G., Jones, D.R. and Milsom, W.K. 1986. Control of breathing in ectothermic vertebrates. In *The Handbook of Physiology, Section 3, The Respiratory System*. Vol. 2. *The Control of Breathing*. Fishman, A.P., Cherniak, N.S., Widdicombe, J.G. and Greger, S.R. (eds), pp. 857–909. American Physiological Society, Bethesda.

Shephard, K.L. 1994. Functions for fish mucus. *Reviews in Fish Biology and Fisheries* 4: 401–429.

Skidmore, J.F. 1970. Respiration and osmoregulation in rainbow trout with gills damaged by zinc sulphate. *The Journal of Experimental Biology* 52: 481–494.

Skidmore, J.F. and Tovell, P.W.A. 1972. Toxic effects of zinc sulphate on the gills of rainbow trout. *Water Research* 6: 217–230.

Smart, G. 1976. The effect of ammonia exposure on gill structure of the rainbow trout (*Salmo gairdneri*). *The Journal of Fish Biology* 8: 471–475.

Smart, G.R. 1978. Investigations of the toxic mechanisms of ammonia to fish-gas exchange in rainbow trout (*Salmo gairdneri*) exposed to acutely lethal concentrations. *The Journal of Fish Biology* 12: 93–104.

Soivio, A. and Tuurala, H. 1981. Structural and circulatory responses to hypoxia in the secondary lamellae of *Salmo gairdneri* gills at two temperatures. *The Journal of Comparative Physiology* 145: 37–43.

Sparling, D.W. and Lowe, T.P. 1996. Environmental hazards of aluminum to plants, invertebrates, fish, and wildlife. *Reviews of Environmental Contamination and Toxicology* 145: 1–127.

Sprague, J.B. 1987. Effects of cadmium on freshwater fish. In *Cadmium in the Environment*. Nriagu, J.O. and Sprague, J.B. (eds), pp. 139–169. John Wiley, New York.

Spry, D.J. and Wiener, J.G. 1991. Metal bioavailability and toxicity to fish in low-alkalinity lakes: a critical review. *Environmental Pollution* 71: 243–304.

Spry, D.J. and Wood, C.M. 1984. Acid–base and plasma ion changes in rainbow trout during short-term toxic zinc exposure. *The Journal of Comparative Physiology* B154: 149–158.

Spry, D.J. and Wood, C.M. 1985. Ion flux rates, acid–base status, and blood gases in rainbow trout exposed to toxic zinc in natural soft water. *The Canadian Journal of Fisheries and Aquatic Sciences* 42: 1332–1341.

Spry, D.J. and Wood, C.M. 1988. Zinc influx across the isolated, perfused head preparation of the rainbow trout (*Salmo gairdneri*) in hard and soft water. *The Canadian Journal of Fisheries and Aquatic Sciences* 45: 2206–2215.

Spry, D.J. and Wood, C.M. 1989a. A kinetic method for the measurement of zinc influx in the rainbow trout and the effects of waterborne calcium. *The Journal of Experimental Biology* 142: 425–446.

Spry, D.J. and Wood, C.M. 1989b. The influence of dietary and waterborne zinc on heat-stable metal ligands in rainbow trout: quantification by [109]Cd radioassay. *The Journal of Fish Biology* 35: 557–576.

Spry, D.J., Hodson, P.V. and Wood, C.M. 1988. Relative contributions of dietary and waterborne zinc in the rainbow trout, *Salmo gairdneri*. *The Canadian Journal of Fisheries and Aquatic Sciences* 45: 32–41.

Stagg, R.M. and Shuttleworth, T.J. 1982a. The effects of copper on ionic regulation by the gills of the seawater-adapted flounder (*Platichthys stellatus* L.). *The Journal of Comparative Physiology* B149: 83–90.

Stagg, R.M. and Shuttleworth, T.J. 1982b. The accumulation of copper in *Platichthys flesus* L. and its effects on plasma electrolyte concentrations. *Journal of Fish Biology* 20: 491–500.

Stagg, R.M. and Shuttleworth, T.J. 1986. Surfactant effects on adrenergic responses in the gills of the flounder (*Platichthys stellatus*). *The Journal of Comparative Physiology* B156: 727–733.

Stagg, R.M. and Shuttleworth, T.J. 1987. Sites of interactions of surfactants with beta-adrenergic responses in trout (*Salmo gairdneri*) gills. *The Journal of Comparative Physiology B* 157: 429–434.

Stagg, R.M., Rankin, J.C. and Bolis, L. 1981. Effect of detergent on vascular responses to noradrenaline in isolated perfused gills of the eel, *Anguilla anguilla* L. *Environmental Pollution (Series A)* 24: 31–37.

Staurnes, M., Sigholt, T. and Reite, O.B. 1984. Reduced carbonic anhydrase and Na-K-ATPase activity in gills of salmonids exposed to aluminium-containing acid water. *Experientia* 40: 226–227.

Stewart, P.A. 1978. Independent and dependent variables of acid–base control. *Respiration Physiology* 33: 9–26.

Stinson, C.M. and Mallatt, J. 1989. Branchial ion fluxes and toxicant extraction efficiency in lamprey (*Petromyzon marinus*) exposed to methylmercury. *Aquatic Toxicology* 12: 237–251.

Stone, H. and Overnell, J. 1985. Non-metallothionein cadmium binding proteins. *Comparative Biochemistry and Physiology* 80C: 9–14.

Sullivan, G.V., Fryer, J.N. and Perry, S.F. 1995. Immunolocalization of proton pumps (H$^+$-ATPase) in pavement cells of rainbow trout gill. *The Journal of Experimental Biology* 198: 2619–2629.

Sullivan, G.V., Fryer, J.N. and Perry, S.F. 1996. Localisation of mRNA for the proton pump (H$^+$-ATPase) and Cl$^-$/HCO$_3^-$ exchanger in the rainbow trout gill. *The Canadian Journal of Zoology* 74: 2095–2103.

Swift, D.J. 1981. Changes in selected blood component concentrations of rainbow trout, *Salmo gairdneri* Richardson, exposed to hypoxia or sublethal concentrations of phenol or ammonia. *Journal of Fish Biology* 19: 45–61.

Taylor, E.W., Beaumont, M.W., Butler, P.J., Mair, J. and Mujallid, M.S.I. 1996. Lethal and sublethal effects of copper upon fish: a role for ammonia. In *Toxicology of Aquatic Pollution – Physiological, Molecular, and Cellular Approaches.* Society for Experimental Biology Seminar Series 57. Taylor, E.W. (ed.), pp. 85–113. Cambridge University Press, Cambridge.

Thomas, D.G., Cryer, A., del G. Solbe, J.F. and Kay, J. 1983. A comparison of the accumulation and protein binding of environmental cadmium in the gills, kidney, and liver of rainbow trout (*Salmo gairdneri*). *Comparative Biochemistry and Physiology* 76C: 241–246.

Thomas, D.G., Brown, M.W., Shurben, D., del G. Solbe, J.F., Cryer, A. and Kay, J. 1985. A comparison of the sequestration of cadmium and zinc in the tissue of rainbow trout (*Salmo gairdneri*) following exposure to the metals singly or in combination. *Comparative Biochemistry and Physiology* 82C: 55–62.

Thomas, S., Fievet, B. and Motais, R. 1988. Adaptive respiratory responses of rainbow trout to pure hypoxia. I. Effects of water ionic composition on blood acid–base response and gill morphology. *Respiration Physiology* 74: 77–90.

Thurberg, P. and Dawson, M.A. 1974. Changes in osmoregulation and oxygen consumption. Chapter III. In *Physiological Response of the Cunner (Tautogolabrus adspersus) to Cadmium.* NOAA Technical Report NMFS SSRF-681.

Thurston, R.V., Russo, R.C. and Vinogradov, G.A. 1981. Ammonia toxicity to fishes: effect of pH on the toxicity of the un-ionized ammonia species. *Environmental Science and Technology* 15: 837–840.

Tietge, J.E., Johnson, R.D. and Bergman, H.L. 1988. Morphometric changes in gill secondary lamellae of brook trout (*Salvelinus fontinalis*) after long-term exposure to acid and aluminum. *The Canadian Journal of Fisheries and Aquatic Science* 45: 1643–1648.

Tomasso, J.R. 1986. Comparative toxicity of nitrite to freshwater fishes. *Aquatic Toxicology* 8: 129–137.

Tomasso, J.R. 1994. Toxicity of nitrogenous wastes to aquaculture animals. *Reviews in Fisheries Science* 2: 291–314.

Tovell, P.W.A., Howes, D. and Newsome, C.S. 1975a. Absorption, metabolism, and excretion by goldfish of the anionic detergent sodium lauryl sulphate. *Toxicology* 4: 17–29.

Tovell, P.W.A., Newsome, C.S. and Howes, D. 1975b. Effect of water hardness on the toxicity of an anionic detergent to fish. *Water Research* 8: 291–296.

Travis, T.W. and Heath, A.G. 1981. Some physiological responses of rainbow trout (*Salmo gairdneri*) to intermittent monochloramine exposure. *Water Research* 15: 977–982.

Tsai, S.-C., Mattice, J.S., Trabalka, J.R., Burch, M.B. and Packard, K.B. 1990. Chlorine sensitivity of early life stages of freshwater fish. In *Water Chlorination: Chemistry, Environmental Impact, and Health Effects*, Vol. 6. Jolley, R.L., Condie, L.W., Johnson, J.D., Katz, S., Minear, R.A., Mattice, J.S. and Jacobs, V.A. (eds), pp. 479–489. Lewis Publishers, Chelsea, MI.

Tuurala, H. and Soivio, A. 1982. Structural and circulatory changes in the secondary lamellae of *Salmo gairdneri* gills after sublethal exposures to dehydroabietic acid and zinc. *Aquatic Toxicology* 2: 21–29.

Tuurala, H. Part, P., Nikinmaa, M. and Soivio, A. 1984. The basal channels of secondary lamellae in *Salmo gairdneri* gills – a non-respiratory shunt. *Comparative Biochemistry and Physiology* 79A: 35–39.

Twitchen, I.D. and Eddy, F.B. 1994. Effects of ammonia on sodium balance in juvenile rainbow trout *Oncorhynchus mykiss* Walbaum. *Aquatic Toxicology* 30: 27–45.

Ultsch, G.R. and Gros, G. 1979. Mucus as a diffusion barrier to oxygen: possible role in O_2 uptake at low pH in carp (*Cyprinus carpio*) gills. *Comparative Biochemistry and Physiology* 62A: 685–689.

USEPA (United States Environmental Protection Agency). 1986. *Quality Criteria for Water*. USEPA Office of Water Regulations and Standards, Washington, DC. EPA440/5-85-001.

Varanisi, U., Robisch, P.A. and Malins, D.C. 1975. Structural alterations in fish epidermal mucus produced by waterborne lead and mercury. *Nature* 258: 431–432.

Verbost, P.M., Flik, G., Lock, R.A.C. and Wendelaar Bonga, S.E. 1987. Cadmium inhibition of Ca^{2+} uptake in rainbow trout gills. *The American Journal of Physiology* 253: R216–R221.

Verbost, P.M., Flik, G., Lock, R.A.C. and Wendelaar Bonga, S.E. 1988. Cadmium inhibits plasma membrane calcium transport. *The Journal of Membrane Biology* 102: 97–104.

Verbost, P.M., Van Rooij, J., Flik, G., Lock, R.A.C. and Wendelaar Bonga, S.E. 1989. The movement of cadmium through freshwater trout branchial epithelium and its interference with calcium transport. *The Journal of Experimental Biology* 145: 185–197.

Verbost, P.M., Lafeber, F.P.J.G., Spanings, F.A.T., Aarden, E.M. and Wendelaar Bonga, S.E. 1992. Inhibition of Ca^{2+} uptake in freshwater carp, *Cyprinus carpio*, during short-term exposure to aluminum. *The Journal of Experimental Zoology* 262: 247–254.

Vogel, W.O.P. 1985. Systemic vascular anastomoses, primary and secondary vessels in fish, and the phylogeny of lymphatics. In *Cardiovascular Shunts – A. Benzon Symposium 21*. Johansen, K. and Burggren, W.W. (eds), pp. 143–159. Munksgaard, Copenhagen.

Waiwood, K.G. and Beamish, F.W.H. 1978a. Effects of copper, pH and hardness on the critical swimming performance of rainbow trout (*Salmo gairdneri* Richardson). *Water Research* 12: 611–619.

Waiwood, K.G. and Beamish, F.W.H. 1978b. The effect of copper, hardness and pH on the growth of rainbow trout, *Salmo gairdneri*. *The Journal of Fish Biology* 10: 591–598.

Walsh, P.J., Danulat, E. and Mommsen, T.P. 1990. Variation in urea excretion in the gulf toadfish, *Opsanus beta*. *Marine Biology* 106: 323–328.

Wang, T., Knudsen, P.K., Brauner, C.J., Busk, M., Vijayan, M.M. and Jensen, F.B. 1998. Copper exposure impairs intra- and extracellular acid–base regulation during hypercapnia in the fresh water rainbow trout (*Oncorhynchus mykiss*). *The Journal of Comparative Physiology B* 168: 591–599.

Watson, T.A. and Beamish, F.W.H. 1980. Effects of zinc on branchial ATPase activity *in vivo* in rainbow trout, *Salmo gairdneri*. *Comparative Biochemistry and Physiology* 66C: 77–82.

Webb, N.A. and Wood, C.M. 1998. Physiological analysis of the stress response associated with acute silver nitrate exposure in freshwater rainbow trout. *Environmental Toxicology and Chemistry* 17: 579–588.

Wedemeyer, G.A. and Yasutake, W.T. 1978. Prevention and treatment of nitrite toxicity in juvenile steelhead trout (*Salmo gairdneri*). *The Journal of the Fisheries Research Board of Canada* 35: 822–827.

Weirich, C.R., Tomasso, J.R. and Smith, T.I.J. 1993. Toxicity of ammonia and nitrite to sunshine bass in selected environments. *The Journal of Aquatic Animal Health* 5: 64–72.

Wendelaar Bonga, S.E. 1997. The stress response in fish. *Physiological Reviews* 77: 592–616.

Wendelaar Bonga, S.E. and Lock, R.A.C. 1992. Toxicants and osmoregulation in fish. *The Netherlands Journal of Zoology* 42: 478–493.

Wendelaar Bonga, S.E. and van der Meij, C.J.M. 1989. Degeneration and death, by apoptosis and necrosis, of the pavement and chloride cells in the gills of the teleost *Oreochromis mossambicus*. *Cell and Tissue Research* 255: 235–243.

Wendelaar Bonga, S.E., van der Meij, C.J.M. and Flik, G. 1984. Prolactin and acid stress in the teleost *Oreochromis* (formerly *Sarotherodon*) *mossambicus*. *General and Comparative Endocrinology* 55: 323–332.

Wendelaar Bonga, S.E., Flik, G., Balm, P.H.M. and Van der Meij, J.C.A. 1990. The ultrastructure of chloride cells in the gills of the teleost *Oreochromis mossambicus* during exposure to acidified water. *Cell and Tissue Research* 259: 575–585.

Wicklund-Glynn, A. 1996. The concentration dependency of branchial intracellular cadmium distribution and influx in the zebrafish (*Brachydanio rerio*). *Aquatic Toxicology* 35: 47–58.

Wicklund-Glynn, A. and Olsson, P.-E. 1991. Cadmium turnover in minnows (*Phoxinus phoxinus*) preexposed to waterborne cadmium. *Environmental Toxicology and Chemistry* 10: 383–394.

Wicklund-Glynn, A., Norrgren, L. and Mussener, A. 1994. Differences in uptake of inorganic mercury and cadmium in the gills of the zebrafish, *Brachydanio rerio*. *Aquatic Toxicology* 30: 13–26.

Wilkie, M.P. 1997. Mechanisms of ammonia excretion across fish gills. *Comparative Biochemistry and Physiology* 118A: 39–50.

Wilkie, M.P. and Wood, C.M. 1991. Nitrogenous waste excretion, acid–base regulation, and ionoregulation in rainbow trout (*Onchorhynchus mykiss*) exposed to extremely alkaline water. *Physiological Zoology* 64: 1069–1086.

Wilkie, M.P. and Wood, C.M. 1994. The effects of extremely alkaline water (pH 9.5) on rainbow trout gill function and morphology. *The Journal of Fish Biology* 45: 87–98.

Wilkie, M.P. and Wood, C.M. 1996. The adaptations of fish to extremely alkaline environments. *Comparative Biochemistry and Physiology* 113B: 665–673.

Wilkie, M.P., Wright, P.A., Iwama, G.K. and Wood, C.M. 1993. The physiological responses of the Lahontan cutthroat trout (*Oncorhynchus clarki henshawi*), a resident of highly alkaline Pyramid Lake (pH 9.4), to challenge at pH 10. *The Journal of Experimental Biology* 175: 173–194.

Wilkie, M.P., Wright, P.A., Iwama, G.K. and Wood, C.M. 1994 The physiological adaptations of the Lahontan cutthroat trout (*Oncorhynchus clarki henshawi*) following transfer from well water to the highly alkaline waters of Pyramid Lake, Nevada (pH 9.4) *Physiological Zoology* 67: 355–380.

Wilkie, M.P., Simmons, H.E. and Wood, C.M. 1996. Physiological adaptations to chronically elevated water pH (pH = 9.5) in rainbow trout (*Oncorhynchus mykiss*). *The Journal of Experimental Zoology* 274: 1–14.

Wilkie, M.P., Laurent, P. and Wood, C.M. 1999. The physiological basis for altered Na^+ and Cl^- movements across the gills of rainbow trout (*Oncorhynchus mykiss*) in alkaline pH (9.5) water. *Physiological Zoology* 72: 360–368.

Wilkinson, K.J. and Campbell, P.G.C. 1993. Aluminum bioconcentration at the gill surface of juvenile Atlantic salmon in acidic media. *Environmental Toxicology and Chemistry* 12: 2083–2095.

Wilkinson, K.J., Campbell, P.G.C. and Couture, P. 1990. Effect of fluoride complexation on aluminum toxicity towards juvenile atlantic salmon (*Salmo salar*). *The Canadian Journal of Fisheries and Aquatic Science* 47: 1446–1452.

Williams, D.R. and Giesy, Jr, J.P. 1978. Relative importance of food and water sources to cadmium uptake by *Gambusia affinis* (Poeciliidae). *Environmental Research* 16: 326–332.

Williams, E.M. and Eddy, F.B. 1986. Chloride uptake in freshwater teleosts and its relationship to nitrite uptake and toxicity. *The Journal of Comparative Physiology* B156: 867–872.

Williams, E.M. and Eddy, F.B. 1988a. Regulation of blood haemoglobin and electrolytes in rainbow trout *Salmo gairdneri* (Richardson) exposed to nitrite. *Aquatic Toxicology* 13: 13–28.

Williams, E.M. and Eddy, F.B. 1988b. Anion transport, chloride cell number and nitrite-induced methaemoglobinaemia in rainbow trout (*Salmo gairdneri*) and carp (*Cyprinus carpio*). *Aquatic Toxicology* 13: 29–42.

Wilson, R.W. 1996. Physiological and metabolic costs of acclimation to chronic sub-lethal acid and aluminium exposure in rainbow trout. In *Toxicology of Aquatic Pollution – Physiological, Molecular, and Cellular Approaches*. Society for Experimental Biology Seminar Series 57. Taylor, E.W. (ed.), pp. 143–167. Cambridge University Press, Cambridge.

Wilson, R.W. and Taylor, E.W. 1992. Transbranchial ammonia gradients and acid–base responses to high external ammonia concentration in rainbow trout (*Oncorhynchus mykiss*) acclimated to different salinities. *The Journal of Experimental Biology* 166: 95–112.

Wilson, R.W. and Taylor, E.W. 1993a. The physiological responses of freshwater rainbow trout, *Oncorhynchus mykiss*, during acutely lethal copper exposure. *The Journal of Comparative Physiology B* 163: 38–47.

Wilson, R.W. and Taylor, E.W. 1993b. Differential responses to copper in rainbow trout (*Oncorhyncus mykiss*) acclimated to sea water and brackish water. *The Journal of Comparative Physiology B* 163: 239–246.

Wilson, R.W., Wright, P.M., Munger, S. and Wood, C.M. 1994a. Ammonia excretion in rainbow trout *Oncorhynchus mykiss*: the importance of gill boundary layer acidification: Lack of evidence for Na^+/NH_4^+ exchange. *The Journal of Experimental Biology* 191: 37–58.

Wilson, R.W., Bergman, H.L. and Wood, C.M. 1994b. Metabolic costs and physiological consequences of acclimation to aluminum in juvenile rainbow trout (*Oncorhynchus mykiss*). 2. Gill morphology, swimming performance, and aerobic scope. *The Canadian Journal of Fisheries and Aquatic Sciences* 51: 536–544.

Witters, H.E., Van Puymbroeck, S. and Vanderbroght, O.L.J. 1991. Adrenergic response to physiological disturbances in rainbow trout, *Oncorhynchus mykiss*, exposed to aluminum at acid pH. *The Canadian Journal of Fisheries and Aquatic Sciences* 48: 414–420.

Witters, H.E., Berckmans, P. and Vangenechten, C. 1996a. Immunolocalization of Na$^+$,K$^+$-ATPase in the gill epithelium of rainbow trout, *Oncorhynchus mykiss*. *Cell and Tissue Research* 283: 461–468.

Witters, H.E., Van Puymbroeck, S., Stouthart, A.J.H.X. and Wendelaar Bonga, S.E. 1996b. Physicochemical changes of aluminium in mixing zones: mortality and physiological disturbances in brown trout (*Salmo trutta* L.). *Environmental Toxicology and Chemistry* 15: 986–996.

Wobeser, G. 1975. Acute toxicity of methyl mercury chloride and mercuric chloride for rainbow trout (*Salmo gairdneri*) fry and fingerlings. *The Journal of the Fisheries Research Board of Canada* 32: 2005–2013.

Wood, C.M. 1974. A critical examination of the physical and adrenergic factors affecting blood flow through the gills of the rainbow trout. *The Journal of Experimental Biology* 60: 241–265.

Wood, C.M. 1989. The physiological problems of fish in acid waters. In *Acid Toxicity and Aquatic Animals*. Society for Experimental Biology Seminar Series 34. Morris, R., Taylor, E.W., Brown, D.J.A. and Brown, J.A. (eds), pp. 125–148. Cambridge University Press, Cambridge.

Wood, C.M. 1991a. Acid–base and ion balance, metabolism, and their interactions after exhaustive exercise in fish. *The Journal of Experimental Biology* 160: 285–308.

Wood, C.M. 1991b. Branchial ion and acid–base transfer – environmental hyperoxia as a probe. *Physiological Zoology* 64: 68–102.

Wood, C.M. 1992. Flux measurements as indices of H$^+$ and metal effects on freshwater fish. *Aquatic Toxicology* 22: 239–264.

Wood, C.M. 1993. Ammonia and urea metabolism and excretion. In *The Physiology of Fishes*. Evans, D.H. (ed.), pp. 379–425. CRC Press, Boca Raton.

Wood, C.M. and McDonald, D.G. 1982. Physiological mechanisms of acid toxicity in fish. In *Acid Rain/Fisheries, Proceedings of an International Symposium on Acidic Precipitation and Fishery Impacts in North-eastern North America*. Johnson, R.E. (ed.), pp. 197–226. American Fisheries Society, Bethesda, MD.

Wood, C.M. and McDonald, D.G. 1987. The physiology of acid/aluminum stress in trout. In *Ecophysiology of Acid Stress in Aquatic Organisms*. Witters, H. and Vanderborght, O. (eds). *Annals Society Royale Zoology Belgium* 117 (Suppl. 1): 399–410.

Wood, C.M. and Marshall, W.S. 1994. Ion balance, acid–base regulation and chloride cell function in the common killifish, *Fundulus heteroclitus* – a euryhaline estuarine teleost. *Estuaries* 17: 34–52.

Wood, C.M. and Munger, R.S. 1994. Carbonic anhydrase injection provides evidence for the role of blood acid–base status in stimulating ventilation after exhaustive exercise in rainbow trout. *The Journal of Experimental Biology* 194: 225–253.

Wood, C.M. and Perry, S.F. 1985. Respiratory, circulatory, and metabolic adjustments to exercise in fish. In *Circulation, Respiration, and Metabolism*. Gilles, R. (ed.), pp. 2–22. Springer-Verlag, Berlin.

Wood, C.M. and Perry, S.F. 1991. A new *in vitro* assay for carbon dioxide excretion by trout red blood cells: effects of catecholamines. *The Journal of Experimental Biology* 157: 349–366.

Wood, C.M., Perry, S.F., Wright, P.A., Bergman, H.L. and Randall, D.J. 1989. Ammonia and urea dynamics in the Lake Magadi tilapia, a ureotelic teleost fish adapted to an extremely alkaline environment. *Respiration Physiology* 77: 1 -20.

Wood, C.M., Hogstrand, C., Galvez, F. and Munger, R.S. 1996. The physiology of waterborne silver toxicity in freshwater rainbow trout (*Oncorhynchus mykiss*). 1. The effects of ionic Ag$^+$. *Aquatic Toxicology* 35: 93–109.

Wood, C.M., Gilmour, K.M., Perry, S.F., Part, P., Laurent, P. and Walsh, P.J. 1998. Pulsatile urea excretion in gulf toadfish (*Opsanus beta*): evidence for activation of a specific facilitated diffusion transport system. *The Journal of Experimental Biology* 212: 805–817.

Wood, C.M., Playle, R.C. and Hogstrand, C. 1999. Physiology and modelling of the mechanisms of silver uptake and toxicity in fish. *Environmental Toxicology and Chemistry* 18: 71–83.

Wren, C.D. and MacCrimmon, H.R. 1983. Mercury levels in the sunfish, *Lepomis gibbosus*, relative to pH and other environmental variables of Precambrian Shield lakes. *The Canadian Journal of Fisheries and Aquatic Sciences* 40: 1737–1744.

Wright, P.A. and Wood, C.M. 1985. An analysis of branchial ammonia excretion in the freshwater rainbow trout: effects of environmental pH change and sodium uptake blockade. *The Journal of Experimental Biology* 114: 329–353.

Wright, P.A., Heming, T.A. and Randall, D.J. 1986. Downstream pH changes in water flowing over the gills of rainbow trout. *The Journal of Experimental Biology* 126: 499–512.

Wright, P.A., Randall, D.J. and Perry, S.F. 1989. Fish gill water boundary layer: a site of linkage between carbon dioxide and ammonia excretion. *The Journal of Comparative Physiology* B158: 627–635.

Wright, P.A., Iwama, G.K. and Wood, C.M. 1993. Ammonia and urea excretion in Lahontan cutthroat trout (*Oncorhynchus clarki henshawi*) adapted to highly alkaline Pyramid Lake (pH 9.4). *The Journal of Experimental Biology B* 175: 153–172.

Yesaki, T.Y. and Iwama, G.K. 1992. Some effects of water hardness on survival, acid-base regulation, ion regulation, and ammonia excretion in rainbow trout in highly alkaline water. *Physiological Zoology* 65: 763–787.

Youson, J.H. and Neville, C.M. 1987. Deposition of aluminum in the gill epithelium of rainbow trout (*Salmo gairdneri* Richardson) subjected to sublethal concentrations of the metal. *The Canadian Journal of Zoology* 65: 647–656.

Zadunaisky, J.A. 1984. The chloride cell: the active transport of chloride and the paracellular pathways. In *Fish Physiology*, Vol. 10B. Hoar, W.S. and Randall, D.J. (eds), pp. 129–176. Academic Press, Orlando.

Zeitoun, I.H., Hughes, L.D. and Ullrey, D.E. 1977. Effect of shock exposures of chlorine on the plasma electrolyte concentrations of adult rainbow trout (*Salmo gairdneri*). *The Journal of the Fisheries Research Board of Canada* 34: 1034–1039.

Zitko, V. and Carson, W.G. 1976. A mechanism of the effects of water hardness on the lethality of heavy metals to fish. *Chemosphere* 5: 299–303.

2 Target organ toxicity in the kidney

Bodil K. Larsen and Everett J. Perkins Jr

Introduction

With more than 20 000 species, fish account for approximately half of the 45 000 vertebrate species living today. Of these, about 95 percent are teleosts (Beyenbach and Baustian, 1989). The teleost group can in terms of species diversity and number be considered the most successful vertebrate group inhabiting greatly diverse environments, indicating their remarkable capacity for adaptation. One of the striking features of fish is their capacity for osmo- and ionoregulation, a phenomenon that has been intensively studied for decades. Teleosts inhabit waters that range from being almost ion free (e.g. Amazon, [ion] < 10 µM, osmolality < 1 mosmol) to full-strength seawater of an osmolality of about 1000 mosmol. Waters of even higher salinities exist (e.g. brine lakes), but relatively few teleosts inhabit waters of more than 1200 mosmol (Rankin and Davenport, 1981).

Despite the extremely variable environments, freshwater and marine species all maintain a plasma osmolality within a rather narrow range, most within the range of 240–400 mosmol, thus being hyperosmotic to freshwater and hypo-osmotic to seawater. Much information has been obtained by studying euryhaline species, which, in contrast to stenohaline species, are capable of living in a wide salinity range although the degree of euryhalinity varies between species. They may live in waters of fluctuating salinity (e.g. estuaries), or it may be species which at some point during their life cycle migrate between freshwater and seawater (e.g. the anadromous salmonids and catadromous eels). These species, thus, face varying and opposing osmotic gradients causing rather complex osmoregulatory problems over a time frame of hours or days. Nevertheless, they handle gradients with relatively small and transient fluctuations in osmolality. In aquatic animals, extrarenal organs, especially the gills in fish, play a quantitatively larger role for ionic regulation. The role of the kidney should not be ignored however. It is the primary organ for elimination of water and is particularly important for freshwater species in which efficient ion reabsorption mechanisms in the kidney minimize the loss of ions. In seawater, the urine flow rate (UFR) is low to minimize water loss, and one of the primary functions is elimination of divalent ions (Hickman and Trump, 1969; Rankin and Davenport, 1981; Nishimura and Imai, 1982). The kidney thereby has distinct functions in the two media; functions which are tightly regulated according to the animals' needs. In euryhaline species, it is of particular importance that kidney function can be rapidly altered. The overall structural and physiologic differences in kidneys of freshwater and marine species and the main changes that occur during freshwater to seawater transfer are quite well understood in a number of species. However, the more specific knowledge about the contribution of each nephron segment with their specific transport mechanisms to the final 'urine output' flow and composition is more limited.

Functional morphology of the teleost kidney

Gross structure

The overall appearance of kidneys from different groups of fish varies greatly from a rather simple-looking string of 'loose' tissue in salmonids to a firm well-defined kidney in carp. The teleost kidney consists of two compartments, which are further divided into the anterior head kidney and the posterior trunk kidney (Figure 2.1). However, in many species, external examination may not be enough to distinguish one part from another because of partial or complete fusion (Hickman and Trump, 1969; Hentschel and Elger, 1989). The head kidney contains chromaffin and inter-renal tissue as well as lymphoid and hematopoietic tissue (Hickman and Trump, 1969; Dantzler, 1989); the last of which may be scattered throughout the whole kidney in some species. The trunk kidney contains the actual renal tubules (nephrons). The embryonic kidney (pronephros) develops in the anterior region where the head kidney later develops and seems to perform excretory function in larvae and remains functional for months in juvenile rainbow trout (Tytler, 1988). The kidney tubules differentiate caudally from the pronephros, and, except for a few advanced species, the nephrons in the anterior part (head kidney) eventually degenerate (Safer *et al.*, 1982; Hentschel and Elger, 1989).

Nomenclature of the nephron

The length of the nephron (kidney unit) and the number of different segments vary among species, and seem to depend on habitat and evolutionary development. Freshwater species and less evolved species (e.g. salmonids, eels, catfish) normally possess the longest, multisegmental nephrons, whereas marine and more advanced species of teleost (e.g. killifish, flounders, sticklebacks) have rather short nephrons and fewer segments, often lacking the distal segment (Hickman and Trump, 1969; Hentschel and Elger, 1987, 1989). The kidney of the aglomerular fish (i.e. lacks glomeruli) represents the fullest expression in this trend and is considered to be the most advanced type (Baustian *et al.*, 1997). Aglomerular species are found in several different orders, such as anglerfish, toadfish, goosefish, sea horses and needlefish (Hickman and Trump, 1969; Hentschel and Elger, 1987, 1989) and comprise about 30 species. Although most of them are marine, a few aglomerular species are found in freshwater. The degeneration of the nephron during evolutionary processes is irreversible. Marine species reinvading freshwater possess the same nephron segments as their marine relatives (Hentschel and Elger, 1987). It is often difficult to distinguish different nephron segments because of a lack of external morphologic markers (Beyenbach, 1995). They can, however, be distinguished by cytologic markers or by segment-specific functional markers, e.g. proximal tubules from marine species can be identified by their ability to secrete magnesium.

Renal corpuscle

The glomerulus (cluster of arterial capillaries) is surrounded by the Bowman's capsule (see below) (Figure 2.2). The size of the renal corpuscle varies, being largest in stenohaline freshwater species and smallest in stenohaline marine species (Nash, 1931; Dantzler, 1989). The glomeruli of euryhaline species undergo alterations in size during transition from one medium to another (see below).

Neck segment

This is a rather short, highly ciliated segment just below the renal corpuscle and appears to be present in most glomerular teleosts (Hickman and Trump, 1969). In some species, it is merely a few rows of undifferentiated cells (Hentschel and Elger, 1989). Ciliar movement may help in transporting the ultrafiltrate down the nephron (Hickman and Trump, 1969).

Proximal tubule (segment PI and PII)

This is characterized by a dense luminal brush border of microvilli, increasing the surface of the apical membrane. PII can frequently be distinguished from segment PI by shorter, slender, and less dense microvilli, and the columnar epithelial cells of PII are usually longer than in PI (Hickman and Trump, 1969; Hentschel and Elger, 1987, 1989; Elger *et al.*, 1998) (Figure 2.2). In marine species, PII is often much longer than PI (Elger *et al.*, 1998), indicating that the need for reabsorption is lower than the need for secretion (see later). Apical coated pits, coated vesicles, large vacuoles, endosomes, as well as a prominent lysosomal apparatus have been identified in PI involved in the endocytosis of macromolecules (Elger *et al.*, 1998). The endocytic apparatus seems to be lacking in PII. The basolateral membrane is elaborated by numerous invaginations in both PI and PII and is associated with numerous mitochondria (Hentschel and Elger, 1989) indicative of high transepithelial transport capacity. Independent of the type of tubule that follows, the end of the proximal tubule is marked by an abrupt disappearance of the brush border.

Intermediate segment

A short, rather narrow segment may be present between the proximal and distal segment. The intermediate segment looks quite similar to the neck segment and may exhibit similar functions in transporting the fluid down the nephron and preventing urine stone formation via ciliary movement (Hentschel and Elger, 1989).

Distal nephron

The distal segment can in some species be distinguished from the proximal segment by a smaller diameter/more narrow lumen and frequently by its limited, short microvilli (Hentschel and Elger, 1987). The basolateral membrane displays extreme amplification which may extend to the apical cell membrane (Stoner, 1985; Hentschel and Elger, 1987; Dantzler, 1989). Numerous, large, elongate, mitochondria indicate a large capacity for ion transport. The zonular occludens between cells in the apical membrane consist of tightly packed strands forming very narrow junctions (Hentschel and Elger, 1987), typical for epithelia with very low water permeability. The distal nephron is absent in many marine teleost species (Hickman and Trump, 1969; Hentschel, 1987; Dantzler, 1989). If a species does possess a distal nephron, it may indicate some degree of euryhalinity, as reabsorption of ions occurs in the distal nephron of freshwater species. Alternatively, other transport mechanisms may have been developed in the distal tubule, for example in chloride cells, which have been identified in the collecting tubule of the marine catfish (see below) which also possess a distal tubule. In freshwater species in which the distal nephron is lacking, its function has probably been assumed by the collecting tubule/duct.

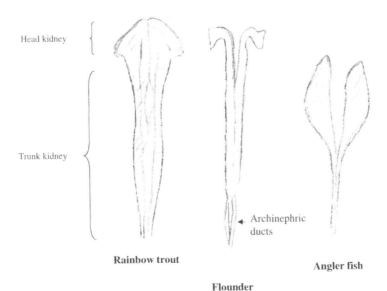

Figure 2.1 Configurational types of kidneys from three different teleost species: rainbow trout (*Oncorhynchus mykis*s), flounder (*Platichthys* sp.) and anglerfish (*Lophius* sp.). Adapted from Hickman and Trump (1969) and Harder (1975).

Collecting tubule/collecting duct/urinary bladder

There seems to be rather few cytologic differences between the distal tubule and the collecting tubule in fish, and it may be referred to as the late distal tubule by some authors (Stoner, 1985; Hentschel and Elger, 1989). Furthermore, it is thought to continue the role of the distal tubule by reabsorbing ions in freshwater and has a high transport capacity with low water permeability (Nishimura and Imai, 1982; Hentschel and Elger, 1987, 1989). Interestingly, chloride cells have been found in the collecting tubule/duct system of the marine catfish *Plotosus lineatus*, which may explain how this species is capable of producing hypertonic urine without possessing Henle's loop (Kowarsky, 1973; Hentschel and Elger, 1987). The unbranched collecting tubules lead into the branched system of collecting ducts which, in turn, run into the two paired archinephric ducts, which unite and open into a widened duct termed the urinary bladder (Hickman and Trump, 1969).

Aglomerular nephrons

In a few species, the proximal end of the tubule does not form glomeruli, or only a few of the nephrons do, and, as will be described in more detail later, these species rely entirely on secretory urine formation (Hickman and Trump, 1969; Hentschel and Elger, 1987, 1989; Dantzler, 1989). An apical endocytic apparatus is lacking in all aglomerular species (Hentschel and Elger, 1987, 1989; Beyenbach and Baustian, 1989; Dantzler, 1989). All degrees of glomerular degeneration are present in marine teleosts, within and between species. It may even be age dependent as in aglomerular anglerfish (*Lophius* spp.), in which the glomeruli present in juveniles degenerates in later stages (Hentschel and Elger, 1989).

Microanatomy of the renal corpuscle

The renal corpuscle consists of a globular cluster of specialized branching and anastomosing capillaries (glomerulus) surrounded by Bowman's capsule, a two-layered pouch-like extension of the nephron. The outer layer (parietal) is continuous with the epithelium of the proximal tubule, whereas the inner (visceral) layer lines the glomerular lobules (Hentschel and Elger, 1989) (Figure 2.2). Marine species generally possess the smallest glomeruli, which may consist of only a few capillary loops (Elger *et al.*, 1998). Furthermore, a decrease in glomerular diameter is frequently observed during seawater acclimation (Nash, 1931; de Ruiter, 1980; Brown *et al.*, 1983; Gray and Brown, 1987; Hentschel and Elger, 1989). A 40 percent decrease in glomerular diameter was observed after long-term acclimation of rainbow trout to seawater (Brown *et al.*, 1983). This response is basically to lower the amount of plasma that is being filtered in order to minimize the loss of water. The barriers which the ultrafiltrate has to pass consist of a capillary endothelial cell layer, a mesangium layer, and the visceral layer with its basement membrane (Figure 2.3).

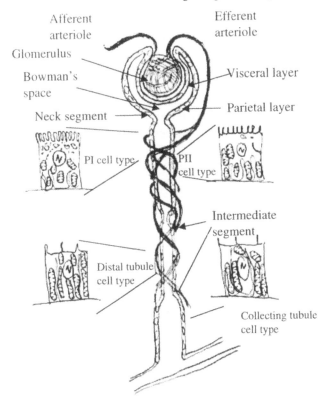

Figure 2.2 Presentation of the nephron with the different cell types existing in PI, PII, distal tubule and collecting tubule. Note the disappearance of the brush border in late segments of the nephron, the lysosomal system in PI, disappearing in PII. Distal tubule and collecting tubule cells are characterized by strongly elaborated basolateral membrane, and elongate mitochondria. Lengths of segments are not relative.

Endothelial cell layer

The capillaries of the glomerulus are lined with large flat endothelial cells, which are fenestrated by pores. The number/size of these pores is highest/largest in freshwater and lowest/smallest in seawater fish, affecting water permeability (de Ruiter, 1980; Hentschel and Elger, 1987, 1989; Brown *et al.*, 1993).

Mesangium layer

This layer can best be described as connective tissue and is thought to have supportive as well as maintenance functions (Yokota *et al.*, 1985). After proliferation of mesangial cells (de Ruiter, 1980; Dantzler, 1989), the thickness of the mesangium layer varies from a few dispersed cells in a matrix (containing collagen) in freshwater to an almost continuous layer in seawater. The layer may then extend in a tree-like manner into the lobules of the glomerulus (Yokota *et al.*, 1985; Hentschel and Elger, 1989; Rankin *et al.*, 1983). Myofibrils are located in the mesangium layer, and their ability to contract may contribute to regulation of the effective filtration area and/or change the blood flow pattern through the glomerulus (Yokota *et al.*, 1985; Dantzler, 1989). Furthermore, an increase in the thickness of this layer possibly reduces the capillary lumen and thereby the endothelial surface area available for filtration (de Ruiter, 1980). Phagocytic capacity of mesangial cells have been observed, enabling the uptake and digestion of trapped substances (de Ruiter, 1980).

Visceral layer

The inner layer of Bowman's capsule consists of large epithelial cells called podocytes. These cells give rise to primary and secondary cell processes which again form foot processes (pedicels) (Figure 2.3). Each podocyte and its pedicels may span over adjacent capillaries, and in freshwater pedicels from different podocytes often interdigitate (Brown *et al.*, 1993). The pedicels terminate in the cement of the basement membrane. The space between adjacent pedicels (filtration slit) are covered by a diaphragm that has a porous substructure (Brown *et al.*, 1993), the pores being too small for macromolecules to pass. During seawater adaptation, the podocytes and their primary processes and pedicels are flattened and broadened and may fuse, reducing the area covered by filtration slit diaphragms (de Ruiter, 1980; Brown *et al.*, 1983; Gray and Brown, 1987). These changes indicate both a lowered hydraulic permeability as well as a reduction in the filtration slit area. Podocytes appear to have phagocytic capacity as well (de Ruiter, 1980).

Glomerular basement membrane

The glomerular basement membrane, a well-defined, three-layered structure secreted by podocytes, and lying above these (Figure 2.3), has important functional properties (de Ruiter, 1980; Hentschel and Elger, 1989; Brown *et al.*, 1993). An increase in thickness has been observed during seawater acclimation (de Ruiter, 1980; Elger and Hentschel, 1981), probably lowering permeability (Figure 2.3). Anionic groups seem to play a role in permselectivity for macromolecules; so not only the size and shape of molecules but also their charge determines whether they are filtered or retained (Elger *et al.*, 1988; Brown *et al.*, 1993).

Water and ions/small molecules pass mainly via an extracellular route: through endothelial fenestrae, basement membrane (possibly via collapsing and reforming water channels), between mesangial cells, and through the pores of filtration slit diaphragms (Brown *et al.*, 1993). Macromolecules can, to some degree, pass though the fenestrated endothelium and the basement membrane but are retained by the filtration slit diaphragm and the mesangium layer.

Blood supply

Arterial system

The renal tissue of glomerular teleosts is supplied with both arterial and venous blood. The arterial blood (directly from the aorta and/or from the trunk) normally enters the renal tissue from the dorsal site via renal arteries into a system of ramifying intrarenal arteries, which then develop into the more or less extensive network of the glomerulus (Hentschel and Elger, 1989). The capillaries of the glomeruli reunite in the efferent arteriole, which again develops into a network of sinusoid capillaries surrounding the kidney tubules.

Renal portal system

A renal portal system has been described in all examined teleosts. It provides a second blood supply for the tubules in addition to the arterial circulation. In marine species, including species with well-developed glomeruli, the venous portal system is the main blood supply, and filtration of arterial blood seems to contribute little to urine formation (Beyenbach and Baustian, 1989). Venous blood from the musculature of the trunk and tail is conducted to the kidney tissue through the intercostal veins and/or the caudal vein, respectively, via afferent renal veins to the afferent intrarenal veins. The afferent intrarenal veins ramify and develop into a system of sinusoid capillaries that surround the renal tubules (Hentschel and Elger, 1989). The venous blood is never fed to the glomerulus; only arterial blood is filtered. The venous blood system surrounding the tubules may unite with the arterial blood supply. Drainage of blood occurs via small intrarenal veins that unite in larger efferent renal veins which join the large cardinal veins (Hentschel and Elger, 1989). In several aglomerular species, the renal portal system is the only blood supply. The control of blood flow through the kidney and its importance

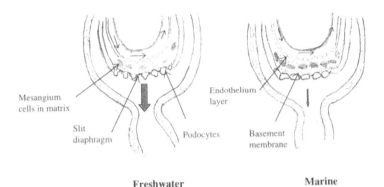

Freshwater   Marine

Figure 2.3 Comparison of glomerular filtration barriers between freshwater- and seawater-adapted fish. Note the reduction in fenestration of the endothelium layer, increased density/thickness of mesangium layer, increased thickness of the basement membrane; podocytes change from being rounded to broad and flat, the filtration slit area is reduced.

for renal function will be discussed below with regard to hormonal control of kidney function.

Overall kidney function

Kidney function/urine flow rate and composition are generally considered a compromise between several factors: the need to eliminate and/or conserve water, metabolic waste/valuable solutes, and ions. However, as the metabolic rate of fish is low and the gills excrete metabolic waste as well, one can assume that teleosts have minimal metabolic clearance requirements, and, therefore, an overall low urine flow rate (Yokota *et al.*, 1985). Glomerular filtration rates in fish are indeed much lower than in endothermic animals with high metabolic rates. Still, fish urine production is greatly influenced by environmental conditions, one of the most important and most studied factors being salinity (Figure 2.4).

Freshwater

With an internal osmolality exceeding the ambient water, freshwater teleosts constantly face the problems of a rather large osmotic influx of water as well as the loss of ions. This influx of water occurs primarily over the large and rather permeable gill epithelium, and it is approximately 50 percent of the body water per hour (Evans, 1993). Additionally, there is an efflux of ions over the gill epithelium. The uptake of ions occurs mainly via ion exchange mechanisms in the gills, although some is gained from food. Large amounts of water are excreted by the kidney, which is the primary organ for elimination of water in all vertebrates (Dantzler, 1989). Urine flow rate and glomerular filtration rate in freshwater species are directly correlated with and determined by the rate of osmotic water influx and are about 10-fold higher than in marine and seawater-acclimated euryhaline species (Table 2.1). As inorganic ions are filtered as well during initial urine formation and occur in the ultrafiltrate in concentrations similar to plasma, very efficient absorption mechanisms have been developed to minimize the loss of ions. Reabsorption of at least 90 percent of filtered Na^+ and Cl^- is commonly measured in freshwater teleosts (Karnaky, 1998; McDonald and Wood, 1998) (Table 2.2). In addition to reabsorption along the nephron itself (including the collecting tubule), a final dilution of the urine and active uptake of at least Na^+ and Cl^- takes place in the urinary bladder, leading to almost 100 percent reabsorption of these ions (Curtis and Wood, 1991). This high reabsorption efficiency can only be accomplished if the last segments of the kidney (distal tubule + collecting tubule/duct system) as well as the urinary bladder is virtually impermeable to water.

Seawater

Marine teleost species face the exact opposite problems of freshwater species: osmotic efflux of water primarily by the gill epithelium and gain of ions. The

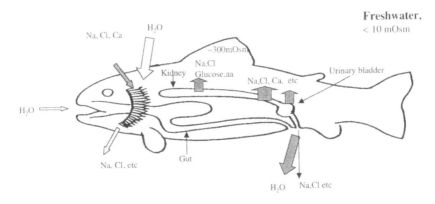

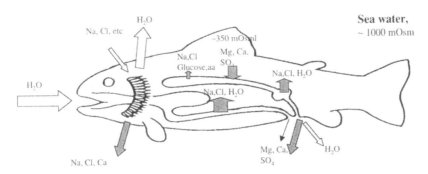

Figure 2.4 Osmoregulatory differences between freshwater- and seawater-acclimated fish. Filled arrows indicate active ion transport, open arrows indicate passive transport.

loading of ions occurs by two mechanisms. First, there is a passive influx of ions over the gill epithelium. Second, to replace osmotically lost water, marine teleosts have to drink large amounts of water; the drinking rate being three to twenty times higher than in freshwater (Karnaky, 1998; Fuentes and Eddy, 1997a). Seawater-adapted rainbow trout drink a volume of seawater each day equivalent to 15 percent of their body weight (Beyenbach *et al.*, 1993). However, this will also increase ion loading as water entering the gastrointestinal tract is absorbed (about 80 percent) secondarily to a Na^+ and Cl^- co-transport (Talbot *et al.*, 1992; Evans, 1993; Karnaky, 1998; Fuentes and Eddy, 1997a). A study on southern flounder showed that over 95 percent of ingested Na^+, K^+ and Cl^+ and 30 percent of Ca^{2+} but less than 15 percent of Mg^{2+} and SO_4^{2-} are absorbed in the intestine (Hickman and Trump, 1969; Björnsson and Nilsson, 1985). Beyenbach *et al.* (1993) report that between 40 percent and 50 percent of ingested Mg^{2+} is absorbed in rainbow trout. A large proportion of the ingested Mg^{2+}, SO_4^{2-} and Ca^{2+} is then excreted rectally, but the excess is excreted by the kidney (Beyenbach *et al.*, 1993; Evans, 1993). The bulk of monovalent ions, primarily Na^+ and Cl^- and variable amounts of Ca^{2+} are excreted by the gills; the rest of the absorbed Ca^{2+} (30–66 percent) is excreted rectally, depending on the species (Björnsson and Nilsson, 1985; Evans, 1993). To minimize water loss, glomerular filtration rate is reduced in marine species (about 10 percent of freshwater), and there is an increased reabsorption of water. Furthermore, fractional tubular water reabsorption is elevated in seawater compared with freshwater (Hickman and Trump, 1969) (Table 2.1).

Urine formation

Fish are the only vertebrates with kidneys known to produce urine by both glomerular filtration and tubular fluid secretion (Beyenbach, 1986; Beyenbach and Baustian, 1989). Evidently, aglomerular species rely entirely on tubular fluid secretion, but several lines of evidence suggest that even glomerular species may also exhibit fluid secretion by the proximal tubule (Beyenbach, 1995).

Glomerular urine formation

Glomerular urine formation is initiated by filtration of plasma in the glomerulus. The driving force for filtration is the hydrostatic pressure of the arterial system, or, more correctly, the hydrostatic pressure difference existing between the glomerular capillary blood and Bowman's space (Figure 2.2) (Yokota *et al.*, 1985; Brown *et al.*, 1993). Glomerular filtration is then energetically inexpensive as the hydrostatic pressure of the arterial system is maintained for other purposes (Yokota *et al.*, 1985; Dantzler, 1989). However, the hydrostatic pressure difference and the colloid osmotic pressure of the blood left behind in the capillaries determine the effective filtration pressure, or net ultrafiltration pressure. As blood flows through the glomerular capillaries and ions and small molecules (with a molecular weight < 30–50 kDa) are filtered – larger molecules being retained – the colloid osmotic pressure of the blood will increase, thereby decreasing the net ultrafiltration pressure (Dantzler, 1989; Brown ,*et al.*, 1993). Filtration in each of the single capillaries will only take place as long as the hydrostatic pressure difference exceeds the colloid osmotic pressure (Rankin *et al.*, 1983; Yokota *et al.*, 1985; Dantzler, 1989; Brown *et al.*, 1993). If equilibrium is reached somewhere along the capillaries, the rate of filtration is a linear function of plasma flow. But if equilibrium does not occur, the rate of filtration is dependent on hydraulic conductivity, filtration area, and net ultrafiltration pressure (Dantzler, 1989; Yokota *et al.*, 1985). The total filtration area for each glomerulus is a function of the total number of capillaries in each glomerulus, and the diameter and length of each of these. For a given net ultrafiltration pressure, the single nephron glomerular filtration rate (SNGFR) (nL min^{-1}) is a function of the ultrafiltration coefficient K_f (hydraulic conductivity / unit area × glomerular filtration area; nL min^{-1} mmHg^{-1}) (Yokota *et al.*, 1985; Brown *et al.*, 1993). Glomerular filtration rate (GFR) is the sum of SNGFR. In most species, urine flow rates (UFR) are directly correlated with GFR, and GFR can be altered by changing the number of filtering nephrons as well as SNGFR itself (Rankin *et al.*, 1983; Brown *et al.*, 1993).

Glomerular filtration rate in freshwater species is relatively high (~ 4–8 mL kg^{-1} h^{-1}; Table 2.1) and UFR is usually less than this (~ 50 percent of GFR in freshwater) (Table 2.1). It has not been possible to measure SNGFR directly by micropuncture methods in any teleost (Brown *et al.*, 1993). An alternative method, measuring filtered radiolabelled ferrocyanide, has been used to estimate SNGFR in freshwater- and seawater-acclimated rainbow trout (see Brown *et al.*, 1993). Three physiologic states of the glomeruli were revealed; the percentage of the nephron population in each state is given in brackets (from Brown *et al.*, 1978, 1980):

- filtering (50 percent in freshwater, 5 percent in seawater);
- non-filtering, but arterially perfused (40–45 percent in both media);
- non-perfused (and therefore not filtering) (~ 10 percent in freshwater, 50 percent in seawater).

Apart from indicating a large capacity for elevation of glomerular filtration rate in both media, these results indicate that alterations in the number of filtering nephrons (glomerular intermittency) is a regulatory mechanism for urine production (Brown *et al.*, 1993). Similarly, the Prussian carp (*Carassius auratius gibelio*) decreases the number of plasma-receiving glomeruli on exposure to saltwater by more than half (Elger *et al.*, 1984a), and after 3 months in seawater, more than 90 percent of the glomeruli disappear (Elger and Hentschel, 1981). Interestingly, although GFR is well known to decrease during seawater acclimation (Brown *et al.*, 1978; Evans, 1993, Karnaky, 1998) (Table 2.1), mean SNGFRs of filtering glomeruli were actually higher in seawater-acclimated trout (3.7 ± 1.1 versus 1.3 ± 0.2 nL min^{-1}; means ± sem) (Brown *et al.*, 1980). Measurements of renal blood flow indicate that both whole kidney and single nephron blood flow are decreased during seawater acclimation (Brown and Oliver, 1985). It is not fully understood why some glomeruli do not filter even though they are perfused and why SNGFRs in filtering glomeruli are higher in seawater than in freshwater. Glomerular filtration pressure is close to equilibrium in trout, so a relatively small increase in the resistance of the glomerular filtration barrier may be enough to cease filtration in some nephrons (Brown *et al.*, 1983). Regulation of glomerular filtration pressure of individual glomeruli or subpopulations of glomeruli is one possibility, as well as restriction of selective, individual afferent arterioles and/or preglomerular sphincters (Elger *et al.*, 1984b). In any case, a 90 percent reduction in GFR during seawater adaptation (Brown *et al.*, 1978, 1980) despite an increase in SNGFR as well as the distribution of filtering/non-filtering–perfused/non-filtering distribution of glomeruli given above implicate glomerular intermittency as an essential part of regulation of GFR and UFR. In some species, glomerular filtration may cease completely (Hickman and Trump, 1969), whereas urine flow continues. UFR higher than GFR is not uncommon in glomerular marine teleost (Hickman and Trump, 1969; Beyenbach and Baustian, 1989; Dantzler, 1989; Beyenbach, 1995) and could be due to significant tubular fluid secretion. This phenomenon may be important in non-filtering nephrons of euryhaline species as well.

Table 2.1 Urine flow rate (UFR), glomerular filtration rate (GFR), fractional reabsorption of water (FR H_2O), plasma osmolality, and urine osmolality in various species, under freshwater and seawater conditions.

Species		UFR (mL h⁻¹ kg⁻¹)	GFR (mL h⁻¹ kg⁻¹)	FR H₂O (%)	Plasma (mosmol)	Urine (mosmol)	Reference
O. mykiss	FW	2.31 ± 0.19	4.55 ± 0.34	48.4 ± 4.2	282 ± 2.1	46.02 ± 6.04	a
SEM (n = 6, 6)	BW	0.56 ± 0.02	1.84 ± 0.11	67.2 ± 3.6	288 ± 1.99	114.2 ± 14.3	
O. mykiss	FW	2.53 ± 0.12	4.40 ± 0.27	42.5			b
SEM (n = 7–11)	FW	2.01 ± 0.17	4.05 ± 0.48	50.4			
O. mykiss	FW	3.3 ± 0.4	7.3 ± 0.8	54.8			c
SEM (n = 7)							
S. salar	FW	1.2 ± 0.22			311 ± 8.6	48 ± 8	d
SEM (n = 6–8)	SW	0.2 ± 0.14			416 ± 8.5	400 ± 8.7	
O. kisutch	FW	4.65 ± 3.66	9.06 ± 4.70	48.7			e
SEM (n = 14–19)	SW	0.406 ± 0.22	1.48 ± 0.72	72.6			
O. mykiss	FW	4.58 ± 0.63	8.42 ± 1.03	45.6			f
SEM (n = 21, 5)	SW	0.32 ± 0.08	1.11 ± 0.35	71.3			
O. mykiss	FW	3.34 ± 0.31	6.84 ± 0.31	51.1			
SEM (n = 5)	SW	0.29 ± 0.06	0.55 ± 0.09	46.8			
A. anguilla	FW	1.1 ± 0.23	1.51 ± 0.18	27.2			g
SD (n = 8, 13)	SW	0.25 ± 0.04	0.43 ± 0.06	41.9			
A. anguilla	FW	3.07 ± 0.25	3.62 ± 0.52	15.4			h
SD (n = 6)	SW	0.62 ± 0.16	0.86 ± 0.23	29.7			

P. stellatus	FW	1.67 (1.05–2.83)		228.4 ± 4.0	78.7 ± 10.3	i
Range (n = 10–12)	SW	0.38 (0.15–0.78)		332.1 ± 4.8	291.6 ± 4.3	
C. commersonii	FW	1.15 (0.76–1.58)	47.7			g
G. morhua	SW	0.73 ± 0.14 s.e.	0.77 ± 0.23 s.e.	5.2	324.9 ± 1.7	328.8 ± 2.0	j
SEM (n = 5–11)						
P. americanus	SW	0.53 ± 0.09	1.35 ± 0.28	60.7		k
SEM (n = 21–37)						
O. tau	FW	3.0 ± 0.28		323 ± 8	325 ± 4	
SEM (n = 5–8)	SW	6.3 ± 0.58		254 ± 3	238 ± 3	l

Sources: Other values are obtained from the following sources:

a Elger et al. (1988).
b Curtis and Wood (1991).
c McDonald and Wood (1998).
d Talbot et al. (1992).
e Miles (1971).
f Brown et al. (1980).
g Hickman and Trump (1969).
h Wales (1984).
i Foster (1975).
j Björnsson and Nilsson (1985).
k Renfro (1980).
l Baustian et al. (1997).

Notes
The values are means ± SEM or SD as indicated in the first column, in which the number of animals is given for each condition, or as a range. FR H_2O is calculated from UFR and GFR as $[1 - (UFR/GFR)] \times 100$ (except for the first example, where it was given).

Modification of the ultrafiltrate along the nephron

As the ultrafiltrate enters Bowman's space, the composition and concentration of inorganic ions is basically similar to plasma, i.e. small organic molecules have been filtered as well, whereas the concentrations of macromolecules is low. The ultrafiltrate will be modified such that valuable nutrients such as sugars, amino acids, and proteins are reabsorbed, and other organic solutes are actively secreted. Inorganic ions are either excreted or reabsorbed.

The role of the proximal tubule

All vertebrates, glomerular and aglomerular, whether in freshwater or seawater, possess a proximal tubule, indicating that it has a key function in kidney physiology (Beyenbach, 1985). Aglomerular species lack PI, indicating that one of PI's primary functions is reabsorption. Isolated tubules from glomerular species do have a higher capacity for reabsorption, probably because of the presence of PI. Transport mechanisms for a variety of small, filtered organic substances, such as glucose and amino acids, which the animal cannot afford to lose, have been identified in the proximal tubule. Physiologic parameters, such as low transepithelial voltages, small transepithelial differences in osmotic pressure and ion concentrations, and histologic characteristics of the tight junctions in the apical membrane indicate that the proximal tubule can be classified as leaky epithelia (Beyenbach, 1985, 1995).

Reabsorption/secretion of organic solutes

MACROMOLECULES

Although the ultrafiltrate is generally considered to contain low amounts of proteins and other macromolecules, the proximal tubule PI segment exhibits an apical endocytic apparatus connected to an elaborate lysosomal system similar to that found in other vertebrates (Hentschel and Elger, 1989). The uptake probably occurs by attachment of the molecules to clatrin-coated pits in the apical membrane, which are ingested and transported to the lysosomal system for digestion (Hentschel and Elger, 1989).

GLUCOSE/AMINO ACIDS

Efficient reabsorption mechanisms for freely filtered, invaluable solutes such as glucose and other sugars as well as amino acids is present in the extensive luminal brush border in proximal tubules from all vertebrate species (Dantzler, 1989; Renfro, 1995). Sugars and most neutral and acidic amino acids cross the luminal membrane in co-transport with Na (Figure 2.5C and D). The mechanisms are saturable and electrogenic, coupled to the movement of Na down its electrochemical gradient, established by Na^+,K^+-ATPase in the basolateral membrane (Lee and Pritchard, 1983; Dantzler, 1989; Renfro, 1995). The transport of basic amino acids seems to be Na independent and may be coupled to a proton gradient from the lumen into the cell (Dantzler, 1989). Exit of sugars and amino acids out of the cell also seems to occur down an electrochemical gradient, possibly by carrier-mediated transport (Dantzler, 1989).

The amino acid taurine can be both secreted and absorbed by kidney tubules. It serves as an intracellular organic osmolyte and is involved in cell volume control during acclimation to different salinities; the intracellular concentration decreases during freshwater acclimation and increases during seawater acclimation (Schröck *et al.*, 1982; King and Goldstein, 1985; Renfro, 1995). The kidney is an important site for regulation of taurine in teleost fish. Secretion across the kidney epithelium is initiated by transport over the basolateral membrane and is dependent on extracellular Na and Cl, indicating co-transport. Inhibition with ouabain demonstrates the necessity of a Na gradient for absorption. Exit across the luminal membrane is possibly by simple diffusion as taurine is accumulated in the renal cells at 500 times the plasma level, or it may be carrier mediated (King and Goldstein, 1985).

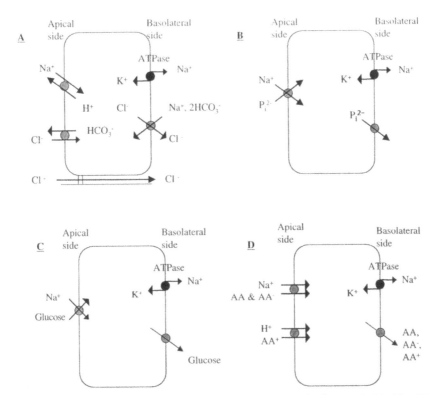

Figure 2.5 Reabsorption mechanisms for various ions. (A) Sodium and chloride; (B) phosphate; (C) glucose; and (D) amino acids. Orientation of arrows indicates either transport against an electrochemical gradient (upward pointed) or transport down an electrochemical gradient (downward pointed). Adapted from Dantzler (1989) and Renfro (1995).

ORGANIC ANION SECRETION

Fish possess a system for secretion of a wide range of hydrophobic organic anions, which include organic acids that exist as anions at physiologic pH. This saturable system excretes a number of endogenous compounds, but it seems to be even more efficient for a number of exogenous compounds, including environmental pollutants and plant and animal toxins (Renfro, 1995; Dantzler, 1996). One possible model for transport is entry at the basolateral membrane occurring in exchange for α-ketoglutarate (α-KG) moving down its electrochemical gradient (α-KG enters the cell in co-transport with Na; Figure 2.6D). Extrusion of the compound through the luminal membrane is possibly carrier-mediated diffusion down an electrochemical gradient, or it may be in exchange for other anions (Dantzler, 1996; Masereeuw *et al.*, 1996; Miller and Pritchard, 1997).

Transepithelial fluxes of inorganic ions in the proximal tubule

The proximal tubule is, at least in freshwater teleosts with high glomerular filtration rates, presented with large amounts of filtered inorganic ions. Both freshwater and marine teleosts need to reabsorb ions but for different purposes. Freshwater species need to conserve the ions and marine species need to conserve water, but this can only be done secondarily to reabsorption of solutes. Na^+ and Cl^- being the most important owing to the large quantities of these ions. Although an overall net absorption of Na^+ and Cl^- by the nephron may be evident in both freshwater and marine species, it does not distinguish net absorption in the whole length of the tubule from net absorption in one portion of the tubule (e.g. PI) and net secretion in another (e.g. PII). Both secretory and absorptive mechanisms for Na^+ and Cl^- and water have been verified in the proximal tubules of several teleost species; they may be present in other species as well (see below), although the relative importance of these fluxes probably varies between species and with environmental factors.

Reabsorption of other inorganic ions is rather poorly investigated in teleosts, but presumably the freshwater teleost nephron exhibits uptake mechanisms for all of them as urine concentrations are extremely low. Reabsorption is not restricted to the proximal tubule, but occurs possibly to a greater extent in later segments of the nephron. Whether reabsorption or secretion is needed depends not only on the concentration of the specific ions in the surrounding water but also on the amounts excreted or taken up by the gills, the electrochemical gradient existing over the gill epithelium, and the gill ion permeability for the specific ions over the gill epithelium (determines direction of passive ion fluxes), the diet, and the amount that is reabsorbed in the gastrointestinal tract. Mg^{2+} reabsorption is known to be regulated, such that an increased reabsorption occurs under conditions in which fish are fed low-Mg diets and during seawater to freshwater transfer (Bijvelds *et al.*, 1996, 1998). Similarly, inorganic phosphate (P_i^{2-}) can be both reabsorbed and secreted, mostly depending on diet, as P_i^{2-} concentration is very low in both freshwater and seawater (see below).

REABSORPTION OF Na⁺ AND Cl⁻ BY THE PROXIMAL TUBULE

Mechanisms by which Na^+ and Cl^- is reabsorbed in the proximal tubule (probably PI) are not clear for teleosts. However, microperfusion and micropuncture studies reveal a lumen-negative transepithelial potential during Na^+ absorption by freshwater-adapted teleosts, indicating active Na^+ uptake, whereas Cl^- uptake may be passive (Beyenbach, 1986; Dantzler, 1989) (Figure 2.5A). As mentioned earlier, Na^+ transport is linked to the uptake of a large number of organic molecules but can also cross the apical membrane in exchange for H^+ (Zonno *et al.*, 1994). Increased activity of the Na^+/H^+ antiporter has been observed with increased salinity, possibly as a result of elevated demand for water reabsorption (Zonno *et al.*, 1994). The transport of Na^+ into the cell from the tubular lumen apparently occurs down an electrochemical gradient established by the basolateral Na^+,K^+-ATPase. Cl^- transport seems to be passive, and may occur either via a paracellular pathway or in exchange for bicarbonate coupled to Na^+/H^+ exchange (Dantzler, 1989).

The leaky epithelium in the proximal tubules allows osmotic water fluxes across the epithelium following the active transport of solutes. The total amount of water reabsorbed by the proximal tubule is a function of the total amount of transported solutes, organic as well as inorganic.

Secretory mechanisms in the proximal tubule

It has become evident that the proximal tubules from several teleost species are capable of active fluid secretion, linked not only to secretion of divalent ions such as Mg^{2+}, SO_4^{2-}, and Ca^{2+} but also to the active secretion of Na^+ and Cl^- (see below). The mechanisms by which this occurs seem identical in aglomerular and glomerular species, although the capacity may differ. The secretory mechanisms are most likely located in PII.

MAGNESIUM SECRETION

In the 1930s, it was demonstrated that marine fish have extremely powerful mechanisms for excretion of magnesium, indicated by very high urine–plasma ratios for Mg^{2+}. Urine–plasma (U/P) ratios less than 1 may indicate net absorption, whereas anything higher than unity indicates net secretion (although the U/P ratio also depends on the tubular reabsorption of water). U/P ratios for Mg^{2+} range between 20 and 132 (see also Table 2.2). Furthermore, plasma Mg^{2+} concentrations are low (~ 1–1.5 mM) (Table 2.2) compared with Mg^{2+} concentrations in the water that is drunk (~ 55 mM), indicating significant regulation of Mg^{2+}. Very small amounts of excreted magnesium originate from filtration (2–5 percent) (Hickman and Trump, 1969; Miles, 1971; Renfro, 1980) with the rest actively secreted by the tubule. Mg^{2+} secretion has been directly correlated with urine production in both glomerular and aglomerular marine teleosts. Infusion of Mg^{2+} in the aglomerular anglerfish, *Lophius piscatorius*, immediately stimulated urine production in a linear manner (Babiker and Rankin, 1979). The exact mechanism by which magnesium is excreted, however, is not known (Beyenbach *et al.*, 1993; Chandra *et al.*, 1997). As an inverse relationship was found between urinary [Mg^{2+}] and [Na^+], it was originally suggested that secretion of Mg^{2+} was coupled to reabsorption of Na^+ (Natochin and Gusev, 1970). However, the original measurements were performed on bladder urine (in contrast to kidney urine, which has not been modified by the bladder). Measurements on kidney urine revealed that the two ions showed a direct relationship (Beyenbach *et al.*, 1993). It was only when the kidney urine had been stored in the bladder and Na^+ and Cl^- had been reabsorbed that the inverse relationship could be found (Beyenbach and Kirschner, 1975; Beyenbach *et al.*, 1993). Mg^{2+} is secreted against an electrochemical gradient (Beyenbach, 1995), and transport shows saturation kinetics with a K_m of 0.22 mM and a V_{max} of 1.5 pmol min^{-1} mm^{-1} in the proximal tubule, indicating high-affinity, low-capacity transport (Cliff *et al.*, 1986). Chandra *et al.* (1997), identifying Mg^{2+} transport sites by ion microscopy imaging stable isotopes, suggested a transcellular pathway, including active transport over the

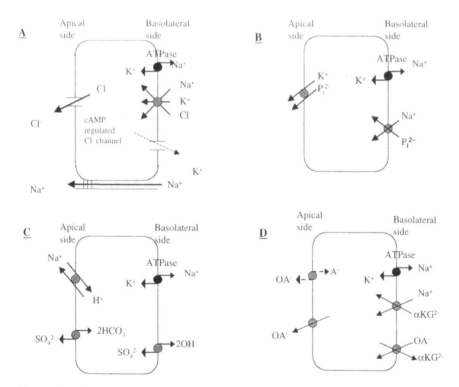

Figure 2.6 Secretory mechanisms for various ions. (A) Sodium and chloride; (B) phosphate; (C) sulfate; and (D) organic acids. Orientations of arrows are as indicated in Figure 2.5. Adapted from Beyenbach (1995) and Renfro (1995).

basolateral membrane, followed by binding to intracellular ligand, leading to intracellular accumulation of the ion.

SULFATE

The seawater ingested by marine teleosts contains approximately 25 mM SO_4^{2-} whereas plasma contains about 0.6 mM, indicating efficient secretion, which is also indicated by relatively high urine–plasma ratios (> 10). Euryhaline species show decreased sulfate excretion during acclimation to lower salinity (Renfro, 1989). Transepithelial unidirectional fluxes have been reported in several studies (Renfro, 1995). A possible mechanism for secretion may be initiated by SO_4^{2-} entry into the cell, driven by a basolateral pH gradient, and may occur as SO_4^-/OH^- exchange or as SO_4^-/H^+ co-transport (Renfro, 1995). The exit over the apical membrane can occur in exchange for HCO_3^- or Cl^- (Renfro and Pritchard, 1983). Similar for most other ions, the transport is dependent on a Na gradient and the activity of the basolateral Na^+,K^+-ATPase (Figure 2.6C).

CALCIUM

Ca^{2+} enters seawater-acclimated fish similarly to magnesium with the water that they drink, and a net absorption of 30–70 percent has been observed in the gut. Teleosts exhibit efficient Ca^{2+} homeostasis, indicated by their stable plasma levels being around 1.5 mM despite a relatively high concentration in seawater (~ 10 mM). Clearance studies (comparing plasma levels with urine levels) indicate that overall net tubular excretion of Ca^{2+} occurs in marine and seawater-acclimated euryhaline teleosts (Table 2.2) (Hickman and Trump, 1969). Teleosts exhibit both renal and extrarenal excretion of Ca^{2+}, and although the gills may excrete a large proportion the kidney possibly participates in the fine control of plasma Ca^{2+} levels and probably responds to the plasma levels itself (Renfro *et al.*, 1982; Bijvelds *et al.*, 1997). Marine and seawater-acclimated euryhaline teleosts produces a urine with Ca^{2+} concentrations well above plasma levels (Bijvelds *et al.*, 1997). However, the mechanisms are unclear. It was suggested that the kidney tubules of seawater-adapted flounder, *Pseudopleuronectus americanus*, possess ATP-driven Ca^{2+} secretion (Renfro *et al.*, 1982). In contrast, Bijvelds *et al.* (1997) suggested that a decrease in Ca^{2+} reabsorption, not an increase in secretion, could account for an overall increase in Ca^{2+} excretion during seawater acclimation. This was suggested because of a decrease in activity of a specific Ca^{2+}-ATPase during seawater acclimation of tilapia (Bijvelds *et al.*, 1997). Whether this apparent disagreement may simply be due to species differences is not known. If tilapia, for example, possess a relatively larger capacity for Ca^{2+} excretion via the gills, this species may not need to rely on active secretory processes by the kidney. Clearly, further investigation is needed.

PHOSPHATE

Teleosts may exhibit both overall reabsorption and secretion of P_i^{2-} (Hickman and Trump, 1969) which is possibly linked to secretory mechanisms in PII and reabsorptive mechanisms in the collecting tubule/duct system, but the precise mechanisms are not known. There are very small amounts of P_i^{2-} in both freshwater and seawater, so net secretion or reabsorption depends on dietary uptake and diffusive losses to the environment.

Reabsorption of P_i^{2-} may occur via apical Na^+,P_i^{2-} co-transport followed by Na^+-independent exit over the basolateral membrane (Figure 2.5B) (Renfro, 1995).

Freshwater-acclimated carp (*Carassius auratus*) respond to an intravenous infusion of P_i^{2-} by a considerable increase in tubular secretion of P_i^{2-} (Kaune and Hentschel, 1987), and most likely the kidney tubule responds to the plasma level of P_i^{2-}. The initial secretory step may be Na-coupled transport over the basolateral membrane (Figure 2.6B) (Renfro, 1995). This correlates well with recent studies in which a Na^+,P_i^{2-} co-transport system was cloned from the seawater-adapted winter flounder (*Pleuronectes americanus*) (Werner et al., 1994) and was later located in the PII and the collecting tubule and collecting duct (CT/CD) system in the same species (Elger et al., 1998). However, the intracelluar distribution was not the same in the two segments. In the PII, the transporter seemed to be located near the basolateral membrane, whereas in the CT/CD system it was observed mainly near the apical membrane. This could explain how animals possess capacity for both secretion (in PII) and reabsorption (in the CT/CD system) (Elger et al., 1998).

Na⁺ AND Cl⁻ SECRETION

A 'secondary' active transport mechanism for secretion of Na^+ and Cl^- has been identified in the proximal tubule in a few species. It includes a furosemide-inhibited Cl^- entry across the basolateral membrane in killifish and shark, indicating the presence of Na^+,K^+,Cl^- co-transport (Beyenbach, 1995). It is dependent on the transmembrane Na^+ gradient generated by the Na^+,K^+-ATPase (Figure 2.6A). Cl^- can then diffuse out over the apical membrane, and down its electrochemical gradient with cAMP increasing the Cl^- permeability of the apical membrane (Beyenbach and Frömter, 1985). Na^+ excretion occurs down an electrochemical gradient, via a paracellular route, with Na^+ and Cl^+ existing in almost equimolar concentrations in the secreted fluid. Fluid transport is not stimulated by cAMP in all species (e.g. winter flounder), indicating that other mechanisms for Na^+ and Cl^- secretion must exist (Beyenbach, 1995).

Aglomerular and glomerular fluid secretion

The secretory mechanisms for secretion of ions in glomerular and aglomerular fish are most likely identical and are the driving forces for aglomerular urine production. Furthermore, tubular fluid secretion has been identified in several glomerular species (Beyenbach, 1995) and contributes significantly to urine production as well. Aglomerular species have very low metabolic rates, and exhibit a sedentary lifestyle (e.g. toadfish) and/or live in very cold environments (e.g. icefish) (Dantzler, 1989). The minimal need to eliminate metabolic waste and the need to conserve water could explain evolution of aglomerular kidneys. Furthermore, in icefish, the need to conserve antifreeze molecules (e.g. glycoproteins), which are relatively small molecules (8–33 kDa) and normally would be filtered, may contribute as well to the evolution of aglomerular kidneys in these animals. The secretory mechanisms in the proximal segment PII, which enable these animals to produce urine, and the fluid secretory mechanisms found in glomerular species seems to be the same, although the capacity probably varies. Indeed, the proximal tubules (PII) in aglomerular and glomerular marine species are structurally so similar that they cannot be distinguished. Aglomerular species are not restricted to seawater despite the lack of a distal tubule. As urine formation is secondary to ion secretion, the ability to survive in freshwater must depend on very efficient reabsorption mechanisms in latter segments of the nephron (collecting tubule) or urinary bladder, in combination with epithelia basically impermeable to water. Transfer of the marine toadfish (*Opsanus tau*) (stenohaline) to 10 percent seawater showed that their limited ability to survive in diluted seawater is related to an inability to produce the necessary diluted urine. They were able to eliminate water but lost large amounts of ions in this process (Table 2.2) (Baustian *et al.*, 1997).

Mechanism for tubular fluid secretion

Fluid secretion in aglomerular as well as in glomerular species was formerly thought to be secondary to transport of Mg^{2+} and SO_4^{2-} alone or possibly involving other divalent ions as well. Secretion of Na^+ and Cl^- did not appear to be involved as the urine concentrations of these ions are frequently low, and net reabsorption is always observed (urine–plasma ratio < 1) (Hickman and Trump, 1969; Cliff and Beyenbach, 1988). Furthermore, proximal tubules of glomerular teleosts were thought to reabsorb Na^+ and Cl^- and fluid, not secrete it. It came as a surprise, therefore, when Beyenbach (1982) elegantly demonstrated that the proximal tubule from winter flounder (*Pleuronectes americanus*) secretes fluid with involvement from Na^+ and Cl^-. As proximal tubules are difficult to identify morphologically in flounder, the researcher used Mg^{2+} secretion as a functional marker. Luminal fluid from freshly dissected tubules was expelled with paraffin and analyzed for Mg^{2+}. The tubules immediately initiated secretion of fluid and thereby expelled the oil (Beyenbach, 1982). Perfusion studies with isolated proximal tubules revealed that Na^+ and Cl^- are the principal electrolytes in the secreted fluid from three unrelated glomerular species: dogfish shark (*Squalus acanthias*), flounder (*Pleuronectes americanus*), and freshwater- and seawater-acclimated killifish (*Fundulus heteroclitus*), as well as the aglomerular toadfish (*Opsanus tau*) (Sawyer and Beyenbach, 1985; Cliff and Beyenbach, 1988; Beyenbach, 1995). Interestingly,

Table 2.2 Plasma ion concentrations, urine ion concentrations, filtration and excretion rate for various ions.

Species		Ion	Plasma conc. (mM)	Urine conc. (mM)	Filtration rate ($\mu mol\ kg^{-1}\ h^{-1}$)	Excretion rate ($\mu mol\ kg^{-1}\ h^{-1}$)	% secretion or reabsorption	Reference
P. americanus SEM (n = 29–37)	SW	Na	180.7 ± 3.4	46.4 ± 7.36	243.95	23.3 ± 6.57	−90.4	a
		Cl	184.3 ± 5.81	182.2 ± 6.88	248.81	106.9 ± 23.37	−51.0	
		K	3.85 ± 0.185	4.07 ± 0.36	5.20	1.81 ± 0.29	−65.2	
		Mg	1.20 ± 0.133	105.40 ± 8.54	1.62	56.87 ± 4.12	+97.2	
		Ca	2.98 ± 0.111	14.07 ± 1.27	4.023	7.45 ± 1.442	+46	
G. morhua SEM (n = 5–11)	SW	Na	170.9 ± 0.8	5.7 ± 1.0	131.6	4.2	−96.8	b
		Cl	135.6 ± 1.2	114.8 ± 5.5	104.4	83.8	−19.7	
		K	4.7 ± 0.2	1.5 ± 0.4	3.6	1.1	−69.4	
		Mg	0.69 ± 0.04	177.0 ± 6.4	0.5	129.2	+99.6	
		Ca	1.93 ± 0.07	114.8 ± 5.5	1.49	4.21	+64.6	
		Pi	0.54 ± 0.02	1.92 ± 0.23	1.94	1.40	+27.8	
P. stellatus SEM (n = 6–21)	FW	Na	149.8 ± 3.3	14.9 ± 2.0		24.9		c
		Cl	121.9 ± 4.9	2.28 ± 0.46		3.81		
		K	2.69 ± 0.07	0.58 ± 0.11		0.97		
		Mg	0.63 ± 0.04	1.06 ± 0.11		1.77		
		Ca	1.79 ± 0.07	1.64 ± 0.24		2.74		
	SW	Na	160.1 ± 2.7	13.9 ± 4.0		5.28		
		Cl	155.6 ± 1.9	87.4 ± 5.4		33.21		
		K	3.34 ± 0.14	1.31 ± 3.2		0.50		
		Mg	0.8 ± 0.13	138.5 ± 6.4		52.63		
		Ca	2.07 ± 0.02	7.7 ± 1.24		2.93		

O. mykiss	FW	Na	154.6 ± 2.4	8.8 ± 1.4	510.2	64.24	−87.4	d
SEM (*n* = 7)		Cl	131.1 ± 2.6	9.3 ± 1.6	432.6	67.89	−84.3	
S. salar	FW	Na	154	12.1		13.7		e
Means taken		Cl	141	12.4		15		
from figures		Mg	0.78	0.54		0.68		
	SW	Na	187	80.6		15		
		Cl	197	207		40		
		Mg	4.5	101		15		

Sources: The values are obtained from the following sources:

a Renfro (1980).
b Björnsson and Nilsson (1985).
c Foster (1975).
d McDonald and Wood (1998).
e Talbot *et al.* (1992).

Notes

The last column (% secreted or reabsorbed) gives either the percentage of the excreted amount which has been secreted by the tubules (positive values) or the percentage of the filtered amount which has been reabsorbed by the tubules (negative values). The amount secreted by the tubules (when excretion rate is greater than filtration rate) has been calculated as [(Excretion rate − filtration rate) / excretion rate] × 100. The amount reabsorbed by the tubules (when filtration rate is greater than excretion rate) has been calculated as [(Filtration rate − excretion rate) / filtration rate] × 100. The values are means ± SEM or SD as indicated in the first column, where the number of animals is also given for each condition, or as a range.

fluid secretion in flounder and toadfish was not stimulated by cAMP or inhibited by loop diuretics, in contrast to that observed in killifish and shark, indicating that the Na^+ and Cl^- secretion differs from the mechanism shown above (Figure 2.6A) (Baustian and Beyenbach, 1993). Moreover, fluid secretion continues in the absence of Mg^{2+} and SO_4^{2-}, whereas the use of Na^+-free choline Ringer or Cl^--free gluconate Ringer solutions in the peritubular bath inhibited fluid secretion (Beyenbach, 1982). Na^+ and Cl^- secretion delivers between four and ten times as many osmolytes to the tubule lumen as does Mg^{2+} and SO_4^{2-} secretion. Moreover, Na^+ and Cl^- secretion drives up to 80 percent of the total fluid volume secreted (Beyenbach, 1986). This could mean that Na^+ and Cl^- provide basal rates of fluid secretion, whereas the secretion of Mg^{2+} (and other divalent ions) causes additive fluid secretion (Beyenbach, 1982; Cliff *et al.*, 1986). Na^+ and Cl^- concentrations in the secreted fluid are close to the peritubular values (approximately plasma values), whereas Mg^{2+} and SO_4^{2-} are found in concentrations less than 30 mM, which is much higher than peritubular values (1 mM) but much lower than Na^+ and Cl^- (Cliff and Beyenbach, 1988, 1992; Beyenbach, 1995). This could be explained by active secretion of Na^+ and Cl^-, possibly at the same site as Mg^{2+} and SO_4^{2-}. However, the proximal tubules show very high permeability to Na^+ and Cl^-, whereas the permeability to secreted Mg^{2+} and SO_4^{2-} is very low, so at least part of the transport could be passive. The relative rates of active secretion of Mg^{2+} and SO_4^{2-} compared with passive diffusion of Na^+ and Cl^- could determine the final fluid composition (Beyenbach, 1995). One advantage of a simultaneous secretion of Na^+ and Cl^- (passive or active) could be that it would prevent development of a large transepithelial concentration gradient, which would increase the energetic cost of Mg^{2+} and SO_4^{2-} transport (Baustian and Beyenbach, 1993).

Beyenbach (1995) suggests that a Donnan-like mechanism could drive the fluid transport. When Mg^{2+} is secreted into the lumen, it behaves like an oncotic agent because it is charged and cannot diffuse back. Na^+ and Cl^- will now be passively distributed in accordance with a Donnan equilibrium. This, however, creates osmotic pressure in the lumen, and it is therefore followed by fluid into the lumen.

It is now clear that proximal tubules of some teleosts species, glomerular as well as aglomerular, whether in freshwater or seawater, undergo tubular fluid secretion. Freshwater-acclimated killifish do exhibit fluid secretion, but only in a few of the tubules (5–10 percent) (Cliff and Beyenbach, 1988, 1992). In seawater-acclimated killifish, 30–70 percent of the tubules secrete fluid. Under normal circumstances, proximal tubular fluid secretion will be masked by the main need for reabsorption of ions and water; it is only when the tubules are not perfused that these small secretory fluxes can be measured. Furthermore, it was observed that not all tubules from aglomerular toadfish secrete fluid *in vitro*, suggesting tubular intermittency, analogous to glomerular intermittency (Beyenbach, 1995).

FLUID SECRETION IN SEAWATER-ADAPTED TELEOSTS

That marine species and seawater-acclimated euryhaline species may exhibit fluid secretion is not surprising as the low GFR (occasionally zero) may necessitate additional routes for excretion of ions, especially divalent ions (Cliff and Beyenbach, 1992). It could also explain why UFRs exceed GFRs in some glomerular species such as long-horn sculpin (*Myoxocephalus octodecimspinosus*) and southern flounder (*Paralichthys lethostigma*) (Beyenbach, 1986). Furthermore, hypertonic urine is not uncommonly observed in marine species, although for many years it was a dogma in kidney physiology that fish are unable to produce hypertonic urine because of the lack of Henle's loop. Interestingly, after abrupt transfer to seawater, killifish (*Fundulus kansae*) produce a urine that is hyperosmotic to plasma by > 100 mosmol (Beyenbach, 1986), which can be explained by Na^+ and Cl^- secretion of the proximal tubule. Although the urine Na^+ and Cl^- concentration is mostly low in marine fish, tubular secretion is not excluded as Na and Cl can be reabsorbed in the collecting duct and urinary bladder.

FLUID SECRETION IN FRESHWATER-ADAPTED TELEOSTS

It may seem surprising that tubular fluid secretion is observed in freshwater species. However, the fluxes may, even though they are small, contribute to formation of urine, for example UFRs greater than GFRs have been measured in the freshwater-adapted American eel (*Anguilla rostrata*), although Na^+ and Cl^- secretion was not thought to be the driving factor (Schmidt-Nielsen and Renfro, 1975). Furthermore, secretion may replace Na^+, Cl^- and water which may be reabsorbed concomitantly with reabsorption of amino acids and glucose. Tubular fluid secretion may also provide a basal volume flow in non-filtering nephrons. As these tubules are not perfused, fluid secretion may provide a method by which other tubule transport mechanisms can be maintained. Finally, the capacity for fluid secretion may in itself be an indicator of euryhalinity (Cliff and Beyenbach, 1992). If freshwater fish do exhibit tubular fluid secretion, it can be expected that Na^+ and Cl^- secretion is the absolute driving force because these animals must conserve Mg^{2+}.

Functions of the distal tubule/collecting tubule/urinary bladder

The urine osmolality of freshwater teleosts is usually less than 20 mosmol, which is about 5 percent of the plasma concentration, indicating very efficient overall reabsorption of ions. Most of this urinary dilution occurs along the distal tubule, collecting tubule system, and urinary bladder (Nishimura *et al.*, 1983; Dantzler, 1989). The high reabsorption efficiency is a combination of high ion transport activity, preferentially Na^+ and Cl^- transport, and an epithelium which is practically impermeable to water (Nishimura *et al.*, 1983; Stoner, 1985; Karnaky, 1998). In the distal tubule, there is a primary active transport of Na^+ out of the cell by the Na^+,K^+-ATPase, which establishes a chemical gradient for coupled, electroneutral movement of Na^+ and Cl^- into cells across the luminal membrane. Exit at the basolateral membrane probably occurs down electrochemical gradients (Stoner, 1985; Dantzler, 1989). One of the most striking features of the distal tubules is its low permeability to water. Nishimura *et al.* (1983) observed that distal tubules from freshwater-acclimated rainbow trout were nearly impermeable to water. As net absorption of Na^+ and Cl^- was demonstrated, this equals a dilution process. A long distal tubule has been observed in several teleost species inhabiting extremely ion-poor water, such as the Amazon rivers (Hochachka *et al.*, 1977), further verifying the distal tubule as a diluting segment in freshwater species. Although most stenohaline do not possess a distal tubule, some do, which may indicate some degree of euryhalinity similar to species which are known to be euryhaline (e.g. salmonids sp.). The distal tubule in marine fish must, however, be modified because reabsorption of ions in this part would only increase ion loading. It is most likely that the permeability to water is increased in these animals as this would allow water to be reabsorbed concomitantly with Na^+ and Cl^-. Furthermore, the function of the distal tubule may even have been reversed to secretion, e.g. in species where chloride cells similar to gill chloride cells have been observed.

The function of the collecting tubule/duct system is probably similar to the function of the distal tubule, but information about this is rather limited.

It is now clear that the urinary bladder plays an important role in teleosts, not only as a storage organ but also for modification of the urine. In earlier studies, implantation of catheters into the bladder meant that kidney urine was not modified by the bladder, and the capacity of the whole kidney/urine bladder system was thereby unknown. In other studies, urine was collected directly from the bladder, which may include, but does not separate, kidney function and bladder function. Curtis and Wood (1991) used a new external catheterization technique to collect naturally discharged urine. They demonstrated that freshwater-acclimated trout urinate in intermittent bursts at 20- to 30-min intervals, allowing the urine to be stored for approximately 25 min. They also demonstrated that natural urine flow is at least 20 percent lower and the excretion rates for Na^+ and Cl^- at least 40 percent lower than those determined by traditional techniques in which catheters implanted into the bladder collect urine continuously. It is clear that Na^+ and Cl^- is absorbed in the urinary bladder, and that in freshwater species this is accompanied by an epithelium with low water permeability, allowing the urine to be further diluted. Furthermore, in freshwater, water permeability seems to be influenced by the hardness of the surrounding water, such that animals in soft water have tighter bladder epithelia than animals in hard water. This may be related to differences in prolactin titers because prolactin is well known to lower epithelial permeability (see below) and the prolactin level is highest in soft water (Marshall, 1995). In marine species, the urinary bladder is not impermeable to water. Beyenbach and Kirschner (1975) compared kidney urine with bladder urine from seawater-acclimated rainbow trout. They observed that both kidney urine and bladder urine contained high amounts of Mg^{2+} ($\sim$ 140 and $\sim$ 170 mM respectively), and that Na^+ concentration was high in kidney urine ($\sim$ 100 mM) but very low in bladder urine ($\sim$ 10 mM). This clearly shows that the urinary bladder reabsorbs Na^+, and that this is followed by osmotic water uptake. The difference in bladder water permeability between freshwater and marine species was also observed when flounder were acclimated to both media, and demonstrated a sixfold decrease in water permeability in flounder acclimated to freshwater (Demarest, 1984). The mechanisms by which Na^+ and Cl^- are reabsorbed seem to differ between media (Figure 2.7). Basolateral Na^+,K^+-ATPase is in both situations the driving force, creating low intracellular Na concentrations. In seawater, Na^+ and Cl^- transport over the apical membrane seems to be coupled, whereas in freshwater (particularly soft water) Na uptake seems to be in exchange for NH_4^+ (or H^+) and Cl^- in exchange for HCO_3^- (Marshall, 1995).

Control of kidney function

In light of how urine is produced in glomerular fish, and where functional/ morphologic changes occur, sites of regulation could include:

- renal blood flow;
- glomerular filtration pressure;
- hydraulic permeability of glomerular filtration barrier;
- tubular transport activity;
- water permeability of tubules/urinary bladder;
- ion permeability of tubules/urinary bladder.

Furthermore, as glomerular and possibly tubular intermittency have been observed in fish, individual nephrons could probably be regulated. How this occurs is not known, but one possibility is that individual or subpopulations of nephrons could have a different sensitivity to circulating hormones, e.g. via different hormone receptor concentrations and/or receptor affinities to hormones. It is, however, very problematic to measure individual nephron responses in fish partly because of the renal anatomy. The dense vasculature surrounding the tubules, and strongly pigmented hemopoietic tissue in some species, makes access to the nephron and histologic preparations difficult (Brown *et al.*, 1993).

Considering the importance of blood pressure to kidney function, it is not surprising that vasoactive hormones have received a lot of attention from researchers in this field. There are numerous vasoactive hormones/compounds, but only three of the most studied hormones will be described here: arginine vasotocin, angiotensin II, and atrial natriuretic peptide. The focus on vasoactive hormones also means that information about possible effects of other hormones in fish kidney physiology is rather limited, so only two of these (cortisol and prolactin, which are known to have osmoregulatory actions in teleosts) will be included here.

Arginine vasotocin (AVT)

AVT is the major neurohypophysial hormone in lower vertebrates and is analogous to mammalian arginine vasopressin, causing constriction of blood vessels. It has long been implicated in the regulation of kidney function in teleosts. Earlier studies indicated that AVT was diuretic in fish (Maetz *et al.*, 1964). However, subsequent studies suggested that low doses are antidiuretic and that higher doses are diuretic

(Henderson and Wales, 1974; Babiker and Rankin, 1978). The development of a sensitive radioimmunoassay made it possible to measure the normal concentration of AVT in plasma (10^{-12} to 10^{-11} M). Balment *et al.* (1993) estimated the plasma level in these earlier studies and found that when a diuretic effect had been observed the plasma levels were far above the physiologic level, suggesting a pharmacologic rather than physiologic effect. AVT levels in the 'low-dose' studies were within the physiologic range, indicating that an antidiuresis is probably the 'normal' response (Table 2.3). As AVT causes vascular pressor action and elevates systemic blood pressure at high doses (Conklin *et al.*, 1997; Warne and Balment, 1997), a possible local antidiuretic effect in the kidney would probably be completely masked (Balment *et al.*, 1993). Administration of AVT to *in situ* perfused trout kidney at physiologic levels elicited an immediate antidiuresis, which could be directly related to a decline in the number of filtering glomeruli (glomerular intermittency) (Amer and Brown, 1995). Having this in mind, comparison of AVT plasma levels showed no consistent difference between freshwater and marine species or between freshwater- and seawater-acclimated euryhaline species (Brown and Balment, 1997). However, the measurements did show that the level measured in freshwater teleosts is similar to the level which was found to elicit antidiuresis *in situ*, which may explain why freshwater fish are in a slight antidiuretic state (i.e. not all nephrons filter in freshwater species). A decrease in AVT could possibly release some of this apparent down-regulation of GFR observed in freshwater-acclimated control animals (Brown and Balment, 1997). Even though no consistent difference in AVT concentrations were found between freshwater- and seawater-acclimated animals, AVT does seem to respond to osmotic stimuli. Both injection of NaCl as well as seawater acclimation of flounder have been found to cause an increase in plasma AVT (Balment *et al.*, 1993). However, the same authors observed that the AVT level decreases again after about a week. In another similar study, no significant changes in plasma AVT were observed after seawater acclimation (Harding *et al.*, 1997). However, a positive correlation between plasma AVT and plasma Na in seawater-acclimated animals was observed. No correlation was noted in any freshwater-acclimated animals. Balment *et al.* (1993) reported the same results. It is possible that a threshold exists for osmotic stimuli to induce AVT secretion (Harding *et al.*, 1997). Furthermore, AVT secretion seems to be sensitive to the volemic status of the animal. In a recent study, flounder were infused with 150 mM Na and Cl (to produce 30 percent increase in blood volume) and were acclimated to seawater (Harding *et al.*, 1997). Non-volume-expanded animals showed a significant correlation between AVT concentration and Na, whereas this correlation was lost in volume-expanded animals (Harding *et al.*, 1997). These results confirm a study in which papaverine treatment (lowers blood pressure) was consistently associated with increased AVT concentration (Brown and Balment, 1997). The sensitivity of AVT secretion to changes in blood pressure is not surprising, considering its local and systemic vasoconstrictor actions. The antidiuretic effect of AVT in the kidney appears to result from activation of glomerular receptors and a vasoconstrictor effect, lowering glomerular filtration rate. In the *in situ* perfused trout kidney, AVT causes increased vascular resistance

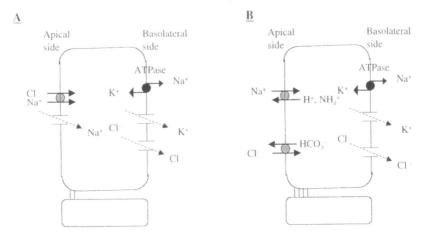

Figure 2.7 Mechanism for reabsorption of sodium and chloride by the urinary bladder in seawater, and possibly in hard freshwater (A) and in soft freshwater (B). Strongly decreased water permeability in soft freshwater is indicated by 'extra' strands in the narrow junction between cells. Adapted from Marshall (1995).

and lowers perfusate flow. The vascular effect is probably via binding to V_1-type receptors in the arterioles, although these have not yet been identified (Brown and Balment, 1997).

In addition to a vascular effect on the glomerulus, AVT may also exhibit a tubular effect. *In situ* perfused trout kidney showed a small but significant reduction in free water clearance in response to 10^{-11} M AVT, indicating that an increase in reabsorption of water possibly contribes to the antidiuretic effect observed during seawater acclimation (Brown and Balment, 1997). In tetrapods, the tubular antidiuretic action (increased water permeability of tubules) is associated with V_2-type receptors coupled to elevated cAMP levels by stimulatory G proteins (Brown and Balment, 1997). V_2-type receptors have not been identified in fish. However, trout tubules respond to AVT in a dose-dependent manner, with increases in the generation of cAMP, indicating that they do possess a V_2-like receptor (Perrott *et al.*, 1993). Significant stimulation of cAMP production was found at an AVT concentration as low as 10^{-12} M in freshwater-adapted rainbow trout (Harding and Balment, 1995; Harding *et al.*, 1997). Interestingly, tubules from 50 percent seawater-acclimated trout showed lower sensitivity to AVT as only AVT concentrations of 10^{-8} M and higher could elicit a significant increase in cAMP level. This may be due to down-regulation of AVT receptors, and this means that AVT can be involved in the adaptive changes in the kidney without changing the circulating level of the hormone AVT (Harding and Balment, 1995). Furthermore, the response to AVT (increased cAMP level) observed in rainbow trout tubules was not observed in tubules from marine species such as plaice, flounder and cod. As these three marine species do not possess a distal tubule, it indicates that the V_2-like receptor is located in the distal tubule, which rainbow trout (and other freshwater teleosts) do possess (Harding and Balment, 1995; Brown and Balment,

1997). Interestingly, AVT could stimulate adenylate cyclase activity in flounder urinary bladder. Urinary bladder is thought to have the same embryologic origin as the distal tubule (Brown and Balment, 1997). The response to AVT correlates with increased water permeability in the bladder during seawater acclimation.

In summary, the antidiuretic effect of AVT seems to occur via a vasoconstrictive effect in the glomerulus possibly via V_1-like receptors, lowering glomerular filtration rate, as well as via tubular effect, increasing water reabsorption, and possibly mediated by V_2-like receptors.

Renin–angiotensin system (RAS)

The renin–angiotensin system (RAS) comprises a complex series of biochemical events which generate active peptides involved in the regulation of a number of physiologic processes (Henderson *et al.*, 1993). Renin is excreted by granular epithelioid cells (differentiated smooth muscle cells in the vasculature of the kidney). It is a proteolytic enzyme and cleaves the substrate angiotensinogen to form a decapeptide, angiotensin I (AI). Angiotensinogen is produced by the liver but is usually found in excess in the circulation and is apparently released from the liver as soon as it is produced. Angiotensin I is then cleaved by the angiotensin-converting enzyme (ACE) to the physiologically active octapeptide form, angiotensin II (AII) (Henderson *et al.*, 1993; Brown and Balment, 1997). Measurement of plasma renin activity is commonly measured by its ability to generate AI *in vitro* from the renin and angiotensinogen present in plasma. However, a RAS system also seems to be present within the kidney itself. ACE is found incorporated in the endothelial plasma membrane but is also found in considerable amounts in kidney tubules (Henderson *et al.*, 1993). Intrarenal RAS physiologic activity was indicated in a study using *in situ* perfused trout kidney preparations, eliminating any systemic effects. In trout kidney preparations using the standard pressure head of 38 cmH_2O, no effect was observed. However, when the kidneys were perfused at a low pressure of 25 cmH_2O (reduces perfusate flow as well), captopril (ACE inhibitor) induced glomerular diuresis, indicating that the kidney has RAS activity (Brown *et al.*, 1995). Furthermore, AII receptors have been localized in glomeruli (Brown *et al.*, 1990; Cobb and Brown, 1994; Sharma *et al.*, 1998).

AII seems to have several physiologic effects, including stimulation of adrenocortical steroidogenesis (Brown *et al.*, 1990), stimulation of drinking (Fuentes and Eddy, 1997b), and regulation of blood pressure (Nishimura and Bailey, 1982). Release of renin by the kidney is related to perfusion pressure (Brown *et al.*, 1990). Inhibition of ACE decreases systemic blood pressure of toadfish, whereas plasma renin concentrations are increased (Nishimura and Bailey, 1982). Generally speaking, AII is a tonic anti-drop regulator of arterial blood pressure in many teleosts, and this seems to be achieved through constrictory actions on the microcirculation (Olson *et al.*, 1994). Its physiologic effect on the kidney seems to be antidiuretic (Table 2.3). AII produces large reductions in GFR and UFR in *in situ* perfused trout kidneys (even if the perfusion pressure is kept constant), as well as in cannulated trout infused with AII (Brown *et al.*, 1980; Rankin *et al.*, 1984; Brown and Balment, 1997). It appears to be an important component of the regulation of glomerular intermittency in teleosts because it reduces the renal tubular transport maximum for glucose (TmG) (Table 2.3) (Brown *et al.*, 1980). This parameter is an estimate for the number of filtering nephrons, which measure glomerular intermittency. In freshwater-acclimated fish, AII infusions which decreased TmG by 50 percent did not alter SNGFR, whereas it reduced SNGFR in seawater-acclimated fish, which could account for the overall decrease in GFR induced by AII (Brown *et al.*, 1980). Application of AII to freshwater- and seawater-acclimated rainbow trout revealed that similar changes in the ultrastructure of the glomerular filtration barrier were observed in seawater-acclimated and AII-infused freshwater animals (i.e. podocyte flattening, primary process broadening and apparent loss of foot processes) (Gray and Brown, 1985). Infusion of AII into seawater-acclimated animals increased the overall thickness of the basement membrane (Gray and Brown, 1985). To exclude a possible indirect effect of AII on ultrastructure via reduced blood pressure, an *in vitro* study was performed with renal slices and isolated glomeruli. Following exposure to AII, morphologic features were investigated and showed that the glomerular filtration area (filtration slit diaphragm area) had been reduced, which is consistent with reduced filtration rates (Brown *et al.*, 1990). Although AII has never been measured during seawater acclimation, these results clearly indicate that the hormone could be involved, based upon morphologic and physiologic changes occurring during acclimation.

Atrial natriuretic peptide (ANP)

Natriuretic peptides, secreted by cardiocytes in all vertebrates, are osmoregulatory and vasoactive hormones (Olson and Duff, 1992) The vascular response to ANP varies greatly among studies, but it has been suggested that ANP controls volume regulation in fish, and has primarily diuretic effects (Table 2.3) (Duff *et al.*, 1997). ANP has been found to increase GFR and to reverse the angiotensin II-induced antidiuresis in the constant pressure *in situ* perfused trout kidney (Dunne and Rankin, 1992). It was not clear whether SNGFR or the number of filtering nephrons was increased.

An increase in atrial filling pressure is a potent and long-lasting stimulus of ANP release by rainbow trout heart (Cousins and Farrell, 1996). A fall in dorsal aortic pressure has been observed in eel and seawater-adapted flounder after administration of human ANP or eel ANP or eel extract (Takei and Balment, 1993). It seems that ANP acts to reduce blood pressure, which could significantly alter kidney function. ANP increased urine output in trout and toadfish (Olson and Duff, 1992; Duff *et al.*, 1997), and an 8-h infusion into conscious rainbow trout decreased both whole body blood volume and extracellular fluid volume (Duff *et al.*, 1997). Additionally, ANP may influence Na^+ balance, possibly via inhibition of proximal tubular Na^+ reabsorption and suppression of renin secretion (Takei and Balment, 1993). A similar diuretic and natriuretic response was evident in the

Table 2.3 Effect of angiotensin II (AII), arginine vasotocin (AVT) and ANP on UFR, GFR, fractional water reabsorption, tubular transport maxima for glucose (TmG), percentage glomeruli filtering, non-filtering but perfused, and non-perfused.

Species	Treatment	M	UFR (mL kg⁻¹ h⁻¹)	GFR (mL kg⁻¹ h⁻¹)	FR H₂O (%)	TmG (mg min⁻¹ kg⁻¹)	Filtration (%)	N-filtration and perfusion	Non-perfusion	Reference
A. anguilla (n = 6, 7)	Control	FW	2.68 ± 0.09	4.51 ± 0.48	40.6	837.2 ± 192.5				a
	0.001 ng AVT kg⁻¹ (2 doses)	FW	1.52 ± 0.10	2.27 ± 0.19	33.0	325.1 ± 62.7				
	Control	FW	1.98 ± 0.17	6.52 ± 0.59	48.4	853.9 ± 256.5				
	100 ng AVT kg⁻¹ (2 doses)	FW	3.32 ± 0.24	6.52 ± 0.77	49.0	1481.5 ± 396.4				
O. mykiss Means taken from figure	Control	FW	3.30	8.5	61.3	162	~77	0	~23	b
	10⁻¹¹ M AVT in perfusate flow	FW	0.96	2.5	61.9	46	~32	~33	~35	
	Control	FW	2.76	10.2	72.9	190	~77	0	~23	
	10⁻⁹ M AVT in perfusate	FW	0.36	1.2	70.0	15	~32	~34	~34	
O. mykiss (n = 6–8)	Control	FW	4.29 ± 0.55	7.58 ± 0.95	43.3	597.1 ± 40.8				c
	150 ng AII min⁻¹ kg⁻¹	FW	1.70 ± 0.39	3.02 ± 0.62	43.7	264.8 ± 49.5				
O. mykiss (n = 21)	Control	FW	4.58 ± 0.63	8.42 ± 1.03	45.6		45	42.2	12.8	d
	150 ng AII min⁻¹ kg⁻¹	FW	1.82 ± 0.34	3.66 ± 0.71	50.4		10	46.4	43.6	
(n = 5)	Control	SW	0.32 ± 0.08	1.11 ± 0.35	71.3		5.1	44.3	50.6	
	600 ng AII min⁻¹ kg⁻¹	SW	0.05 ± 0.01	0.19 ± 0.08	75.9		5.6	21.6	72.8	
O. mykiss (n = 9–15)	Control	FW	2.9 ± 0.4	5.7 ± 0.1	49.1					e
	300 ng ANP h⁻¹ kg⁻¹	FW	6.3 ± 0.8	12.7 ± 2.3	50.4					

Sources: Values are means ± SEM and are obtained from the following papers [a]Henderson and Wales (1974); [b]Amer and Brown (1995); [c]Brown et al., (1990); [d]Brown et al. (1980); [e]Duff et al. (1997).

Note
FR H₂O is calculated from means of UFR and GFR as $[1 - (UFR/GFR)] \times 100$.

aglomerular toadfish (Lee and Malvin, 1987). However, a parallel relationship between urine flow and glomerular filtration rate suggests that the primary site of ANP activity in trout is the glomerulus (Duff *et al.*, 1997). Although these studies suggest involvement of ANP in teleost renal function, knowledge is still rather limited. An increase in ANP production could be a proper adaptive response during freshwater acclimation as an increased passive water influx could possibly stimulate ANP production and thereby urine production, but this has not been investigated. Furthermore, the precise action of ANP on the nephron is not known. A simple explanation could be dilation of the afferent arteriole or efferent vasoconstriction (Brown *et al.*, 1993).

Cortisol

Although there are large amounts of data suggesting that cortisol is one of the most important hormones in fish osmoregulation, there is rather limited and not very consistent information regarding kidney physiology. It is well known that cortisol has a positive effect on hypo-osmoregulation in seawater (Bern and Madsen, 1992), but it has also been shown to improve hyperosmoregulation in ion-poor freshwater (Laurent and Perry, 1990). One of the few reported effects of cortisol on kidney was the observation of enhanced sulfate secretion by the proximal tubule (Renfro, 1989). However, cortisol administered directly into the renal portal system of the aglomerular anglerfish (*Lophius piscatorius* L.) did not alter urine flow or urinary electrolyte concentration, indicating that cortisol is not involved in tubular functions in this species (Babiker and Rankin, 1979). In contrast, cortisol increased net Na^+ and Cl^- fluxes and osmotic water permeability in the bladder of *Gillichthys mirabilis* (Doneen, 1976). A well-known effect of cortisol is stimulation of gill and intestine Na^+,K^+-ATPase activity (Madsen, 1990; Bern and Madsen, 1992; McCormick, 1995), whereas its effect on kidney ATPase is not clear. Cortisol treatment of brown trout (*Salmo trutta*) showed no effect on either Na^+,K^+-ATPase mRNA or activity in the kidney (Madsen *et al.*, 1995). Furthermore, Na^+,K^+-ATPase activity in the kidney itself does not show a consistent pattern in seawater- and freshwater-acclimated animals (Madsen *et al.*, 1994, 1995; Zonno *et al.*, 1994; Fuentes *et al.*, 1996, 1997). Considering the numerous transport mechanisms that involve the basolateral Na^+,K^+-ATPase, it may not be surprising that no consistent pattern is seen. Furthermore, the enzyme is involved in both secretion and reabsorption of ions, and data may be obscured by species and fish size differences. In addition, in these studies, Na^+,K^+-ATPase activity was measured in a homogenate of kidney tissue, which makes it impossible to say whether increases and decreases in enzyme activity could have occurred concomitantly in different parts of the nephron. In summary, these studies do not indicate that cortisol is implicated in the control of kidney function.

Prolactin

Prolactin is generally recognized as a 'freshwater hormone', essential for maintaining hydromineral balance in freshwater (Borski *et al.*, 1992; Ayson *et al.*, 1993). The plasma level of prolactin decreases during seawater acclimation (Ayson *et al.*, 1993; Yada *et al.*, 1994) and increases after transfer to freshwater (Ayson *et al.*, 1993). Furthermore, treatment with prolactin limits seawater adaptability and antagonizes the seawater adaptive effect of cortisol and growth hormone (Hirano *et al.*, 1987; Madsen and Bern, 1992; Sakamoto *et al.*, 1997; Seidelin and Madsen, 1997). Most importantly, it is thought to lower epithelial permeability (Nishimura, 1985; Hirano, 1986). Prolactin treatment of seawater-adapted starry flounder (*Platichthys stellatus*) increased urine production and urinary Na$^+$ excretion and decreased urine osmolality, indicating that it decreases water reabsorption via a decrease in water permeability (Foster, 1975). Furthermore, the profound changes in water permeability of the distal tubule/urinary bladder during freshwater/seawater acclimation of teleosts correlates well with changes in prolactin levels. The mechanisms by which prolactin lowers permeability is not clear, but it may be via ultrastructural changes of the nephron and urinary bladder (Nishimura, 1985). It has been observed that prolactin greatly stimulates growth rates of the tubular epithelium, especially in the later parts of the nephron (i.e. distal tubule/collecting duct) (Wendelaar Bonga, 1976), and it restores the decrease in height of tubular cells in hypophysectomized eels (Olivereau and Olivereau, 1977). Furthermore, prolactin causes an extreme amplification of the basolateral membrane, which can possibly be related to higher capacity for reabsorption of ions (Wendelaar Bonga, 1976). In contrast, the structural changes occurring in the glomerulus during freshwater acclimation could not be related to prolactin (Wendelaar Bonga, 1976).

Concluding remarks on teleost kidney physiology

A summary of teleost renal function in freshwater and marine species is presented in Figure 2.8. The major differences between the two situations are:

- high UFR and GFR in freshwater, low in seawater;
- efficient dilution of urine in freshwater through high reabsorption activity, particularly by the distal tubule, collecting tubule/system, urinary bladder and by extremely low water permeability;
- tubular secretion of inorganic ions in PII (divalent as well as monovalent) in seawater, followed by:
- tubular reabsorption of Na$^+$ and Cl$^-$ in collecting tubule and urinary bladder, which is the main driving force for reabsorption of water.

Although teleost kidney function has been rather intensively examined, the majority of studies have focused on osmotic and ionic regulation. Information about the role of the kidney in other aspects of the animal's physiology, for example acid–base regulation and excretion of metabolic waste, is more limited. Although teleosts do excrete metabolic waste (e.g. ammonia, urea) via the kidney, the majority seems to be excreted via the gills under most circumstances (Wood,

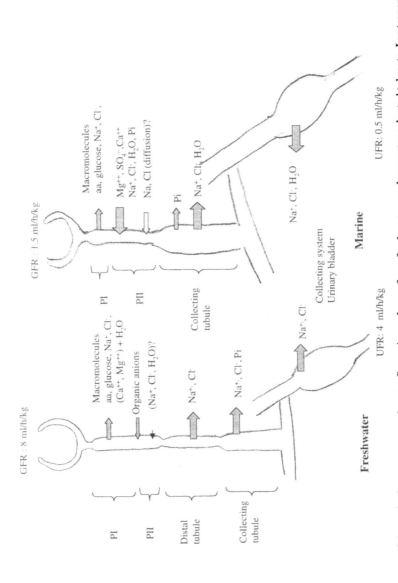

Figure 2.8 Summary of the major ion transport and water fluxes in nephrons from freshwater- and seawater-adapted teleosts. Ion transport may exist in other sites not shown here.

1993). Furthermore, in marine species, the very low urine flow rate limits that capacity for excretion, and increases in urine flow rate could cause osmoregulatory problems. Similarly, the kidney seems to play a rather limited role in acid–base regulation under most circumstances; again, the gills being more important (Heisler, 1986; Dantzler, 1989). Although the total amount of excreted acid–base load by the kidney may be limited, it plays a role at least in reabsorbing HCO_3^-. During acidotic stress, teleosts commonly accumulate bicarbonate, and plasma concentrations of more than 50 mM have been measured during hypercapnia in freshwater-acclimated rainbow trout (Larsen, 1997). As bicarbonate is freely filtered, the kidney must possess very efficient mechanisms for this ion. Interactions between various kidney functions have been poorly investigated (e.g. acid–base regulation and ammonia excretion is possibly linked), and hormonal influences on kidney functions are unknown. We are only beginning to understand the hormonal regulation of kidney function, and the precise mechanisms at the cellular level are largely unknown. Here, we have only presented information about the most commonly investigated hormones; other hormones are also being investigated as well as the cellular mechanisms by which they act.

Nephrotoxicity

The role of the teleost kidney in osmoregulation, cortisol synthesis, and elimination of waste may all be affected by toxicant-induced cellular damage. A decrease in the efficiency of any of these renal processes may, in turn, have a significant effect on the ability of the exposed individual to survive, grow, or reproduce normally. Therefore, the fish kidney can be an important site of toxic action for some xenobiotics.

Most descriptions of nephrotoxicity in fish have focused on histopathology and enzyme alterations. Relatively few attempts have been made to elucidate the biochemical mechanisms underlying the observed pathologies or to characterize the resulting effect on the animal's overall health. The paucity of mechanistic nephrotoxicity studies in fish may stem in part from an assumption that the nephrotoxic mechanisms of most chemicals are the same for fish as for mammals. The similarity of basic nephron structure between fish and mammals lends credence to this assumption. There are, however, some unique features of the teleost kidney that may present special problems.

Mechanisms of nephrotoxicity

The typical teleost kidney receives a large fraction of the cardiac output because of the extensive renal portal system (Pritchard and Miller, 1980; Pritchard and Bend, 1984). High blood flow to the renal tissues allows for high levels of renal exposure to circulating compounds and may combine with other factors to lead to xenobiotic accumulation in the kidneys. Among the physiologic properties contributing to bioconcentration of xenobiotics in the kidney are glomerular filtration, active secretion and reabsorption of chemicals in the tubules, and extracellular pH. It should be remembered, however, that some marine species do not exhibit true glomeruli. The lack of glomerular filtration can increase the significance of active secretion in the tubule. As a consequence, toxicants that disrupt the normal rates of filtration, secretion, or reabsorption in mammals might effect very different responses in fish of varying glomerular morphology.

Cellular injury

Damage to renal cells can be initiated through a wide variety of mechanisms, including alterations in cell signaling, gene regulation, protein synthesis, membrane transport, ionic regulation, enzyme function, and cytoskeletal integrity. The normal cell maintains a certain capacity for compensation and repair of injury, depending on the type and extent of injury. However, when the damage exceeds the cell's capability for repair, cell death may occur – either through the protective pathway of apoptosis or by necrosis.

Apoptosis is the innate ability of many cell types to self-destruct in a controlled manner. This process is considered to be a means of suppressing the extent of toxic damage by lessening the effect on adjacent cells. Necrotic cell death, a more destructive process, is more frequently the result of toxic insult. Whereas apoptosis can protect neighboring cells by preventing inflammation and release of cellular debris, necrosis is typified at the tissue level by localized areas of cellular degeneration and inflamed tissue. Necrotic lesions are caused by the swelling and eventual breakdown of cell membranes, leading to rupture of the cell and release of its contents into the extracellular space. Reports of cell death in fish kidneys generally focus on necrotic degeneration, although apoptosis almost certainly occurs to some extent in teleost renal cells.

Kidney repair and compensation

Vertebrate kidneys often respond to toxic insult by initiating a series of compensatory mechanisms, including hypertrophy and cellular regeneration. The hypertrophy of proximal tubules gives the kidney increased tubular capacity, which can make up for decreases in GFR resulting from nephron loss. Increases in single nephron GFR and renal plasma flow may accompany increases in tubule size. In rats, functional compensation for nephron loss occurs before structural hypertrophy and appears to be linked to prostaglandin-induced vasodilation (Pelayo and Shanley, 1990). Therefore, both structural and functional changes may work together to offset decreased renal function. This process has not been closely examined in fish, however, and may be less important in piscine renal compensation.

The occurrence of cellular regeneration and its role in piscine kidney compensation is well established. This process occurs through replication of healthy epithelial cells and appears to be common in damaged nephrons of most vertebrates. As the remaining healthy cells multiply, they begin to migrate along the basement membrane and reconstruct the damaged portions of the tubule

(Reimschuessel and Williams, 1995; Augusto *et al.*, 1996). The repaired tubules can then regain their normal function.

Mammals have a remarkable ability to repair and compensate for cellular damage to the nephron, but they lack the ability to produce new nephrons after infancy (Reimschuessel *et al.*, 1990). Fish, however, retain renal stem cells and undergo some level of nephrogenesis for their entire lifespan. Formation of new nephrons is stimulated after exposure of fish to various toxic agents, and it plays an important role in renal recovery. Nephrogenesis in mature animals is apparently a unique regenerative response in fish. The nephrogenic pathway originates in small, darkly staining clusters of basophilic cells, from which crude, small glomeruli attached to simple tubular structures begin to appear. These simple structures apparently become fully formed and functional nephrons.

Localization of renal damage

In teleosts, as in mammals, renal lesions are observed most frequently in the proximal tubules. Tubular necrosis is reported in almost all histologic examinations of xenobiotic-induced kidney damage in fish, although glomerular and inter-renal alterations occur with some toxicants. The propensity of proximal tubule effects is probably related to the high capacity for membrane transport in the tubular epithelial cells and the concentration of toxic compounds in the tubular lumen. These factors can combine to increase the intracellular concentrations of circulating xenobiotics, thereby increasing the potential for cellular injury.

Nephrotoxic agents

Metals

The ubiquitous occurrence of metals in aquatic systems is a continuing environmental concern. Metals constitute one of the most-studied groups of aquatic toxicants and present a multiplicity of toxicologic effects. However, the complex environmental chemistry of some metals makes assessments of exposure and associated risk difficult. In nature, metals can exist in multiple electron spin states and can form both inorganic and organic complexes. Site-specific parameters such as pH, hardness, dissolved organic carbon, and salinity determine the metal species present in a given waterway (Roesijadi and Robinson, 1994), and the speciation of the metal, in turn, has a great effect on its bioavailability and ultimate toxicity.

A key issue in the nephrotoxic nature of many heavy metals is their ability to accumulate in renal tissue (Table 2.4). Metal accumulation in fish kidneys is driven in large part by toxicokinetic factors such as high renal blood flow and plasma protein binding. The affinity of some toxic metals for sulfhydryl groups on endogenous compounds, such as the peptide glutathione (GSH), often plays a major role in renal transport. Sulfhydryl binding is also thought to contribute significantly to the ultimate cellular injury resulting from exposure to some metals. By binding to sulfhydryl groups on cellular proteins, metals are able to disrupt enzyme function, alter membrane integrity, and reduce cytoprotective ability.

Some metals, such as iron (Fe^{2+}/Fe^{3+}) and copper (Cu^+/Cu^{2+}) are implicated in the initiation of oxidative stress through the production of reactive oxygen species (ROS). The generation of ROS, such as superoxide radicals (O_2^-), hydrogen peroxide (H_2O_2), and hydroxyl radicals (OH$\cdot$), is a normal consequence of cellular metabolism. Transition metals participate in ROS production through the single electron reduction of O_2 and H_2O_2, leading to formation of OH$\cdot$. Of the ROS produced physiologically, OH$\cdot$ is the most reactive, and is capable of oxidative damage to proteins, lipids and DNA. Excessive oxidative damage to biomolecules can result in severe cellular dysfunction. The cell is protected, however, against ROS-induced damage by a variety of enzymatic processes. One such protective mechanism is the conversion of O_2^- to H_2O_2 by superoxide dismutase (SOD), which is followed by glutathione peroxidase-catalyzed reduction of H_2O_2 to H_2O. Non-enzymatic processes also serve to protect from oxidative damage by reducing or scavenging free radicals. Together, these biochemical reactions make up the antioxidant defenses of the cell. Oxidative stress occurs when ROS production within the cell overwhelms the capacity of antioxidant defenses and causes structural and functional damage.

CADMIUM

The widely occurring industrial contaminant cadmium (Cd) is known to exhibit significant toxicity to most vertebrates including fish. Cadmium has a high potential for bioconcentration in fish and is accumulated in multiple organs. In most teleost species examined, the kidney is the primary site of accumulation, but substantial differences exist in the tissue distribution of Cd between species. For example, a comparison of Cd distribution in roach (*Rutilus rutilus*), stone loach (*Noemacheilus barbatulus*), and rainbow trout (*Oncorhynchus mykiss*) demonstrated a preferential storage of Cd in kidney of roach and trout, while liver was the main organ of deposition in stone loach (Brown *et al.*, 1986).

The toxic effects of Cd range from osmoregulatory to death, but the mechanisms of many of the observed effects are not well understood (Glynn *et al.*, 1992; Melgar *et al.*, 1997; Scherer *et al.*, 1997). Two important features of the toxicity of Cd are its ability to: (1) displace essential metals such as Zn and Fe from sulfhydryl ligands and (2) bind to critical cellular proteins. Both of these processes can result in alterations in normal enzymatic processes, and could ultimately lead to cell death. Cadmium has also been shown to produce lipid peroxidation with subsequent decreases in GSH, GSH peroxidase, and GSH reductase levels (Stohs and Bagchi, 1995). Because Cd does not produce free radicals directly, the resulting oxidative stress is thought to result from displacement of endogenous Fe, which generates reactive oxygen species as described earlier in this chapter.

Not only is the kidney a primary target organ for Cd in many fish species, but renal Cd concentrations may remain elevated for extended periods after exposure. An early examination of Cd distribution in channel catfish (*Ictalurus punctatus*) demonstrated kidney concentrations of over 4 mg kg^{-1} Cd after a 3-week exposure to 800 µg L^{-1} Cd (Smith *et al.*, 1976). More recently, Schultz *et al.* (1996) examined the detailed tissue distribution of Cd in channel catfish after intravascular injection

Table 2.4 Effects of organochlorine insecticides on renal enzymes of teleosts.

Compound	Enzyme	Effect	Species	Reference
Aldrin	Na⁺,K⁺-ATPase	↓ in vitro	*Labeo rohita*	Verma *et al.* (1978)
	Mg²⁺-ATPase	↓ in vitro	*L. rohita*	
Dieldrin	Na⁺,K⁺-ATPase	↓ in vitro	*L. rohita*	Verma *et al.* (1978)
	Mg²⁺-ATPase	↓ in vitro	*L. rohita*	
Toxaphene	Na⁺,K⁺-ATPase	↓ in vitro	*Ictalurus punctatus*	Desaiah and Koch (1975)
	Mg²⁺-ATPase	↓ in vitro	*I. punctatus*	
p,p′-DDT	Na⁺,K⁺-ATPase	↓ in vivo	*Oncorhynchus mykiss*	Campbell *et al.* (1974)
Chlordecone	Ca²⁺-ATPase	↓ in vitro	*Saccobranchus fossilis*	Bansal and Chandra (1985)
Mirex	Ca²⁺-ATPase	↓ in vitro	*S. fossilis*	Bansal and Chandra (1985)
Endrin	Acid phosphatase	↑ in vivo	*Channa gachua,* *Channa punctatus*	Sastry and Sharma (1978)
	Alkaline phosphatase	↑ in vivo	*C. gachua* *C. punctatus*	Sastry and Sharma (1979a)
	Glucose-6-phosphatase	↑ in vivo	*C. gachua* *C. punctatus* *Ophiocephalus punctatus*	Sastry and Sharma (1979b)
Endosulfan	Acid phosphatase	↓ in vitro	*C. gachua*	Sharma (1988)
	Alkaline phosphatase	↓ in vitro	*C. gachua*	

and found a similar pattern of renal accumulation. Preferential storage of Cd (30–37 percent injected dose) was noted in trunk kidney, with tissue concentrations remaining unchanged for 335 days. Common carp fingerlings (*Cyprinus carpio*), exposed for up to 30 days at both lethal and sublethal Cd concentrations, demonstrated time- and concentration-dependent accumulation in the kidney (Suresh *et al.*, 1993). Other teleosts show the same combination of high Cd loading in the kidney and prolonged Cd retention, suggesting that renal cells may be especially at risk of damage during chronic exposure (Thomas *et al.*, 1985; Brown *et al.*, 1986; Glynn *et al.*, 1992; Allen, 1995; Melgar *et al.*, 1997; Scherer *et al.*, 1997) (Table 2.4).

The distribution and toxicity of Cd is affected greatly by the metal-binding protein metallothionein (MT) (Figure 2.9). Metallothionein is a low molecular weight, cysteine-rich protein that can efficiently sequester metals, thereby reducing their toxicity. Because MT is induced by exposure to metals, sublethal doses may serve to protect the organism when faced with a subsequent high exposure. In the case of Cd, however, MT becomes directly involved in mediating renal toxicity, at least in mammals (Maitani *et al.*, 1988). Metallothionein binds circulating Cd and transports it to the mammalian kidney, where the Cd–MT complex is freely filtered by the glomerulus. As the Cd–MT is reabsorbed, free Cd is released via the lysozomes inside the tubule cells, where it induces MT. If Cd levels continue to increase, intracellular MT capacity is overwhelmed and free Cd wreaks havoc on the intracellular machinery. The existence of MT in fish is well established, and induction of piscine MT after Cd exposure is commonly observed (Roesijadi and Robinson, 1994; Olsson *et al.*, 1995; Zhang and Schlenk, 1995; Schlenk and Rice, 1998). Also, MT is involved in Cd transport to the kidney in some fish, but its involvement in a mammalian-type Cd transport/release mechanism has not been established.

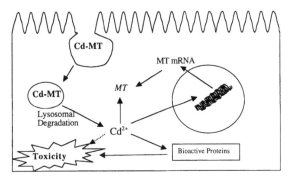

Figure 2.9 Simplified mechanism of Cd–MT renal toxicity. Cd–MT is filtered by the glomerulus and is then transported into the renal cell via endocytosis. Free Cd^{2+} is released during lysosomal degradation, and stimulates the synthesis of MT within the tubule cell. As intracellular Cd concentrations increase, the capacity for MT production is overwhelmed, and binding to non-MT proteins can occur. Modified from Goldstein and Schnelmann (1996).

In both terrestrial and aquatic species, cadmium is inhibitory (*in vitro* and *in vivo*) to numerous enzymes, including acetylcholinesterase (AChE), acid phosphatase, alkaline phosphatase, Na⁺,K⁺-ATPase, and lactate dehydrogenase. In some instances of *in vivo* exposure, however, increases in the activity of select kidney enzymes have been described. An example is the case of the rosy barb (*Barbus conchius*), which shows significant Cd-related inhibition (*in vitro* and *in vivo*) of acid phosphatase and glutamic pyruvic transaminase in kidney homogenates, whereas kidney alkaline phosphatase activity is increased by more than 100 percent (Gill *et al.*, 1991a). The enzymatic inhibitory activity of Cd is probably due in most cases to either binding of Cd to the enzyme or displacement of a cofactor, but species and tissue differences exist for the inhibitory mechanism of some enzymes (e.g. Na⁺,K⁺-ATPase) (Kinne-Saffran *et al.*, 1993).

Cadmium also affects kidney-related steroidogenesis in some piscine species by altering cortisol production, a function of the inter-renal tissue. The inter-renal cells of rainbow trout particularly seem to respond sensitively to Cd exposure, with cortisol levels increasing rapidly and remaining elevated during exposure to sublethal concentrations (Hontella *et al.*, 1996). Whether Cd directly mediates this effect or whether it is an indirect effect of osmoionic disruption is unclear. However, it is possible that Cd-induced increases in cortisol are actually a result of calcium (Ca^{2+}) channel blockade at the gill, leading to hypocalcemia and stimulation of inter-renal cells. Another proposed mechanism is that impaired glucose metabolism, related to renal toxicity, may alter cortisol production. This idea is supported by findings of decreased kidney glycogen after Cd and Hg exposure (Srivastava, 1982). The long-term consequences of altered cortisol regulation may include effects on other hormonal systems and decreased responsiveness to stress.

Chronic exposure to environmentally relevant concentrations of $CdCl_2$ often lead to extensive renal pathologies in fish. Common carp (*Cyprinus carpio*) display progressive tubular necrosis at exposures of 58 µg L⁻¹ $CdCl_2$ for several weeks (Singhal and Jain, 1997). These necrotic changes (nuclear pyknosis, karyohexis, and karyolysis) are also associated with glomerular alterations, loss of brush border in PI and PII segments, and increases in intertubular space. Similar tubular necrosis and glomerular collapse have been documented in *Puntius conchonius* after sublethal exposures (Gill *et al.*, 1989). However, a study in which stickleback (*Gasterosteus aculeatus*) were exposed to Cd^{2+} concentrations as high as 6 mg L⁻¹ described only moderate changes in tubular cells, consisting primarily of vacuolization and granulation (Oronsaye, 1989). The variability in response between these species is unclear, but may be related to differences in MT induction, drinking rates, or renal transport processes.

MERCURY

Mercury (Hg) exists naturally in high concentrations in some waterways, but its toxicity and bioaccumulative nature raise concerns for the safety of exposed organisms. In 1997, 1782 mercury advisories existed for waterways in the USA

(USEPA, 1998). These advisories are established to warn the public of potential fish consumption hazards based on fish muscle residue analysis. Although the primary regulatory agenda for Hg in the USA has been driven by human heath concerns, bioconcentration and biomagnification of Hg by teleosts pose legitimate threats to the health of aquatic systems receiving inputs of Hg.

Mercury may exist in its elemental state (Hg^0_2), as inorganic salts ($HgCl_2$), and in organic (primarily methylated) forms. The inorganic salt is primarily implicated in mammalian nephrotoxic effects, although nephrotoxicity has been noted in fish exposed to organomercurials (Gill *et al.*, 1988). In aquatic systems, only 5–20 percent of the total mercury exists as methylmercury (MeHg). Yet, bioaccumulation of Hg in fish occurs almost completely (> 95 percent) through the methylated forms (MeHg), with extensive MeHg deposition in the kidneys (Watras and Huckabee, 1994).

In vitro experiments with flounder (*Pseudoplectonectes americanus*) tubules have shown a concentration-dependent increase in cellular injury with short-term exposure to $HgCl_2$, MeHgCl, and parachloromercuribenzenesulfonate (PCMBS) (Trump and Jones, 1975). Of these compounds, MeHg causes the most rapid histologic progression of tubule cell injury. However, the inorganic form is responsible for the most substantial decrease in chlorophenol red uptake, a measure of active transport capacity. The histopathologic effects of Hg in fish kidney are similar to those in mammals – typified by tubular epithelial granulation and hyperplasia, glomerular shrinkage, enlargement of Bowman's space, pyknotic nuclei, and progressive necrotic degeneration of tubule cells (Pandey, 1994; Reimschuessel and Gonzalez, 1998). In *Puntius conchonius*, sublethal exposure (20–33 percent of 96-h LC_{50}) to ethoxy ethyl mercuric chloride (MEMC), an antifungal seed dressing, has been shown to cause notable tubular pathologies in exposures of 7 days. Cellular damage has been noted in mullet (*Liza parsia*) kidney after only 4 days of inorganic Hg exposure (20 µg L^{-1}), with severe tubulonecrosis occurring by day 15. Mercury-induced damage is generally localized in the first and second segments of the proximal tubule of teleost nephrons, which is similar to the proximal tubule-localized toxicity in mammals (Trump and Jones, 1975), but damage to the head kidney has also been described. In *Channa punctatus*, sublethal Hg exposure led to a time- and concentration-dependent necrosis of inter-renal, chromaffin, and hematopoietic tissues in the head kidney (Banerjee and Bhattacharya, 1994). This was accompanied by the typical tubular degeneration. Toxic damage to the head kidney may lead to hematologic effects, but the functional changes related to head kidney necrosis have not been examined in Hg-exposed fish. In fact, few examples of functional effects related to Hg-induced renal damage have been described in teleosts. But some functional changes have been noted, such as lysozymuria in $HgCl_2$-exposed plaice (*Pleuronectus platessa*) (Fletcher and White, 1986). A similar form of proteinuria also occurs in humans experiencing Hg toxicity.

The renal toxicity of Hg is thought to be related to binding of Hg^{2+} to sulfydryl groups on membrane proteins, leading to alterations in cellular permeability and subsequent swelling and lysis. One of the primary cellular targets for mercuric ion is Na^+,K^+-ATPase (Werner and Costa, 1995). Inhibition of sodium reabsorption in tubule cells is apparently responsible for the well-established diuretic effect of low doses of Hg^{2+} in mammals. At high concentrations of Hg^{2+}, severe ionic disruptions can occur accompanied by protein denaturation and decreased protein synthesis. Lipid peroxidation also occurs in piscine renal cells after Hg exposure. However, direct production of ROS by Hg does not appear to be involved (Rana *et al.*, 1995). A complex sequence of events, including disruption of oxidative phosphorylation and calcium homeostasis, depletion of GSH and antioxidant enzymes (SOD, catalase, etc.), and loss of mitochondrial membrane potential is implicated in the oxidative stress produced by Hg in mammalian kidneys. A similar mechanism is likely involved in fish nephrotoxicity.

The repair of teleost kidneys after Hg toxicity includes both nephrogenesis and tubular regeneration. Cell growth is accompanied by overall increases in mRNA and protein synthesis, as is documented by the dramatically enhanced activity of polyamine biosynthetic enzymes in flounder kidney following acute MeHg toxicity. Because polyamines are required for cellular growth and development, the resulting increases in putrescine, spermidine, and spermine may also be considered part of the regenerative process (Manen *et al.*, 1976).

Pesticides

Several classes of pesticides have been shown to produce nephrotoxicity in aquatic species. Because the numerous compounds classified as pesticides have such diverse structural and functional features, the renal toxicity of pesticides is likely to occur through myriad mechanisms. However, inhibition of renal enzymes is a common feature among nephrotoxic pesticides.

ACETYLCHOLINESTERASE-INHIBITING INSECTICIDES

The organophosphorous (OP) and carbamate insecticides are among the world's most heavily used pesticides. Although developed as safe alternatives (i.e. non-persistent) to organochlorine insecticides, these compounds often present significant acute toxicity to non-target species. The major toxic concern for aquatic species exposed to OPs or carbamates is mortality resulting from acute cholinergic poisoning – a situation caused by uncontrolled stimulation of acetylcholine (ACh) receptors in synapses and neuromuscular junctions. There is concern, however, that sublethal exposures to these chemicals may lead to a multitude of other problems, including neurologic, reproductive, and respiratory/osmoregulatory impairment.

The majority of renal lesions in fish related to AChE-inhibiting pesticides are similar in nature to other nephrotoxicants. Glomerular shrinkage, tubular dilation, vacuolization of epithelial cells, and progressive epithelial necrosis of the proximal tubules are all commonly observed (Kumar and Pant, 1984; Srivastava *et al.*, 1990; Gill *et al.*, 1991b). In addition to the expected AChE inhibition, other enzymes can be affected in fish renal tissues (Table 2.4).

Organophosphate- or carbamate-induced kidney lesions are almost always linked to either prolonged or very high-level exposures. The typically low degree of water solubility and short half-lives of these compounds in the natural environment probably make these long-term exposures rare outside of the laboratory, but the possibility of sublethal renal toxicity in fish may exist in situations with periods of heavy agricultural application.

Of particular interest are a few highly water-soluble AChE-inhibiting pesticides, which may have greater bioavailability in natural waterways than similar compounds of low solubility. An example is the carbamate aldicarb (Temik®), which is soluble in water at concentrations over 5 g L^{-1}. Aldicarb is used heavily around the world to control insect and nematode infestations of crops, but is toxic to fish, with LC$_{50}$ values of 1 mg L^{-1} or less for most species examined (Spradley, 1991; Perkins *et al.*, 1999). At sublethal concentrations, gill and kidney lesions may occur in some teleosts. The kidneys of rosy barb (*Barbus conchonius* Hamilton) are noted to display hemorrhage and congestion after only 2 days of sublethal aldicarb exposure (Kumar and Pant, 1984). Tubular necrosis and substantial degeneration of the surrounding lymphoid tissue is evident after 7 days of exposure. The renal toxicity of aldicarb has not been examined in other teleosts, but the outward signs of sublethal toxicity exhibited by the rosy barb, such as rapid overproduction of mucus at the gills, is similar to that noted in other species (Schlenk *et al.*, 1992).

Also of concern is the deliberate use of AChE-inhibiting pesticides on fish in aquacultural applications (Burka *et al.*, 1997). Unregistered use of the organophosphate malathion, for instance, as a selective aquacultural piscicide has been documented in the popular press and in trade publications. The utility of this approach stems from a wide differential in acute toxicity of this compound between fish species, allowing for survival of desired species (particularly channel catfish) with subsequent death of undesired species. Malathion, however, has been shown to cause mild renal lesions, decreased kidney glycogen and protein content, and increased kidney cholesterol in some species (Awasthi *et al.*, 1984; Dutta and Marcelino, 1990). This sublethal effect could decrease the fitness of the desired species without causing immediate outward signs of toxicity.

Another example of deliberate exposure is the use of OPs for treatment of ectoparasites in aquaculture. The OP Nuvan (dichlorvos), is also relatively water soluble, and is used in some countries as an antiparasitic treatment for fish (Ross, 1989). Despite its effectiveness against parasites such as sea lice, long-term exposure to Nuvan causes notable tubular alterations, including epithelial vacuolation, dilation of the tubular lumen, and necrotic cell death in gray mullet (*Liza parsia*) (Mophatra and Noble, 1992). It is doubtful, however, that the extended exposure periods needed to produce these lesions (45 days) would occur in actual applications.

Other OP insecticides cause similar renal pathology. Long-term exposure of *Ophiocephalus punctatus* to diazinon, at a concentration causing 10–20 percent mortality in 30 days, has been shown to cause glomerular shrinkage, necrosis of proximal tubules, and sloughing of tubular epithelial cells into the tubular lumen

(Sastry and Sharma, 1981). Rupture of the distal tubules was also described. In *Channa punctatus*, a similar effect was seen after extended exposure to Elsan [*O,O*-dimethyl *S*-(*a*-ethoxicarbonyl benzyl)phosphorodithioate]. Effects on the cells of the head kidney also accompanied Elsan-induced tubular damage (Banerjee and Bhattacharya, 1994).

ORGANOCHLORINE INSECTICIDES

Commercial use of organochlorine insecticides (OC), such as DDT (dichloro-diphenyl-trichloroethane), has been banned in the USA because of the environmental persistence and demonstrated biomagnification of these compounds. But in many parts of the world, OCs are still widely used because of their effectiveness and low acute toxicity to humans. The hydrophobicity of most OC pesticides keeps dissolved concentrations low in aquatic systems. Yet, it is the tendency of these chemicals to be stored for long periods in biologic tissues that make them a concern for piscine species. Exposure of fish to OCs occurs through both direct exposure to water and sediments as well as through the diet, and deposition of OCs is noted in most tissues.

Effects of OCs on renal enzyme activities have been documented in a variety of teleosts (Table 2.4). Adenosine triphosphatases (ATPases) are especially sensitive to inhibition by OC pesticides. ATPases are membrane-bound ion transporters involved in ionic regulation and energy production of the cell. Long-term inhibition of these transporters could lead to loss of cellular function as a result of decreased adenosine triphosphate (ATP) in the cell. Because of the importance of ion transport in renal cells, disruption of this function could also affect renal function directly. The effect of OCs on ATPase activity may be related to disruption of cellular membranes, which would also affect the efficiency of kidney function. Studies of chlorophenol red uptake (an anionic dye) in isolated goldfish tubules have demonstrated time- and dose-dependent decreases in active tubular transport processes (Gruppuso and Kinter, 1973). Species and tissue differences exist in the sensitivity to ATPase inhibition, but some level of inhibition has been noted in the kidney of all fish species examined.

Herbicides

The acute toxicity of agricultural herbicides to fish is generally much lower than that of insecticides. However, more total pounds of herbicides are applied than any other group of pesticides in the USA (USGS, 1997). The commonly used herbicides, such as atrazine and 2,4-dichlorophenoxyacetic acid (2,4-D), are frequently found as major contaminants in surface water (Oulmi *et al.*, 1995a; Poleksic' *et al.*, 1997). Because of their common occurrence, the chronic effects of herbicides on fish are of great concern.

Atrazine is one of the most prominent aquatic contaminants and is among the best-studied nephrotoxic herbicides. Although atrazine is selective for plants, it demonstrates a rather high potential for toxicity in fish, with LC_{50} values as low as 1 mg L^{-1} (Oulmi *et al.*, 1995a). In addition to acute mortality, renal and hepatic toxicity occurs in both fish and mammals (Fisher-Sherl *et al.*, 1991). The toxicity of atrazine to plants is mediated through photosynthetic inhibition, but its toxicologic mechanism in animals is not well understood.

Atrazine causes a variety of cytologic changes in the kidneys of rainbow trout, including mitochondrial alterations, peroxisomal and lysosomal proliferation, fragmentation of endoplasmic reticulum, and disruption of Golgi fields and brush border (Oulmi *et al.*, 1995a). The severity of atrazine toxicity increases with exposure levels, and the ultrastructural changes are selective for specific tubular segments. The PI and PII segments of the proximal tubule are the most sensitive to atrazine, with a variety of responses occurring at concentrations as low as 10 μg L^{-1} in the PII segments. Some less significant cellular damage occurs in the distal tubules. Tubular effects are also described along with decreases in renal alkaline phosphatase activity in carp (*Cyprinus carpio*) exposed to non-lethal atrazine concentrations. Other herbicides, including linuron, 2,4-D, and dichlobinil, also cause similar tubular alterations in fish (Oulmi *et al.*, 1995b; Poleksic' *et al.*, 1997). However, because the mechanism of action in plants is different between these compounds, the specific subcellular targets in fish kidneys are likely to be diverse.

Antibiotics

Exposure of fish to medicinal antimicrobials probably occurs exclusively in aquaculture applications, although there is the potential for movement of significant amounts of some antibiotic compounds into natural waterways as a result of land-based agriculture, such as poultry farms and cattle feed lots. Antibiotics are widely used as feed additives to increase growth and prevent diseases in high-density livestock operations (Addison, 1984). An additional input of antibiotics into the aquatic environment may come from municipal sewage outflows, although concentrations would presumably be very low (Hirsch *et al.*, 1999).

The nephrotoxicity of some classes of antibiotics is a well-known problem for mammals. The aminoglycosides, a *Streptomyces*-derived class of antibacterials, are particularly well-studied nephrotoxicants. Aminoglycosides are used in humans, livestock, and fish to control Gram-negative infections, but the severe renal toxicity of most of the compounds in this class requires a great deal of precaution in their administration.

Glomerular filtration is responsible for most of the renal elimination of aminoglycosides – a factor that makes their response in aglomerular fish particularly interesting. Despite being freely filtered by the glomerulus, these compounds tend to be concentrated in renal tissues. This is apparently due, in mammals, to endocytosis of aminoglycoside–phospholipid complexes that form in the brush border membranes of the tubular lumen after glomerular filtration (Dekant and Vamvakas, 1996). It is assumed that the mechanism of aminoglycoside tubular reuptake in fish is similar, but this has not been well substantiated.

Once inside tubule cells, aminoglycosides affect phospholipid metabolism and perturb lysosomal and mitochondrial function. The disturbance of lipid metabolism in lysosomes is related to inhibition of phospholipase activity (specifically phosphatidylinositol phospholipase C), which causes phospholipidosis in the cell. In mitochondria, some aminoglycosides can also cause inhibition of Na^+,K^+-ATPase, leading to severe intracellular electrolyte alterations and inhibition of oxidative phosphorylation. These cellular alterations can cause tubular necrosis and eventual renal failure. However, the exact steps in the progression of aminoglycoside-induced tubulonecrosis are unclear.

The histologic effects of aminoglycosides in fish are very similar to those described in mammals and are generally characterized by massive necrosis of the proximal tubules. Gentamicin has been examined in various teleost species because of both its use in aquaculture and its utility as a tool for the study of nephrogenesis. In goldfish (*Carassius auratus*) and tilapia (*Oreochromis nilotica*) species, acute tubulonecrosis is seen within a few days at high (25–50 mg kg^{-1}) doses. Pyknotic nuclei and degeneration of proximal tubule epithelial cells characterize these lesions. However, regeneration of damaged cells and formation of new nephrons begins within a week of dosing (Reimschuessel and Williams, 1995; Augusto *et al.*, 1996).

Although gentamicin shows some degree of toxicity in all piscine species examined, wide variability exists in sensitivity between species. Aglomerular fish seem to be especially susceptible to gentamicin-induced renal effects. The prime example is the toadfish (*Opsanus tau*), a marine fish which lacks glomeruli, and subsequently has a limited capacity for renal excretion of aminoglycosides. Despite the lack of glomerular filtration, gentamicin causes extensive tubulonecrosis at relatively low doses (Jones *et al.*, 1997). Because aminoglycosides produce their toxicity intracellularly, basolateral transport must occur for these effects to be manifest. Tubular damage in toadfish is essentially irreversible, with little or no cellular regeneration and no nephrogenesis. The lack of repair may be due to the extended plasma half-life of gentamicin in this species, which prolongs the effective duration of toxicity and negates any attempts for regeneration.

Hexachlorobutadiene

The chlorinated hydrocarbon hexachlorobutadiene (HCBD) is a widespread environmental contaminant originating from the manufacture of chlorinated ethylene solvents. Examination of the renal toxicity of HCDB in fish has focused on the use of this compound as a tool for studying nephrogenesis in goldfish (*Carassius auratus*), but exposure of feral fish to HCBD (and similar compounds) may occur as a result of its widespread environmental distribution (Reimschuessel *et al.*, 1989).

The morphologic changes induced by HCBD in goldfish kidney include glomerular shrinkage, vacuolization, and widespread necrosis of the tubular epithelium. This tubular degeneration begins soon after dosing with HCBD, and is primarily localized in the PII segment of the proximal tubule. Within 3–4 days of a single intraperitoneal injection, cellular regeneration begins. The initial regenerative phase consists of repopulating of the damaged tubules with flattened epithelial cells. A secondary vacuolization of these newly developing cells occurs during the second week after exposure. However, many basophilic clusters of nephrogenic cells also begin to appear during the second week. These clusters of cells begin developing into new nephrons and are still present, in reduced numbers, at 10 weeks after exposure. The specific cell signals that trigger the nephrogenic response to HCBD are unknown.

The related compound tetrachloroethylene (PCE) also causes tubular degeneration followed by nephrogenesis in rainbow trout (Reimschuessel *et al.*, 1993). This response has only been described after a large-scale contamination of a hatchery pond. The concentration of PCE in the pond is unknown, but approximately 50 percent of the fish were dead within a few days. The tubular effects were localized in the initial segments of the proximal tubule and no glomerular changes were noted, and nephrogenesis was evident after 10 days.

Concluding remarks on nephrotoxicity

Nephrotoxicants are diverse in terms of both structure and mechanism of toxicity. The resulting effects of nephrotoxic insult in fish, however, are often quite similar among classes of toxicants. Glomerular perturbations and necrosis of proximal tubular epithelium are the apparent hallmarks of piscine renal toxicity. Ultrastructural alterations of the distal tubules and cells of the head kidney have also been described, and may also be common. There are many known mechanisms of nephrotoxicants in vertebrates, ranging from membrane disruption to metabolic inhibition. But the specifics of renal toxicity in fish are largely unknown.

Functional effects of renal toxicity have seldom been measured in teleosts. This may be due to difficulty in obtaining good measurements of renal function in fish, or it may stem from a general lack of concern. Most of the published piscine nephrotoxicity studies describe their purpose as either the development of histologic biomarkers for certain types of exposure or as an alternative model for better understanding of mammalian renal function. It is reasonable to expect, however, that processes such as glomerular filtration, active secretion, and urine formation will be affected when severe tubulonecrosis and glomerular collapse occur. From the few studies that examined functional alterations, such as disruption of ion transport or occurrence of proteinuria, it is evident that significant effects do accompany renal damage in fish. The full extent of adaptation to losses in renal function remains to be determined, but the evidence suggests that teleosts have a great capacity for withstanding renal injury. The processes of nephrogenesis, renal hypertrophy, and tubular regeneration aid in renal compensation. It is the gill, however, that may take over much of the kidney's role in osmoregulation and excretion after renal damage. Unfortunately, most chemicals that cause major renal damage also create serious lesions in the gill, complicating the situation further. The interplay between renal and branchial toxicity, the underlying mechanisms of tubular cell death, and the environmental significance of teleost nephrotoxicity are not well understood – but it is clear that a number of common environmental contaminants do cause significant renal toxicity in both freshwater and marine species.

References

Addison, J.B. 1984. Antibiotics in sediments and run-off waters from feedlots. *Residue Reviews* 92: 1–28.

Allen, P. 1995. Long-term mercury accumulation in the presence of cadmium and lead in *Oreochromis aureus* (Steindachner). *Journal of Environmental Science and Health* B30: 549–567.

Amer, S. and Brown, J.A. 1995. Glomerular actions of arginine vasotocin in the in situ perfused trout kidney. *American Journal of Physiology* 269: R775–R780.

Augusto, J., Smith, B., Smith, S., Robertson, J. and Reimschuessel, R. 1996. Gentamicin-induced nephrotoxicity and nephrogenesis in *Oreochromis nilotica*, a tilapian fish. *Diseases of Aquatic Organisms* 26: 49–58.

Awasthi, M., Shah, P., Dubale, M.S. and Gadhia, P. 1984. Metabolic changes induced by organophosphates in the piscine organs. *Environmental Research* 35: 320–325.

Ayson, F.G., Kaneko, T., Tagawa, M., Hasegawa, S., Grau, E.G., Nishioka, R.S., King, D.S., Bern, H.A. and Hirano, T. 1993. Effects of acclimation to hypertonic environment on plasma and pituitary levels of two prolactins and growth hormone in two species of Tilapia, *Oreochromis mossambicus* and *Oreochromis niloticus*. *General and Comparative Endocrinology* 89: 138–148.

Babiker, M.M. and Rankin, J.C. 1978. Neurohypophysial hormonal control of kidney function in the European Eel *Anguilla anguialla* L. adapted to sea-water or fresh water. *Journal of Endocrinology* 76: 347–358.

Babiker, M.M. and Rankin, J.C. 1979. Factors regulating the functioning of the in vitro perfused aglomerular kidney of the anglerfish, *Lophius piscatorius* L. *Comparative Biochemistry and Physiology* 62: 989–993.

Balment, R.J., Warne J.M., Tierney M. and Hazon N. 1993. Arginine vasotocin and fish osmoregulation. *Fish Physiology and Biochemistry* 11 (1–6): 189–194.

Banerjee, S. and Bhattacharya, S. 1994. Histopathology of kidney of *Channa punctatus* exposed to chronic nonlethal level of Elsan, mercury, and ammonia. *Ecotoxicology and Environmental Safety* 29: 265–275.

Bansal, S.K. and Chandra, S.V. 1985. The in vitro effect of chlordecone and mirex on Ca^{2+}-activated ATPase in the teleost *Saccobranchus fossilis*. *Aquatic Toxicology* 6: 37–44.

Baustian, M.D. and Beyenbach, K.W. 1993. Isolated renal proximal tubules: Studies of transepithelial secretory mechanisms in teleost and elasmobranch fish. In *New Insights in Vertebrate Kidney Function*. Brown, J.A., Balment, R.J. and Rankin, J.C. (eds), pp. 45–63. Cambridge University Press, Cambridge, UK.

Baustian, M.D., Wang, S.Q. and Beyenbach, K.W. 1997. Adaptive responses of aglomerular toadfish to dilute sea water. *Journal of Comparative Physiology B* 167: 61–70.

Bern, H.A. and Madsen, S.S. 1992. A selective survey of the endocrine system of the rainbow trout *Oncorhynchus mykiss* with emphasis on the hormonal regulation of ion balance. *Aquaculture* 100: 237–262.

Beyenbach, K.W. 1982. Direct demonstration of fluid secretion by glomerular renal tubules in a marine teleost. *Nature* (London) 299: 54–56.

Beyenbach, K.W. 1985. Comparative physiology of the renal proximal tubule. *Renal Physiology* 8: 222–236.

Beyenbach, K.W. 1986. Secretory NaCl and volume flow in renal tubules. *American Journal of Physiology* 250: R753–R763.

Beyenbach, K.W. 1995. Secretory electrolyte transport in renal proximal tubules of fish. In *Cellular and Molecular Approaches to Fish Ionic Regulation, Fish Physiology*, Vol. 14. Wood, C.M. and Shuttleworth, T.J. (eds), pp. 85–105. Academic Press, San Diego.

Beyenbach K.W. and Baustian, M.D. 1989. Comparative physiology of the proximal tubule. from the perspective of aglomerular urine formation. In *Structure and Function of the Kidney*. Kinne, R.K.H. (ed.), pp. 103–142. Karger, Basel.

Beyenbach, K.W. and Frömter, E. 1985. Electrophysiological evidence for Cl secretion in shark renal proximal tubules. *American Journal of Physiology* 248: F282–F295.

Beyenbach, K.W. and Kirschner, L.B. 1975. Kidney and urinary bladder functions of the rainbow trout in Mg and Na excretion. *American Journal of Physiology* 229: 389–383.

Beyenbach, K.W., Freire, C.A., Kinne, R.K.H. and Kinne-Saffran, E. 1993. Epithelial transport of magnesium in the kidney of fish. *Mineral Electrolyte Metabolism* 19: 241–249.

Bijvelds, M.J.C., Flik, G., Kolar, Z.I. and Wendelaar Bonga, S.E. 1996. Uptake, distribution and excretion of magnesium in *Oreochromis mossambicus*: dependence on magnesium in diet and water. *Fish Physiology and Biochemistry* 15: 287–298.

Bijvelds, M.J.C., Van der Heiden, A.J.H., Flik, G., Verbost, P.M., Kolar, Z.I., Wendelaar Bonga, S.E. 1997. Calcium pump activities in the kidneys of *Oreochromis mossambicus*. *Journal of Experimental Biology* 198: 1351–1357.

Bijvelds, M.J.C., Van der Velden, J.A., Kolar, Z.I. and Flik, G. 1998. Magnesium transport in freshwater teleosts. *Journal of Experimental Biology* 201: 1981–1990.

Björnsson, B.T. and Nilsson, S. 1985. Renal and extra-renal excretion of calcium in the marine teleost, *Gadus morhua*. *American Journal of Physiology* 248: R18–R22.

Borski, R.J., Hansen, M.U., Nishioka, R.S. and Grau, E.G. 1992. Differential processing of the two prolactins of the Tilapia *Oreochromis mossambicus* in relation to environmental salinity. *Journal of Experimental Zoology* 264: 46–54.

Brown, J.A. and Balment, R.J. 1997. Teleost renal function: regulation by arginine vasotocin and by angiotensins. In *Ionic Regulation in Animals: a Tribute to Professor W.T.W. Potts*. Hazon, N., Eddy, F.B. and Flik, G. (eds), pp. 150–165. Springer-Verlag, Berlin.

Brown, J.A. and Oliver, J.A. 1985. The renin angiotensin system and single nephron glomerular structure and function in the trout, *Salmo gairdneri*. In *Current Trends in Comparative Endocrinology: Proceedings of the Ninth International Symposium on Comparative Endocrinology*, Vol. I. Lofts, B. and Holmes, W.N. (eds), pp. 905–909. Hong Kong University Press, Hong Kong.

Brown, J.A., Jackson, B.A., Oliver, J.A. and Henderson, I.W. 1978. Single nephron filtration rate SNGFR in the trout, *Salmo gairdneri*. Validation of the use of ferrocyanide and the effect of environmental salinity. *Pflügers Archives* 377: 101–108.

Brown, J.A., Oliver, J.A., Henderson, I.W. and Jackson, B.A. 1980. Angiotensin and single nephron glomerular function in the trout *Salmo gairdneri*. *American Journal of Physiology* 239: R509–R514.

Brown, J.A., Taylor, S.M. and Gray, G.J. 1983. Glomerular ultrastructure of the trout, *Salmo gairdneri*. *Cell Tissue Research* 230: 205–218.

Brown, J.A., Gray, C.J. and Taylor, S.M. 1990. The renin–angiotensin system and glomerular function of teleost fish. *Progress in Clinical and Biological Research* 342: 528–533.

Brown, J.A., Rankin, J.C. and Yokota, S.D. 1993. Glomerular haemodynamics and filtration in single nephrons of non-mammalian vertebrates. In *New Insights in Vertebrate Kidney Function*. Brown, J.A., Balment, R.J. and Rankin, J.C. (eds), pp. 1–44. Cambridge University Press, Cambridge.

Brown, J.A., Paley, R.K., Amer, S. and Aves, S.J. 1995. Evidence for an intrarenal renin–angiotensin system in the rainbow trout. *Journal of Endocrinology* 147 (Suppl.): 79.

Brown, M.W., Thomas, D.G., Shurben, D., de L.G. Solbe, J.F., Kay, J. and Cryer, A. 1986. A comparison of the differential accumulation of cadmium in the tissues of three species of freshwater fish, *Salmo gairdneri, Rutilus rutilus,* and *Noemacheilus barbatulus. Comparative Biochemistry and Physiology* 84C: 213–217.

Burka, J.F., Hammell, K.L., Horsberg, T.E., Johnson, G.R., Rainie, D.J. and Speare, D.J. 1997. Drugs in salmonid aquaculture – a review. *Journal of Veterinary Pharmacology and Therapeutics* 20: 333–349.

Campbell, R.D., Leadem, T.P. and Johnson, D.W. 1974. The in vivo effect of p,p'DDT on Na+-K+-activated ATPase activity in rainbow trout (*Salmo gairdneri*). *Bulletin of Environmental Contamination and Toxicology* 11: 425–428.

Chandra, S., Morrison, G.H. and Beyenbach, K.W. 1997. Identification of Mg-transporting renal tubules and cells by ion microscopy imaging of stable isotopes. *American Journal of Physiology* 273: F939–F948.

Cliff, H. and Beyenbach, K.W. 1992. Secretory renal proximal tubules in seawater- and freshwater-adapted killifish. *American Journal of Physiology* 262: F108–F116.

Cliff, W.H. and Beyenbach, K.W. 1988. Fluid secretion in glomerular renal proximal tubules of freshwater-adapted fish. *American Journal of Physiology* 254: R154–R158.

Cliff, W.H., Sawyer, D.B. and Beyenbach, K.W. 1986. Renal proximal tubule of flounder. II. Transepithelial Mg secretion. *American Journal of Physiology* 250: R616–R624.

Cobb, C.S. and Brown, J.A. 1994. Characterisation of Angiotensin II binding to glomeruli from Rainbow Trout *Oncorhynchus mykiss* adapted to fresh eater and seawater. *General and Comparative Endocrinology* 94: 104–112.

Conklin, D.J., Chavas, A., Duff, D.W., Weaver, L., Zhang, Y. and Olson, K.R. 1997. Cardiovascular effects of arginine vasotocin in the rainbow trout *Oncorhynchus mykiss. Journal of Experimental Biology* 200: 2821–2832.

Cousins, K.L. and Farrell, A.P. 1996. Stretch-induced release of atrial natriuretic factor from the heart of rainbow trout *Oncorhynchus mykiss. Canadian Journal of Zoology* 74: 380–387.

Curtis, B.J. and Wood, C.M. 1991. The function of the urinary bladder in vivo in the freshwater rainbow trout. *Journal of Experimental Biology* 155: 567–583.

Dantzler, W.H. 1989. *Comparative Physiology of the Vertebrate Kidney*. Springer-Verlag, Berlin.

Dantzler, W.H. 1996. Comparative aspects of renal organic anion transport. *Cell Physiology and Biochemistry* 6: 28–38.

Dekant, W. and Vamvakas, S. 1996. Biotransformation and membrane transport in nephrotoxicity. *Critical Reviews in Toxicology* 26: 309–334.

Demarest, J.R. 1984. Ion and water transport by the flounder urinary bladder: salinity dependence. *American Journal of Physiology* 246 (4 pt 2): F395–401.

Desaiah, D. and Koch, R.B. 1975. Toxaphene inhibition of ATPase activity in catfish, *Ictalurus punctatus*, tissues. *Bulletin of Environmental Contamination and Toxicology* 13: 238–244.

Doneen, B.A. 1976. Water and ion movements in the urinary bladder of the gobiid teleost Gillichthys mirabilis in response to prolactins and to cortisol. *General and Comparative Endocrinology* 28: 33–41.

Duff, D.W, Conklin, D.J and Olson, K.R. 1997. Effect of Atrial Natriuretic Peptide on fluid volume and glomerular filtration in rainbow trout. *Journal of Experimental Zoology* 278: 215–220.

Dunne, J.B. and Rankin J.C. 1992. Effects of atrial natriuretic peptide and angiotensin II on salt and water excretion by the perfused rainbow trout kidney. *Journal of Physiology* 446: 92.

Dutta, H.M. and Marcelino, J. 1990. Effects of malathion on kidney and skin of bluegill fish, *Lepomis macrochromis. Journal of Freshwater Biology* 2: 77–88.

Elger, B., Ruehs, H. and Hentschel, H. 1988. Glomerular permselectivity to serum proteins in rainbow trout *Salmo gairdneri*. *American Journal of Physiology* 255: R418–R423.

Elger, M. and Hentschel, H. 1981. The glomerulus of a stenohaline fresh-water teleost, *Carassius auratus gibelio*, adapted to saline water. *Cell Tissue Research* 220: 73–85.

Elger, M., Kaune, R. and Hentschel, H. 1984a. Glomerular intermittency in a freshwater teleost, *Carassius auratius gibelio*, after transfer to salt water. *Journal of Comparative Physiology B* 154: 225–231.

Elger, M., Wahlqvist, I. and Hentschel, H. 1984b. Ultrastructure and adrenergic innervation of preglomerular arterioles in the euryhaline teleost, *Salmo gairdneri*. *Cell Tissue Research* 237: 451–458.

Elger, M., Werner, A., Herter, P., Kohl, B., Kinne, R.K.H. and Hentschel, H. 1998. Na–Pi cotransport sites in proximal tubule and collecting tubule of winter flounder *Pleuronectes americanus*. *American Journal of Physiology* 274: F374–F383.

Evans, D.H. 1993. Osmotic and ionic regulation. In *The Physiology of Fishes*. Evans, D.H. (ed.), pp. 315–341. CRC Press, Boca Raton.

Fisher-Sherl, T., Veeser, A., Hoffmann, R.W., Kuhnhauser, C., Negele, R.D. and Ewringmann, T. 1991. Morphological effects of acute and chronic atrazine exposure in rainbow trout (*Oncorhynchus mykiss*). *Archives of Environmental Contamination and Toxicology* 20: 454–461.

Fletcher, T.C. and White, A. 1986. Nephrotoxic and haematological effects of mercuric chloride in the plaice (*Pleuronectes platessa* L.). *Aquatic Toxicology* 8: 77–84.

Foster, R.C. 1975. Changes in urinary bladder and kidney function in the starry flounder *Platichthys stellatus* in response to prolactin and to freshwater transfer. *General and Comparative Endocrinology* 27: 157–161.

Fuentes, J. and Eddy, F.B. 1997a. Drinking in Atlantic salmon presmolts and smolts in response to growth hormone and salinity. *Comparative Biochemistry Physiology* 117A: 487–491.

Fuentes, J. and Eddy, F.B. 1997b. Effect of manipulation of the renin–angiotensin system in control of drinking in juvenile Atlantic salmon *Salmo salar* in fresh water and after transfer to sea water. *Journal of Comparative Physiology B* 167: 438–443.

Fuentes, J., Soengas, J.L., Buceta, M., Otero, J., Rey, P. and Rebolledo, E. 1996. Kidney ATPase response in seawater-transferred rainbow trout *Oncorhynchus mykiss*. Effect of salinity and fish size. *Journal of Physiology and Biochemistry* 52: 231–238.

Fuentes, J., Soengas, P.R. and Rebolledo, E. 1997. Progressive transfer to seawater enhances intestinal and branchial Na+-K+-ATPase activity in non-anadromous rainbow trout. *Aquaculture International* 5: 217–227.

Gill, T.S., Pant, J.C. and Tewari, H. 1988. Branchial and renal pathology in the fish exposed chronically to methoxy ethyl mercuric chloride. *Bulletin of Environmental Contamination and Toxicology* 41: 241–246.

Gill, T.S., Pant, J.C. and Tewari, H. 1989. Cadmium nephropathy in a freshwater fish, *Puntius conchonius*, Hamilton. *Ecotoxicology and Environmental Safety* 18: 165–172.

Gill, T.S., Tewari, H. and Pande, J. 1991a. *In vivo* and *in vitro* effects of cadmium on selected enzymes in different organs of the fish *Barbus conchonius* Ham. (rosy barb). *Comparative Biochemistry and Physiology C* 100C: 501–505.

Gill, T.S., Pande, J. and Tewari, H. 1991b. Hemopathological changes associated with experimental aldicarb poisoning in fish (*Puntius conchonius* Hamilton). *Bulletin of Environmental Contamination and Toxicology* 47: 628–633.

Glynn, A.W., Haux, C. and Hogstrand, C. 1992. Chronic toxicity and metabolism of Cd and Zn in juvenile minnows (*Phoxinus phoxinus*) exposed to Cd and Zn mixture. *Canadian Journal of Fisheries and Aquatic Science* 49: 2070–2079.

Goldstein, R.S. and Schnelmann, R.G. 1996. Toxic responses of the kidney. In *Cassaert and Dou's Toxicology*, 5th edn. Klaassen, C. (ed.), pp. 417–442. McGraw Hill, New York.

Gray, C.J. and Brown, J.A. 1985. Renal and cardiovascular effects of angiotensin II in the rainbow trout, *Salmo gairdneri*. *General and Comparative Endocrinology* 59: 375–381.

Gray, C.J. and Brown, J.A. 1987. Glomerular ultrastructure of the trout, *Salmo gairdneri*: effects of angiotensin II and adaptation to seawater. *Cell Tissue Research* 249: 437–442.

Gruppuso, P.A. and Kinter, L.B. 1973. DDT inhibition of active chlorophenol red transport in goldfish (*Carassius auratus*) renal tubules. *Bulletin of Environmental Contamination and Toxicology* 10: 181–186.

Harder, W. 1975. *Anatomy of Fishes*. E. Schweizerbart'sche Verlagsbuchhandlung, Stuttgart.

Harding, K.E. and Balment, R.J. 1995. Effects of AVT on cAMP accumulation in renal tubules of teleost fish. *Journal of Endocrinology* 147 (Suppl.): 60.

Harding, K.E., Warne, J.M., Hyodo, S. and Balment, R.J. 1997. Pituitary and plasma AVT content in the flounder *Platichthys flesus*. *Fish Physiology and Biochemistry* 17: 357–362.

Heisler, N. 1986. Acid–base regulation in fishes. In *Acid–base regulation in Animals*, pp. 309–356. Elsevier Science Publishers, Amsterdam.

Henderson, I.W. and Wales, N.A.M. 1974. Renal diuresis and antidiuresis after injections of arginine vasotocin in the freshwater eel *Anguilla anguilla*. *Journal of Endocrinology* 61: 487–500.

Henderson, I.W., Brown, J.A. and Balment, R.J. 1993. The renin–angiotensin system and volume homeostasis. In *New Insights in Vertebrate Kidney Function*. Brown, J.A., Balment, R.J. and Rankin, J.C. (eds), pp. 311–350. Cambridge University Press, Cambridge.

Hentschel, H. and Elger, M. 1987. The distal nephron in the kidney of fishes. In *Advances in Anatomy, Embryology and Cell Biology Series*, Vol. 108. Beck, F., Hild, W., Kriz, W., Ortmann, R., Pauly, J.E. and Schiebler, T.H. (eds). Springer-Verlag, Berlin.

Hentschel, H. and Elger, M. 1989. Morphology of glomerular and aglomerular kidneys. In *Structure and Function of the Kidney*. Kinne, R.K.H. (ed.), pp. 1–72. Karger, Basel.

Hickman, C.P. 1968. Ingestion, intestinal absorption and elimination of sea water and salts in the southern flounder, *Paralichthyes lethostigma*. *Canadian Journal of Zoology* 46: 457–466.

Hickman, Jr, C.P. and Trump B.F. 1969. The kidney. In *Fish Physiology*, Vol. 1. Hoar, W.S. and Randall, D.J. (eds), pp. 91–239. Academic Press. New York.

Hirano, T. 1986. The spectrum of prolactin action in teleosts. In *Comparative Endocrinology: Development and Directions*. Ralph, C.L. (ed.), pp. 53–74. A.R. Liss. New York.

Hirano, T., Ogasawara, T., Bolton, J.P., Collie, N.L., Hasegawa, S. and Iwata, M. 1987. Osmoregulatory role of prolactin in lower vertebrates. In *Comparative Physiology of Environmental Adaptations*. 1. *Adaptations to Salinity and Dehydration*. Kirsch, R. and Lahlou, B. (eds), pp. 112–124. Karger, Basel.

Hirsch, R., Ternes, T., Haberer, K. and Kratz, K.-L. 1999. Occurrence of antibiotics in the aquatic environment. *The Science of the Total Environment* 225: 109–118.

Hochachka, P.W., Moon, T.W., Bailey, J. and Hulbert, W.C. 1977. The osteoglossid kidney: correlations of structure, function and metabolism with transition to air breathing. *Canadian Journal of Zoology* 56: 820–832.

Hontella, A., Daniel, C. and Ricard, A.C. 1996. Effects of acute and subacute exposures to cadmium on the interrenal and thyroid function in rainbow trout, *Oncorhynchus mykiss*. *Aquatic Toxicology* 35: 171–182.

Jones, J., Kinnel, M., Christenson, R. and Reimschuessel, R. 1997. Gentamicin concentrations in toadfish and goldfish serum. *Journal of Aquatic Animal Health* 9: 211–215.

Karnaky, Jr., K.J. 1998. Osmotic and ionic regulation. In *The Physiology of Fishes*, 2nd edn. Evans, D.H. (ed.), pp. 157–176. CRC Press, Boca Raton.

Kaune, R. and Hentschel, H. 1987. Stimulation of renal phosphate secretion in the stenohaline freshwater teleost *Carassius auratus* gibelio bloch. *Comparative Biochemistry and Physiology* 87A: 359–362.

King, P.A. and Goldstein, L. 1985. Renal excretion of nitrogenous compounds in vertebrates. *Renal Physiology* 8 (4–5): 261–278.

Kinne-Saffran, E., Hulseweh, M., Pfaff, C. and Kinne, R.K. 1993. Inhibition of Na,K-ATPase by cadmium: different mechanisms in different species. *Toxicology and Applied Pharmacology* 121: 22–29.

Kowarsky, J. 1973. Extra-branchial pathways of salt exchange in a teleost fish. *Comparative Biochemistry and Physiology* 46A: 477–486.

Kumar, S. and Pant, S.C. 1984. Organal damage caused by aldicarb to a freshwater teleost *Barbus conchonius* Hamilton. *Bulletin of Environmental Contamination Toxicology* 33: 50–55.

Larsen, B.K. and Jensen, F.B. 1997. Influence of ionic composition on acid–base regulation in rainbow trout (*Oncorhynchus mykiss*) exposed to environmental hypercapnia. *Fish Physiology and Biochemistry* 16: 157–170.

Laurent, P. and Perry S.F. 1990. Effects of cortisol on gill chloride cell morphology and ionic uptake in the freshwater trout, *Salmo gairdneri*. *Cell Tissue Research* 259: 429–442.

Lee, J. and Malvin, R.L. 1987. Natriuretic response to homologous heart extract in aglomerular toadfish. *American Journal of Physiology* 252: R1055–R1058.

Lee, S.-H. and Pritchard, J.B. 1983. Role of the electrochemical gradient for Na$^+$ in D-glucose transport by mullet kidney. *Journal of Membrane Biology* 75: 171–178.

McCormick, S.D. 1995. Hormonal control of gill Na$^+$, K$^+$, ATPase and chloride cell function. In *Cellular and Molecular Approaches to Fish Ionic Regulation*, *Fish Physiology*, Vol. 14. Wood, C.M. and Shuttleworth, T.J. (eds), pp. 285–315. Academic Press, San Diego.

McDonald, M.D. and Wood, C.M. 1998. Reabsorption of urea by the kidney of the freshwater rainbow trout. *Fish Physiology and Biochemistry* 18: 375–386.

Madsen, S.S. 1990. Enhanced hypoosmoregulatory response to growth hormone after cortisol treatment in immature rainbow trout, *Salmo gairdneri*. *Fish Physiology and Biochemistry* 8 (4): 271–279.

Madsen, S.S. and Bern, H.A. 1992. Antagonism of prolactin and growth hormone: impact on seawater adaptation in two salmonids, *Salmo trutta and Oncorhynchus mykiss*. *Zoological Science* 9 (4): 775–784.

Madsen, S.S., McCormick, S.D., Young, G., Endersen, J.S., Nishioka, R.S. and Bern, H.A. 1994. Physiology of seawater acclimation in the striped bass, *Morone saxatilis* Walbaum. *Fish Physiology and Biochemistry* 13: 1–11.

Madsen, S.S., Jensen, M.K., Noehr, J. and Kristiansen, K. 1995. Expression of Na$^+$-K$^+$-ATPase in the brown trout, *Salmo trutta*: in vivo modulation by hormones and seawater. *American Journal of Physiology* 269: R1339–R1345.

Maetz, J., Bourguet, J., Lahlou, B. and Hourdry, J. 1964. Peptides neurohypophysaires et osmoregulation chez *Carassius auratus*. *General and Comparative Endocrinology* 4: 508–522.

Maitani, T., Cuppage, F.E. and Klassen, C.D. 1988. Nephrotoxicity of intravenously injected cadmium-metallothionein: critical concentration and tolerance. *Fundamental and Applied Toxicology* 10: 98–108.

Manen, C., Scmidt-Nielson, B. and Russell, D.H. 1976. Polyamine synthesis in liver and kidney of flounder in response to methylmercury. *American Journal of Physiology* 231: 560–564.

Marshall, W.S. 1995. Transport processes in isolated teleost epithelia: Opercular epithelium and urinary bladder. In *Cellular and Molecular Approaches to Fish Ionic Regulation, Fish Physiology*, Vol. 14. Wood, C.M. and Shuttleworth, T.J. (eds), pp. 1–23. Academic Press, San Diego.

Masereeuw, R., Russel, F.G.M. and Miller, D.S. 1996. Multiple pathways of organic anion secretion in renal proximal tubule revealed by confocal microscopy. *American Journal of Physiology* 271 (*Renal Fluid Electrolyte Physiology* 40): F1173–F1182.

Melgar, M.J., Perez, M., Alonso, J. and Miguez, B. 1997. The toxic and accumulative effects of short-term exposure to cadmium in rainbow trout (*Oncorhynchus mykiss*). *Veterinary and Human Toxicology* 39: 79–83.

Miles, H.M. 1971. Renal function in migrating adult Coho salmon. *Comparative Biochemistry and Physiology* 38A: 787–826.

Miller, D.S. and Pritchard, J.B. 1997. Dual pathways for organic anion secretion in renal proximal tubule. *Journal of Experimental Zoology* 279: 462–470.

Mophatra, B.C. and Noble, A. 1992. Liver and kidney damage in grey mullet *Liza parsia* (Hamilton and Buchanan) on exposure to an organophosphate 'Nuvan'. *Journal of the Marine Biology Association, India* 34: 218–221.

Nash, J. 1931. The numbers and size of glomeruli in the kidneys of fishes, with observations on the renal morphology of the renal tubules of fish. American Journal of Anatomy 47: 425–445.

Natochin, Y.V., and Gusev, G.P. 1970. The coupling of magnesium secretion to sodium reabsorption in the kidney of teleost. *Comparative Biochemistry and Physiology* 37: 107–111.

Nishimura, H. 1985. Endocrine control of renal handling of solutes and water in vertebrates. *Renal Physiology* 8: 279–300.

Nishimura, H. and Bailey, J.R. 1982. Intrarenal renin–angiotensin system in primitive vertebrates. *Kidney International* 22 (Suppl. 12): S185–S192.

Nishimura, H. and Imai, M. 1982. Control of renal function in freshwater and marine teleost. *Federation Proceeding* 41: 2355–2360.

Nishimura, H., Imai, M. and Ogawa, M. 1983. Sodium chloride and water transport in the renal distal tubule of the rainbow trout. *American Journal of Physiology* 244: F247–F254.

Olivereau, M. and Olivereau, J. 1977. Effects of hypophysectomy and prolactin replacement in eel kidney structure during adaptation to sea water. *Acta Zoologica* 58: 103–115.

Olson, K.R. and Duff, D.W. 1992. Cardiovascular and renal effects of eel and rat atrial natriuretic peptide in rainbow trout, *Salmo gairdneri. Journal of Comparative Physiology B* 162: 408–415.

Olson, K.R., Chavez, A., Conklin, D.J., Cousins, K.L., Farrell, A.P., Ferlic, R., Keen, J.E., Kne, T., Kowalski, K.A. and Veldman, T. 1994. Localization of angiotensin II responses in the trout cardiovascular system. *Journal of Experimental Biology* 194: 117–138.

Olsson, P., Kling, P., Peterson, C. and Silversand, C. 1995. Interaction of cadmium and oestradiol-17β on metallothionein and vitellogenin synthesis in rainbow trout (*Oncorhynchus mykiss*). *Biochemical Journal* 307: 197–203.

Oronsaye, J.A.O. 1989. Histological changes in the kidneys and gills of the stickleback, *Gasterosteus aculeatus* L., exposed to dissolved cadmium in hard water. *Ecotoxicology and Environmental Safety* 17: 279–290.

Oulmi, Y., Negele, R.D. and Braunbeck, T. 1995a. Segment specificity of the cytological response in rainbow trout (*Oncorhynchus mykiss*) renal tubules following prolonged exposure to sublethal concentrations of atrazine. *Ecotoxicology and Environmental Safety* 32: 39–50.

Oulmi, Y., Negele, R.D. and Braunbeck, T. 1995b. Cytopathology of liver and kidney in rainbow trout (*Oncorhynchus mykiss*) after long-term exposure to sublethal concentrations of linuron. *Diseases of Aquatic Organisms* 21: 35–52.

Pandey, A.K. 1994. Branchial and renal lesions in the estuarine mullet, *Liza parsia*, exposed to sublethal concentration of mercury. *Proceedings of the National Academy of Science, India* 64B: 281–287.

Pelayo, J.C. and Shanley, P.F. 1990. Glomerular and tubular adaptive responses to acute nephron loss in the rat. Effect of prostaglandin synthesis inhibition. *Journal of Clinical Investigation* 85 (6): 1761–1769.

Perkins, E.J., El-Alfy, A. and Schlenk, D. 1999. *In Vitro* sulfoxidation of aldicarb by hepatic microsomes of channel catfish, *Ictalurus punctatus*. *Toxicological Sciences* 48: 67–73.

Perrott, M.N., Sainsbury, R.J. and Balment, R.J. 1993. Peptide hormone-stimulated second messenger production in the teleostean. *General Comparative Endocrinology* 89: 387–395.

Poleksic', V., Karan, V., Dulic', Z., Elezovic', I. and Naskovic, N. 1997. Herbicide toxicity to fish: histopathological effects. *Pesticides* 12: 257–268.

Pritchard, J.B. and Bend, J.R. 1984. Mechanisms controlling the renal excretion of xenobiotics in fish: effects of chemical structure. *Drug Metabolism Reviews* 15: 655–671.

Pritchard, J.B. and Miller, D.S. 1980. Teleost kidney in evaluation of xenobiotic toxicity and elimination. *Federation Proceedings* 39: 3207–3212.

Rana, S.V.S., Singh, R. and Verma, S. 1995. Mercury-induced lipid peroxidation in the liver, kidney, brain, and gills of a fresh water fish, *Channa punctatus*. *Japanese Journal of Ichthyology* 42: 255–259.

Rankin, J.C. and Davenport, J. 1981. *Animal Osmoregulation*. Wiley and Sons, New York.

Rankin, J.C., Henderson, I.W. and Brown, J.A. 1983. Osmoregulation and the control of kidney function. In *Control Processes in Fish Physiology*. Rankin, J.C., Pitcher, T.J. and Duggan, R.T. (eds), pp. 66–88. Wiley and Sons. New York.

Rankin, J.C., Wahlqvist, I. and Wallace, B. 1984. Antidiuretic actions of angiotensin II, catecholamines and neurohypophysial hormones in the in situ perfused rainbow trout kidney. *General and Comparative Endocrinology* 53: 442.

Reinschuessel, R. and Gonzalez, C.M. 1998. Renal alterations following sublethal mercury toxicity: a fish model for aquatic environmental contamination. In *Advances in Animal Alternatives for Safety and Efficacy Testing*. Salem, H. and Katz, S.A. (eds), pp. 399–401. Taylor & Francis, Washington, DC.

Reimschuessel, R. and Williams, D. 1995. Development of new nephrons in adult kidney following gentamicin-induced nephrotoxicity. *Renal Failure* 17: 101–106.

Reimschuessel, R., Bennett, R.O., May, E.B. and Lipsky, M.M. 1989. Renal histopathological changes in the goldfish (*Carassius auratus*) after sublethal exposure to hexachlorobutadiene. *Aquatic Toxicology* 15: 169–180.

Reimschuessel, R., Bennett, R.O., May, E.B. and Lipsky, M.M. 1990. Development of newly formed nephrons in the goldfish kidney following hexachlorobutadiene-induced nephrotoxicity. *Toxicology and Pathology* 18: 32–38.

Reimschuessel, R., Bennett, R.O., May, E.B. and Lipsky, M.M. 1993. Pathological alterations and new nephron development in rainbow trout (*Oncorhynchus mykiss*) following tetrachloroethylene contamination. *Journal of Zoo and Wildlife Medicine* 24: 503–507.

Renfro, J.L. 1980. Relationship between renal fluid and Mg secretion in a glomerular marine teleost. *American Journal of Physiology* 238 (7): F92–F98.

Renfro, J.L. 1989. Adaptability of marine teleost renal inorganic sulfate excretion: evidence for glucocorticoid involvement. *American Journal Physiology* 257: R511–R516.

Renfro, J.L. 1995. Solute transport by flounder renal cells in primary culture. In *Cellular and Molecular Approaches to Fish Ionic Regulation, Fish Physiology*, Vol. 14. Wood, C.M. and Shuttleworth, T.J. (eds), pp. 147–171. Academic Press, San Diego.

Renfro, J.L. and Pritchard, J.B. 1983. Sulfate transport by flounder renal tubule brush border: Presence of anion exchange. *American Journal of Physiology* 243: F150–F159.

Renfro, J.L., Dickman, K.G. and Miller, D.S. 1982. Effect of sodium ion and ATP on peritubular calcium transport by the marine teleost renal tubule. *American Journal of Physiology* 243: R34–R41.

Roesijadi, G.K. and Robinson, W.E. 1994. Metal regulation in aquatic animals: mechanisms of uptake, accumulation, and release. In *Toxicology Aquatic: Molecular, Biochemical, and Cellular Perspectives*. Malins, D.C. and Ostrander, G. (eds), pp. 387–420. Lewis Publishers, Boca Raton.

Ross, A. 1989. Nuvan use in salmon farming: the antithesis of the precautionary principle. *Marine Pollution Bulletin* 20: 372–374.

de Ruiter, A.J.H. 1980. Changes in glomerular structure after sexual maturation and seawater adaptation in males of the euryhaline teleost *Gasterosteus aculeatus* L. *Cell Tissue Research* 206: 1–20.

Safer, A.M.A., Tytler, P. and El-Sayed, N. 1982. The structure of the head kidney in the Mudskipper, *Periophthalmus koelreuteri* Pallas. *Journal of Morphology* 174: 121–131.

Sakamoto, T., Shepard, B.S., Madsen, S.S., Nishioka, R.S., Siharath, K., Richman, N.H., Bern, H.A. and Grau, E.G. 1997. Osmoregulatory actions of growth hormone and prolactin in an advanced teleost. *General and Comparative Endocrinology* 106: 95–101.

Sastry, K.V. and Sharma, S.K. 1978. The effect of in vivo exposure of endrin on the activities of acid, alkaline and glucose-6-phosphatases in liver and kidney of *Ophiocephalus* (Channa) *punctatus*. *Bulletin of Environmental Contamination and Toxicology* 20: 456–460.

Sastry, K.V. and Sharma, S.K. 1979a. Endrin toxicosis on few enzymes in liver and kidney of *Channa punctatus* (Bloch). *Bulletin of Environmental Contamination and Toxicology* 22: 4–8.

Sastry, K.V. and Sharma, S.K. 1979b. In vivo effect of endrin on three phosphatases in kidney and liver of the fish *Ophiocephalus punctatus*. *Bulletin of Environmental Contamination and Toxicology* 21: 185–189.

Sastry, K.V. and Sharma, K. 1981. Diazinon-induced histopathological and hematological alterations in a freshwater teleost, *Ophiocephalus punctatus*. *Ecotoxicology and Environmental Safety* 5: 329–340.

Sawyer, D.B. and Beyenbach, K.W. 1985. Mechanism of fluid secretion in isolated shark renal proximal tubules. *American Journal of Physiology* 249: F884–F890.

Scherer, R., McNicol, R.E. and Evans, R.E. 1997. Impairment of lake trout foraging by chronic exposure to cadmium: a black-box experiment. *Aquatic Toxicology* 37: 1–7.

Schlenk, D. and Rice, C.D. 1998. Effect of zinc and cadmium treatment on hydrogen peroxide-induced mortality and expression of glutathione and metallothionein in a teleost hepatoma cell line. *Aquatic Toxicology* 43: 121–129.

Schlenk, D., Erickson, D.A., Lech, J.L. and Buhler, D.R. 1992. The distribution, elimination, and *in vivo* biotransformation of aldicarb in the rainbow trout (*Oncorhynchus mykiss*). *Fundamental and Applied Toxicology* 18: 131–136.

Schmidt-Nielsen, B. and Renfro, J.L. 1975. Kidney function of the American Eel *Anguilla anguilla*. *American Journal of Physiology* 228: 420–431.

Schröck, H., Forster, R.P. and Goldstein, L. 1982. Renal handling of taurine in marine fish. *American Journal of Physiology* 242: R64–R69.

Schultz, I.R., Peters, E.L. and Newman, M.C. 1996. Toxicokinetics and disposition of inorganic mercury and cadmium in channel catfish after intravascular administration. *Toxicology and Applied Pharmacology* 149: 39–50.

Seidelin, M. and Madsen, S.S. 1997. Prolactin antagonizes the seawater-adaptive effect of cortisol and growth hormone in anadromous brown trout *Salmo trutta*. *Zoological Science* 14: 249–256.

Sharma, M., Sharma, R., Greene, A.S., McCarthy, E.T. and Savin, V.J. 1998. Documentation of angiotensin II receptors in glomerular epithelial cells. *American Journal of Physiology* 274 *(Renal Physiology* 43): F623–F627.

Sharma, R.M. 1988. Effect of endosulfan on adenosine triphosphatase (ATPase) activity in liver, kidney, and muscles of *Channa gachua*. *Bulletin of Environmental Contamination and Toxicology* 41: 317–323.

Singhal, R.N. and Jain, M. 1997. Cadmium-induced changes in the histology of kidneys in common carp, *Cyprinus carpio* (Cyprinidae). *Bulletin of Environmental Contamination and Toxicology* 58: 456–462.

Smith, B.P., Hejtmanick, E. and Camp, B.J. 1976. Acute effects of cadmium on *Ictalurus punctatus* (catfish). *Bulletin of Environmental Contamination and Toxicology* 15: 271–277.

Spradley, J.P. 1991. *Toxicity of Pesticides to Fish*. Arkansas Cooperative Extension Service. MP330-3M-7-91. University of Arkansas, Fayetteville, AR.

Srivastava, D.K. 1982. Comparative effects of copper, cadmium and mercury on tissue glycogen of the catfish, *Heteropneustes fossils* (Bloch). *Toxicology Letters* 11: 135–139.

Srivastava, S.K., Tiwari, P.R. and Srivasty, A.K. 1990. Effects of chlorpyrifos on the kidney of freshwater catfish, *Heteropneustes fossilis*. *Bulletin of Environmental Contamination and Toxicology* 45: 748–751.

Stohs, S.J. and Bagchi, D. 1995. Oxidative mechanisms in the toxicity of metal ions. *Free Radical Biology and Medicine* 18: 321–336.

Stoner, L.C. 1985. The movement of solutes and water across the vertebrate distal nephron. *Renal Physiology* 8: 237–248.

Suresh, A., Sirvaramakrishna, B. and Radhakrishnaiah, K. 1993. Patterns of accumulation in the organs of fry and fingerlings of freshwater fish *Cyprinus carpio* following cadmium exposure. *Chemosphere* 26: 945–953.

Takei, Y. and Balment, R.J. 1993. Natriuretic factors in non-mammalian vertebrates. In *New Insights in Vertebrate Kidney Function*. Brown, J.A., Balment, R.J. and Rankin, J.C. (eds), pp. 350–385. Cambridge University Press, Cambridge.

Talbot, C., Stagg, R.M. and Eddy, F.B. 1992. Renal, respiratory and ionic regulation in Atlantic salmon *Salmo salar* L. kelts following transfer from fresh water to seawater. *Journal of Comparative Physiology B* 162: 358–364.

Thomas, D.G., Brown, M.W., Shurben, D., Solbe, J.F., Cryer, A. and Kay, J. 1985. A comparison of the sequestration of cadmium and zinc in the tissues of rainbow trout *(Salmo gairdneri)* following exposure to the metals singly or in combination. *Comparative Biochemistry and Physiology C* 82: 55–62.

Trump, B.F. and Jones, R.T. 1975. Cellular effects of mercury on fish kidney tubules. In *Proceedings of the Pathology of Fishes Conference*, pp. 585–612. University of Wisconsin Press, Wisconsin.

Tytler, P. 1988. Morphology of the pronephros of the juvenile brown trout, *Salmo trutta*. *Journal of Morphology* 195: 189–204.

USEPA (United States Environmental Protection Agency). 1998. Update: Listing of fish and wildlife advisories. United States Environmental Protection Agency Fact Sheet EPA-823-F-98-009.

USGS (US Geological Survey). 1997. National and study unit pesticide use. National Water Quality Assessment Pesticide National Synthesis Project. US Geological Survey, Department of the Interior, Washington, DC.

Verma, S.R., Gupta, A.K., Bansal, S.K. and Dalela, R.C. 1978. In vitro disruption of ATP dependent active transport following treatment with aldrin and its epoxy analog dieldrin in a fresh water teleost, *Labeo rohita*. *Toxicology* 11: 193–201.

Wales, N.A.M. 1984. Vascular and renal actions of salmon calcitonin in freshwater- and seawater-adapted European eels *(Anguilla anguilla)*. *Journal of Experimental Biology* 113: 381–387.

Warne, J.M. and Balment, R.J. 1997. Vascular actions of neurohypophysial peptides in the flounder. *Fish Physiology and Biochemistry* 17: 313–318.

Watras C.J. and Huckabee J.W. 1994. *Mercury Pollution: Integration and Synthesis.* CRC Press, Boca Raton, FL.

Wendelaar Bonga, S.E. 1976. The effect of prolactin on kidney structure of the euryhaline teleost *Gasterosteus aculeatus* during adaptation to freshwater. *Cell Tissue Research* 166: 319–338.

Werner, A., Murer, H. and Kinne, R. 1994. Cloning and expression of a renal Na–Pi cotransport system from flounder. *American Journal of Physiology* 267 (*Renal Fluid Electrolyte Physiology*): F311–F317.

Werner, M. and Costa, M.J. 1995. Nephrotoxicity of xenobiotics. *Clinica Chimica Acta* 237: 107–154.

Wood, C.M. 1993. Ammonia and urea metabolism and excretion. In *The Physiology of Fishes.* Evans, D.H. (ed.), pp. 379–425. CRC Press, Boca Raton.

Yada T., Hirano, T. and Grau, E.G. 1994. Changes in plasma levels of the two prolactins and growth hormone during adaptation to different salinities in the euryhaline Tilapia, *Oreochromis mossambicus. General and Comparative Endocrinology* 93: 214–223.

Yokota, S.D., Benyajati, S. and Dantzler, W.H. 1985. Comparative aspects of glomerular filtration in vertebrates. *Renal Physiology* 8: 193–221.

Zhang, Y.S. and Schlenk, D. 1995. Induction and characterization of hepatic metallothionein expression from cadmium-induced channel catfish (*Ictalurus punctatus*). *Environmental Toxicology and Chemistry* 14: 1425–1431.

Zonno, V., Vilella, S. and Storelli, C. 1994. Salinity dependence of Na^+/H^+ exchange activity in the eel *Anguilla anguilla* renal brush border membrane vesicles. *Comparative Biochemistry Physiology* 107A: 133–140.

3 Toxic responses of the skin

James M. McKim and Gregory J. Lien

Introduction

The importance of fish skin is realized when one considers that it is the interface between the external and internal environment of the animal. Forming the body's first line of defense, it comes into direct contact with all waterborne toxic chemicals, parasites, and disease organisms. Skin is one of the largest organs in a fish, making up approximately 10 percent of the body weight, and offers a complex surface that is responsible for maintaining the integrity and constancy of the *milieu interieur*. Morphologic and functional differences can be great as there are 20 000 species of fish occupying a multitude of habitats in both fresh and saltwater. Many variations in skin composition have evolved to fulfill specific needs of each species, such as sensory organs within the skin, special epidermal and dermal structures and cell types, along with the unique biochemical machinery of the various cell types. Regardless of these many subtle and not so subtle differences, there are basically two major skin layers, the epidermis and dermis with an underlying hypodermis or subcutis recognized in most species. In contrast to the dry, hard, and rather impermeable keratinized skin of mammals, fish skin is continually hydrated, unkeratinized and covered completely by a layer of slimy mucus. Because of the unkeratinized, hydrated nature of fish skin, it can be quite sensitive to waterborne chemicals and physical stressors.

At this time, there are no toxicity tests designed specifically for fish skin as there are for mammals, and therefore no routine use of fish skin as a target organ or endpoint in the development of water quality criteria or environmental risk assessments. In most instances with fish it becomes difficult to assess accurately the acute (i.e. 96-h LC_{50}), local dermal toxicity of waterborne irritants or caustic chemicals (e.g. metals, detergents, chlorine, acid, etc.) because the gills are also exposed and the thin lamellae of the respiratory surface and its function in gas exchange are, in most cases, more sensitive to acutely lethal concentrations of these types of chemicals. Even though many of the same gill epidermal responses to toxic chemicals also occur within the skin epithelium, most acute lethal toxicity (96-h LC_{50}) of waterborne caustic chemicals has been linked to damage of the respiratory surface and loss of respiratory and osmoregulatory function (Heath, 1987). However, skin can be the primary target organ in aqueous exposures of air-breathing fish of the tropics, larval fish utilizing skin respiration during development, and bottom-dwelling species where their skin comes into direct contact with contaminated sediment and its pore water. The toxic impacts of non-caustic chemicals on fish skin, both directly and indirectly, are not well understood and remain an important area for future research. These relationships must be considered when evaluating the impact of waterborne chemicals and determining toxicity and target organ sensitivity.

The lethal and sublethal toxicity of chemicals to fish skin or any other target organ, with the exception of fish gills, has not been examined extensively by aquatic toxicologists for use in risk assessment. Up to this point, work in aquatic toxicology has been tied more to impacts on the survival, growth, and reproductive aspects of the life cycle of a species, with little effort on a specific target organ or organs directly or indirectly responsible for the toxic effect. Historically, more emphasis on target organ toxicity and its use in risk assessment has been in the field of mammalian toxicology.

Target organ toxicity studies in fish are dominated by work on the sensitive gill epithelium responsible for the vital function of respiratory gas exchange and the regulation of ion balance and acid/base control (see Chapter 1, this volume). However, because of its direct contact with the environment, many important functions, sensitivity, and visibility to toxicologists, the skin has received considerable attention in structure–function laboratory studies and in field studies on skin tumor epizootics. Unfortunately, at this time, the use of the current skin toxicity data set is limited by the lack of a well-developed skin dose–response paradigm.

As will be pointed out in this chapter, fish skin has a number of vital functions, many of which could be life threatening if perturbed beyond certain limits. These limits must be better explored and understood with respect to this organ's normal function before any consistent use of skin as a target organ in aquatic toxicity and environmental risk assessment. We hope this chapter will begin to lay a foundation from which this understanding of skin as a target organ and chemical exchange surface in aquatic toxicology can grow.

Fish skin's contribution to toxic responses can occur in several ways: (1) skin can itself be the major target organ and receive direct acute or chronic damage; (2) skin can act primarily as an exchange surface that is responsible for facilitating the entry of chemicals by diffusion directly across the skin into the capillary system of the dermis where it can enter the systemic circulation and be distributed to a specific target organ(s); (3) chemoreceptors within the epidermis of fish are in many cases very sensitive to waterborne chemicals and can have an ecologic impact on the ability of certain fish to locate food (taste and smell), to reproduce (pheromones), or to detect clues necessary for migration; (4) direct damage to specific epidermal areas that are in continuous direct contact with contaminated sediment and its pore water; and (5) direct toxicity can occur by a combination of any two or more of the above possibilities.

The first goal of this chapter is to acquaint the reader with the normal histologic structure–function of fish skin and its physiologic and toxicokinetic functions. The second goal is to present what is known about the toxicity of specific chemicals, the types of lesions that they cause, and the impact that they have on normal skin structure–function. This will be followed by a brief discussion of future skin toxicology research needs.

Structure–function of fish skin

Structure

Many styles, variations, and adaptations in the structure of fish skin are observed across the numerous fish species that make up this highly diverse group of aquatic vertebrates. However, there is a basic structural design for fish skin that seems to hold across the many species and their unique life histories. A number of extensive research papers and reviews are available that cover the detailed histologic structure and chemistry of the epidermis and dermis of fish skin, that describe how the cell types in these major skin layers differ between species (Henrickson and Matoltsy, 1968a,b,c; Bullock and Roberts, 1975; Leonard and Summers, 1976; Whitear, 1986a,b), and that characterize the dermatopathology of diseases of fish integument (Roberts and Bullock, 1976). However, for our purposes here, we will provide a general overview of fish skin structure that can be used by aquatic toxicologists in evaluating the impact of waterborne chemicals and other environmental stressors on fish skin. For this overview, we selected the rainbow trout (*Oncorhynchus mykiss*) and the channel catfish (*Ictalurus punctatus*) on the basis of: (1) common use in aquatic toxicity tests, (2) basic descriptions of skin histology are available (Hawkes, 1974a,b; Yokote, 1982; Grizzle and Rogers, 1976), and (3) trout are scaled and catfish are scaleless.

The skin of both rainbow trout and channel catfish is composed of two distinct layers (epidermis and dermis) overlying the muscle, as reported for other vertebrates (Hawkes, 1974a,b; Grizzle and Rogers, 1976). In detail, the skin of these two species differs considerably; however, the general structure is quite similar (Figures 3.1 and 3.2). Fish skin varies in thickness according to age, sex, maturation, and the part of the body that it covers. The combined thickness of the epidermis and dermis is similar in a 1-kg trout (1.12 mm) and a channel catfish (1.02 mm); however, the catfish epidermis (0.50 mm) is thicker than that of trout (0.06 mm), whereas trout dermis (1.06 mm) is thicker than that of catfish (0.52 mm) (McKim *et al.*, 1996). Because of the protection received by the scales, scaled fish have less need of a thick epidermis, yet the scale pockets in the dermis require more space. Conversely, the scaleless catfish requires a thicker epidermis with a large supply of mucous goblet cells and club cells for better protection at the skin surface, and a thinner dermis is acceptable as no scale formation or regeneration is required.

Epidermis

The outer skin layer or epidermis of both species consists of squamous or cuboidal cells that contain filamentals called tonofibrils that tie the epithelial cells together

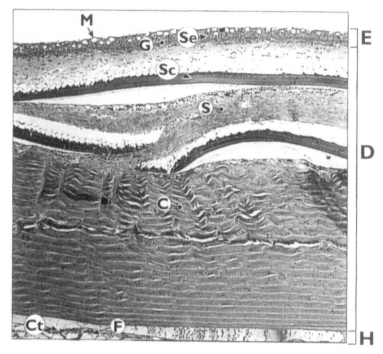

Figure 3.1 Skin from the dorsal–lateral surface of a 1-kg rainbow trout. The skin is made up of three distinct zones: (1) epidermis (E), composed of squamous epithelium (Se), stratum germinativum (G), and mucous cells (M); (2) dermis (D), composed of stratum spongiosum (S), stratum compactum (C), and scales (Sc); and (3) hypodermis (H), consisting of loose connective tissue (Ct) and fat cells (F) (125×, stained with hematoxylin and eosin) (from McKim *et al.*, 1996). (Reproduced with permission from Academic Press.)

and give rigidity to the epidermis. The basal layer or stratum germinativum (basal lamina), made up of a layer of columnar epithelial cells, is responsible for initiating epidermal cell differentiation and maintaining the continuous upward movement of newly differentiating mucous cells (goblet cells), club cells (alarm substance cells), and granulocytes (secretory cells) within the epidermal layer (Figures 3.1 and 3.2).

Epithelial cells in the outer layer are usually polygonal in shape with raised microridges on the surface in both scaled and scaleless species, as demonstrated for rainbow trout in Figure 3.3. These ridges are a fingerprint for the species and are thought to aid in defense and as an anchor for the surface coat of mucus present on the surface of all fish (Hawks, 1974a). Mucous cells are numerous and are in various stages of development throughout the epidermis in both the trout and catfish (Figures 3.1 and 3.2). They begin their differentiation in the stratum germinativum and migrate upward through the epidermis at a rate required to replace mature cells lost at the surface. As they move upward through the middle layer of the epidermis, they actively synthesize mucin packets and store them within the cell. On reaching the skin surface, they force themselves between surface

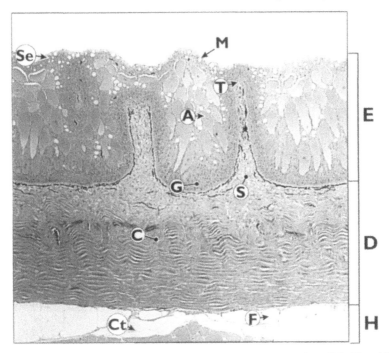

Figure 3.2 Skin from the dorsal–lateral surface of a 1-kg channel catfish. The skin is made up of three distinct zones: (1) epidermis (E), composed of squamous epithelium (Se), stratum germinativum (G), mucous cells (M), and alarm substance cells (A); (2) dermis (D), composed of stratum spongiosum (S), stratum compactum (C), and taste buds (T); and (3) hypodermis (H), consisting of loose connective tissue (Ct) and fat cells (F) (125×, stained with hematoxylin and eosin) (from McKim *et al.*, 1996). (Reproduced with permission from Academic Press.)

epithelial cells, release their contents to the surface and die, as shown for rainbow trout in Figure 3.4. The debris from these dying cells also becomes part of the surface coat of mucus. Mucous cells are present in the epidermis of both species, but catfish have more than trout (Figures 3.1 and 3.2). The catfish epidermis shows a gradual transition between the small newly differentiated mucous cells and the larger mature cells at the surface preparing to eject their contents onto the skin surface. In addition, the catfish epidermis contains a number of large irregularly shaped club or alarm substance cells that are not as prevalent in trout. In contrast to the goblet cells, the club cells only release their contents when the skin surface is damaged.

It should be pointed out at this time that cell renewal in the teleost epidermis, which includes epithelial cells, club cells, and goblet mucous cells, is rapid (4 days) with continuous translocation of cells from the lower to the upper levels of the epidermis (Tsai, 1996). Club cells and goblet cells are both derived from the epithelial cells during differentiation. This rapid renewal of epithelial cells is a key part in the functioning of the outer skin of fish and is much faster than epidermal renewal in terrestrial mammals (Tsai, 1996).

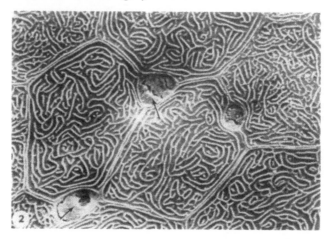

Figure 3.3 Scanning electron micrograph of the epidermal surface of a rainbow trout. Individual filament-containing cells have microridges on their surface, and their boundaries are delineated by a continuous circumferential microridge. The tops of the mucous glands (arrows) squeeze between the filament-containing cells to gain access to the surface (2500×, from Hawkes, 1983). (US Government publication.)

Dermis

The dermis of fish skin lies between the epidermis and the underlying muscle (Figures 3.1 and 3.2). It is made up of two layers. The outer layer or stratum spongiosum consists of collagen, fibroblasts and pigment cells. The lower layer, the stratum compactum, is a non-cellular layer of orthogonal bands of collagen required for structural rigidity and attachment of the skin to the underlying muscle (Hawkes, 1974a; Whitear, 1986b). Collagen fibers in sheets are attached at right angles, which allows them to bend and move horizontally during swimming without wrinkling the skin surface (Figures 3.1 and 3.2). The dermis contains the primary vasculature of the skin, composed of an extensive network of blood vessels concentrated in the upper stratum spongiosum of the dermis (Figure 3.5A) just below the epidermis with branches reaching upward into the epidermis that covers the scales (Figure 3.5B) (Jakubowski, 1982). On the distal part of the scale (covered with epidermis), shown in Figure 3.5B, the topography of the capillary network follows closely the sculpture of the scale.

Scales are flexible, translucent, calcified plates that provide further surface protection and are located in pockets obliquely inserted within the superficial dermis or stratum spongiosum (Van Oosten, 1957). The outermost edges of the scales are covered with epidermis and a fold of dermal tissue, as shown in Figure 3.6. For a more in depth account of the development and regeneration of scales, the reader is referred to a recent review by Bereiter-Hahn and Zylberberg (1993). Trout possess scales originating from the stratum spongiosum of the dermis, whereas the catfish are scaleless and have many taste bud papillae emanating from the stratum spongiosum (Figures 3.1 and 3.2). The epidermis of the catfish

Figure 3.4 Electron micrograph of a mucous cell from an adult rainbow trout. The membrane-bound packets of mucin break down as the contents of the cell are emptied (6400×, from Hawkes, 1983). (US Government publication.)

at the point where taste bud papillae near the surface is only 0.08 mm thick, but is 0.50 mm thick where there are no taste buds. The rather uneven surface of the catfish skin in Figure 3.2 is probably caused by the large number of taste buds forming pits on the epidermal surface. In those fish that are scaleless, the epidermis is thicker than in scaled species and often has a more abundant supply of mucous and alarm substance cells.

The color of fish skin comes from an interaction between three major types of pigment cells (melanophores, xanthophores, and iridophores) that are located within the dermis forming a chromatophore unit (Hawks, 1974b). The melanophores contain melanin granules which absorb light in the visible range and appear black. The xanthophores have drosopterin and carotenoid granules, which are yellow. Iridophores contain platelets of guanine or hypoxanthine, which reflect light and give fish a shiny appearance. The xanophores lie above the melanophores and are surrounded by cytoplasmic extensions of the melanophores. Environmental or physiologic stimuli can cause melanin to move in and out of the cytoplasmic extensions of the melanophores, thereby darkening or lightening the fish color as required for its camouflage protection (Hawks, 1974b).

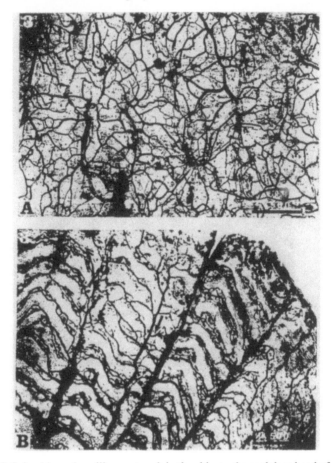

Figure 3.5 Subepidermal capillary network in the skin on the scaleless head of the Amur (*Ctenopharyngodon idella*) (A), and on the scales (B). Vessels are stained with an india ink injection (from Jakubowski, 1982). (Reproduced with permission from Polish Academy of Science.)

The layer beneath the dermis, the subcutis or hypodermis, is mostly loose connective tissue. Pigment cells, fat cells, small blood vessels, and nerves are also numerous within the hypodermis, which rests on the muscle layer beneath (Figures 3.1 and 3.2) (Hawkes, 1974a).

Function

The function of fish skin is unique in that it must provide an effective protective barrier against physical abrasion, parasites, disease organisms and environmental chemicals, while simultaneously acting as a sensitive, interactive, exchange surface between the internal and external environments of the fish. To perform this task successfully, fish skin must rely on a number of capabilities that involve morphologic (structural), physiologic, biochemical, and sensory processes.

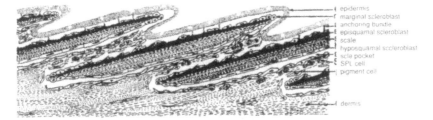

Figure 3.6 Longitudinal section perpendicular to the skin of a typical teleost fish showing the scales in their pockets obliquely inserted in the superficial dermis. The inner surface of the scale pocket is lined by the SPL (scale-pocket lining). The scleroblasts are attached to the scales (from Bereiter-Hahn and Zylberberg, 1993). (Reproduced with permission from Elsevier.)

The protective barrier is provided by: epidermal and dermal thickness, scales arising from within the dermal layer, copious amounts of mucus synthesized within the goblet cells of the epidermal layer, and pigment cells within the dermis that camouflage the fish.

The skin, as an interactive, semipermeable exchange surface, is responsible for cutaneous respiration, absorption of exogenous chemicals, biotransformation and excretion of endogenous and exogenous chemicals, continuous synthesis of mucus onto the surface of the skin, and synthesis and secretion of pheromones. Sense organs for taste, odor, pressure and temperature are also an integral part of the skin, are highly developed in fish and are important in their ability to find food and escape predators.

Mucus production

Using the work of Allen (1978) on mammalian gastrointestinal mucus glycoproteins as a model, the structure of fish mucus was visualized by Satchell (1984) as: (1) glycoprotein of high molecular weight (2×10^6), (2) high carbohydrate to protein ratio, (3) carbohydrate 65 percent of dry weight, and (4) a molecular structure resembling a 'bottle brush' (Figure 3.7). The two major amino acids cited by Allen (1978) as making up the protein core of mammalian glycoprotein were serine and threonine, both of which have been identified in the skin mucus of plaice (*Pleuronectes platessa*) by Fletcher and Grant (1968) and in the Atlantic salmon (*Salmo salar*) by Harris and Hunt (1973). The carbohydrate side-chains that are attached to the protein core of Allen's 'bottle brush,' which form the bristles of the brush, contain up to five different monosaccharides with sialic acid as *N*-acetylneuraminic acid (NANA) attached as the terminal group (Figure 3.7). The necessary carbohydrate side-chains for the model glycoprotein were shown to be available in fish mucus from char (*Salvelinus alpinus*) and Atlantic salmon. This included four different monosaccharides or amino sugars – fucose, galactose, *N*-acetylglucosamine and *N*-acetylgalactosamine for char, and mannose, fucose, glucose and galactose for Atlantic salmon – plus the very

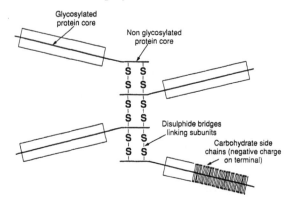

Figure 3.7 Diagrammatic representation of an epithelial mucous glycoprotein molecule (modified from Satchell, 1984).

important monosaccharide sialic acid in the form of NANA as the terminating side-chain in both species (Harris and Hunt, 1973; Wold and Selset, 1977). *N*-acetylneuraminic acid has also been identified in the skin mucus of European eel, *Anguilla anguilla* (Olivereau and Lemoine, 1971), Japanese eel, *Anguilla japonica* (Asakawa, 1974), loach, *Misgurnus* sp. (Enmoto *et al.*, 1964), char (Wold and Selset, 1977), brown trout, *Salmo trutta* (Pickering, 1974), and Atlantic salmon (Harris and Hunt, 1973). The percentage composition of the major components of surface mucus from a sample of six different species of marine fish is also given in Table 3.1. These components were generally similar, but sialic acid seemed less important in these species than in salmonids.

After infusing brown trout with radioactive sialic acid as NANA, Pickering (1976) found that the kinetics of sialic acid secretion in the epithelial mucus occurred in less than 10 h, which seemed to indicate that mature cells near the surface of the epithelium still had the capability of synthesizing this monosaccharide. These carbohydrate side-chains can and will vary between species and for mucins from different locations on the fish's body, but the sialic acid would always be terminal on the side-groups. As the pK_a of sialic acid is 3, it would be fully ionized at environmental pH. That would make this large glycoprotein molecule with many terminal sialic acid side-groups strongly negative in its charge. At low ionic strengths, such as freshwater, there would be little charge shielding by counter ions (positive ions) and the negatively charged side-chains would repulse each other, expanding the tertiary structure of the molecule and increasing its viscous properties (Allen, 1978). Therefore, mucus close to the surface would be quite viscous, and at greater distances from the skin surface mucus would be less viscous as it became less concentrated. Alterations in the mucin structure brought on by a pathologic or environmental change could have an effect on the viscosity and protective function of this mucous layer (Allen, 1978).

Besides the glycoproteins in fish skin surface mucus, there is a considerable variety of biologic materials arising from the secretions of other gland cells (e.g.

Table 3.1 Composition of the non-diffusible macromolecular components of the surface mucus of some marine fishes (from Fletcher, 1978).

	Percentage composition (w/w)					
	Plaice	Dab	Lemon sole	Cod	Ray	Sea trout
Protein	58	57	63	58	48	60
Hexosamine	10.5	10.9	10.8	5.0	15.0	8.0
Hexose (as galactose)	10.0	12.3	12.0	9.0	18.8	12.1
Fucose	7.8	8.0	7.6	1.0	4.0	1.0
Sialic acid (as NANA)	0.6	0.9	0.5	2.0	Nil	1.4
Uronic acid	<0.3	<0.3	<0.3	Nil	0.5	–
Sulfate	11.0	7.5	11.3	6.2	8.0	4.0
Phosphate	0.8	1.0	2.0	1.5	–	–
Ash	9.8	–	—	10.1	–	–

Note

–, not determined; nil, not detectable; NANA, *N*-acetylneuraminic acid.

sacciform cells, club cells, ionocytes, transudate from epithelial cells and debris from dead surface epithelial cells). Many of these secretions contain not only acidic and neutral mucopolysaccharide but also lipids, including phospholipids. The general pattern of phospholipid composition in fish skin secretions was similar to that observed in membrane lipids (Mittal and Nigam, 1986), i.e. phosphatidylcholine and phosphatidylethanolamine made up approximately two-thirds of the total phospholipids in fish skin mucus. Mittal and Nigam (1986) also suggested that these lipids are derived from surface membranous profiles (fragments) in the secretions of skin glands, extrusions of membrane-bound vesicles from surface epithelial cells, and exfoliated cells.

Generally, the surface of fish skin is rough or ridge-like with a variable covering of a slippery mucoid secretion, which is slowly but continually sloughing into the surrounding aqueous environment. The mucous layer is a dynamic, fragile, mobile structure, breaking up and being replaced intermittently from the basal laminae (stratum germinativum) below. Its texture varies considerably from a fluid to a highly viscous, gel-like consistency depending on the species and the environmental situation.

The suggested functional significance of fish epidermal mucus includes osmoregulation, protection from abrasions, entanglement of particulate materials, defense against pathogens and parasites, reduction of swimming drag or friction, and protection against environmental contaminants. The character of the mucus must be such that while performing these functions it must also maintain flexibility, allow the penetration of respiratory gases and excretory products, and provide for the easy movement of olfactory molecules through to olfactory cells of the sensory system.

The mucous layer is an important barrier that helps to control water and solute movement across the skin of fish. Shephard (1981) also determined that mucus alone did not seem to have much greater than a 10 percent reduction in water movement. However, his later work showed that the action of the skin mucous layer in reducing the inflow of water in freshwater fish and the loss of water in saltwater fish was in part due to a reduction in the osmotic gradient set up in the unstirred mucous layer (Shephard, 1982, 1984). The strong negative charge on the mucus glycoproteins sets up a Na^+, K^+, and Ca^{2+} ion concentration gradient within the unstirred mucous layer which in freshwater is highest next to the skin surface. This shifts the osmotic gradient in favor of less water movement into the fish across the skin (hydration). In saltwater fish, the process is reversed making the ion gradient higher on the water side of the mucous layer and lower next to the skin, which helps to protect the fish from dehydration.

The same osmotic gradient that controls water movement also helps to control the diffusion of salts, both in and out, across a fish's skin. In addition, the skin of freshwater fish also has the capability of actively taking up ions needed by the animal to maintain its normal salt and acid balance. Recently, the skin of rainbow trout was shown to contribute to systemic Ca^{2+} balance through active transport by mitochondria-rich epithelial cells (Marshall *et al.*, 1992) and to acid–base balance through Cl^-/HCO_3^- exchange (Ishimatsu *et al.*, 1992).

The fish epidermal mucous layer contains natural antibodies, lysozymes, and bactericidins which are thought to provide a defense system against aquatic microorganisms and parasites (Diconza, 1970; Fletcher and White, 1973). Cutaneous mucous antibodies have also been induced by several investigators against specific bacteria and erythrocyte antigens (Fletcher and Grant, 1969; Bradshaw *et al.*, 1971; Fletcher and White, 1973; Ourth, 1980). St. Louis *et al.* (1984) used fluorescent antibody techniques to demonstrate the presence of immunoglobulin (Ig)-producing cells in both the dermis and overlying mucous layer. Their close association with the epithelial mucus suggested an active secretory immune defense mechanism that probably functions along with the circulating serum Ig system.

Fish mucus is known to reduce drag in swimming fish by at least 50 percent (Daniel, 1981). Rosen and Cornford (1971), using turbulent flow rheometry, found that mucus scraped from freshly caught fish had a friction-reducing capacity of 50–70 percent in narrow pipes. Larger fish have higher mucous drag-reducing activity and burst swimmers also seemed to have more drag reduction activity than cruising or bottom-maneuvering species (Bernadsky *et al.*, 1993).

Respiration

In addition to the protective function, fish skin also possesses an array of physiologic functions, one of which is respiration. Cutaneous respiration will be discussed here because of the possible impact that chemical toxicity to the skin could have on this critical function. The skin of larger fish has been shown to consume oxygen, but in some species (non-exchangers) the skin takes up only enough to satisfy its own respiratory demands (Kirsch and Nonnotte, 1977; Nonnotte and Kirsch, 1978; Nonnotte, 1981, 1984). However, in many species, variable amounts of oxygen taken up across the skin go into the systemic circulation and are used by other organs and tissues (Figure 3.8; Nonnotte and Kirsch, 1978). Most of the freshwater fish in Figure 3.8 are not oxygen exchangers, whereas most of the saltwater fish are, and take up more oxygen across the skin than is required for the skin only.

Based upon morphometric and anatomic information, researchers have long suggested that cutaneous oxygen flux contributes significantly to total respiration in larval fish (McDonald and McMahon, 1977; McElman and Balon, 1980; Oikawa and Itazawa, 1985). Because the gill epithelium of both small and large fish consists of one or a few cell layers, its thickness does not change much with fish size. In contrast, skin thickness tends to decrease with decreasing fish size, and in small fish may approach the thickness of the gill epithelium. In newly hatched larval chinook salmon (*Oncorhynchus tshawytscha*), as much as 80 percent of oxygen uptake takes place across the skin (Rombough and Moroz, 1990). Half-way through the alevin stage of development, only 50 percent of the total oxygen consumed was taken up across the skin. At swim-up, skin oxygen uptake had declined to 40 percent of the total oxygen consumed, and shortly thereafter appeared to stabilize in the early fry stage at about 30 percent (Rombough and Ure, 1991). In adult rainbow trout, oxygen uptake across the skin as a percentage of total oxygen consumption was reduced further and ranged from 15 to 25 percent (Kirsch and Nonnotte, 1977). This certainly emphasizes the importance of skin in the respiration of newly hatched fish that must be considered when determining the impact of chemical contaminants on the skin or when working with predictive toxicity models dealing with young or larval stages of fish.

Biotransformation and excretion

The excretion of lead, cadmium, and naphthalene across the skin of juvenile rainbow trout has been demonstrated previously (Varanasi and Markey, 1978; Varanasi *et al.*, 1978). These studies suggested that lead and cadmium diffuse from the plasma across the skin and into the mucous layer which is then sloughed from the fish. Naphthalene was recovered in mucus immediately after intraperitoneal (i.p.) injection or dietary exposure, but was later replaced by naphthalene metabolites. Whether or not these metabolites arose from the direct transport via the blood to the skin and/or from biotransformation of naphthalene by the mixed function oxidases that may be present in the skin was not determined. Both of these studies point to the skin as an exchange surface for the excretion of environmental chemicals and their metabolites and they suggest a possible role of the skin in the biotransformation of xenobiotics.

The skin of fish is highly metabolic in nature, with the presence of steroid-metabolizing enzymes observed in both the epidermal and dermal layers of several species. Histochemical staining was used by Ali *et al.* (1987) to show the locations of specific metabolizing enzymes within the two major layers of the skin (Figure 3.9). *In vitro* incubations with fish skin strips from various areas of the body have demonstrated the biotransformation of endogenous steroids to the same degree in

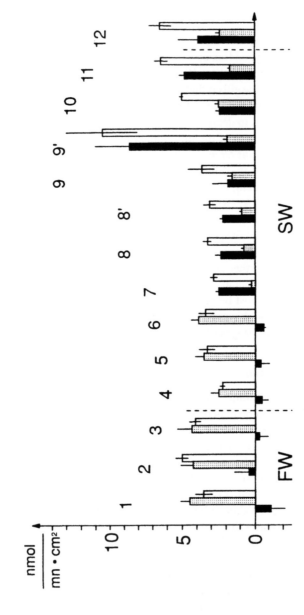

Figure 3.8 Transcutaneous oxygen net flux (filled bars), cutaneous oxygen consumption (dotted bars) and cutaneous oxygen uptake (open bars) in freshwater teleosts (FW), seawater teleosts (SW) and the frog. 1, freshwater eel (*Anguilla anguilla*); 2, rainbow trout (*Oncorhynchus mykiss*); 3, tench (*Tinca tinca*); 4, butterfish (*Pholis gunnellus*); 5, cod (*Gadus morhua*); 6, five-bearded rockling (*Ciliata mustela*); 7, shanny (*Blennius pholis*); 8, flounder (ES) and 8′ flounder (BS) (*Platichthys flesus*); 9, sole (ES) and 9′ (BS) (*Solea solea*); 10, seawater-adapted eel (*Anguilla anguilla*); 11, cultured seawater eel (*Anguilla anguilla*); 12, frog (*Rana temporaria*). BS, blind side; ES, eyed side (from Nonnotte and Kirsch, 1978). (Reproduced with permission from Elsevier.)

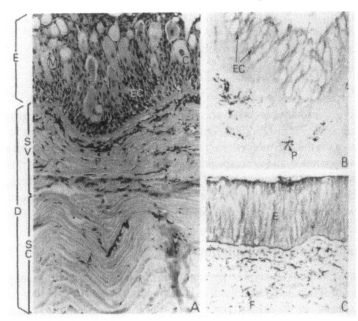

Figure 3.9 Histologic sections of the skin of *Clarias gariepinus*. (A) General structure. Hemalumeosin (230×). (B) 17β-Hydroxysteroid dehydrogenase, with a weak enzyme activity in epithelial cells (145×). (C) UDP-glucose dehydrogenase. Enzyme activity is demonstrated in the epidermis and fibroblasts (145×). C, club cells; D, dermis; E, epidermis; EC, epithelial cells; F, fibroblasts; M, mucous gland; P, pigment; SC, stratum compactum; SV, stratum vasculare (from Ali *et al.*, 1987). (Reproduced with permission from Academic Press.)

both males and females. Rainbow trout and brown trout were shown to contain 3α-hydroxysteroid dehydrogenase (HSD), 17β-HSD, and 5α-reductase (Hay *et al.*, 1976; Soivio *et al.*, 1982), whereas the skin of the scaleless African catfish (*Clarias gariepinus*) contained the same enzymes as the trout plus 5β-reductase, 11β-HSD, and 20β-HSD (Ali *et al.*, 1987). These same authors also found considerable quantities of water-soluble compounds, mainly steroid glucuronide conjugates, in the skin of brown trout, rainbow trout, and African catfish. The presence of UDP-glucose dehydrogenase, an enzyme required for glucuronic acid synthesis, was also found in African catfish skin, which indicated its capability to conjugate endogenous compounds and perhaps exogenous xenobiotics. Although no records of xenobiotic metabolism by fish skin are currently available, at least some of the necessary enzymes (phase I and phase II) for xenobiotic metabolism seem to be present in fish skin.

Early work by Todd *et al.* (1967), using behavioral observations, suggested that the skin mucus of the yellow bullhead (*Ictalurus natalis*) contained pheromones that allowed them to make conspecific identifications. Later efforts by Richards (1974), using skin extracts, further demonstrated that channel catfish could discriminate between different conspecifics. Most recently, steroid glucuronides produced in the skin were shown to be highly effective as pheromones important in the reproductive behavior of the African catfish (Lambert *et al.*, 1986), the zebra fish (*Brachydanio rerio*) (Van den Hurk and Lambert, 1983), and the black goby (*Gobius jozo*) (Colombo *et al.*, 1980).

Sensory receptors

Sensory receptors (taste, odor, pressure, temperature) are important attributes of fish skin as they translate subtle environmental signals into important behavioral responses via the nervous system. These behavioral responses are critical to secure food, to reproduce, and to escape predators. Because of the location of this sensory system in the skin, environmental chemicals can disrupt their normal function and impact survival. The skin sensory system and other aspects of nueroendrocrine behavior are discussed in further detail in Chapter 4 of Volume 2.

Skin absorption of xenobiotics

Fish skin functions as a typical membrane in that it prohibits, permits, or promotes passage of fluids, ions, and dissolved chemicals. The rate or extent of passage of molecules across the skin is typically referred to as permeability. Permeability is the number of molecules crossing a unit area of membrane per unit time when a unit concentration difference exists across the membrane. As described above, the semipermeability of the cutaneous surface is necessary to maintain the homeostatic physiology of the animal. However, as a result of this semipermeability, the skin is only a partial barrier to penetration by many xenobiotic compounds. The permeability of the cutaneous surface to toxic compounds is of interest and concern, from both toxicologic and environmental viewpoints. In aquatic toxicology, xenobiotic absorption potential via the skin is usually concomitant with gill exposure; only with specialized apparatus are gill and skin exposure to the permeant separated from one another. In the environment, whole body exposure and potential absorption via the cutaneous route is naturally the rule. Therefore, there is a need to ascertain the capacity of xenobiotics to penetrate the skin. Understanding and quantifying the percutaneous penetration of xenobiotic chemicals in aquatic organisms will aid in predicting the toxicologic consequences of exposure to xenobiotic compounds and in reducing uncertainty in environmental risk assessments.

Uptake by the skin can be the result of adsorption and/or absorption of the compound. Adsorption is defined as the attachment of one substance to another. In the present context, adsorption pertains to the binding of a xenobiotic chemical to the surface of the skin. Absorption refers to the uptake process by which compounds penetrate living membranes and enter the systemic system via the blood. Absorption is measured as a net increase in the content of a substance in the blood occurring when uptake exceeds loss. Percutaneous absorption is the process whereby xenobiotics traverse the epidermis and enter the microcirculation of the dermis. The quantification of the absorption process, which is referred to as pharmacokinetics (kinetics of therapeutic chemical concentrations) or toxicokinetics (kinetics of toxic chemical concentrations), is essential for a complete toxicity assessment of a compound.

Toxicants may adversely affect the structure and/or function of one or more target organs. The adverse effect may be local or systemic. Local effect is a toxic effect of the chemical at the site of first contact with the organism. Local effects on fish apply to a few exchange surfaces, gills, gastrointestinal tract (GI) tract, and skin. Systemic effect refers to an adverse effect at a site distant from the exchange surface of the chemical. This implies that the chemical has been absorbed and transported by the blood from the exchange surface. The magnitude of the adverse effect is related to the concentration, potency, and persistence of the toxic compound. Typical measurements of the capacity of an exposure to produce a toxicologic effect include concentration of the toxic material in the target organ, accumulated dose, area under the concentration–time curve, and threshold or maximum concentration. These measures of toxicant exposure require a quantification of the rate of absorption. A quantitative assessment of percutaneous absorption of xenobiotic chemicals in fish will allow the prediction of skin and systemic toxicity that results from an exposure of the skin.

Efforts to understand the mechanisms and to predict the kinetics of xenobiotic absorption in fish have focused primarily on the gill as the major site of exchange. Various authors have investigated the mechanisms recognized to control xenobiotic flux at the gills (Boddington *et al.*, 1979; McKim and Goeden, 1982; Piiper and Scheid, 1984; McKim *et al.*, 1985, 1994; Barber *et al.*, 1988; Black and McCarthy, 1988; Erickson and McKim, 1990a,b; Hayton and Barron, 1990; Black *et al.*, 1991; McKim and Erickson, 1991; Nichols *et al.*, 1991, 1993; Streit *et al.*, 1991; Schmieder and Weber, 1992; Lien and McKim, 1993; Sijm *et al.*, 1993, 1994; Streit and Sire, 1993; Lien *et al.*, 1994; McKim, 1994). Efficient diffusion of xenobiotic chemicals across the gills is facilitated by a large surface area, short diffusion distance, large volume of respiratory water, majority of cardiac output flowing through gills, and countercurrent flow of blood and water. However, absorption of xenobiotics across the skin may also be an important route of uptake for some fish and for some chemicals. The skin provides an extensive surface area for potential absorption. Diffusion distances from the external surface to the plasma may be only a few cell layers thick in some instances (Grizzle, 1979). The importance of the cutaneous surface as a respiratory gas and ion exchange surface in teleosts is described above. The relative importance of transcutaneous absorption of xenobiotics in fish, in comparison with branchial absorption, and the factors controlling the rate of absorption across this surface are only beginning to be understood.

In mammals, the stratum corneum is considered the principal barrier to percutaneous penetration by xenobiotics. This thin layer is composed of densely packed, keratinized epithelial cells and provides a good physical barrier. This is evidenced in part by the fact that removal or otherwise compromising the stratum corneum can result in a greater absorption rate. Diffusion across the biologically inactive stratum corneum can also be clearly differentiated from diffusion across the viable epidermis (Guy *et al.*, 1985). Diffusion across this 'dead' layer is several orders of magnitude less than that across viable epidermis, despite the fact that the typical thickness of the stratum corneum is an order of magnitude less than the viable epidermis (Scheuplein and Blank, 1971). A major function of the stratum corneum in mammals is to prevent dehydration of underlying tissues. The lipid-rich nature of this layer is an effective barrier to the flow of water. The chemical barrier effect of the stratum corneum in mammals is also due in part to the partially hydrated nature of this layer. The relatively dry (the degree of hydration is not a fixed quantity but varies with the relative humidity of the surrounding air) nature of these cells and the intercellular space is an effective barrier to compounds of a hydrophilic nature. Indeed, increasing the moisture content of the stratum corneum greatly enhances the uptake of these materials applied to the skin (Scheuplein and Blank, 1971). The lipid–protein matrix of the cells and the lipid-rich amorphous material of the intercellular area of the stratum corneum makes the skin a substantial depository for lipophilic compounds. The prolonged saturation phase of the stratum corneum with lipophilic compounds (e.g. in drug patch therapy) provides a temporary depot for these compounds, thereby reducing initial kinetics of systemic delivery. This depot also prevents the occurrence of large peak plasma concentrations of an absorbed chemical following a loading dose and provides for steady dosing into plasma for long periods after the applied exposure is ceased. This property has been recently exploited in the cutaneous delivery of pharmaceuticals in humans.

Percutaneous absorption kinetics

The molecular basis of the movement of xenobiotics through fish skin is not well defined. Several important postulates regarding movement of xenobiotics across skin are widely accepted, however, and form the basis of our current understanding and description of percutaneous absorption. First, chemicals must be dissolved before they can diffuse across membranes. In addition, only the free or unbound form of the chemical is able to diffuse. The free form of the chemical is typically referred to as the permeant or effective concentration in the context of absorption. Processes that reduce environmental bioavailability or the effective concentration of the permeant may reduce the rate of absorption (Landrum *et al.*, 1994).

No active transport process(es) for the movement of unionized xenobiotics across fish skin has (have) been documented. This is of utmost importance because the laws of physics pertaining to passive diffusion can then be applied to permeation of the skin and aid in its description. Generally, if the solution is dilute and if tissue damage does not occur as a result of the sorbed chemical, Fick's law applies. Fick's first law states that the unit area steady-state flux (J) of solute normal to the direction of diffusion is proportional to the activity gradient ($C_1 - C_2$) and inversely proportional to the distance (x).

$$J = \frac{D(C_1 - C_2)}{x}$$

(3.1)

where D is the diffusion coefficient.

Diffusion across fish skin can be characterized practically by liquid-state diffusion in contrast to the tortuous path between tightly stacked hexagonal cells of the stratum corneum. The diffusion of chemicals in fish skin is likely to resemble the diffusion in viable epidermis of mammals. The skin of fish is completely hydrated. The diffusion of xenobiotics therefore occurs in the aqueous cytoplasm of fish skin cells versus the lipid-rich cells of the stratum corneum in mammals. As mentioned above, the diffusion across viable epidermis of mammals is several orders of magnitude greater than that across the stratum corneum. Fish skin also contains relatively large amounts of unsaturated fatty acids. Membranes become more fluid-like when they contain more unsaturated fatty acids, and a fluid-like membrane in turn facilitates greater diffusion of xenobiotics (Rozman and Klaassen, 1996).

Another generalization is that the more soluble the compound is in the cellular membrane then the greater the membrane permeability. This is only true up to a limiting molecular volume. The affinity of a compound for the skin is generally proportional to the octanol–water partition coefficient, a widely used analogue for lipid solubility. This relationship is not perfect, however, because tissues have separate aqueous and lipid regions providing an affinity for both water-soluble and lipid-soluble compounds, something a non-polar solvent such as octanol cannot duplicate.

A layer of mucus on the skin may affect permeability through ionic interactions or a thickening of the 'unstirred layer.' However, the diffusion coefficients of ions through mucus associated with fish skin are similar to those through control saline solutions (Marshall, 1978). Shephard (1981) states that a layer of dilute mucus presents no more of a permeability barrier than does an unstirred layer of water of equal thickness. The cell walls of fish skin are also thin in comparison with the thick keratinized cells of the stratum corneum of mammals. All of these attributes of fish skin increase the potential for greater absorption of xenobiotics and increase the importance of the skin as a route of uptake for hazardous materials.

The extensive area of the cutaneous surface is another factor which could affect the relative importance of cutaneous uptake in small fish. It has been estimated that, for fish weighing less than 2.5–4.0 g, the cutaneous surface area is slightly greater than the total lamellar surface area (Oikawa and Itazawa, 1985; Rombough and Moroz, 1990). Figure 3.10 shows the relationship between calculated cutaneous and branchial surface areas for fish weighing up to 10 g. The surface area–volume ratio of branchial surfaces for 30-day-old fathead minnows (*Pimephales promelas*) is approximately 15:1; the surface area–volume ratio for the cutaneous surface is approximately 25:1 (Lien and McKim, 1993). When only the functional or effective surface area of the gill lamellar surface of the fathead is considered, the relative area of the skin becomes even more significant. Cutaneous surface area exceeds functional respiratory area by more than 5:1. Surface area–volume ratios (cm^2/cm^3) for cutaneous and functional respiratory area are approximately 25:1 and 4:1, respectively, for the average fathead minnow used by Lien and McKim (1993).

Experimental evidence also suggests that transcutaneous absorption of xenobiotic chemicals may be noteworthy in fish, especially in small fish. As early

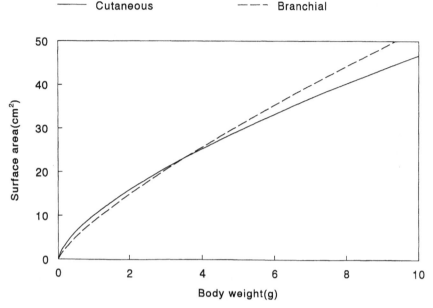

Figure 3.10 Calculated area of cutaneous and branchial surfaces in small fish (from Lien and McKim, 1993). (Reproduced with permission from Elsevier.)

as 1975, Tovell and co-workers (1975) reported that ~ 100-g goldfish (*Carassina auratus*) absorbed sodium lauryl sulfate across the skin. Approximately 20 percent of total absorption was attributed to cutaneous absorption in these fish. Table 3.2 includes estimated *in vivo* percutaneous absorption coefficients for six compounds ranging in log K_{ow} from 0.79 to 6.0 and for five species of freshwater fish ranging in weight from < 1 g to > 1 kg. Saarikoski *et al.* (1986) reported that the primary route of uptake for 17 phenols, anisoles, and carboxylic acids in guppies (*Poecilia reticulata*) was across the gill epithelium, but that 25–40 percent of the total absorption was across the skin. Lien and McKim (1993) suggested that cutaneous absorption of 2,2′,5,5′-tetrachlorobiphenyl may be as high as 50 percent of total absorption in fathead minnows and Japanese medaka (*Oryzias latipes*). Cutaneous absorption of pentachloroethane and hexachloroethane were estimated to contribute 20 percent and 30 percent, respectively, of total absorption in fathead minnows (Lien *et al.*, 1994). In small fish, absorption across the skin may be a major route of uptake for mid-hydrophobic compounds (i.e. log*P* 3–6). Sijm and Van der Linde (1995) also consider that skin uptake and elimination may contribute considerably towards bioconcentration in small fish.

Toxicokinetic studies suggest that cutaneous absorption may be especially important in small fish but that diffusion of xenobiotic compounds across the skin of larger fish also occurs (Table 3.2). A detailed examination has been made of the absorption dynamics of a homologous series of chlorinated ethanes across the gills (Nichols *et al.*, 1990, 1991, 1993; McKim *et al.*, 1994) and skin (McKim

Table 3.2 Exchange coefficients (L h⁻¹ kg⁻¹) for cutaneous absorption of organic chemicals by five species of fish.

	Guppy	Fathead minnow	Japanese medaka	Channel catfish	Rainbow trout
Butyric acid (log K_{ow} = 0.79)	2.1[a]				
1,1,2,2-Tetrachloroethane (log K_{ow} = 2.39)		7.5[b]		0.6[d]	0.1[d]
Pentachloroethane (log K_{ow} = 3.06)		23.1[b]		0.8[d]	0.1[d]
Hexachloroethane (log K_{ow} = 4.04)		41.5[b]		0.7[d]	0.3[d]
Pentachlorophenol (log K_{ow} = 5.15) (pH 5)	33[a]				
2,5,2,5-Tetrachlorobiphenyl (log K_{ow} = 6.0)		67.6[c]	39.6[c]		

Sources:
a   Saarikoski *et al.* (1986).
b   Lien *et al.* (1994).
c   Lien and McKim (1993).
d   Nichols *et al.* (1996).

et al., 1996; Nichols *et al.*, 1996) of rainbow trout and channel catfish. These studies examined, independently, uptake across the gills and skin in a system that can isolate gill from skin exposure. These studies revealed that cutaneous absorption in large fish (≈ 1 kg) contributed 2–4 percent of the total absorbed dose in rainbow trout and 7–8 percent of the total dose in channel catfish (Nichols *et al.*, 1996).

Techniques for quantifying skin exchange

Only recently have attempts been made to measure the rate of percutaneous absorption of xenobiotics in fish. The major difficulty has been separating gill absorption from that of the skin. Several *in vitro, in vivo,* and mathematical modeling techniques have been developed to overcome this difficulty. Some of these techniques were originally developed to measure percutaneous oxygen exchange whereas others were developed solely for studying xenobiotic exchange.

In vitro *techniques*

In vitro techniques have been widely used in the assessment of percutaneous penetration in mammals. Diffusion rates can be measured more accurately in *in vitro* systems without the interference of other complicating factors such as blood transport.

With the exception of anesthetics, *in vitro* absorption studies with fish have not included organic xenobiotic diffusion across skin. The uptake of the anesthetic benzocaine hydrochloride across the skin of three freshwater fish was determined using skin pouches (Ferreira *et al.*, 1984). Ussing-type chambers have been used with isolated skin (Marshall *et al.*, 1992) opercular epithelium (Wood and Marshall, 1994; Verbost *et al.*, 1997), urinary bladder epithelium (Marshall, 1988), monolayer cultures of renal proximal tubules (Dawson and Renfro, 1990; Lu *et al.*, 1994), sperm duct epithelium (Marshall *et al.*, 1989), and intestine (Musch *et al.*, 1987), but these have been restricted to the study of ion transport, amino acids and acid–base regulation. *In vitro* methods with fish skin could also be very useful for determining biotransformation in skin. *In vivo* data are difficult to interpret because of the possibility of further systemic metabolism of absorbed chemicals. Similarly, *in vitro* techniques may be incomplete if biotransformation is not considered when measuring skin penetration.

In vivo *techniques*

In vivo studies include blood flow to the skin, which maintains the skin as a viable organ with all of its normal physiologic properties. Physiologic properties such as blood flow, capillary bed structure, and blood lipoprotein binding can have a strong effect on skin absorption and cannot be accurately duplicated *in vitro*.

A method involving a rubber dam to isolate water flowing over the gills from water in contact with the cutaneous surface has been used to estimate percutaneous oxygen uptake (Kirsch and Nonotte, 1977; Nonnotte, 1981; Rombough and Ure, 1991; Wells and Pinder, 1996) and xenobiotic uptake across the skin (Saarikoski *et al.*, 1986).

In an experimental system that combines the rubber dam with an oral membrane to separate cutaneous and branchial routes of exposure, *in vivo* estimates of xenobiotic chemical flux across the cutaneous surface of intact fish were obtained (McKim *et al.*, 1996). Figure 3.11 shows chloroethane mass balance during a cutaneous-only exposure which emphasizes those factors that control the mass balance. The kinetics of percutaneous absorption varied among 1,1,2,2-tetrachloroethane (TCE), pentachloroethane (PCE), and hexachloroethane (HCE), but the absorption of each chloroethane was similar across species. The flux of TCE, PCE, and HCE across the skin was 1.4 percent, 1.8 percent, and 1.4 percent of total (i.e. hypothetical combined gill and skin) flux in the rainbow trout, respectively, and 2.8 percent, 3.6 percent, and 3.2 percent of total flux in the channel catfish respectively. The cutaneous absorption of TCE, PCE, and HCE was two to four times greater in catfish than in trout. McKim *et al.* (1996) suggest that the differences in cutaneous absorption were due to the differences in the structure of the skin between these two fish, and/or the changes in diffusivity caused by experimental temperatures.

As an alternative to the rubber dam or diaphragm technique, microelectrodes have been used to estimate flux rates of inorganic ions (Smith *et al.*, 1994) and Po_2 gradients in the boundary layer adjacent to the outer surface of the skin

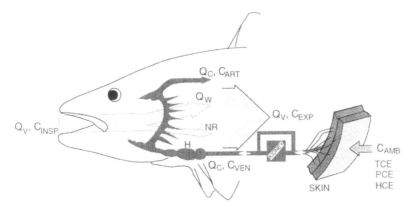

Figure 3.11 Diagram showing flows and concentrations used to construct a mass balance for a cutaneous exposure. Q_V and Q_W are ventilation volume and respiratory water in contact with perfused lamellae respectively; C_{INSP}, C_{EXP}, and C_{AMB} are concentration in inspired water, expired water, and exposure water respectively; NR, non-respiratory water; Q_C, cardiac output; C_{ART} and C_{VEN} are concentration in arterial and venous blood respectively; TCE, tetrachloroethane; PCE, pentachloroethane; HCE, hexachloroethane (from McKim *et al.*, 1996). (Reproduced with permission from Academic Press.)

(Rombough, 1998). To our knowledge, this technique has not been used for xenobiotic uptake, but may have application to skin absorption in the future.

Mathematical models

Acquiring *in vitro* and *in vivo* percutaneous absorption kinetics in fish is often difficult and time-consuming. Consequently, there is a need to predict reliably the kinetics of percutaneous absorption of xenobiotics in fish. The mathematical modeling of percutaneous absorption in fish has been the objective of several studies (Lien and McKim, 1993; Lien *et al.*, 1994; Nichols *et al.*, 1996). Insight into the mechanisms and the dynamics of percutaneous xenobiotic absorption can be gained by theoretical treatments such as physiologically based toxicokinetic (PBTK) model simulations (McKim and Nichols, 1994). The strategy of physiologically based models is to incorporate relevant principles or physical laws of diffusion, fluid dynamics, and thermodynamic equilibrium into a mathematical model that simulates chemical interactions with biologic systems. These models can be helpful in identifying key physical and biologic processes and rate-limiting processes in skin exchange (McKim, 1994). They can also be very useful in separating skin absorption from absorption by other routes (i.e. gill, oral) and correct for normal excretion (i.e. urine, bile, and gill for skin-only exposure) and biotransformation processes. The ability to predict absorbed dose accurately on a mechanistic basis provides a basis for extrapolation of toxicity data across species, chemicals, exposure regimes, and environmental conditions.

The dynamics of organic chemical exchange across the cutaneous surface in small fish has been described previously (Lien and McKim, 1993; Lien *et al.*,

1994) in terms of a few fundamental physiological, morphological, and physicochemical parameters using a mathematical model (Figure 3.12). The following simplifying assumptions were adopted for this model. First, chemical adsorption to the skin surface, or accumulation of chemical in the skin tissue, does not retard the kinetics of absorption significantly. Second, water renewal at the cutaneous surface does not limit chemical flux. Third, the diffusivity of the xenobiotic in tissue would be proportional to the aqueous diffusivity. Last, the diffusion distance could be approximated from morphologic measurements. Therefore, blood flow from the skin and the diffusion barrier become the potential rate-controlling parameters.

The parameters used in the small fish cutaneous exchange model include blood/water equilibrium partition coefficients (K_{BW}), blood flow from the skin (Q_B^S), chemical diffusion coefficients for the cutaneous barrier (D^S), the cutaneous surface area (A^S), effective diffusion distance across the cutaneous barrier (T^S), and wet weight of the fish (W). An outline of the cutaneous exchange model is provided below.

Percutaneous flux (F^S) is described mathematically using a first-order rate constant (k_X^S) referenced to the difference between the free concentration in exposure water ($f_W \cdot C_W$) and arterial blood ($f_B \cdot C_B^{Art}$):

$$F^S = k_X^S \cdot (f_W \cdot C_W - f_B \cdot C_B^{Art}) \tag{3.2}$$

The first-order rate constant for cutaneous exchange (k_X^S) was calculated from the exchange capacities of the blood flow from the skin (k_B^S) and the cutaneous diffusion barrier (k_D^S) according to the following equation (Lien and McKim, 1993):

$$k_X^S = k_X^S \cdot \left(1 - e^{-k_D^S / - k_B^S}\right) \tag{3.3}$$

This equation is analogous to the treatment of exchange at the gill given by Erickson and McKim (1990b) when water flow is not limiting. The perfusion layer (i.e. dermis) is simplified to a network of parallel tubes to represent capillaries at an average distance from the exposure water. The effect of chemical concentration gradients transverse and longitudinal to the capillaries are considered minimal. The capacity of blood flow from the cutaneous surface to support exchange of chemical (k_B^S) may be estimated using the following relationship:

$$k_B^S = Q_B^S \cdot K_{BW} \tag{3.4}$$

The capacity of cutaneous surfaces to support diffusion of chemical (k_D^S) is then calculated as follows:

Exposure water

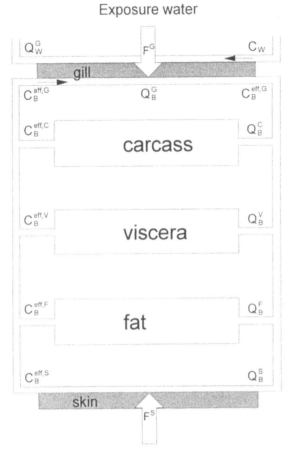

Figure 3.12 A physiologically based toxicokinetic model for the absorption and accumulation of non-metabolized, waterborne organic compounds in small fish. F^G and F^S are flux of chemical across branchial and cutaneous surfaces respectively; $Q_B{}^G$ and $Q_B{}^S$ are flow of blood to the branchial and cutaneous surfaces respectively; $C_B{}^{aff,G}$ and $C_B{}^{eff,G}$ are concentration in blood afferent and efferent to the gill respectively; $Q_B{}^C$, $Q_B{}^V$ and $Q_B{}^L$ are flow of blood to the carcass, viscera and fat compartments respectively; $C_B{}^{eff,L}$, $C_B{}^{eff,V}$ and $C_B{}^{eff,C}$ are concentration of chemical in blood efferent to the fat, viscera and carcass compartments respectively (from Lien *et al.*, 1994). (Reproduced with permission from SETAC Press.)

$$k_D^S = \frac{D^S \cdot A^S}{T^S} \bigg/ W \tag{3.5}$$

Based on model simulations, Lien and McKim (1993) demonstrated that the uptake of 2,2′,5,5′-tetrachlorobiphenyl (TCB) by fathead minnows cannot be accounted for solely by gill absorption (Figure 3.13). The model predicts that branchial and cutaneous surfaces have approximately equal capacity to support exchange of TCB. Lien and McKim (1993) concluded that the combined effect of

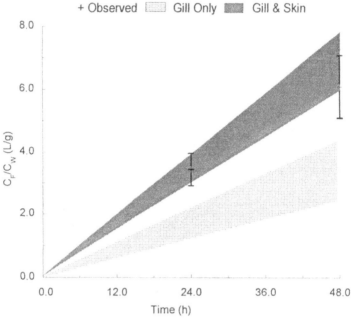

Figure 3.13 Predicted (heavy shaded area) and observed (+ = mean ± SD) accumulation of 2,5,2',5'-tetrachlorobiphenyl via branchial and cutaneous exchange surfaces in fathead minnows and predicted accumulation for gill uptake only (light shaded area). Upper and lower limits for each shaded area result from the use of total ($A^G = A^{G'}$) and functional respiratory areas ($A^G = A^{G'}$) respectively (from Lien and McKim, 1993). (Reproduced with permission from Elsevier.)

a greater cutaneous surface area compared with branchial area, a large surface area–volume ratio for either barrier, and a short diffusion distance across the skin in these small fish could result in the substantial percutaneous absorption.

Subsequently, Lien *et al.* (1994) evaluated the small fish model by exposing fathead minnows and Japanese medaka to three chlorinated ethanes, TCE, PCE, and HCE. They found that the cutaneous uptake of TCE, PCE, and HCE was approximately 9 percent, 21 percent, and 32 percent, respectively, of the combined branchial and cutaneous initial uptake rate constant. They also report that the percutaneous absorption of these three chlorinated ethanes appears to be regulated primarily by blood binding and flow. Blood exiting the skin has less capacity to transport these compounds from the site of entry (i.e. skin) than the capacity to diffuse across the membrane. The results of this study also suggest that there is an increase in the initial uptake rate constant with an increase in the molecular volume of the compounds. Lien *et al.* (1994) deduced that if diffusion were limiting in this case one would expect a decrease in the initial uptake rate constant with an increase in the molecular volume of the compounds. The reverse was observed in this instance. They also propose that cutaneous absorption may contribute significantly to total (i.e. combined branchial and cutaneous) absorption in small fish, depending on the lipophilicity of the compound.

In fish with thick skin, the epidermis and dermis may serve as a sink for chemicals to accumulate before systemic transport, thereby initially retarding the kinetics of uptake. The steady-state model described above for small fish would not be appropriate for fish with thick skin. As an alternative, Nichols *et al.* (1996) created a discrete skin compartment which accumulates chemical depending on the balance between diffusion across the skin surface and the exchange between skin tissues and blood (assumed in this case to be determined by an equilibrium distribution). The skin compartment constituted 10 percent of the volume of the fish and consisted of epidermis and dermis (Figure 3.14). Absorption across the skin was assumed to be by passive diffusion and to be a function of chemical permeability (i.e. combined chemical diffusivity and skin thickness) and the concentration gradient. A mass balance of the skin compartment was described as follows:

$$V_S dC_S/dt = Q_S(C_{ART} - C_{VEN}) + K_p A_S(\Delta C_S) \tag{3.6}$$

where V_S, C_S, Q_S, and A_S refer to the volume, chemical concentration, blood flow rate, and area of the skin, respectively, C_{ART} and C_{VEN} refer to the chemical concentration in arterial and venous blood of the skin, respectively, and K_p and ΔC refer to the permeability coefficient and concentration gradient respectively. Second, skin absorption was incorporated into an established model with branchial flux and distribution components (Nichols *et al.*, 1991, 1993). This allowed the authors to fit the value of apparent dermal permeability coefficients by modeling to chemical concentrations in arterial blood and expired branchial water. The PBTK model was used to simulate cutaneous-only exposure in specially designed respirometer/metabolism chambers. Rainbow trout and channel catfish were exposed to waterborne chlorinated ethanes to evaluate the model.

Nichols *et al.* (1996) found that fitted permeability coefficients increased slightly with the number of chlorine substituent groups among the chlorinated ethanes tested, but not in the manner expected from a direct proportionality between permeability and skin/water equilibrium partitioning of the chemicals. They also concluded that absorption of the chlorinated ethanes was limited more by diffusion across the skin than by the capacity of blood flow through the skin to transport the chemicals away from the site of entry. The skin of the rainbow trout and channel catfish that were used had approximately the same surface area and thickness, yet they found that the permeability estimates for catfish were 2.4 (HCE) to 5.1 (PCE) times higher than the corresponding values for rainbow trout. The higher temperature of the catfish exposure and the structural differences of catfish skin may have contributed to the higher permeability observed in the catfish experiments. These studies also suggest that cutaneous absorption of the chloroethanes (with a log K_{OW} of 2.6–4.6) would contribute less than 10 percent toward a hypothetical combined inhalation and dermal exposure for a 1-kg fish of either species.

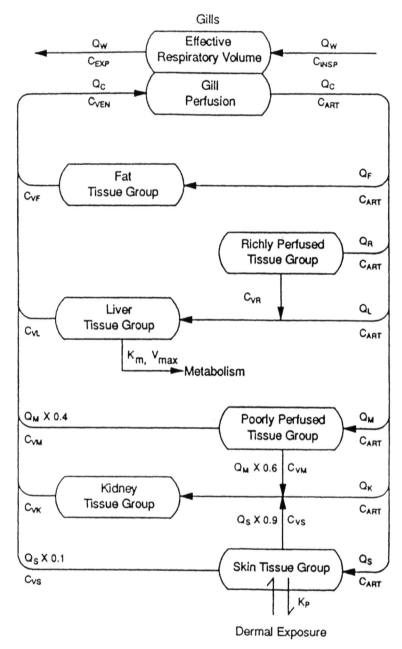

Figure 3.14 A physiologically based toxicokinetic model for the percutaneous absorption of waterborne organic chemicals in fish. Q_w, Q_C and Q_I are effective respiratory volume, cardiac output, and arterial blood flow to the compartment respectively; C_{EXP} and C_{INSP} are concentration in expired water and inspired water respectively; C_{ART}, C_{VEN} and C_{VI} are concentration in arterial blood, venous blood, and blood exiting the compartment respectively; subscripts for compartments: F, fat; R, richly perfused; L, liver; M, poorly perfused; K, kidney; S, skin (from Nichols *et al.*, 1996). (Reproduced with permission from Academic Press.)

Rate-limiting factors in skin absorption

If significant advances are to be made in predictive toxicology and environmental risk assessment, it is critical to understand the factors controlling the absorption of organic compounds across the skin of fish. A variety of biologic, chemical and physical factors are known to affect gill flux of contaminants between fish and the aquatic environment (McKim and Erickson, 1991; McKim, 1994). Some of these factors also pertain to percutaneous absorption. Understanding the relationships between the physiologic and morphologic properties of the skin of fish and the structural/electronic properties of various toxic compounds will provide better estimates of absorbed dose and more complete risk assessments.

Morphometric factors

The conductance capacity of the skin is directly proportional to the surface area. Therefore, a determination of the surface area of the skin is crucial to skin absorption studies. Surface area of fish can be estimated using the allometric equation presented by Schmidt-Nielson (1984):

$$A_s(\text{cm}^2) = 10.0 \cdot Wt(\text{g})^{0.67} \tag{3.7}$$

Other methods include: (1) removing the skin and tracing its outline on graph paper, (2) projecting the image onto a digitizing tablet, or (3) multiplying body length by the average circumference. Oikawa and Itazawa (1985) found that the slope of the regression analysis of log body surface area (mm^2) to log body mass (g) for juvenile and later stages of carp (*Cyprinus carpio*) was 0.664. Rombough and Moroz (1990) report a mean allometric exponent of 0.632 for total cutaneous area of young chinook salmon weighing 0.32–13.4 g. These relationships compare favorably with the allometric equation presented by Schmidt-Nielson (1984).

Physiologic factors

Percutaneous absorption implies that the chemical has penetrated the skin and the capillaries and is transported to the rest of the body. Blood flow to the skin may affect the absorption rate of xenobiotics across the skin. Removal of xenobiotic from the dermis by the blood may limit the percutaneous absorption rate for very hydrophilic compounds in fish with thin skin (Lien *et al.*, 1994). Direct measurements of the volume flow of blood to the skin in fish is lacking. As a first approximation, blood flow to the skin has been estimated from the relative blood volume of the skin (Lien and McKim, 1993; Lien *et al.*, 1994). The blood flow rate to the skin of a ~ 1-kg rainbow trout and a channel catfish was estimated to be 7.5 percent of cardiac output, based on the assumption that on a volume-weighted basis the blood perfusion of skin equals that of white muscle (Nichols *et al.*, 1996). The vascularization of fish skin has been investigated and quantified (Jakubowski and Rembiszewski, 1974; Jakubowski, 1977, 1982). Capillary density, diameter, and distance information may be useful in developing mathematical models for skin absorption.

If lipid solubility is an important determinant of the absorption rate then the lipid content of the tissue is equally important to the overall kinetics of absorption. Skin with a total lipid content of 11.43 percent, 2.68 percent, and 2.77 percent was reported for *Cyprinus carpio*, *Oreochromis mossambicus*, and *Salmo gairdneri* respectively (Ferreira *et al.*, 1984). They report that the highest percutaneous uptake of the anesthetic benzocaine hydrochloride was found in the species with the highest skin lipid content. More lipids in the skin could maintain a gradient of the freely diffusible form for a longer period of time and therefore increase the overall mass of xenobiotic transported over time. Ferreira *et al.* (1984) also stated that factors such as skin thickness and scale coverage were also determinants in the percutaneous uptake rate in fish.

The lipid content of skin has also been measured for other species of fish. Hoffman *et al.* (1999) reported a total lipid content of 6.9 percent for skin of lake trout (*Salvelinus namaycush*). The skin of channel catfish, fathead minnows, and rainbow trout contains 5.7 percent, 4.6 percent, and 2.9 percent total lipid respectively (Bertelsen *et al.*, 1998).

Environmental factors

Environmental factors such as temperature, pH, and organic macromolecules may affect the percutaneous absorption rate of xenobiotics.

The effect of temperature on percutaneous absorption can involve several possible mechanisms. First, aqueous diffusivity is directly proportional to temperature (Wilke and Chang, 1955), the same is supposed for tissue. Therefore, an increase in temperature can increase the permeability rate of a xenobiotic across skin. The calculated diffusivity at 21°C is 30 percent greater than at 12°C (Nichols *et al.*, 1996). Second, temperature can affect the lipophilicity of the compound. A 10°C change in temperature can effect a 25 percent change in the octanol/water partition coefficient of compounds (Chiou *et al.*, 1977). Third, the temperature dependence of cardiac output has been established (Barron *et al.*, 1987). As discussed above, cardiac output may have an influence on the rate of percutaneous absorption.

The percutaneous absorption rate of ionizable compounds can be affected in part by environmental pH. As discussed below, the lipid-soluble unionized form of the xenobiotic is much more readily absorbed than the ionized form. The proportion of the ionizable xenobiotics that is non-ionized can be determined using the Henderson–Hasselbalch equation.

According to Fick's law of diffusion, the rate of absorption is directly proportional to the concentration gradient of the permeant across the membrane. This gradient applies to the free or unbound concentration. Xenobiotics bound to dissolved, colloidal, particulate matter or to organic macromolecules will reduce the freely available concentration and hence the bioavailability of those compounds.

Physicochemical factors

Physicochemical properties of the xenobiotic, such as lipid solubility, molecular volume, and ionization, can greatly affect the rate of percutaneous absorption.

The most important physicochemical property affecting xenobiotic permeability is lipid solubility. The significance of lipid solubility in cellular permeability was established by Collander (1949). This phenomenon follows the simple physicochemical precept that 'like dissolves like.' The most familiar measure of the lipophilicty of a compound is its octanol/water partition coefficient or K_{ow}. The use of the octanol/water partition coefficient as a predictor of chemical partitioning to fish tissues has been described by Bertelsen *et al.* (1998). More direct *in vitro* methods for the determination of chemical partition coefficients for fish plasma and tissues have been published (Schmieder and Henry, 1988; Hoffman *et al.*, 1992). Percutaneous absorption data in Table 3.2 illustrate the influence of lipophilicity on these rates. There is, however, a limit to the impact that lipophilicty exerts on skin permeability (Gillette, 1987). Highly lipophilic compounds such as 3,4,7,8-tetrachloro-dibenzodioxin (TCDD) are very soluble in triglycerides but not in phospholipids and other lipids with polar heads (Rozman and Klaassen, 1996).

The diffusivity of a xenobiotic is inversely related to the molecular volume of the compound. Diffusivity is a measure of permeability by diffusion and is inversely related to the radius of the molecule. Generally, for diffusion in a lattice, up to a molecular weight of 100 the diffusion coefficient is inversely proportional to the square root of the molecular weight; for molecules of molecular weight of more than 100, the cube root of the molecular weight applies (Stein, 1967). Diffusivity in tissue is also dependent on the characteristics of the tissue (e.g. tortuosity, fractional volume) and temperature.

The property of a weak acid or base that has the greatest impact on dermal uptake is the extent to which it is ionized at the pH of the exposure water. The ionization constant is the molar ratio of ionized to unionized molecules of a weak acid or base. Passive diffusion of the relatively lipid-soluble unionized form of the xenobiotic is the major chemical process for percutaneous absorption. The rate of absorption of the unionized form can be related directly to its solubility in lipid. As a consequence, the ionization of the compound in water and the presence of acid or base groups and their pK is of importance in the absorption process. Ionized compounds generally have[a] low lipid solubility. Ferreira *et al.* (1984) concluded that the differences observed between the uptake of the anesthetic benzocaine hydrochloride and neutralized form by the gills and skin of three freshwater fish could be accredited to the degree of ionization of the compound and hence the lipid solubility. Organic anions and cations of small radii may diffuse through pores in a membrane; however, this diffusion is very slow compared with the diffusion of lipid-soluble neutral compounds.

In addition to percutaneous absorption rate and absorbed dose, distribution and biotransformation of xenobiotics are essential to predicting localized or systemic toxicity and to characterize fully the potential hazard of waterborne xenobiotics. Compounds absorbed via the cutaneous route may also have a more direct (or first pass) route to elimination by the kidney and the gills (see Figure 3.11).

Toxic responses of the skin

The heavily keratinized layer on the surface of the skin of terrestrial animals is quite different from the hydrated, mucus-covered skin surface of most aquatic vertebrates. The skin of a fish and its mucous coat is the major barrier between the cells, tissues and organs of the fish's body and the aqueous environment. Damage to either the mucous coat or the underlying epidermal or dermal layers can rapidly disrupt the integrity of the fish's internal environment, causing abnormal behavior, disease or death. Being on the outside of the fish, the mucus is the first line of defense, and, as seen in Appendix 1, it responds quickly to chemical and physical perturbations. The composition of fish skin mucus (see Structure–function of fish skin, p. 153) is such that it can bind metals and particulates, fight disease organisms, absorb organics, reduce the inward and outward flux of water and ions, and protect against abrasions of all kinds. The underlying epidermis is responsible for the synthesis of the mucous layer and, as such, becomes highly active under the influence of a variety of environmental stressors as well as endogenous hormones. The dermis is the vascular zone supplying blood, nutrients, and other raw materials to the highly metabolic epidermis. The dermal layer also supports the scales and the necessary nutrients and metabolic equipment for their development in those species with scales. For those species without scales, the dermis usually supports a thicker epidermis and a larger number of chemosensory papillae (see Structure and function of skin).

Skin toxicity

The acute, subacute, and chronic fish skin toxicity data presented in Appendix 1 are divided into nine categories that represent the major environmental stressors on which skin toxicity data exist (heavy metals, acidic pH, inorganic chemicals, detergents, organic chemicals, chemical mixtures, ultraviolet radiation, pulp mill effluent, and temperature). The combining of these nine categories into one table allows comparisons to be made within and among the categories. The observed toxic impacts on the skin by various environmental stressors summarized in Appendix 1 emphasize the species, type of exposure, and exposure concentrations, summarize the cellular and physiologic changes that take place in the skin in response to a particular stress, and provide a reference for the work. In most cases, the exposure concentrations involved in toxic responses of the skin summarized in Appendix 1 are considerably lower than the lethal concentration (96-h LC_{50}) and provide a unique set of sublethal effects on the largest target organ in fish.

Data on species that represent branchial respiration as well as those representing air-breathers are included, and no great difference in skin toxicity is observed between the two modes of respiration. It should be noted at this point that there are seven studies reported in Appendix 1 that deal with air-breathing fish of the tropics, four that deal with larval fish which utilize the skin for a major part of their respiratory needs, and twenty-two studies that deal with juvenile–adult fish with branchial respiration. Supplemental information on skin responses to environmental chemicals is also available in a review article by Banerjee (1993).

Metals

The skin toxicity of cadmium, copper, lead, zinc, and chromium was studied by several researchers at concentrations well below the 96-h LC_{50} for each metal, and for exposure times of 7–60 days (Appendix 1). The responses to these sublethal chronic metal concentrations followed a similar sequence of architectural changes in the surface epidermal cells: the release of copious amounts of mucus onto the surface of the skin and the continual cycling of mucous cells upward from the stratum germinativum through the middle and upper layers of the epidermis to the surface, where the newly synthesized mucus is released. This mucous cell cycling event involved both hypertrophy and hyperplasia along with a thickening and thinning of the skin. The dermal layer also showed increased metabolic activity with more lymphocytes, chloride cells and the formation of new capillaries near the basal lamina of the epidermis, with blood cells, in some cases, invading the dermal layer. As this response is seen at sublethal levels, it can be considered as an attempt at compensation; however, the cost of this adjustment may over time cause deterioration and death. Scaled and scaleless fish showed the same general pattern of response, as did those with gills compared with those with accessory breathing organs.

Higher acutely lethal metal concentrations such as the mercury studies summarized in Appendix 1 indicate more drastic changes in the skin, with sloughing of the epidermal layer and a loss of skin architecture accompanied by degeneration of the deeper layers. Of course, this will impact all of the skin's many functions, including its ability to protect against pathogens and parasites, that are always present in the aqueous environment. Recovery of the epidermis after an acute exposure to 1.0 mg L^{-1} mercury required 7–10 days before normal cell architecture was again present.

The organometallic triphenyltin was extremely toxic to the skin and caused sublethal degenerative changes (i.e. vaculation in the cytoplasm of epidermal cells) in the skin of larval fish at 0.005 mg L^{-1}, whereas erosion of the epidermis occurred at 0.018 mg L^{-1} (Appendix 1).

Certain exogenous chemicals released into the aquatic environment such as heavy metals can, on binding to the negative glycoproteins that make up the normal mucin covering of the epidermis, change the structure of the glycoproteins and thereby alter their function. Varanasi *et al.* (1975) showed through electron spin resonance studies that waterborne lead and mercury do alter the gel structure of the glycoprotein. The overall effect of both metals was to make the gel coat more fluid and less viscous. The increased fluidization was probably caused by the counter ion effect of the positive metal ions neutralizing the repulsion of the sialic acid side-groups on the glycoprotein, effectively reducing its viscosity (Allen, 1978). These types of structural alterations in the mucous gel coat could certainly impact hydrodynamic aspects of swimming behavior and life-threatening functions such as the acquisition of respiratory gases, the movement of essential ions, and protection from abrasion (Varanasi *et al.*, 1975; Crowther and Marriott, 1984). The ability of the mucus to recover from these structural changes caused by lead and mercury was slow, even though both metals were continually being removed from the skin in sloughed mucus (Varanasi *et al.*, 1975). Removal of the protective mucous layer or a change in its structure can bring about significant changes in the behavior of an aquatic animal by allowing chemicals to reach the sensitive chemoreceptors of taste and odor located in the surface of the epidermis, effectively eliminating or blocking their function (see Chapter 4 of Volume 2).

Acid pH

pH levels of approximately 5.5 and lower caused both hyperplasia and hypertrophy of epithelial goblet cells (mucous cells) in all species and life stages regardless of test temperatures (Appendix 1). This resulted in a thicker epidermis and a copious layer of mucus. In the scaleless brown bullhead at a pH of 5.5, there was a doubling in the number of mucous cells with no changes in cell diameter or volume. At pH values below 5.0, there was epidermal necrosis and a constant sloughing of epithelial cells.

Detergents

Exposure of three fish species to sublethal concentrations of linear alkylate sulfonate (LAS) detergents caused copious mucus secretion, cycling of mucous cells, and high metabolic activity in the skin, as described previously for heavy metals. There was also a shift in mucus composition from acidic to neutral glycoprotein then a shift back to acidic glycoprotein and acidic glyco-saminoglycans, a combination known to be suggestive of pathologic conditions in mammals (Appendix 1). This accelerated metabolic activity was shown to increase the levels of glucose-6-phosphate and lactate dehydrogenase which were correlated with increased mucus synthesis. It was suggested that the changes in enzyme levels stimulated by waterborne detergent exposure to the skin might be investigated as a new model for the effects of pollutant stress on fish skin.

Inorganics

Exposure of juvenile catfish to ammonium sulfate caused an increase in size and density of mucous cells, accompanied by increased mucus secretion. Vacuoles, epidermal sloughing and other more advanced necrotic changes progressed throughout the 45-day catfish exposure to ammonium sulfate (Appendix 1). Larval fish exposed to ammonia (unionized) were much more sensitive than the juvenile catfish, and the larval skin seemed to lack the ability to cycle epithelial cells or mobilize its metabolic reserves. This may be due to their dependence on the skin for respiration as the gills are not yet fully functional.

Organics

The organics are the first group of chemicals to deviate substantially in their toxic skin responses from the previous chemical groups (Appendix 1). In general, the effects of organics seem to involve a breakdown of the mucous barrier by solubilizing and removing it with little to no stimulation of mucous cell cycling noted previously with the other chemical groups. The removal of this barrier seemed to make the skin of the fish more susceptible to various pathogens that resulted in eroded fins, fungal problems on the skin surface, and a darkening of skin color. The penetration of these lipophilic chemicals into the deeper layers of the skin are more pronounced probably because of their high lipid solubility. Some of the chemicals listed here under organics have also been shown to cause skin tumors of several different types.

Mixtures

Because the mixtures examined are a mix of heavy metals, organics, inorganics, and detergents, it would be expected that the responses produced by these mixtures would resemble all three of these chemical groups, and in fact this was the case (Appendix 1). For most of the mixtures studied, there was increased mucus secretion, mucous cell cycling, and increased metabolic activity in the epidermal layer. This occurred in all the fish exposed to sublethal mixtures except for the chlorinated sewage pond mixture. The chlorinated chemicals and polycyclic aromatic hydrocarbons (PAHs) in the pond mixture were linked to the numerous dermal papillomas found on the fish. Several chlorinated compounds and PAHs within the organics group were shown to cause dermal papillomas.

Ultraviolet B radiation

Because of its recent emergence as an important environmental stressor in aquatic systems, the impact of ultraviolet B (UV-B) radiation on fish skin has been well documented in various situations in both marine and freshwater (Appendix 1). The general progression involves an initial darkening of the skin on the back of the fish followed by a thickening of the epidermis, filament epithelial cells becoming necrotic in appearance, vacuolation of the epidermis with the appearance of sunburn cells containing large pale-staining nuclei, and, finally, the separation of the skin from the underlying muscle. The skin of animals can undergo considerable oxidative stress because oxygen is always present and the skin is often exposed to UV-B radiation. Nakano *et al.* (1993) found that the skin of plaice (*Paralichthys olivaceus*) and coho salmon (*Oncorhynchus kisutch*) has the capability of synthesizing the enzyme superoxide dismutase (SOD). Nakano *et al.* (1993) suggest that the distribution of higher SOD activity in the dark parts of the skin might be related to melanization and to regulation of reactive oxygen species. Most of the problems with UV-B described to date have involved holding fish in shallow raceways or ponds unprotected from the sun. However, environmental situations involving loss of protective cover in shallow streams and lakes could initiate this problem. Schindler (1998) reviewed the effect of climatic warming and pH on the dissolved organic carbon content (DOC) of the surface waters of boreal lakes in Canada and the resulting impacts on UV penetration. The loss of DOC caused a seven- to eightfold increase in UV penetration, with the resulting loss of both invertebrate and vertebrate fauna.

Temperature

Acute temperature stress elicited a general increase in metabolic activity that included mucous cell cycling, increased mucin production, apoptotic cells, cells with peroxidase activity, and lymphocytes and pigment cells extending into dermis and epidermis. The overall response was very similar to many of the other stressors described previously and seemed to define generally the response of the skin of fish to a majority of the stressors investigated to date.

The skin responses to stressors described in Appendix 1 represent sublethal and chronic responses to a wide range of toxicants which will vary in severity with the amount of stress encountered. Acute exposures involve more extensive damage to the epidermal layer, with considerably more degeneration, sloughing, and scab formation that consists of cellular debris under which epithelial cells are rapidly laid down with a complete loss of cellular architecture. Under the same conditions, the dermal layer becomes heavily infiltrated with erythrocytes, lymphocytes, and macrophages. Dermal connective tissue shows vacuolation and degeneration with the formation of 'dermal herrings' or collagen bundles.

Appendix 1 has summarized the major studies dealing with localized acute and sublethal skin toxicity. If it were not for the well-developed structure and function of fish skin and its rapid response to stress, these aquatic vertebrates would be far less resistant to environmental chemicals.

General skin stress response syndrome

The impact of various stressors on the skin of fish in Appendix 1 demonstrated a general skin stress response syndrome (GSSRS). The details of this response syndrome can vary with species, life stage, type of stressor, and exposure concentration, but in general the skin goes through a general set of responses for many of the stressors reviewed in Appendix 1. It is described here as a possible new toxic endpoint for fish skin. The initiation and/or intensity of this toxic syndrome might be further developed in the future for dose–response studies with fish skin for use in environmental risk assessment.

This stress syndrome is a linked series of events that occur in response to direct exposure to individual or mixtures of physical and/or chemical stressors.

The initial response of fish skin to sublethal stress caused by surface-acting chemicals or physical stressors is the release of copious amounts of mucus from goblet cells onto the surface of the epithelium. This rapidly forms a protective barrier over the entire epithelial surface. This mucous layer forms an unstirred layer next to the skin and allows the creation of a concentration gradient within the mucous layer, which helps retard the diffusion of environmental chemicals. The composition of the mucins secreted (glycoproteins and mucopolysaccharides) not only limits chemical entry but, because of their unique structure (see Structure and function of skin), also binds inorganic and organic chemicals and moves them off the skin surface as the mucus is continually being dissipated. As the mucous cells at the surface initially release their contents, the epithelial layer has fewer mucous cells and the epidermis becomes thinner. Increased metabolic activity in the basal lamina of the epidermis initiates the differentiation of new mucous goblet cells which move rapidly upward through the lower and middle epithelial layers, growing larger as they move to the surface to release their newly synthesized load of mucus. This cycling of mucous cells continues at a rapid pace throughout the stressful encounter, with increases in mucous cell size (hypertrophy), mucous cell numbers (hyperplasia), and total epidermal thickness (Ingersoll *et al.*, 1990). The extremely high metabolic activity of the skin is evidenced by the rapid incorporation of intravenously injected [^{14}C]-sialic acid into mucous cells of rainbow trout within only 4 h, whereas complete replacement of mucous cells required about 10 days (Pickering, 1976). The differentiation of club cells and other epithelial cells with secretory granules occurs continuously. They contain mucopolysaccharides and phospholipids for release at the skin surface as part of the mucous coat. Accompanying all of this metabolic activity that composes the GSSRS is a shift from acidic glycoproteins to neutral ones and an increase in certain enzymes, such as glucose-6-phosphate and lactate dehydrogenase that are required for the increased metabolic activity of the skin (Zaccone *et al.*, 1985).

While the events described above are taking place in the epidermis, the dermal layer of the skin is also undergoing a rapid change. Chloride cells appear along with lymphocytes and erythrocytes that migrate in from the capillaries in the dermis. Finally, pigment cells in the dermis extend cytoplasmic extensions up into the epidermis. These extensions of the pigment cells contain melanin granules that cause a darkening of the exposed skin.

It should also be noted that the skin of fish seems able to adapt to low-level stressors, as described by Agrawal *et al.* (1979) for salinity stress in the freshwater air-breathing catfish (*Heteropneustes fossilis*). When exposed to a 1 percent increase in salinity, the skin went through the GSSRS that included the mucous cell cycling events discussed previously. After about 14 days, the animals' osmoregulatory systems had compensated and the cell cycling diminished to normal. We assume that fish skin compensation to mild stressors can also occur for other environmental stressors, but the increased cost in energy required to adapt may limit the skin's ability to compensate to multiple stressors.

Epizootic skin neoplasms

Over the last 30 years there has been an increasing number of reported epizootics of spontaneous tumors reported in specific geographic areas of North America. According to Harshbarger and Clark (1990), this may be in direct association with an increase in waterborne contaminants representing the impact of increased industrialization and urbanization. Although several studies have suggested that the appearance of certain types of feral fish tumors are related to environmental contaminants, there is presently no direct proof of this hypothesis (Black, 1983). Many biologic, chemical, and physical factors can combine to impact the formation of neoplastic disease in feral fish, as shown in Figure 3.15 (Mawdesley-Thomas, 1975). These feral fish epizootics involve various kinds of tissue neoplasms, but those of interest to us here involve only neoplastic diseases of the skin. For those interested, a complete listing of all neoplasms of bony fish found in forty-one different locations in North America up through 1990 were compiled by Harshbarger and Clark (1990).

An in depth critical review by Mix (1986) describes all of the skin tumor studies, including epidermal papillomas (EP) and oral papillomas (OP) in Pacific rim flatfish and pseudobranchial tumors (PB) in Pacific cod. He points out that the etiology for these Pacific rim neoplasms was not caused by pollution, as first suspected, but was shown to be unicellular parasites (Dawe *et al.*, 1979; Dawe, 1981), and the papillomas were pseudotumors caused by the parasitic disease.

The remaining epizootics involving skin neoplasia are primarily found in freshwater and are listed in Table 3.3. Of the species listed in Table 3.3, the brown bullhead and the white sucker seem to be the most likely candidates for skin and lip tumors because of their apparent sensitivity and bottom-feeding habits that place them in close contact with contaminated bottom sediments. In a study by Smith *et al.* (1996), the exchange of chemicals from sediments to fish was clearly demonstrated by the uptake of benzo(a)pyrene from spiked sediments placed in direct contact with the skin of anesthetized channel catfish for 6 h.

The frequency of epidermal papillomas observed in feral white sucker populations in Ontario, Canada, was greater in industrialized areas receiving heavy inputs of contaminated water (Sonstegard, 1977). However, no direct correlations with specific chemicals was possible. Later studies on the brown bullhead were carried out by Baumann *et al.* (1987), who surveyed for skin, lip, and liver tumors in fish sampled from the highly contaminated Black River, Ohio, and the pristine Buckeye Lake, Ohio. They found PAH concentrations in fish from the Black River to be 100 times higher than in fish from Buckeye Lake, where tumor frequency was significantly lower (Figure 3.16). This study certainly supported the idea that contaminated areas and specific chemicals or groups of chemicals predisposed the bullhead to neoplasia, but no single chemical could be identified. Black (1983) took this one step further by extracting sediment PAHs from both the Black River, Ohio, and the Buffalo River, New York, and directly exposing brown bullheads to these extracts by painting the lips and skin of the head. After 18 months, a high frequency of lip and skin papillomas were induced (Figures 3.17 and 3.18). This

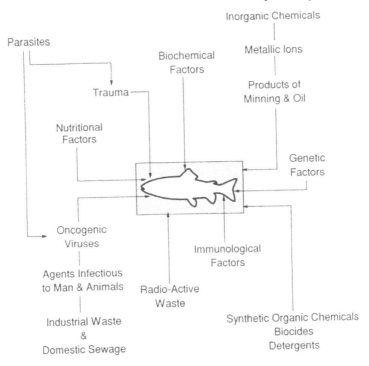

Figure 3.15 Possible etiologies of neoplasia in feral fish (from Mawdesley-Thomas, 1975). (Reproduced with permission from University of Wisconsin, Madison, Wisconsin.)

showed, for the first time, that a mixture of known carcinogens in the sediment had the capability of inducing skin tumors when in direct contact with the surface epidermis of fish (Figure 3.19). This direct relationship of high PAH concentrations in bottom sediments to the higher frequency of external (dermal) tumors is further demonstrated in Table 3.4. The higher the PAH concentration in the sediment the higher the tumor frequency (Black and Baumann, 1991). It is also important to note that the frequency of skin tumors collaborates in most cases with the frequency of hepatic neoplasms in these contaminated river systems. According to Baumann *et al.* (1996), there are presently no clear-cut delineations between tumor frequencies in feral fish from urban and industrialized areas in comparison with uncontaminated sites. However, Baumann *et al.* (1996) state that cutaneous tumor prevalences > 20 percent were never found at uncontaminated or reference sites for either the brown bullhead or the white sucker, and that prevalences this high and higher should be 'interpreted as an indicator of environmental degradation' in the area where the fish were surveyed.

A multidisciplinary approach will be required to understand finally the many biologic, chemical and physical factors involved in causing epizootics of neoplastic disease in marine and freshwater fish. This will certainly include: (1) monitoring of specific chemicals and/or groups of chemicals in sediments and pore waters,

Table 3.3 Epizootic neoplasms in freshwater fishes (from Black and Bauman, 1991)

Species	Organs site/neoplasm type[a]	Geographic location	Reference
Black bullhead	Oral papilloma	Sewage pond, Tuskagee, Alabama	Grizzle et al. (1984)
Brown bullhead	Oral papilloma	Schuylkill River, Delaware[b]	Lucke and Schlumberger (1941)
	Oral papilloma	Lakes in Polk County, Florida	Harshbarger and Clark (1990)
	Oral and epidermal papilloma	Western Great Lakes, inland lakes, New York	Black (1983); Baumann et al. (1987); Smith et al. (1989)
	Dermal melanoma	Sundbury River, Massachusetts, E. Lake Erie and Upper Niagara River	Harshbarger and Clark (1990)
Freshwater drum	Dermal chromatoblastoma, hepatocellular	Sudbury River, Massachusetts, E. Lake Erie, and Upper Niagara River	Black (1983)
White sucker	Epidermal papilloma, hepatocellular, cholangiocellular	Great Lakes system	Black (1983); Sonstegard (1977); Smith et al. (1989); Hayes et al. (1990)

Notes

a   Neoplasms exhibit a range of invasive potential from non-invasive to locally extensive invasion to an occasional neoplasm exhibiting metastatic growth.

b   Historical data (1941); 166 tumor-bearing fish were studied (160/166 had oral tumors), indicating the tumors must have been common in these populations. Fish (40/166) also had epidermal tumors located at sites other than the mouth (e.g. barbels, head, body).

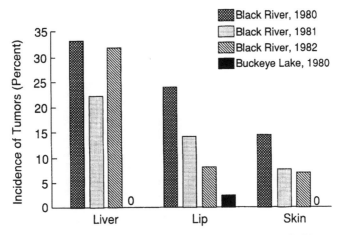

Figure 3.16 Incidence of grossly visible tumors (percent) in brown bullhead age 3 and older from the Black River, Ohio, a PAH-contaminated system, and from Buckeye Lake, Ohio, a reference location (from Baumann and Whittle, 1988). (Reproduced with permission from Elsevier.)

(2) biologic monitoring of tumor frequency in various tissues, (3) exposure to specific chemicals or groups of chemicals found in the environment, and (4) correlation of specific chemicals found in the tissues with the type and frequency of the tumors found.

Research needs

The following discussion describes what we believe to be some of the more outstanding research needs in a continuing effort to understand skin as an important exchange surface, and to make skin toxicology useful in the development of environmental risk assessments.

Skin dose–response paradigm

To use the present data from studies on the toxic responses of the skin there must be a dose–response paradigm established. Presently, toxicologists and risk assessors can indicate that there are alterations in the skin of chemically exposed fish, but little information is available on what this means in terms of their survival, normal physiologic–biochemical function or behavior. More emphasis needs to be applied to the effects of non-caustic chemicals on fish skin and the development of new toxic endpoints for these chemicals indicative of exposure and effects. There are also few or no data currently available regarding how sensitive or responsive fish skin is to environmental chemicals in comparison with other commonly used toxicity test endpoints or target organs, or regarding how certain groups of chemicals may have a greater impact on skin. Therefore, the development and use of toxicity tests that provide these types of data would make toxic responses

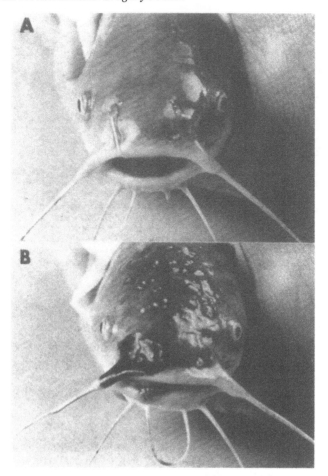

Figure 3.17 After 18 months, control fish (A) painted with solvent only and (B) experimental fish painted with the residue of river sediment extracts showing multiple skin papillomas superimposed upon epidermal hyperplasia (from Black, 1982). (Reproduced with permission from Batelle Press.)

of the skin, as described in Appendix 1, important toxicity test endpoints that could be more effectively used in predictive toxicology and risk assessment.

Causative agents for skin tumor epizootics

In the past, a considerable effort has gone into determining the causative agents for several major skin tumor epizootics in both fresh and saltwater environments. Yet, so far in most cases, no specific chemical cause can be linked to these lesions. Because these may be indicative of large-scale environmental contamination and possible human exposure, further investigations are warranted. These early research efforts on skin tumor epizootics involved sampling and analysis of feral fish

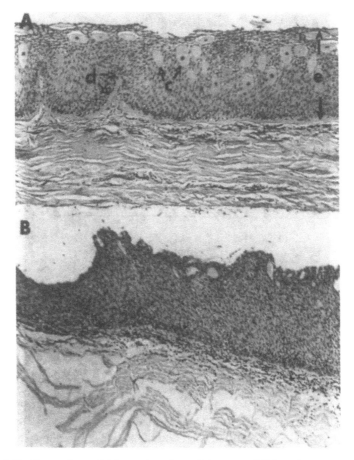

Figure 3.18 Comparison of skin sections from (A) a control fish and (B) a hyperplastic river sediment extract-treated fish. Note irregular thickening of the epidermis (e), loss of alarm substance cells (c), and loss of dermal papillae (d) in the treated fish (120×, from Black, 1982). (Reproduced with permission from Batelle Press.)

populations that involved costly and difficult analytical work on many different environmental chemicals in water, sediment, and tissue. The complexity of these efforts made it difficult to acquire definitive answers on cause and effect. Because of these early difficulties and the continuing need to understand the causative agents for skin tumor epizootics, more research is needed on laboratory studies that deal with specific suspect chemicals chronically applied directly to fish skin at varying concentrations.

Skin absorption, biotransformation, and excretion

The current emphasis on the use of fish early life stage toxicity testing to predict chronic toxicity and to screen for chemicals causing endocrine disruption,

Figure 3.19 Chromatophoromas on the skin and fins of the croaker, *Nibea mitsukurii*, thought to be caused by environmental carcinogens (e.g. nifurpirinol) (from Kimura *et al.*, 1984) (US government publication).

developmental abnormalities, and cancer suggests a strong need for a well-parameterized toxicity model for fish early life stages. To accomplish this goal will require a major research effort in the area of skin absorption, biotransformation, and excretion in fish early life stages.

An accurate description of the absorption of parent chemicals across fish skin is critical to modeling the disposition of various chemicals taken up across the large skin exchange surface. To accomplish this more effectively, research efforts need to focus on *in vitro* percutaneous absorption with skin flaps to determine the permeability of chemicals with varying structure (unionized and ionized) across the skin of larval, juvenile, and adult fish. The emphasis on skin absorption should definitely include new *in vivo* methods of determining the absorption across the skin of larval and early juveniles, such as microdialysis (McKim, Jr. *et al.*, 1993), separation of gill water exposure from skin water exposure by a latex dam (Saarikoski *et al.*, 1986), and the development of new microrespirometer chambers for larval and early juvenile fish (Rombough and Moroz, 1990; Rombough and Ure, 1991; Rombough, 1998). In addition, *in vitro* skin methods utilizing diffusion cells with patches of excised skin (Scheuplein and Blank, 1971) could be developed for fish to provide valuable information on the diffusion rates of chemicals across the skin of both scaled and scaleless fish.

The skin of larval stages acts as the primary respiratory surface, whereas in the

Table 3.4 Some observed tumor frequencies in brown bullhead in relation to aromatic hydrocarbon pollution (from Black and Baumann, 1991).

Waterway	Hepatic neoplasia (%)			External tumors[a]	Sediment PAH[b]
	Altered foci	Hepatocellular	Bile ductular (%)		
Black River, 1982	19.2	32.2	37.2	13.5	81.0
n	121	121	121	193	
Cuyahoga River, 1984–7	23.1	11.5	11.5	12.4	9.4
n	52	52	52	145	
Buffalo River, 1983, 1986	38.8	5.5	11.1	22.6	6.7
n	36	36	36	53	
St Mary's River, 1984	16.7	8.3	8.3	0.0	1.8
n	12	12	12	68	
Old Woman's Creek, 1984–8	10.0	0.0	0.0	1.7	0.45
n	10	10	10	230	
Buckeye Lake, 1982	0.0	0.0	0.0	1.5	0.07
n	10	10	10	78	

Notes
a Epidermal and oral neoplasms.
b Combined values of benzanthracene, benzofluoranthenes, 3,4-benzo(a)pyrene in microgram per gram dry weight of sediment.

early juvenile stages the skin remains quite thin and is more readily penetrated by chemicals than at later stages. Determining chemical absorption rates of fish skin should also include additional work on understanding the elimination rates of parent chemicals and major metabolites to allow the more effective development of useful predictive toxicity models. Finally, detailed morphologic studies of the skin vascular system, including measurements of blood flow, would greatly improve the parameterization and accuracy of the present skin exchange models.

The skin of fish was shown earlier in this chapter, through *in vitro* studies of skin strips, to have the necessary phase I and II enzymes within the epidermis and dermis to metabolize endogenous steroids. However, this has not been demonstrated for exogenous chemicals and is important in the further development of predictive toxicity models. Because of the difficulties involved in the homogenization of fish skin, the use of skin slices would probably result in the best data set for demonstrating the potential for skin biotransformation of xenobiotics. If fish skin has the proper enzymes to biotransform xenobiotics, then are chemicals taken up across the skin biotransformed prior to entering the systemic blood? If so, how do the biotransformation rates compare with other tissues, such as the liver?

The skin excretion of steroids and steroid conjugates by fish has been previously reported along with their ability to excrete other chemical substances into the environment that are known to have hormonal effects or act as chemical signals. Further investigations into these fish pheromones produced by the skin are encouraged to provide new insights into highly sensitive, subtle, endpoints that may have a sensitive, chronic impact on fish survival, behavioral interactions and reproduction.

Appendix 1 A summary of lethal, sublethal and chronic fish skin toxicity for eight different categories of waterborne stressors.

Chemical	Fish species	Mean size (g)	Exposure concentration (mg L⁻¹)	Toxic responses of the skin	References
1. Metals					
Cadmium [Cd(NO₃)₂]	*Cyprinus carpio* (carp) (freshwater) (scaled)	22	0.560, 0.022 static renewal	All dead in 8 days at 0.560, whereas fish at 0.022 survived 21 days. **Epidermis:** copious mucus on surface and many mucous cells near surface; necrotic pavement cells; highly active filament cells; mature club cells migrated out, many were newly differentiated; chloride cells appeared; massive extravasation of leukocytes; mast cells appeared; **Dermis:** dermal 'herrings' (bundles of collagen fibers) common; formation of new capillaries; melanosomes became common, extended into epidermis; erythrocytes in matrix	Iger *et al.* (1994c)
Copper (CuSO₄)	*Heteropneustes fossilis* (freshwater, air-breathing catfish) (scaleless)	28	0.32 static renewal	7 days' sublethal exposure; architectural abnormalities in the surface epidermal cells; copious mucus secretion; loss of shape, size, and structure of epidermal cells and mucous goblet cells; increase in tubular dilated mucous cells	Khangarot and Tripathi (1991)

Appendix 1 Continued

Chemical	Fish species	Mean size (g)	Exposure concentration (mg L^{-1})	Toxic responses of the skin	References
Lead (PbNO$_4$)	Cyprinus carpio (carp) (freshwater) (scaled)	250	1.5–6.0 (static renewal)	2 months' sublethal exposure. **Epidermis:** lymphocytes common near basal lamina and throughout; granulocytes and macrophages appeared; mucous cells releasing contents at surface decrease in number with time; chloride cells appear. **Dermis:** capillaries observed near basal lamina; diapedesis occurs in endothelium; erythrocytes invading the dermis	Iger and Abraham (1989)
Zinc (ZnSO$_4$)	Notopterus notopterus (freshwater) (scaled)	200	39.85 (static renewal)	60 days' sublethal exposure at 20°C; copious mucous secretions on surface of skin; tumorous growth at base of pectoral fin; epidermis thickens; mucous cells show hyperplasia and hyperactivity; club cells show hypertrophy with shrinkage of cytoplasm; necrotic basement membrane; dermal connective tissue shows degenerative changes	Roy et al. (1993)
Chromium (K$_2$Cr$_2$O$_7$)	Heteropneustes fossilis (freshwater, air-breathing catfish) (scaleless)	28	5.6 (static renewal)	7 days' sublethal exposure at 26°C; mucus cells loose and have hexagonal or polygonal shape; mucous cells at surface of skin acquired a dilated flask or cylindrical shape; some focal necrosis in mucous cells; hypersecretion in mucous cells; dilated mucous cells are needle-shaped at their tips	Khangarot and Tripathi (1992)

Contaminant	Species		Concentration	Effects	Reference
Mercury ($HgCl_2$)	*Heteropneustes fossilis* (freshwater, air-breathing catfish) (scaleless)	34	0.30 (static renewal)	10 days' exposure at 96-h LC_{50} concentration. **General appearance:** copious amounts of mucus over surface of fish; Hg very corrosive; erosion of barbels causing them to break off at tip and at base; many hemorrhages and ulcers noted on lateral sides of body; degeneration and shedding of fins and their distal extremities. **Cellular changes:** hyperactivity of goblet mucous cells, discharge contents and degenerate; degenerating club cells are sloughed at the epidermal surface to form a crust or scab of cellular debris; simultaneous regeneration from lower levels of the epidermis and dermis ongoing with degeneration in lower levels; space left behind by degeneration is quickly filled with haphazardly arranged polygonal epithelial cells	Rajan and Banerjee (1991)
Mercury ($HgCl_2$)	*Heteropneustes fossilis* (freshwater, air-breathing catfish) (scaleless)	21	0.1–1.0 (static)	The acute toxic effects of Hg in this study follow those described above for 0.3 mg L^{-1}; higher concentration of 1.0 mg L^{-1} caused sloughing of the epidermal layer; vacuolization in the muscle bundles and subcutis; destruction of all cell types. Recovery in contaminant-free water occurred in a week in most cases with the thickened epidermal layer thinning to normal; deeper skin lesions required up to 10 days	Prasad and Shil (1993)
Triphenyltin	*Phoxinus phoxinus* (European minnow) (freshwater) (scaled)	Fry	0.005–0.018	5 days' exposure of newly hatched larvae at 16°C; degenerative changes in the skin such as hydropic vacuolation in the cytoplasm of epithelial cells at 0.005 mg L^{-1}; at higher concentrations, nuclear alterations and erosion of the epithelium also occurred	Fent and Meier (1994)

Appendix 1 Continued

Chemical	Fish species	Mean size (g)	Exposure concentration (mg L⁻¹)	Toxic responses of the skin	References
2. Acid pH					
Acid pH stress	*Ictalurus nebulosus* (brown bullhead) (freshwater) (scaleless)	200	pH 5.7–6.15 (static)	5 days' sublethal exposure to low pH; mucous cell volume density increased; no increase in cell diameter or volume; number of mucous cells doubled	Zuchelkowski *et al.* (1981)
Acid pH stress	*Salvelinus fontinalis* (brook trout) (freshwater) (scaled)	Finger-lings	pH 2.2–10.8 (flow-through)	166 h lethal and sublethal exposure to low pH at 10°C and 20°C; temperature had no effect on degree or form of tissue injury; mucous cells of gills, nares, and skin exhibited progressive degrees of hypertrophy and excessive secretion of mucus with increased pH stress; epithelial necrosis and sloughing extensive on gills, corneae, and skin; thresholds for tissue and cellular changes were pH 5.2 and 9.0.	Daye and Garside (1976)
Acid pH stress	*Salmo salar* (Atlantic salmon) (freshwater) (scaled)	Alevin	pH 3.7–6.8 (static)	Effects noted in skin of alevins continuously exposed from fertilization to various levels of acidic pH at 5–6°C; pH 5.0 and below caused cell dysplasia; pH 4.2 and below caused epithelial sloughing; pH 4.0 and below caused nuclear pyknosis and epithelial and mucous cell hyperplasia; skin damage at this stage could be cause of death as involved in ion regulation and respiration prior to functional gills	Daye and Garside (1980)

Acid pH stress	*Salvelinus fontinalis* (brook trout) (freshwater) (scaled)	Swim-up stage	pH 4.3–6.3	40 days' exposure of eggs through swim-up at 10°C; pH 4.3 and 5.1 caused an increase in mucous cell volume density, an increase in mucous cell diameter, and an increase in the number of mucous cells	Ingersoll *et al.* (1990)
3. Detergents					
Linear alkyl benzene sulfonate (LAS detergent)	*Cirrhina mrigala* (freshwater) (scaled)	2 cm	0.005	30 days' exposure to 25% of the LC_{50} at 25°C; copious amounts of mucus were secreted by the epithelial mucous cells; clumping of mucus on the skin surface indicated the likely interaction between constituents of mucus and LAS	Misra *et al.* (1987)
Dodecylbenzene sodium sulfonate (C_{12}-LAS detergent)	*Rita rita* (freshwater) (scaled)	14 cm	6.9	45 months' exposure caused a doubling of the mucous cells in the opercular epithelium; this was followed by a rapid reduction in mucous cells as mucus was released to the surface of the skin; this cycling continued during exposure; opercular mucus composition changed from acidic glycoprotein to neutral glycoprotein; however, at a later stage the mucus shifts to acidic glycoprotein and acidic glycosaminoglycans, which represents a pathologic condition of the skin, as observed in mammals	Roy (1988)

Appendix 1 Continued

Chemical	Fish species	Mean size (g)	Exposure concentration (mg L^{-1})	Toxic responses of the skin	References
Linear alkyl benzene sulfonate (LAS detergent)	*Heteropneustes fossilis* (freshwater air-breathing catfish) (scaleless)	55	1.0–2.5 (static renewal)	24–192 h of sublethal exposures at 25°C; marked increase in mucous cells that produce acidic glycoproteins with a shift later to sialic acids; loss of mitochondrial enzymes in upper levels; glucose-6-phosphate and lactate dehydrogenase increased considerably; these changes were correlated with metabolic requirements for enhanced production of mucus under stress; changes in enzyme activity may provide new experimental model for effects of pollutant stress	Zaccone *et al.* (1985)
4. Inorganics Ammonium sulfate (dissolves in water to NH$_3$ and NH$_4^+$)	*Heteropneustes fossilis* (freshwater, air-breathing catfish) (scaleless)	38	200.0 (static renewal)	45 days' sublethal exposure at 10% of 96-h LC$_{50}$. **Outer opercule epidermis:** increased density and size of mucous cells; mucous cell cycling in cell size and density; vacuoles appear in necrotic cells; club cells exhibit vacuolization; epidermis sloughs off in time. **Inner opercular epidermis:** less necrotic change than outer; hyperplasia and vacuolization at various stages; more mucoprotein secreted was thought to give better protection of epidermis and respiratory surface	Paul and Banerjee (1996)

Compound	Species	Age/size	Concentration	Effects	Reference
Ammonia (un-ionized)	*Nibea japonica* (Japanese Croaker Larvae) (saltwater) (scaled)	3 days' old	0.26 (flow-through)	24-h exposure of 3-day-old larvae to the 24-h LC_{50} at 15°C; very thin epidermis in control larvae; mucous cells reduced; mucous cells showed no hyperplastic ability, which may explain greater susceptibility of fish larvae to ammonia; rough apical surface of epithelial cells appeared as irregular foldings; cytoplasmic vacuolation of epithelial cells; disorganization of mitochondrial cristae in chloride cells; vacuolation in the chondrocytes	Guillen *et al.* (1994)
5. Organics					
Malachite green	*Heteropneustes fossilis* (freshwater air-breathing catfish) (scaleless)	28	0.1–0.5 (static)	96-h LC_{50} of 0.2 mg L^{-1}; the authors used a decrease in the number of mucous cells to quantitate the impact of malachite green on the epidermis; mucocyte density bioassay showed malachite green causes a significant decrease in the number of mucous cells of dorsal and opercular epidermis, causing a rapid breakdown in the mucous barrier	Rajan and Banerjee (1994)
Chloramine-T	*Oncorhynchus mykiss* (rainbow trout) (freshwater) (scaled)	16.5	5.0, 10.0, and 20.0 (1-h stat-bath, two per week for 1 month)	10.0 and 20.0 mg L^{-1} caused a predisposition to an erosive dermatitis of the caudal fin; appeared to be caused by opportunistic pathogens of the genus *Pseudomonas* spp. and *Flavobacter* spp.; actual etiology not known	Powell *et al.* (1994)

Appendix 1 Continued

Chemical	Fish species	Mean size (g)	Exposure concentration (mg L⁻¹)	Toxic responses of the skin	References
Organic manure (fertilizer)	*Cyprinus carpio* (carp) (freshwater) (scaled)	175	3000.0	1–30 days' exposure at 20°C; thickness of epidermis increased from 140 to 180 μm; pigment cells extended into cytoplasmic extensions making the fish very dark (thought to be a stress response); copious amounts of mucus accompanied by a drop in the number of mature mucous cells and an increase in number of newly differentiated mucous cells; thick mucous coat continuous; pavement cell secretions formed glycocalyx over body; lymphocytes and macrophages infiltrated the epidermis; despite high bacterial count in water, no bacteria were observed in skin or in the mucous coat	Iger *et al.* (1988)
Polycyclic aromatic hydrocarbons (PAH)	*Ictalurus nebulosus* (brown bullhead) (freshwater) (scaleless)	Adult (20 cm)	Skin painting with 5.0% sediment extract once per week	18 months' exposure at 25°C to 18 PAHs ranging from 700 to 28 000 ng g⁻¹ in sediment extracts from Black and Buffalo rivers. Painted skin changes were: 3 months, blanching and coarsening of skin, slight hyperemia on lips; 10–14 months, skin darkening, irregular pitting of skin due to hyperplasia of epidermis (Figure 3.17), increased basophilia, increase in cell number and thickening of basal layer of epidermis and loss of normal skin architecture (i.e. loss of dermal papillae and club cells; Figure 3.18); at termination, half of surviving fish had papillomas (Figure 3.17)	Black (1983) and Black *et al.* (1985)

Note: in the superscript mg L⁻¹ and ng g⁻¹ the units are $mg\ L^{-1}$ and $ng\ g^{-1}$.

Compound	Species	Life stage/size	Exposure	Effects	Reference
Nifurpirinol (NP)	*Nibea mitsukurii* (croaker) (saltwater) (scaled)	Juvenile (13 cm)	Injected (s.c., i.p.) and static water	1-h water exposure of 0.5 and 1.0 mg L^{-1} NP per month; caused 71% and 100% incidence, respectively, of chromatophoromas (melanotic lesion) on the skin and fins of the croaker (Figure 3.19)	Kimura *et al.* (1984)
3-Methylcholanthrene	*Gasterosteus aculeatus* (stickleback) and *Rhodeus amarrus* (bitterling, freshwater, scaled)	Juvenile to adult	Skin painted with 0.5 mg twice per week	3- to 6-month induction of epitheliomas on the painted skin area	Ermer (1970)
Benzo(a)pyrene	*Gasterosteus aculeatus* (stickleback) and *Rodeus amarrus* (bitterling, freshwater, scaled)	Juvenile to adult	Skin painted with 0.5 mg twice per week	3- to 7-month induction of epitheliomas on the painted skin area	Ermer (1970)
6. Mixtures					
Chlorinated sewage pond	*Ictalurus melas* (black bullhead) (scaleless)	Adult	Fish taken from pond	Collections of fish made from December, 1979, through July, 1980; prevalence of oral papillomas was 65–75%; suggested chlorination of sewage formed carcinogenic organic compounds	Grizzle *et al.* (1981)
Pulp mill waste	*Perca fluviatilis* and *Carassius auratus* (perch and goldfish) (freshwater) (scaled)	100–300	Fish taken from river in gill nets	Histopathology of the fins showed acute fin erosion; edema and hyperemia in dermis; necrosis of the distal part of the fins; healed fin erosion was also noted in a number of individuals; no disease organisms noted; etiology, chemicals in waste stream	Lindesjoo and Thulin (1994)

Appendix 1 Continued

Chemical	Fish species	Mean size (g)	Exposure concentration (mg L^{-1})	Toxic responses of the skin	References
Rhine river water (metals, salts, organics, pathogens)	*Oncorhynchus mykiss* (rainbow trout) (freshwater) (scaled)	70	Filtered river water (flow-through)	7 days, fish skin became dark; 14–24 days, fish appeared pale and slightly green. **Epidermis:** 4–7 days, 30% decrease in thickness; 24 days, 30% increase in thickness; filament cells contain phagosomes and are very mitotic; pavement cells increase synthetic activity, many vesicles with high electron density; apoptotic pavement and filament cells; mucous cells elongated rather than ovoid with high electron–dense vesicles; a few apoptotic mucous cells. **Dermis:** 4 days, many fibroblasts in outer layer near basal lamina; outer dermis with many collagen fibers, which form dermal 'herrings' at 14–24 days; pigment cells move into cytoplasmic extensions; extensions extended into epidermis; some pigment cells degenerating at 14 days	Iger *et al.* (1994a)
7. UV-B radiation					
UV-B radiation	*Pleuronectes platessa* (plaice) (saltwater) (scaled)	Juveniles	18 mJ cm^{-2} UV-B at 30 μW cm^{-2}	Flagellate infestations on plaice skin caused a greater susceptibility to wavelengths of radiation in the UV-B bandwidth; non-infested skin showed moderate edema, whereas infested skin was strongly necrotic	Bullock (1985)

| UV-B radiation | *Oncorhynchus mykiss* (rainbow trout) (freshwater) (scaled) | Brood-stock | Field study | Broodstock at a commercial trout hatchery was observed; grayish focal thickening of the outermost zone of the dorsal fin web; this progressed to fin erosion and sloughing of the entire fin; malpighian cells lack organization; nuclear droplets present in central zone of epidermis; loss of attachment between the basal zone and underlying basement membrane; sunburn cells present in the epidermis; stratum compactum contains large numbers of leukocytes and macrophages; melanocytes no longer discrete units but sporadic with distended dendrites | Bullock and Coutts (1985) |
| UV-B radiation | *Pleuronectes platessa* (plaice) (saltwater) (scaled) | Juveniles | Field study | Juvenile plaice from shallow pools were monitored for sunburn just prior to emigration to deeper water; sunburn in some individuals showed thickening of epidermis; filament cells appeared necrotic with large pale-staining nuclei; dorsal epidermis had a vacuolated appearance and lacked structural organization; a frequent feature of traumatized skin was its separation from the underlying musculature; extreme separation also had an amorphous exudate in the resultant space | Berghahn *et al.* (1993) |

Appendix 1 Continued

Chemical	Fish species	Mean size (g)	Exposure concentration (mg L^{-1})	Toxic responses of the skin	References
UV-B radiation	*Oncorhynchus mykiss* (rainbow trout) *Oncorhynchus apache* (apache trout) *Oncorhynchus clarki henshawi* (lahontan cutthroat trout)	0.53 g 1.4 g 1.3 g		7 days' exposure to a high (357 μW cm^{-2} or 6.43 J cm^{-2} day^{-1}) and a low (190 μW cm^{-2} or 3.42 J cm^{-2} day^{-1}) UV-B; sunburn caused a darkening of the dorsal skin; degradation of the epidermal layer and a decrease in mucus in the darkened area just posterior to the head and anterior to and around the dorsal fin extending to the lateral line; loss of mucous protection allowed infection by opportunistic fungus; 93% of the fish affected in 7 days with no difference between high and low UV-B exposure. Apache trout were not affected	Little and Fabacher (1994)
UV-B radiation	*Polyodon spathula* (paddlefish) (freshwater) (scaleless)	Juveniles Covered vs. uncovered		51 days of exposure to UV-B in covered and uncovered raceways; first documentation of sunburn in paddlefish; covered, no skin lesions; uncovered, burned with skin lesions and sloughing; rostrum curved upward; raceways must be covered to avoid mortality and lesions in juvenile paddlefish	Ramos *et al.* (1994)

8. Temperature
Temperature stress

Oncorhynchus mykiss (rainbow trout) (freshwater) (scaled)	276	7°C rise in temperature	3-h exposure to a 7°C rise in temperature (from 15°C to 22°C); fish sampled over 14 days; skin thickness decreased at 3 h; after 24 h, skin thickness increased above controls because of increased mitotic activity; enhanced aging indicated by appearance of apoptotic cells; filament cells and mucous cell vesicles with peroxidase activity; epidermis and dermis invaded by many lymphocytes and macrophages, the latter containing peroxidase; pigment cells extend into epidermis; synthetic activity stimulated in fibroblasts; response same as for other stressors	Iger *et al.* (1994b)

References

Agrawal, S.K., Ral, A.K., Banerjee, T.K. and Mittal, A.K. 1979. Histophysiology of the epidermal mucous cells in relation to salinity in a freshwater teleost, *Heteropneustes fossilis* (Bloch) (Heteropneustidae, Pisces). *Zoologische Beitraege* 25: 403–410.

Ali, S.A., Schoonen, W.G.E.J., Lambert, J.G.D., Van den Hurk, R. and Van Oordt, P.G.W.J. 1987. *General and Comparative Endocrinology* 66: 415–424.

Allen, A. 1978. Structure of gastrointestinal mucous glycoproteins and the viscous and gel forming properties of mucous. In *Chemical Aspects of Mucus*. Clamp, J.R., Allen, A., Gibbons, R.A. and Roberts, G.P. (eds). *British Medical Bulletin* 34: 28–33.

Asakawa, M. 1974. Sialic acid containing glycoprotein in the external mucus of eel, *Anguilla japonica* Temminck et Shlegel-II carbohydrate and amino acid composition. *Bulletin of the Japanese Society of Scientific Fisheries* 40: 303–308.

Banerjee, T.K. 1993. Response of fish skin to certain ambient toxicants. In *Advances in Fish Research*, Vol. I. Singh, B.R (ed.), pp. 185–192. Narendra Publishing House, Delhi.

Barber, M.C., Suarez, L.A. and Lassiter, R.R. 1988. Modeling bioconcentration of nonpolar organic pollutants by fish. *Environmental Toxicology and Chemistry* 7: 545–558.

Barron, M.G., Tarr, B.D. and Hayton, W.L. 1987. Temperature-dependence of cardiac output and regional blood flow in rainbow trout, *Salmo gairdneri* (Richardson). *Journal of Fish Biology* 31: 735–744.

Baumann, P.C. and Whittle, D.M. 1988. The status of selected organics in the Laurentian Great Lakes: An overview of DDT, PCB, dioxins, furans, and aromatic hydrocarbons. *Aquatic Toxicology* 11: 241–257.

Baumann, P.C., Smith, W.D. and Parland, W.K. 1987. Tumor frequencies and contaminant concentrations in brown bullheads from an industrialized river and a recreational lake. *Transactions of the American Fisheries Society* 116: 79–86.

Baumann, P.C., Smith, I.R. and Metcalfe, C.D. 1996. Linkages between chemical contaminants and tumors in benthic great lakes fish. *Great Lakes Research* 22: 131–152.

Bereiter-Hahn, J. and Zylberberg, L. 1993. Regeneration of teleost fish scale. *Comparative Biochemistry and Physiology* 105A: 625–641.

Berghahn, R., Bullock, A.M. and Karakiri, M. 1993. Effects of solar radiation on the population dynamics of juvenile flatfish in the shallows of the Wadden Sea. *Journal of Fish Biology* 42: 329–345.

Bernadsky, G., Sar, N. and Rosenberg, E. 1993. Drag reduction of fish skin mucus: Relationship to mode of swimming and size. *Journal of Fish Biology* 42: 797–800.

Bertelsen, S.L., Hoffman, A.D., Gallinat, C.A., Elonen, C.M. and Nichols, J.W. 1998. Evaluation of log K_{ow} and tissue lipid content as predictors of chemical partitioning to fish tissues. *Environmental Toxicology and Chemistry* 17: 1447–1455.

Black, J.J. 1982. Epidermal hyperplasia and neoplasia in brown bullheads (*Ictalurus nebulosus*) in response to repeated applications of a PAH containing extract of polluted river sediment. In *Polynuclear Aromatic Hydrocarbons: Formation, Metabolism, and Measurement*. Cooke, M.W. and Dennis, A.J. (eds), pp. 149–211. Batelle Press, Columbus, OH.

Black, J.J. 1983. Field and laboratory studies of environmental carcinogenesis in Niagara River fish. *Journal of Great Lakes Research* 9: 326–334.

Black J.J. and Baumann, P.C. 1991. Carcinogens and cancers in freshwater fishes. *Environmental Health Perspectives* 90: 27–33.

Black, M.C. and McCarthy, J.F. 1988. Dissolved organic macromolecules reduce the uptake of hydrophobic organic contaminants by the gills of rainbow trout (*Salmo gairdneri*). *Environmental Toxicology and Chemistry* 7: 593–600.

Black, J., Fox, H., Black, P. and Bock, F. 1985. Carcinogenic effects of river sediment extracts in fish and mice. In *Water Chlorination: Chemistry, Environmental Impact, and Health Effects*, Vol. 5. Jolley, R.L., Condie, L.W., Johnson, J.D., Katz, S., Minear, R.A., Mattice, J.S. and Jacobs, V.A. (eds), Ch. 33. CRC Press, New York.

Black, M.C., Millsap, D.S. and McCarthy, J.F. 1991. Effects of acute temperature change on respiration and toxicant uptake by rainbow trout, *Salmo gairdneri* (Richardson). *Physiological Zoology* 64: 145–168.

Boddington, M.J., Mackenzie, B.A. and DeFreitas, A.S.W. 1979. A respirometer to measure the uptake efficiency of waterborne contaminants in fish. *Ecotoxicology and Environmental Safety* 3: 383–393.

Bradshaw, C.M., Richard, A.S. and Sigel, M.M. 1971. Igm antibodies in fish mucus. *Proceedings of the Society for Experimental Biology and Medicine* 136: 1122.

Bullock, A.M. 1985. The effect of ultraviolet-B radiation upon the skin of the plaice, *Pleuronectes platessa* L., infested with the bodonid ectoparasite *Ichthyobodo necator* (Henneguy, 1883). *Journal of Fish Diseases* 8: 547–550.

Bullock, A.M. and Coutts, R.R. 1985. The impact of solar ultraviolet radiation upon the skin of rainbow trout, *Salmo gairdneri* Richardson, farmed at high altitude in Bolivia. *Journal of Fish Diseases* 8: 263–272.

Bullock, A.M. and Roberts, R.J. 1975. The dermatology of marine teleost fish. I. The normal integument. *Oceanography and Marine Biology* 13: 383–411.

Chiou, C.T., Freed, V.H., Schmedding, D.W., and Kohnert, R.L. 1977. Partition coefficient and bioaccumulation of selected organic chemicals. *Environmental Science and Technology* 11: 475–478.

Collander, R. 1949. The permeability of plant protoplasts to small molecules. *Physiology Plantarum* 2: 300–311.

Colombo, L., Marconato, A., Colombo Belvedere, P. and Frisco, C. 1980. Endocrinology of teleost reproduction: A testicular steroid pheromone in black goby, *Gobius jozo* L. *Bollettino di Zoologia* 47: 355–364.

Crowther, R.S. and Marriott, C. 1984. Counter-ion binding to mucus glycoproteins. *Journal of Pharmacy and Pharmacology* 36: 21–26.

Daniel, T.L. 1981. Fish mucus: *In situ* measurements of polymer drag reduction. *Biological Bulletin: Marine Biological Laboratory, Woods Hole* 160: 376–382.

Dawe, C.J. 1981. Polyoma tumors in mice and X cell tumors in fish, viewed through telescope and microscope. In *Phyletic Approaches to Cancer*. Dawe, C.J., Harshbarger, J.C., Kondo, S., Sugimura, T. and Takayama, S. (eds), pp. 19–49. Japan Science Society Press, Tokyo.

Dawe, C.J., Bagshaw, J. and Poore, C.M. 1979. Amebic pseudotumors in pseudobranches of Pacific cod, *Gadus macrocephalus* (Abstract). *Proceedings of the American Association for Cancer Research* 20: 245.

Dawson, M.A. and Renfro, J.L. 1990. Organic ion secretion by winter flounder renal proximal tubule primary monolayer cultures. *Journal of Pharmacology and Experimental Therapeutics* 254: 39–44.

Daye, P.G. and Garside, E.T. 1976. Histopathologic changes in surfical tissues of brook trout, *Salvelinus fontinalis* (Mitchill), exposed to acute and chronic levels of pH. *Canadian Journal of Zoology* 54: 2140–2155.

Daye, P.G. and Garside, E.T. 1980. Structural alterations in embryos and alevins of the Atlantic salmon, *Salmo salar* (L.), induced by continuous or short-term exposure to acidic levels of pH. *Canadian Journal of Zoology* 58: 27–43.

Diconza, J.J. 1970. Some characteristics of natural haemagglutinins found in serum and mucus of the catfish, *Tachysurus australis. Australian Journal of Experimental Biology and Medical Science* 48: 515.

Enmoto, N., Nakagawa, H. and Tomiyasu. Y. 1964. Studies on the external mucous substance of fish. IX. Preparation of crystalline n-acetyl neuraminic acid from the external mucous substance of the loach. *Bulletin of the Japanese Society of Scientific Fisheries* 30: 495–499.

Erickson, R.J. and McKim, J.M. 1990a. A simple flow-limited model for exchange of organic chemicals at fish gills. *Environmental Toxicology and Chemistry* 9: 159–165.

Erickson, R.J. and McKim, J.M. 1990b. A model for exchange of organic chemicals at fish gills: Flow and diffusion limitations. *Aquatic Toxicology* 18: 175–198.

Ermer, M. 1970. Versuche mit cancerogenen Mitteln bei kurzlebigen Fischarten. *Zoologischer Anzeiger* 184: 175–193.

Fent, K. and Meier, W. 1994. Effects of triphenyltin on fish early life stages. *Archives of Environmental Contamination and Toxicology* 27: 224–231.

Ferreira, J.T., Schoonbee, H.J. and Smit, G.L. 1984. The uptake of the anaesthetic benzocaine hydrochloride by the gills and the skin of three freshwater fish species. *Journal of Fish Biology* 25: 35–41.

Fletcher, T.C. 1978. Defence mechanisms in fish. In *Biochemical and Biophysical Perspectives in Marine Biology*, Vol. IV. Malins, D.C. and Sargent, J.R. (eds), pp. 189–222, Academic Press, New York.

Fletcher T.C. and Grant, P.T. 1968. Glycoproteins in the external mucous secretions of the plaice, *Pleuronectes platessa*, and other fishes. *Biochemical Journal* 106: 205–224.

Fletcher, T.C. and Grant, P.T. 1969. Immunoglobulin in the serum and mucus of the plaice (*Pleuronectes platessa*). *Biochemical Journal* 115: 65.

Fletcher, T.C. and White, A. 1973. Lysozyme activity in the plaice (*Pleuronectes platessa* L.). *Experientia* 29: 1283.

Gillette, J.R. 1987. Dose, species, and route extrapolation: general aspects. In *Pharmacokinetics in Risk Assessment*, Vol. 8. National Academy Press, Washington, DC.

Grizzle, J. 1979. *Anatomy and Histology of the Golden Shiner and Fathead Minnow*. Special Publication, National Technical Information Service (NTIS) PB-294219. Auburn University Press, Auburn, AL.

Grizzle, J.M. and Rogers, W.A. 1976. *Anatomy and Histology of the Channel Catfish*. Special Publication (Project 2-187-R). Auburn University Press, Auburn, AL.

Grizzle, J., Schwedler, T.E. and Scott, A.L. 1981. Papillomas of black bullheads, *Ictalurus melas* (Rafinesque), living in a chlorinated sewage pond. *Journal of Fish Diseases* 4: 345–351.

Grizzle, J.M., Melius, M.P. and Strength, D.R. 1984. Papillomas on fish exposed to chlorinated wastewater effluent. *Journal of the National Cancer Institute* 73: 1133–1142.

Guillen, J.L., Endo, M., Turnbull, J.F., Kawatsu, H., Richards, R.H. and Aoki, T. 1994. Skin responses and mortalities in the larvae of Japanese croaker exposed to ammonia. *Fisheries Science* 60: 547–550.

Guy, R.H., Hadgraft, J. and Maibach, H.I. 1985. Percutaneous absorption in man: A kinetic approach. *Toxicology and Applied Pharmacology* 78: 123–129.

Harris, J. and Hunt, S. 1973. Epithelial mucins of the Atlantic salmon (*Salmo salar* L.). *Transactions of the Biochemical Society* 1: 153–155.

Harshbarger, J.C. and Clark, J.B. 1990. Epizootiology of neoplasms in bony fish of North America. *The Science of the Total Environment* 94: 1–32.

Hawkes, J.W. 1983. Skin and scales. In *Microscopic Anatomy of Salmonids: An Atlas* (Resource Publication 150). Yasutake, W.T. and Wales, J.H. (eds), pp. 14–23. US Department of the Interior, Fish and Wildlife Service, Washington, DC.

Hawkes, J.W. 1974a. The structure of fish skin. I. General organization. *Cell and Tissue Research* 149: 147–148.

Hawkes, J.W. 1974b. The structure of fish skin. II. The chromatophore unit. *Cell and Tissue Research* 149: 159–172.

Hay, J.B., Hodgins, M.B. and Roberts, R.J. 1976. Androgen metabolism in skin and skeletal muscle of the rainbow trout (*Salmo gairdnerii*) and in accessory sexual organs of the spur dogfish (*Squalus acanthias*). *General and Comparative Endocrinology* 29: 402–413.

Hayes, M.A., Smith, I.R., Crane, T.L., Rushmore, T.H., Thorn, C., Kocal, T.E. and Ferguson, H.W. 1990. Pathogenesis of skin and liver neoplasms in white suckers (*Catostomus commersoni*) from industrially polluted areas in Lake Ontario. *Science of the Total Environment* 94: 105–123.

Hayton, W.L. and Barron, M.G. 1990. Rate-limiting barriers to xenobiotic uptake by the gill. *Environmental Toxicology and Chemistry* 9: 151–158.

Heath, A.G. 1987. Respiratory cardiovascular responses. In *Water Pollution and Fish Physiology*. pp. 31–49, CRC Press, Boca Raton.

Henrickson, R.C. and Matoltsy, A.G. 1968a. The fine structure of teleost epidermis. I. Introduction and filament containing cells. *Journal of Ultrastructure Research* 21: 194–212.

Henrickson, R.C. and Matoltsy, A.G. 1968b. The fine structure of teleost epidermis. II. Mucus cells. *Journal of Ultrastructure Research* 21: 213–221.

Henrickson, R.C. and Matoltsy, A.G. 1968c. The fine structure of teleost epidermis. III. Club cells and other cell types. *Journal of Ultrastructure Research* 21: 222–232.

Hoffman, A.D., Bertelsen, S.L. and Gargas, M.L. 1992. An in vitro gas equilibration method for determination of chemical partition coefficients in fish. *Comparative Biochemistry and Physiology* 101A: 47–51.

Hoffman, A.D., Jenson, C.T., Lien, G.J. and McKim, J.M. 1999. Individual tissue weight to total body weight relationships and total, polar, and nonpolar lipids in tissues of hatchery lake trout. *Transactions of the American Fisheries Society* 128: 178–181.

Iger, Y. and Abraham, M. 1989. Effects of lead pollution on carp skin. In *Proceedings of the 1989 European Aquaculture Society Conference (Special publication, No. 10)*. Billard, R. and Pauw, N. (eds), pp. 131–133. European Aquaculture Society Press, Bordeaux, France.

Iger, Y., Abraham, M., Dotant, A., Fattal, B. and Rahamim, E. 1988. Cellular responses in the skin of carp maintained in organically fertilized water. *Journal of Fish Biology* 33: 711–720.

Iger, Y., Jenner, H.A. and Wendelaar Bonga, S.E. 1994a. Cellular responses in the skin of rainbow trout (*Oncorhynchus mykiss*) exposed to Rhine water. *Journal of Fish Biology* 45: 1119–1132.

Iger, Y., Jenner, H.A. and Bonga, S.E.W. 1994b. Cellular responses in the skin of the trout (*Oncorhynchus mykiss*) exposed to temperature elevation. *Journal of Fish Biology* 44: 921–935.

Iger, Y., Lock, R.A.C., Van der Meij, J.C.A. and Wendelaar Bonga, S.E. 1994c. Effects of water-borne cadmium on the skin of the common carp (*Cyprinus carpio*). *Archives of Environmental Contamination and Toxicology* 26: 342–350.

Ingersoll, C.G., Sanchez, D.A., Meyer, J.S., Gulley, D.D. and Tietge, J.E. 1990. Epidermal response to pH, aluminum, and calcium exposure in brook trout (*Salvelinus fontinalis*) fry. *Canadian Journal of Fisheries and Aquatic Sciences* 47: 1616–1622.

Ishimatsu, A., Iwama, G.K., Bentley, T.B. and Heisler, N. 1992. Contribution of the secondary circulatory system to acid–base regulation during hypercapnia in rainbow trout (*Oncorhynchus mykiss*). *Journal of Experimental Biology* 170: 43–56.

Jakubowski, M. 1977. Size and vascularization of respiratory surfaces of gills and skin in some Cobitids (Cobitidae, Pisces). *Bulletin De L'Academie Polonaise Des Sciences, Serie des sciences biologiques* Cl. II. 25: 307–316.

Jakubowski, M. 1982. Size and vascularization of the gill and skin respiratory surfaces in the white amur, Ctenopharyngodon idella (Val.) (Pisces, Cyprinidae). *Acta Biologica Cracoviensia, Series: Zoologia*, 24: 93–104.

Jakubowski, M. and Rembiszewski, J.M. 1974. Vascularization and size of respiratory surfaces of gills and skin in the antarctic fish Gymnodraco acuticeps Boul. (Bathydraconidae). *Bulletin De L'Academie Polonaise Des Sciences, Serie des sciences biologiques* Cl. II. 22: 305–313.

Khangarot, B.S. and Tripathi, D.M. 1991. Changes in humoral and cell-mediated immune responses and in skin and respiratory surfaces of catfish, *Saccobranchus fossilis*, following copper exposure. *Ecotoxicology and Environmental Safety* 22: 291–308.

Khangarot, B.S. and Tripathi, D.M. 1992. The stereoscan observations of the skin of catfish, *Saccobranchus fossilis*, following chromium exposure. *Journal of Environmental Science and Health (Part A)* 27: 1141–1148.

Kimura, I., Taniguchi, N., Kumai, H., Tomita, I., Kinae, N., Yoshizaki, K., Ito, M. and Ishikawa, T. 1984. Correlation of epizootiological observations with experimental data: Chemical induction of chromatophoromas in the croaker, *Nibea mitsukurii*. In *Use of Small Fish Species in Carcinogenicity Testing, National Cancer Institute Monograph* 65, pp. 139–154, NIH Publication No. 84-2653, National Cancer Institute, Bethesda, MD.

Kirsch, R. and Nonnotte, G. 1977. Cutaneous respiration in three fresh-water teleosts. *Respiration Physiology* 29: 339–354.

Lambert, J.G.D., Van den Hurk, R., Schoonen, W.G.E.J., Resink, J.W. and Van Oordt, P.G.W.J. 1986. Gonadal steroidogenesis and the possible role of steroid glucuronides as sex pheromones in two species of teleosts. *Fish Physiology and Biochemistry* 2: 101–107.

Landrum, P.F., Hayton, W.L., Lee, II, H.L., McCarty, L.S., Mackay, D. and McKim, J.M. 1994. Synopsis of discussion sessions on the kinetics behind environmental bioavailability. In *Bioavailability: Physical, Chemical, and Biological Interactions*. Hamelink, J.L., Landrum, P.F., Bergman, H.L. and Benson, W.H. (eds). CRC Press, Boca Raton.

Leonard, J.B. and Summers, R.G. 1976. The ultrastructure of the integument of the American eel, *Anguilla rostrata. Cell and Tissue Research* 171: 1–30.

Lien, G.J. and McKim, J.M. 1993. Predicting branchial and cutaneous uptake of 2,2′,5,5′-tetrachlorobiphenyl in fathead minnows (*Pimephales promelas*) and Japanese medaka (*Oryzias latipes*): Rate limiting factors. *Aquatic Toxicology* 27: 15–32.

Lien, G.J., Nichols, J.W., McKim, J.M. and Gallinat, C.A. 1994. Modeling the accumulation of three waterborne chlorinated ethanes in fathead minnows (*Pimephales promelas*): A physiologically based approach. *Environmental Toxicology and Chemistry* 13: 1195–1205.

Lindesjoeoe, E. and Thulin, J. 1994. Histopathology of skin and gills of fish in pulp mill effluents. *Diseases of Aquatic Organisms* 18: 81–93.

Little, E.E. and Fabacher, D.L. 1994. Comparative sensitivity to rainbow trout and two threatened salmonids, Apache trout and Lahontan cutthroat trout, to ultraviolet-B radiation. *Archives of Hydrobiology* 43: 217–226.

Lu, M., Wagner, G.F. and Renfro, J.L. 1994. Stanniocalcin stimulates phosphate reabsorption by flounder renal proximal tubule in primary culture. *American Journal of Physiology* 267: R1356–R1362.

Lucke, B. and Schlumberger, H. 1941. Transplantable epitheliomas of the lip and mouth of catfish. *Journal of Experimental Medicine* 74: 397–408.

McDonald, D.G. and McMahon, B.R. 1977. Respiratory development in Arctic char *Salvelinus alpinus* under conditions of normoxia and chronic hypoxia. *Canadian Journal of Zoology* 55: 1461–1467.

McElman, J.F. and Balon, E.K. 1980. Early ontogeny of white sucker, *Catostomus commersoni*, with steps of saltatory development. *Environmental Biology of Fishes* 5: 191–224.

McKim, J.M. 1994. Physiological and biochemical mechanisms that regulate the accumulation and toxicity of environmental chemicals in fish. In *Bioavailability: Physical, Chemical, and Biological Interactions*. Hamelink, J.L., Landrum, P.F., Bergman, H.L. and Benson, W.H. (eds), pp. 179–203. CRC Press, Boca Raton.

McKim, J.M. and Erickson, R.J. 1991. Environmental impacts on the physiological mechanisms controlling xenobiotic transfer across fish gills. *Physiological Zoology* 64: 39–67.

McKim, J.M. and Goeden, H.M. 1982. A direct measure of the uptake efficiency of a xenobiotic chemical across the gills of brook trout (*Salvelinus fontinalis*) under normoxic and hypoxic conditions. *Comparative Biochemistry and Physiology* 72C: 65–74.

McKim, J.M. and Nichols, J.W. 1994. Use of physiologically based toxicokinetic models in a mechanistic approach to aquatic toxicology. In *Aquatic Toxicology: Molecular, Biochemical, and Cellular Perspectives*. Malins, D.C. and Ostrander, G.K. (eds), pp. 469–519, CRC Press, Boca Raton.

McKim, J.M., Schmieder, P.K. and Veith, G. 1985. Absorption dynamics of organic chemical transport across trout gills as related to octanol–water partition coefficient. *Toxicology and Applied Pharmacology* 77: 1–10.

McKim, J.M., Jr, McKim, J.M., Sr, Naumann, S., Hammermeister, D.E., Hoffman, A.D. and Klaassen, C.D. 1993. *In vivo* microdialysis sampling of phenol and phenyl glucuronide in the blood of unanesthetized rainbow trout. *Fundamental and Applied Toxicology* 20: 190–198.

McKim, J.M., Nichols, J.W., Lien, G.J. and Bertelsen, S.L. 1994. Respiratory-cardiovascular physiology and chloroethane gill flux in the channel catfish, *Ictalurus punctatus. Journal of Fish Biology* 44: 527–547.

McKim, J.M., Nichols, J.W., Lien, G.J., Hoffman, A.D., Gallinat, C.A. and Stokes, G.N. 1996. Dermal absorption of three waterborne chloroethanes in rainbow trout (*Oncorhynchus mykiss*) and channel catfish (*Ictalurus punctatus*). *Fundamental and Applied Toxicology* 31: 218–228.

Marshall, W.S. 1978. On the involvement of mucous secretion in teleost osmoregulation. *Canadian Journal of Zoology* 56: 1088–1091.

Marshall, W.S. 1988. Passive solute and fluid transport in brook trout (*Salvelinus fontinalis*) urinary bladder epithelium. *Canadian Journal of Zoology* 66: 912–918.

Marshall, W.S., Bryson, S.E. and Idler, D.R. 1989. Control of ion transport by the sperm duct epithelium of brook trout (*Salvelinus fontinalis*). *Fish Physiology and Biochemistry* 7: 331–336.

Marshall, W.S., Bryson, S.E. and Wood, C.M. 1992. Calcium transport by isolated skin of rainbow trout. *Journal of Experimental Biology* 166: 297–316.

Mawdesley-Thomas, L.E. 1975. Neoplasia in fish. In *The Pathology of Fishes*. Ribelin, W.E. and Migaki, G. (eds), pp. 805–870. University of Wisconsin Press, Madison.

Misra, V., Chawla, G., Kumar, V., Lal, H. and Viswanathan, P.N. 1987. Effect of linear alkyl benzene sulfonate in skin of fish fingerlings (*Cirrhina mrigala*): Observations with scanning electron microscope. *Ecotoxicology and Environmental Safety* 13: 164–168.

Mittal, A.K. and Nigam, G.D. 1986. Fish skin surface lipids: Phospholipids. *Journal of Fish Biology* 29: 123–138.

Mix, M.C. 1986. Cancerous diseases in aquatic animals and their association with environmental pollutants: A critical literature review. In *Marine Environmental Research, Special Issue*, Vol. XX, nos 1 and 2. Heath, G.W. and US Regional (eds). Roesijadi, G. and Spies, R.B. (eds). Elsevier Applied Science Press, Essex, UK.

Musch, M.W., McConnell, F.M., Goldstein, L. and Field, M. 1987. Tyrosine transport in winter flounder intestine: Interaction with NA+-K+-2Cl-cotransport. *American Journal of Physiology* 253: R264–R269.

Nakano, T., Sato, M. and Takeuchi, M. 1993. Superoxide dismutase activity in the skin of fish. *Journal of Fish Biology* 43: 492–496.

Nichols, J.W., McKim, J.M., Andersen, M.E., Gargas, M.L., Clewell, III, H.J. and Erickson, R.J. 1990. A physiologically-based toxicokinetic model for the uptake and disposition of waterborne organic chemicals in fish. *Toxicology and Applied Pharmacology* 106: 433–447.

Nichols, J.W., McKim, J.M., Lien, G.J., Hoffman, A.D. and Bertelsen, S.L. 1991. Physiologically-based toxicokinetic modeling of three waterborne chloroethanes in rainbow trout (*Oncorhynchus mykiss*). *Toxicology and Applied Pharmacology* 110: 374–389.

Nichols, J.W., McKim, J.M., Lien, G.J., Hoffman, A.D., Bertelsen, S.L. and Gallinat, C.A. 1993. Physiologically-based toxicokinetic modeling of three waterborne chloroethanes in channel catfish (*Ictalurus punctatus*). *Aquatic Toxicology* 27: 83–112.

Nichols, J.W., McKim, J.M., Lien, G.J., Hoffman, A.D., Bertelsen, S.L. and Elonen, C.M. 1996. A physiologically based toxicokinetic model for dermal absorption of organic chemicals by fish. *Fundamental and Applied Toxicology* 31: 229–242.

Nonnotte, G. 1981. Cutaneous respiration in six freshwater teleosts. *Comparative Biochemistry and Physiology* 70A: 541–543.

Nonnotte, G. 1984. Cutaneous respiration in the catfish *Ictalurus melas. Comparative Biochemistry and Physiology* 78A: 515–517.

Nonnotte, G. and Kirsch, R. 1978. Cutaneous respiration in seven seawater teleosts. *Respiration Physiology* 35: 111–118.

Oikawa, S. and Itazawa, Y. 1985. Gill and body surface areas of the carp in relation to body mass, with special reference to the metabolism–size relationship. *Journal of Experimental Biology* 117: 1–14.

Olivereau, M. and Lemoine, A.M. 1971. Teneur en acide N-acetyl-neuraminique de la peau chez l'anguille apres autotransplantation de l'hypophyse. *Zeitschrift fuer Vergleichnende Physiologie (Journal of Comparative Physiology)* 73: 44–52.

Ourth, D.D. 1980. Secretory IgM, lysozyme and lymphocytes in the skin mucus of the channel catfish, *Ictalurus punctatus. Developmental and Comparative Immunology* 4: 65.

Paul, V.I. and Banerjee, T.K. 1996. Ammonium sulphate induced stress related alterations in the opercular epidermis of the live fish *Heteropneustes* (Saccobranchus) *fossilis* (Bloch). *Current Science (Bangalore)* 70: 1025–1029.

Pickering, A.D. 1974. The distribution of mucous cells in the epidermis of the brown trout *Salmo trutta* (L.) and the char *Salvelinus alpinus* (L.). *Journal of Fish Biology* 6: 111–118.

Pickering, A.D. 1976. Synthesis of N-acetyl neuraminic acid from [14C] glucose by the epidermis of the brown trout, *Salmo trutta* (L.). *Comparative Biochemistry and Physiology* 54B: 325–328.

Piiper, J. and Scheid, P. 1984. Model analysis of gas transfer in fish gills. In *Fish Physiology, Gills – Anatomy, Gas Transfer and Acid–base Regulation*, Vol. XA. Hoar, W.S. and Randall, D.J. (eds), pp. 229–262. Academic Press, New York.

Powell, M.D., Speare, D.J. and MacNair, N. 1994. Effects of intermittent chloramine-T exposure on growth, serum biochemistry, and fin condition of juvenile rainbow trout (*Oncorhynchus mykiss*). *Canadian Journal of Fisheries and Aquatic Sciences* 51: 1728–1736.

Prasad, M. and Shil, M. 1993. Histopathology of skin of the catfish, *Heteropneustes fossilis*: Short-term effects of mercuric chloride. *Journal of Fish Diseases*, 16: 797–800.

Rajan, M.T. and Banerjee, T.K. 1991. Histopathological changes induced by acute toxicity of mercuric chloride on the epidermis of freshwater catfish, *Heteroneustes fossilis. Ecotoxicology and Environmental Safety* 22: 139–152.

Rajan, M.T. and Banerjee, T.K. 1994. Toxicity of a fish disinfectant, malachite green to *Heteropneustes fossilis* and induced changes in the mucocyte density of its skin. *Journal of Freshwater Biology* 6: 177–182.

Ramos, K.T., Fries, L.T., Berkhouse, C.S. and Fries, J.N. 1994. Apparent sunburn of juvenile paddlefish. *Progressive Fish Culture* 56: 214–216.

Richards, I.S. 1974. Caudal neurosecretory system: Possible role in pheromone production. *Journal of Experimental Zoology* 187: 405–408.

Roberts, R.J. and Bullock, A.M. 1976. The dermatology of marine teleost fish. II. Dermatopathology of the integument. *Oceanography and Marine Biology an Annual Review* 14: 227–246.

Rombough, P.J. 1998. Partitioning of oxygen uptake between the gills and skin in fish larvae: A novel method for estimating cutaneous oxygen uptake. *The Journal of Experimental Biology* 201: 1763–1769.

Rombough, P.J. and Moroz, B.M. 1990. The scaling and potential importance of cutaneous and branchial surfaces in respiratory gas exchange in young chinook salmon (*Oncorhynchus tshawytscha*). *Journal of Experimental Biology* 154: 1–12.

Rombough, P.J. and Ure, D. 1991. Partitioning of oxygen uptake between cutaneous and branchial surfaces in larval and young juvenile chinook salmon (*Oncorhynchus tshawytscha*). *Physiological Zoology* 64: 717–727.

Rosen, M.W. and Cornford, N.E. 1971. Fluid friction of fish slimes. *Nature (London)* 234: 49–51.

Roy, D. 1988. Statistical analysis of anionic detergent-induced changes in the goblet mucous cells of opercular epidermis and gill epithelium of *Rita rita* (Ham.) (Bagridae: Pisces). *Ecotoxicology and Environmental Safety* 15: 260–271.

Roy, U.K., Gupta, A.K. and Chakrabarti, P. 1993. Deleterious effect of zinc on the skin of *Notopterus notopterus* (Pallas). *Journal of Freshwater Biology* 5: 191–196.

Rozman, K.K. and Klaassen, C.D. 1996. Absorption, distribution, and excretion of toxicants. In *Casarett and Doull's Toxicology: the Basic Science of Poisons*. Klaassen, C.D. (ed.), pp. 91–112. McGraw-Hill, New York.

Saarikoski, J., Lindstrom, R., Tyynela, M. and Viluksela, M. 1986. Factors affecting the absorption of phenolics and carboxylic acids in the guppy (*Poecilia reticulata*). *Ecotoxicology and Environmental Safety* 11: 158–173.

St. Louis, C.E.A., Oterland, C.K. and Anderson, P.D. 1984. Evidence for cutaneous secretory immune system in rainbow trout (Salmo gairdneri). *Developmental and Comparative Immunology* 8: 71–80.

Satchell, G.H. 1984. Respiratory toxicology of fishes. In *Aquatic Toxicology*, Vol. II. Weber, L.J. (ed.) pp. 1–50. Raven Press, New York.

Scheuplein, R.J. and Blank, I.H. 1971. Permeability of the skin. *Physiological Reviews* 51: 702–745.

Schindler, D.W. 1998. A dim future for boreal waters and landscapes. *BioScience* 48: 157–164.

Schmidt-Nielson, K. 1984. *Scaling: Why is Animal Size so Important?* Cambridge University Press, Cambridge, UK.

Schmieder, P.K. and Henry, T.R. 1988. Plasma binding of 1-butanol, phenol, nitrobenzene and pentachlorophenol in the rainbow trout and rat: A comparative study. *Comparative Biochemistry and Physiology* 91C: 413–418.

Schmieder, P.K. and Weber, L.J. 1992. Blood and water flow limitations of gill uptake of organic chemicals in the rainbow trout (*Onchorynchus mykiss*). *Aquatic Toxicology* 24: 103–122.

Shephard, K.L. 1981. The influence of mucus on the diffusion of water across fish epidermis. *Physiological Zoology* 54: 224–229.

Shephard, K.L. 1982. The influence of mucus on the diffusion of ions across the oesophagus of fish. *Physiological Zoology* 55: 23–34.

Shephard, K.L. 1984. The influence of mucus on the diffusion of chloride ions across the oesophagus of the minnow *Phoxinus phoxinus* (L.). *Journal of Physiology (London)* 346: 449–460.

Sijm, D.T.H.M. and Van der Linde, A. 1995. Size-dependent bioconcentration kinetics of hydrophobic organic chemicals in fish based on diffusive mass transfer and allometric relationships. *Environmental Science Technology* 29: 2769–2777.

Sijm, D.T.H.M., Part, P. and Opperhuizen, A. 1993. The influence of temperature on the uptake rate constants of hydrophobic compounds determined by the isolated perfused gills of rainbow trout (*Oncorhynchus mykiss*). *Aquatic Toxicology* 25: 1–14.

Sijm, D.T.H.M., Verberne, M.E., Part, P. and Opperhuizen, A. 1994. Experimentally determined blood and water flow limitations for the uptake of hydrophobic compounds using perfused gills of rainbow trout (*Oncorhynchus mykiss*): Allometric applications. *Aquatic Toxicology* 30: 325–341.

Smith, A.A., New, R.A., Wiles, J.E. and Kleinow, K.M. 1996. Effect of varying sediment organic carbon content upon the dermal bioavailability and disposition of benzo(a)pyrene in the catfish, *Ictalurus punctatus*. *Marine Environmental Research* 42: 87–91.

Smith, I.R., Ferguson, H.W. and Hayes, M.A. 1989. Histopathology and prevalence of epidermal papilloma epidemic in brown bullhead, *Ictalurus nebulosus* (Lesueur), and white sucker, *Catostomus commersoni* (Lacepede) populations from Ontario, Canada. *Journal of Fish Diseases* 12: 373–388.

Smith, P.J.S., Sanger, R.H. and Jaffe, L.F. 1994. The vibrating Ca^{2+} electrode: a new technique for detecting plasma membrane regions of Ca^{2+} influx and efflux. *Methods in Cell Biology* 40: 115–134.

Soivio, A., Pesonen, S., Teravainen, T. and Nakari, T. 1982. Seasonal variations in oestrogen levels in the plasma of brown trout (*Salmo trutta lacustris*) and in the metabolism of testosterone in the skin. *Annals Zoologici Fennici* 19: 53–59.

Sonstegard, R.A. 1977. Environmental carcinogenesis studies in fishes of the Great Lakes of North America. *Annals of the New York Academy of Sciences* 298: 261–269.

Stein, W.D. 1967. *The Movement of Molecules Across Cell Membranes*. Academic Press, New York.

Streit, B. and Sire, E.O. 1993. On the role of blood proteins for uptake, distribution, and clearance of waterborne lipophilic xenobiotics by fish: A linear system analysis. *Chemosphere* 26: 1031–1039.

Streit, B., Sire, E.O., Kohlmaier, G.H., Badeck, F.W. and Winter, S. 1991. Modelling ventilation efficiency of teleost fish gills for pollutants with high affinity to plasma proteins. *Ecological Modelling* 57: 237–262.

Todd, J.H., Atema, J. and Bardach, J.E. 1967. Chemical communication in social behavior of a fish, the yellow bullhead (*Ictalurus natalis*). *Science* 158: 672–673.

Tovell, P.W.A., Howes, D. and Newsome, C.S. 1975. Absorption, metabolism and excretion by goldfish of the anionic detergent sodium lauryl sulphate. *Toxicology* 4: 17–29.

Tsai, J.C. 1996. Cell renewal in the epidermis of the loach *Misgurnus anguillicaudatus* (Cypriniformes). *Journal of Zoology* 239: 591–599.

Van den Hurk, R. and Lambert, J.G.D. 1983. Ovarian steroid glucuronides function as sex pheromones for male zebrafish, *Brachydario rerio*. *Canadian Journal of Zoology* 61: 2381–2387.

Van Oosten, J. 1957. The skin and scales. In *The Physiology of Fishes*, Vol. I. Brown, M.E. (ed.) pp. 207–244. Academic Press, New York.

Varanasi, U. and Markey, D. 1978. Uptake and release of lead and cadmium in mucus and skin of coho salmon (*Oncorhynchus kisutch*). *Comparative Biochemistry and Physiology* 60C: 187–191.

Varanasi, U., Robisch, P.A. and Malins, D.C. 1975. Structural alterations in fish epidermal mucus produced by water-borne lead and mercury. *Nature (London)* 258: 431–432.

Varanasi, U., Uhler, M. and Stranahan, S.I. 1978. Uptake and release of naphthalene and its metabolites in skin and epidermal mucus of salmonids. *Toxicology and Applied Pharmacology* 44: 277–289.

Verbost, P.M., Bryson, S.E., Wendelaar Bonga, S.E. and Marshall, W.S. 1997. Na^+ dependent Ca^{2+} uptake in isolated opercular epithelium of *Fundulus heteroclitus*. *Journal of Comparative Physiology* 167: 205–212.

Wells, P.R. and Pinder, A.W. 1996. The respiratory development of Atlantic salmon. II. Partitioning of oxygen uptake among gills, yolk sac and body surfaces. *Journal of Experimental Biology* 199: 2737–2744.

Whitear, M. 1986a. The skin of fishes including cyclostomes-epidermis. In *Biology of the Integument. Vol. II. Vertebrates.* Bereiter-Hahn, J., Maltoltsy, A.G. and Richards, K.S. (eds), pp. 8–38. Springer-Verlag, Heidelberg.

Whitear, M. 1986b. The skin of fishes including cyclostomes-dermis. In *Biology of the Integument. Vol. II. Vertebrates.* Bereiter-Hahn, J., Maltoltsy, A.G. and Richards, K.S. (eds), pp. 39–64. Springer-Verlag, Berlin.

Wilke, C.R. and Chang, P. 1955. Correlation of diffusion coefficients in dilute solutions. *American Institute Chemical Engineering Journal* 1: 264–270.

Wold, J.K. and Selset, R. 1977. Glycoproteins in the skin mucous of the char (*Salmo alpinus* L.). *Comparative Biochemistry and Physiology* 56B: 215–218.

Wood, C.M. and Marshall, W.S. 1994. Ion balance, acid–base regulation, and chloride cell function in the common killifish, *Fundulus heteroclitus* – a euryhaline estuarine teleost. *Estuaries* 17(1A): 34–52.

Yokote, M. 1982. Skin. In *An Atlas of Fish Histology, Normal and Pathological Features.* Hibiya, T. (ed.) pp. 8–15. Kodansha, Tokyo.

Zaccone, G., Lo Cascio, P., Fasulo, S. and Licata, A. 1985. The effect of an anionic detergent on complex carbohydrates and enzyme activities in the epidermis of the catfish *Heteropneustes fossilis* (Bloch). *Histochemical Journal* 17: 453–466.

Zuchelkowski, E.M., Lantz, R.C. and Hinton, D.E. 1981. Effects of acid-stress on epidermal mucous cells of the brown bullhead *Ictaluru nebulosus* (LeSeur): A morphometric study. *Anatomical Record* 200: 33–39.

Further reading

Agarwal, S.K. and Shah, K.K. 1987. Histochemistry of the epidermis of a hill-stream fish, *Garra gotyla. Indian Journal of Animal Sciences* 57: 908–914.

Anders, K. and Yoshimizu, M. 1994. Role of viruses in the induction of skin tumors and tumor-like proliferations of fish. *Diseases of Aquatic Organisms* 19: 215–232.

Bowser, P.R., Wolf, M.J., Reimer, J. and Shane, B.S. 1991. Epizootic papillomas in brown bullheads *Ictalurus nebulosus* from Silver Stream Reservoir, New York. *Diseases of Aquatic Organisms*, 11: 117–127.

Brown, E.R., Hazdra, J.J., Keith, L., Greenspan, I., Kwapinski, J.B.G. and P. Beamer. 1973. Frequency of fish tumors found in a polluted watershed as compared to nonpolluted Canadian waters. *Cancer Research* 33: 189–198.

Bullock, A.M., Marks, R. and Roberts, R.J. 1978. The cell kinetics of teleost fish epidermis: mitotic activity of the normal epidermis at varying temperatures in plaice (*Pleuronectes platessa*). *Journal of Zoology (London)* 184: 423–428.

Cho, S.Y., Mohri, S., Endo, Y. and Fujimoto, K. 1992. Characterization of lipid prooxidants in sardine skin. *Bulletin of the Korean Fisheries Society* 25: 501–510.

Crouse-Eisnor, R.A., Cone, D.K. and Odense, P.H. 1985. Studies on relations of bacteria with skin surface of *Carassius auratus* L, and *Poecilia reticulata. Journal of Fish Biology* 27: 395–402.

Davidson, G.A., Ellis, A.E. and Secombes, C.J. 1993. Novel cell types isolated from the skin of rainbow trout, *Oncorhyncus mykiss. Journal of Fish Biology*, 42: 301–306.

Diconza, J.J. and Halliday, W.J. 1971. Relationship of catfish serum antibodies to immunoglobulin in mucus secretions. *Australian Journal of Experimental Biology and Medical Science* 49: 517–525.

Earnest-Koons, K., Wooster, G.A. and Bowser, P.R. 1996. Invasive walleye dermal sarcoma in laboratory-maintained walleyes *Stizostedion vitreum. Diseases of Aquatic Organisms* 24: 227–232.

Eddy, F.B. and Fraser, J.E. 1982. Sialic acid and mucus production in rainbow trout *Salmo gairdneri* in response to zinc and seawater. *Comparative Biochemistry and Physiology* 73C: 357–359.

Fauconneau B. and Saglio, P. 1984. Protein-bound and free amino acid content in the skin mucus of the European eel *Anguilla anguilla* (L.). *Comparative Biochemical Physiology* 77B: 513–516.

Finn, J.P. and Nielson, N.O. 1971. The inflammatory response of rainbow trout. *Journal of Fish Biology* 3: 463–478.

Fletcher, T.C. and White, A. 1973. Antibody production in the plaice (*Pleuronectes platessa* L.) after oral and parenteral immunization with *Vibrio anguillarum* antigens. *Aquaculture* 1: 417–420.

Fromm, P.O. 1980. A review of some physiological and toxicological responses of freshwater fish to acid stress. *Environmental Biology of Fishes* 5: 79–93.

Gona, O. 1979. Mucous glycoproteins of teleostean fish: A comparative histochemical study. *Histochemical Journal* 11: 709–718.

Goto-Nance, R., Watanabe, U., Kamiya, H. and Ida, H. 1995. Characterization of lectins from the skin mucus of the loach *Misgurnus anguillicaudatus*. *Fisheries Science* 61: 137–140.

Hamelink, J.L., Landrum, P.F., Bergman, H.L. and Benson, W.H. (eds). 1994. *Bioavailability: Physical, Chemical, and Biological Interactions*. CRC Press, Boca Raton, FL.

Handy, R.D. and Eddy, F.B. 1989. Surface absorption of aluminum by gill tissue and body mucus of rainbow trout, *Salmo gairdneri*, at the onset of episodic exposure. *Journal of Fish Biology* 34: 865–874.

Handy, R.D. and Eddy, F.B. 1990. The interactions between the surface of rainbow trout, *Oncorhynchus mykiss*, and waterborne metal toxicants. *Functional Ecology* 4: 385–392.

Harris, J.E., Watson, A. and Hunt, S. 1973. Histochemical analysis of mucus cells in the epidermis of brown trout *Salmo trutta* (L.). *Journal of Fish Biology* 5: 345–351.

Hawkins, W.E., Jacobs, A.D., Gregory, D.G. and Ostrander, G.K. 1996. Poorly differentiated mesenchymal neoplasm in the dermis of a white bass. *Journal of Aquatic Animal Health* 8: 150–154.

Iger, Y., Jenner, H.A. and Bonga, S.E.W. 1994. Cellular responses in the skin of carp (*Cyprinus carpio*) exposed to copper. *Aquatic Toxicology (Amsterdam)* 29: 49–64.

Iger, Y., Balm, P.H.M., Jenner, H.A. and Bonga, S.E.W. 1995. Cortisol induces stress-related changes in the skin of rainbow trout (*Oncorhynchus mykiss*). *General and Comparative Endocrinology* 97: 188–198.

Imaki, H. and Chavin, W. 1984. Ultrastructure of mucous cells in the sarocopterygian integument. *Scanning Electron Microscopy* 1: 409–422.

Ingale, S.R. and More, N.K. 1984. Histochemical analysis and significance of the mucopolysaccharides from the skin of three species of *Mystus*. *Comparative Physiology and Ecology* 9: 191–196.

Ingram, G.A. 1980. Substances involved in the natural resistance of fish to infection: A review. *Journal of Fish Biology* 16: 23–60.

Jakowska, S. 1963. Mucus secretion in fish (A note). *Annals of the New York Academy of Sciences* 106: 458–462.

Kimura, M., Hama, Y., Sumi, T., Asakawa, M., Rao, B.N.N., Horne, A.P., Li, S.C., Li, Y.T. and Nakagawa, H. 1994. Characterization of a deaminated neuraminic acid-containing glycoprotein from skin mucus of the loach, *Misgurnus anguillicaudatus*. *Journal of Biological Chemistry* 269: 32138–32143.

Lasker, R. and Threadgold, L.T. 1968. 'Chloride cells' in the skin of the larval sardine. *Experimental Cell Research* 52: 582–590.

Leppi, T.J. 1968. Morphochemical analysis of mucous cells in the skin and slime gland of Hag fishes. *Histochemie* 15: 68–78.

Lobb, C.J. and Clem, L.W. 1981. The metabolic relationships of the immunoglobulins in fish serum, cutaneous mucus, and bile. *Journal of Immunology* 127: 1525–1530.

Lobb, C.J. and Clem, L.W. 1981. Phylogeny of immunoglobulin structure and function. XI. Secretory immunoglobulins in the cutaneous mucus of the sheepshead, *Archosargus probatocephalus*. *Developmental and Comparative Immunology* 5: 587–590.

Lock, R.A.C. and Van Overbeeke, A.P. 1981. Effects of mercury chloride and methylmercuric chloride on mucous secretion in rainbow trout, *Salmo gairdneri* Richardson. *Comparative Biochemistry and Physiology* 69C: 67–73.

McCormick, S.D., Hasegawa, S. and Hirano, T. 1992. Calcium uptake in the skin of a freshwater teleost. *Proceedings of the National Academy of Sciences of the United States of America* 89: 3635–3638.

McKone, C.E., Young, R.G., Bache, C.A. and Lisk, D.J. 1971. Rapid uptake of mercuric ions by goldfish. *Environmental Science and Technology* 5: 1138–1139.

Marshall, W.S. 1979. Effects of salinity acclimation, prolactin, growth hormone, and cortisol on the mucous cells of *Leptocottus armatus* (Teleostei, cottidae). *General and Comparative Endocrinology* 37: 358–368.

Marshall, W.S. and Bern, H.A. 1979. Teleostean urophysis: Urotensin II and ion transport across the isolated skin of a marine teleost. *Science* 204: 519–521.

Mattheij, J.A.M. and Stroband, H.W.J. 1971. The effects of osmotic experiments and prolactin on the mucous cells in skin and ionocytes in the gill of the teleost, *Cichlasoma biocellatum*. *Zeitschrift fuer Zellforschung und Mikroskopische Anatomie* 121: 93–101.

Mattheij, J.A.M., Stroband, H.W.J., Kingma, F.J. and Van Oordt, P.G.W.J. 1972. Prolactin and osmoregulation in the cichlid fish, *Cichlasoma biocellatum*, and the effect of this hormone on the thyroid, gills and skin. *General and Comparative Endocrinology* 18: 607.

Miller, T.G. and Mackay, W.C. 1982. Relationship of secreted mucus to copper and acid toxicity in rainbow trout. *Bulletin of Environmental Contamination and Toxicology* 28: 68–74.

Mittal, A.K. 1968. Studies on the structure of the skin of *Rita rita* (Ham) (Bagridae, Pisces) in relation to its age and regional variations. *Indian Journal of Zoology* 9: 61–78.

Mittal, A.K. and Banerjee, T.K. 1975. Histochemistry and the structure of the skin of a murrel, *Channa striata* (Bloch, 1797) (Channiformes, Channidae). I. Epidermis. *Canadian Journal of Zoology* 53: 833–843.

Mittal, A.K. and Garg, T.K. 1994. Effect of an anionic detergent–sodium dodecyl sulphate exposure on club cells in the epidermis of Clarias batrachus. *Journal of Fish Biology* 44: 857–875.

Mittal, A.K., Agarwal, S.K. and Banerjee, T.K. 1976. Protein and carbohydrate histochemistry in relation to the keratinization in the epidermis of *Barbus sophor* (Cyrinidae, Pisces). *Journal of Zoology (London)* 179: 1–17.

Mittal, A.K., Whitear, M. and Agarwal, S.K. 1980. Fine structure and histochemistry of the epidermis of the fish (*Monopterus cuchia*). *Journal of Zoology (London)* 191: 107–125.

Nairn, R.S., Morizot, D.C., Kazianis, S., Woodhead, A.D. and Setlow, R.B. 1996. Nonmammalian models for sunlight carcinogenesis: genetic analysis of melanoma formation in Xiphorphorus hybrid fish. *Photochemistry and Photobiology* 64: 440–448.

Nakano, T., Ono., K. and Takeuchi, M. 1992. Levels of zinc, iron, and copper in the skin of abnormally pigmented Japanese flounder. *Bulletin of the Japanese Society of Scientific Fisheries* 58: 2207.

Octotake, M., Iwama, G.K. and Nakanishi, T. 1996. The uptake of bovine serum albumin by the skin of bath-immunised rainbow trout (*Oncorhynchus mykiss*). *Fish Shellfish Immunology* 6: 321–333.

Olson, K.R. 1996. Secondary circulation in fish: Anatomical organization and physiological significance. *Journal of Experimental Zoology* 275: 172–185.

Ostrander, G.K., Hawkins, W.E., Kuehn, R.L., Jacobs, A.D., Berlin, K.D. and Pigg, J. 1995. Pigmented subcutaneous spindle cell tumors in native gizzard shad (*Dorosoma cepedianum*). *Carcinogenesis* 16: 1529–1535.

Pfeiffer, W., Sasse, D. and Arnold, M. 1971. The alarm substance cells of *Phoxinus phoxinus* and *Morulius chrysophakedion* (Cyprinidae, Ostariophysi, Pisces). Histochemical and electron microscopical study. *Zeitschrift fuer Zellforschung und Mikroskopische Anatomie* 118: 203–213.

Phromsuthirak, P. 1977. Electron microscopy of wound healing in the skin of (*Gasterosteus aculeatus*). *Journal of Fish Biology* 3: 463–478.

Pickering, A.D. and Macey, D.J. 1977. Structure, histochemistry and effect of handling stress on the mucous cells of the epidermis of the char, *Salvelinus alpinus* (L.). *Journal of Fish Biology* 10: 505–512.

Pickering, A.D., Pottinger, T.G. and Christie, O. 1982. Recovery of the brown trout, *Salmo trutta* L., from acute handling stress: A time-course study. *Journal of Fish Biology* 20: 229–244.

Pohla-Gubo, G. and Adam, H. 1982. Influence of the anionative detergent Na-alkyl-benzenesulfonate (LAS) on the head-epidermis of juvenile rainbow trout *Salmo gairdneri* (Richardson). *Zoologischer Anzeiger* 209: 97–110.

Poulet, F.M., Casey, J.W. and Spitsbergen, J.M. 1993. Response of fish skin to certain ambient toxicants. *Diseases of Aquatic Organisms* 16: 97–104.

Powell, M.D., Speare, D.J. and Burka, J.F. 1992. Preservation of mucous biofilms of fish gills and skin: Potential diagnostic and research applications. In *Proceedings of the 1992 Meeting of the Aquaculture Association of Canada*, Vol. 92, pp. 64–66. University of British Columbia, Vancouver, BC.

Premdas, P.D. and Metcalfe, C.D. 1996. Experimental transmission of epidermal lip papillomas in white sucker, *Catostomus commersoni*. *Canadian Journal of Fisheries and Aquatic Sciences* 53: 1018–1029.

Roberts, R.J., Bell, M. and Young, H. 1973. Studies on the skin of plaice (*Plueronectes platessa* L.). II. The development of larval plaice skin. *Journal of Fish Biology* 5: 103–108.

Roberts, R.J., Shearer, W.M., Elson, K.G.R. and Munroe, A.L.S. 1973. Studies on ulcerative dermal necrosis of salmonids. I. The skin of the normal salmon head. *Journal of Fish Biology* 2: 223–229.

Rodger, H.D., Inglis, V. and Richards, R.H. 1995. Effect of trenbolone acetate on development of Atlantic salmon skin and the lack of epidermal protection against laboratory-induced furunculosis. *Journal of Aquatic Animal Health* 7: 50–53.

Rombough, P.J. 1992. Intravascular oxygen tensions in cutaneously respiring rainbow trout (*Oncorhynchus mykiss*) larvae. *Comparative Biochemistry and Physiology* 101A: 23–27.

Rydevik, M. 1988. Epidermal thickness and secondary sexual characters in mature male and immature Baltic salmon, *Salmo salar* L., parr: Seasonal variations and effects of castration and androgen treatment. *Journal of Fish Biology* 33: 941–944.

Segner, H., Linnenbach, M. and Marthaler, R. 1979. Towards a use of epidermal mucous cells in the field assessment of acid stress in fish. *Applied Ichthyology* 3: 187–190.

Segner, H., Marthaler, R. and Linnenbach, M. 1988. Growth, aluminum uptake and mucous cell morphometrics of early life stages of brown trout, *Salmo trutta*, in low pH water. *Environmental Biology of Fishes* 21: 153–159.

Shelbourne, J.E. 1957. Site of chloride regulation in marine fish larvae. *Nature (London)* 180: 920–922.

Smith, I.R., Ferguson, H.W. and Hayes, M.A. 1989. Histopathology and prevalence of epidermal papilloma epidemic in brown bullhead, *Ictalurus nebulosus* (Lesueur), and white sucker, *Catostomus commersoni* (Lacepede) populations from Ontario, Canada. *Journal of Fish Diseases* 12: 373–388.

Stiffler, D.F. 1988. Cutaneous exchange of ions in lower vertebrates. *American Zoologist* 28: 1019–1029.

Stiffler, D.F., Graham, J.B., Dickson, K.A. and Stockmann, W. 1986. Cutaneous ion transport in the freshwater teleost *Synbranchus marmoratus. Physiological Zoology* 59: 406–418.

Takeda, T. 1996. Effects of experimental ventilation and ambient PO_2 on O_2 uptake of the isolated cutaneous tissue in the carp, *Cyprinus carpio. Comparative Biochemistry and Physiology* 113A: 107–111.

Takemura, A., Kanematsu, M. and Oka, M. 1996. Early sex distinction in greater amberjack *Seriola dumerili* using skin mucus. *Nippon Suisan Gakkaishi* 62: 62–67.

Teunis, P.F.M., Vredovoogd, W., Weterings, C., Bretschneider, F. and Peters, R.C. 1991. The emergence of electroreceptor organs in regenerating fish skin and concurrent changes in their transduction properties. *Neuroscience* 45: 205–212.

Toda, M., Goto-Nance, R., Muramoto, K. and Kamiya, H. 1996. Characterization of the lectin from the skin mucus of the kingklip *Genypterus capensis. Fisheries Science* 62: 138–141.

Uchida, N., Nakano, H., Hagimoto, R. and Anzai, H. 1992. Biological activities of fish skin collagen-thrombocyte aggregation-inducing activity and affinity for fibronectin. *Bulletin of the College of Agriculture and Veterinary Medicine Nihon University* 49: 111–116.

Van de Winkle, J.G.J., van Kuppevelt, T.H.M.S.M., Jannsen, H.M.J. and Lock, R.A.C. 1986. Glycosaminoglycans in the skin mucus of rainbow trout (*Salmo gairdneri*). *Comparative Biochemistry and Physiology* 85B: 473–475.

Vethaak, A.D. and Jol, J.G. 1996. Diseases of flounder *Platichthys flesus* in Dutch coastal and estuarine waters, with particular reference to environmental stress factors. I. Epizootiology of gross lesions. *Diseases of Aquatic Organisms* 26: 81–97.

Wendelaar Bonga, S.E. 1978. The effect of changes in external sodium, calcium, and magnesium concentrations on prolactin cells, skin and plasma electrolytes of *Gasterosteus aculeatus. General and Comparative Endocrinology* 34: 265–275.

Westfall, B.A. 1945. Coagulation film anoxia in fishes. *Ecology* 26: 283–287.

Whitear, M. 1993. Epithelial sensory cells in fish. In *Advances in Fish Research*, Vol. I. Singh, B.R. (ed.), pp. 169–184. Narendra Publishing House, Delhi.

Whitear, M. and Mittal, A.K. 1986. Structure of the skin of *Agonus cataphractus* (Teleostei). *Journal of Zoology (London)* A210: 551–574.

Wiklund, T. and Bylund, G. 1993. Skin ulcer disease of flounder *Platichthys flesus* in the northern Baltic Sea. *Diseases of Aquatic Organisms* 17: 165–174.

Yamashita, E., Arai, S. and Matsuno, T. 1996. Metabolism of xanthophylls to vitamin A and new apocarotenoids in liver and skin of black bass, *Micropterus salmoides. Comparative Biochemistry and Physiology* 113B: 485–489.

Zaccone, G. 1980. Structure, histochemistry and effect of stress on the epidermis of *Ophisurus serpens* (L) (Telesoteiophichtidae). *Cellular and Molecular Biology* 26: 663–674.

Zuchelkowski, E.M., Lantz, R.C. and Hinton, D.E. 1986. Skin mucous cell response to acid stress in male and female brown bullhead catfish, *Ictalurus nebulosus* (LeSeur). *Aquatic Toxicology* 8: 39–148.

4   Toxic responses of the liver

David E. Hinton, Helmut Segner,
and Thomas Braunbeck

Introduction

This chapter is about the liver as a target organ of toxic chemicals. It is a target organ because of: its large blood supply leading to pronounced toxicant exposure and accumulation; its clearance function involving microvasculature, hepatocytes, possibly phagocytic cells, and intrahepatic biliary system; and its pronounced metabolic capacity, critical for internal homeostasis and for survival of the organism.

The liver serves three main functions. (1) *Uptake, metabolism, storage and redistribution of nutrients and other endogenous molecules.* The liver has a central function in maintaining homeostasis of the organism by synthesis and secretion of molecules into the blood as well as by removal, metabolism, and eventually excretion of compounds. For example, this organ governs blood glucose levels, it synthesizes and releases hormones such as somatomedins, it removes hormones from the blood circulation, it produces and secretes proteins (serum albumins, vitellogenin, choriogenin or zona radiata protein) as well as serum cholesterol and lipids. Many hepatic products are released into the hepatic venous blood for use by distant organs and tissues. (2) *Metabolism of xenobiotics.* Biotransformation catalyzes the conversion of poorly excretable, lipophilic chemicals into more readily excretable water-soluble compounds. In general, hepatic biotransformation is a detoxification process in that it enhances excretion of toxic substances and thereby decreases toxic body burden. However, during the biotransformation process, many electrophilic reactive species are generated which readily interact with basic cellular constituents such as DNA and proteins. This can lead to disruption of normal cellular function and result in toxicity including carcinogenesis. (3) *Formation and excretion of bile.* Bile excretion is important for the elimination of degradation products of endogenous compounds such as heme or steroid hormones, and, in addition, for elimination of xenobiotics and their metabolites and some metals such as copper and mercury.

All of these functions of the liver, synthesis and redistribution of nutrients and intermediary metabolites, biotransformation, and bile formation, may be involved in the generation of hepatic lesions. It is this great metabolic capacity of the liver that makes it both a target and an organ to protect itself. When 'toxic hits' on the target occur, they may lead to alterations or injury in liver structure and function.

Because of the multiple physiologic functions of the liver and its considerable plasticity, the liver responds to toxic insults in many different ways. Thus, there exists no prototype reaction in hepatotoxicity. Rather, a combination of morphologic pattern, functional alteration, and mechanisms is used to classify hepatotoxicants (Vandenberghe, 1996).

Two further aspects of liver toxicity have to be emphasized. First, deterioration of hepatic structure and function is not only relevant for the liver itself but may also lead to sequella aberrations in other organs and to death of the organism. Second, the liver possesses a pronounced capacity to acclimate to toxic stress: its great metabolic capacity makes it a target of toxic substances while its capacity to produce protective molecules or to perform efficient repair and recovery and acclimation provide for survival and production of a state that is altered, but compatible with life.

An understanding of chemical hepatotoxicity requires an appreciation of anatomic and physiologic features of the liver. With respect to fish liver, it is important to emphasize that although fish and rodent liver agree in many features there are still many differences that may strongly influence chemical toxicity. In this chapter, therefore, we seek to: (1) provide a short review of the available knowledge on the anatomy and physiology of the piscine liver, (2) illustrate toxic mechanisms in this organ using studies of reference hepatotoxicants, (3) review model systems for investigations of hepatotoxicity, (4) review cellular targets, early, intermediate, and endstage aspects of carcinogenesis as an example of chronic toxicity, and (5) summarize information gaps and future directions.

Important aspects of fish liver anatomy related to interpretation of toxicity

Certain features of the hepatic gross and microscopic anatomy of fish are recognized as being different from those of mammalian liver (for detailed descriptions, see Hampton *et al.*, 1985, 1989; reviewed in Hinton, 1998), and they are important considerations in a text of target organ toxicity. Whereas chapters in conventional toxicology texts illustrate important vascular and parenchymal relationships of mammalian liver (for recent examples, see Moslen, 1995; Vandenberghe, 1996), the assumption that these are relevant to livers of fish may lead to confusion when attempting to interpret toxic manifestations in the latter vertebrates. Table 4.1 provides a comparison of rodent and fish hepatic anatomic features and Table 4.2 reviews quantitative aspects of liver in mammalian and fish species. First, gross inspection of livers of fish reveals that most are in the form of a single lobe, whereas livers of mammals have multiple lobes. Second, the classic lobule, a consistent microscopic feature of mammalian liver and represented herein with sections of mouse liver (Figure 4.1), is not present in livers of fish (Figure 4.2). Portal tracts found at corners of the classic lobule in mammals (Figure 4.3) are delimited by perilobular connective tissue and contain bile duct, portal venule, and hepatic arteriole. Portal tracts indicate sites where arterial blood enters hepatic sinusoids supplying hepatocytes at the periphery of

Table 4.1 Comparison of rodent and fish hepatic anatomical features.

Anatomic entity	Rodent	Fish
Liver lobes	Multiple	Single
Lobules	Distinct	Indistinct or absent
Portal tracts	At corners of the classic lobule; contain bile duct, portal venule, hepatic arteriole; define sites where arterial blood enters hepatic microcirculation	Absent; larger bile ducts may coexist with hepatic artery in so-called 'biliary arterial tract' (BAT); no portal vein branches are in BATs
Blood supply	Hepatic artery and portal vein: capillary-like sinusoids contain arterial and venous blood	Same
Venous drainage	Hepatic veins to caudal vena cava	Hepatic veins to sinus venosus
Architecture	Laminae as single layer between adjacent sinusoids	Tubules
Biliary system	Canaliculi and hierarchy of passageways: cholangioles, small and large bile ducts	Same; short canaliculi enter bile preductules or cholangioles in center of hepatic tubules
Kupffer cells	Present	Absent, Ictaluridae – exceptions
Perisinusoidal macrophages	Present	Present
Macrophage aggregates	Absent	Present

hepatic lobules with oxygenated blood. Cytologic features of hepatocytes at the periphery (portal zone) of lobules differ from those of hepatocytes at the center of the classic lobule (compare Figures 4.1 and 4.3). This has led to investigations in mammalian species revealing a lobular zonation. Analysis of the venular profiles in trout liver (Figure 4.2) reveals no bile ducts or hepatic arterioles in close association and no tissue component analogous to portal tracts of rodent liver (Figure 4.3).

Why is this important in our considerations of liver toxicity in fish? For mammals, on liver sections we can differentiate portal or afferent (i.e. relatively more oxygenated) regions of parenchyma from centrolobular or efferent regions, but such differentiation in fish is not possible. Also, exposure of rodents to certain reference hepatotoxicants results in a mosaic of altered and unaltered parenchyma, depending on the targeting of afferent or efferent zones by the toxicant. No such pattern has been observed in fish studied with these compounds (see section on reference hepatoxicants below). Heterogeneity of mammalian hepatocytes has been demonstrated with regard to oxygen saturation, cell volume, glycogen, isozymes of cytochrome P450 and other enzymes, and cell shape. This has led to

Table 4.2 Volume densities of cells and spaces in liver parenchyma.

Component	*% volume*				
	Rat[a]	*Dog[b]*	*Trout[c]*	*Ide[d]*	*Brown trout[e]*
Hepatocytes	77.80	84.40	84.50	88.90	87.3
Non-hepatocytes	22.20	15.60	15.50	11.10	15.18
Endothelial cells	2.80		1.80	1.00	1.35
Fat-storing cells	1.40		0.70	0.10	0.6
Kupffer cells/ macrophages	2.10	7.80	0.20	0.10	0.7
Bile epithelial cells			1.30	2.10	1.32
Spaces					
Sinusoidal lumen	10.60	4.30	9.40	6.60	6.23
Space of Disse	4.90	3.00	1.00	0.50	4.53
Bile canaliculi	0.40	0.50	1.10	0.70	0.45

Notes
Reference volume, liver parenchyma.
a Blouin (1977).
b Hess *et al.* (1973).
c Hampton *et al.* (1985).
d Segner, unpublished ($n = 4$ animals, with ten measurements per animal).
e Rocha *et al.* (1997).

the concept of metabolic zonation within the mammalian liver lobule (Jungermann and Sasse, 1978). Less (Schar *et al.*, 1985), little (Segner and Braunbeck, 1990), or no (Hampton *et al.*, 1985) evidence exists, however, for hepatic metabolic zonation in trout and, apparently, other fish.

Analysis of mouse liver sections (Figure 4.1 and, especially, Figure 4.3) reveals features of hepatocyte arrangement in these vertebrates. Hepatocytes are typically arranged as laminae, one hepatocyte thick, resulting in both lateral plasma membranes (those facets facing sinusoids) of individual hepatocytes in close proximity to the microcirculation (Figure 4.3) being separated from sinusoids by the space of Disse (Elias and Sherrick, 1969) and communicating with sinusoidal lumens by fenestrae in the endothelium (the last two features are not shown in the figures).

Microscopic analysis of trout liver sections (Figure 4.4) reveals a pattern of arrangement of hepatocytes quite different from that of mammals. When viewed in longitudinal orientation and proceeding from one sinusoidal profile, across the hepatocytes, to the next sinusoidal profile (Figure 4.4), a double row of hepatocytes is obvious. In addition, rounded nuclei of hepatocytes contrast with flattened and elongated nuclei of a different cell type, the bile preductular epithelial cells, found between rows of hepatocytes in an individual tubule (Figure 4.4). Using high-resolution light microscopy (Figure 4.5), transects of hepatocyte arrays resemble tubules. Basal aspects of trout hepatocytes project toward sinusoids or adjacent hepatocytes of neighboring tubules (Figure 4.5), while cellular apices are directed toward the center of the tubule. It is also important to realize that the cells with

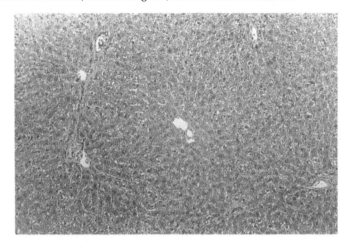

Figure 4.1 Classic lobule of mouse liver. The central venule (initial tributary of hepatic venous system) is at the center of the lobule and is surrounded only by large rounded hepatocytes, whose cytoplasm stains throughout. Portal tracts containing bile ducts, hepatic arterioles, and portal venules mark four corners at the periphery of the lobule and identify portal venule profiles by their association with portal tract structures. Hepatocytes at the periphery of the lobule show unstained areas within the cytoplasm (glycogen depots). Hematoxylin and eosin stain (375×).

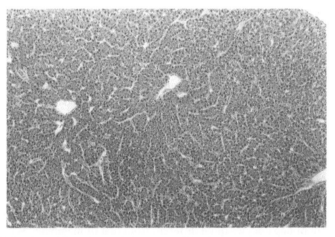

Figure 4.2 Section of liver from rainbow trout. Fixation was by portal venous perfusion. Note the absence of accompanying structures near venous profiles and the absence of a lobular pattern. Hematoxylin and eosin stain (200×).

flattened nuclei and high nuclear to cytoplasmic ratios (Figure 4.4) are commonly encountered in the centers of tubules (Figure 4.5), and, by electron microscopic analysis, these were confirmed as biliary epithelial cells and they formed junctional complexes with each other and with hepatocytes (Hinton and Pool, 1976; Hampton *et al.*, 1988, 1989). When first described (Hinton and Pool, 1976), the name 'bile

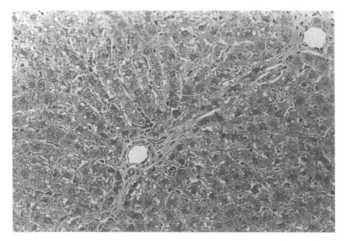

Figure 4.3 Higher magnification view of the portal tract of mouse liver showing the bile duct and hepatic arteriole with the portal venule. Note the sinusoids, with erythrocytes in the lumens, positioned on either side of the hepatic lamina. Clear areas within hepatocytes are glycogen depots. Hematoxylin and eosin stain (750×).

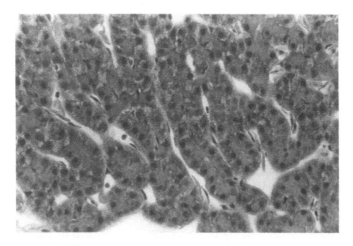

Figure 4.4 Higher magnification view of rainbow trout liver showing hepatic tubules in predominately longitudinal orientation. Each tubule shows a double row of hepatocytes. In between the rows of hepatocytes in an individual longitudinally sectioned tubule are small cells with large nuclear–cytoplasmic ratios. Hematoxylin and eosin (750×).

preductules' was used to avoid confusion with mammalian terms and to signify that a different location for transitional and larger elements of the bile passageways were present in the parenchyma of fish liver. Whereas the mammalian liver has only canaliculi as biliary passageways of the parenchymal units, fish livers are arranged in such a way that hepatocytes and biliary epithelial cells are in close proximity.

The importance of this observation is that, for a given volume of liver, fish

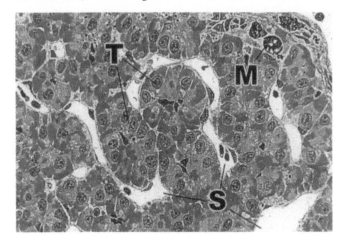

Figure 4.5 Higher magnification of a glycol methacrylate-embedded section of perfusion-fixed rainbow trout liver showing three hepatic tubules in cross section. S, hepatic sinusoid; T, tubules with biliary epithelial cell nuclei at the center; M, macrophage. Liver was fixed by perfusion of fixative through the hepatic portal vein and pieces were embedded in glycol methacrylate and stained by hematoxylin and eosin (1500×).

have significantly more biliary epithelial cells than do mammals (Table 4.2). When Hampton *et al.* (1988) performed enzyme histochemical studies on trout liver, they localized Mg^{2+}-dependent ATPase to plasma membranes of hepatocytes at the bile canaliculi and lumenal plasma membranes of biliary epithelial cells. In this manner, three successively larger diameters for bile passageways were shown in hepatic parenchyma, and, specifically, at the center of hepatic tubules.

The two-dimensional features of hepatic tubules (Figures 4.4 and 4.5) suggest that individual tubules curve, anastomose and/or branch, thereby forming a complex continuum of parenchyma tunneled by an extensive microcirculation. As stated above, this large blood supply no doubt leads to intensive toxicant exposure and accumulation while hepatocytes and biliary epithelial cells, i.e. the tubular elements, perform clearance functions.

To facilitate understanding of the architectural pattern of parenchyma and microvasculature of piscine livers, Figure 4.6 is used to illustrate features and relationships. With the longitudinal orientation of tubules, only one surface (basal surface) of the teleost hepatocyte will face the sinusoid. In addition, at those sites where curving and bending result in adjacent tubular lengths in close proximity, sinusoids will not intervene (Figures 4.5 and 4.6) and component hepatocytes will be separated by some distance from the nearest sinusoid. When compared with mammalian liver, this tubular architectural pattern of fish may result in different (lower) rates of uptake of endogenous compounds and xenobiotics.

Another difference between mammalian liver and that of fish is the fixed macrophage of the mammalian sinusoids, the Kupffer cell, containing the majority of lysosomal enzymes in the liver. As noted in Table 4.1, the Kupffer cells are absent in most teleosts. The exception appears to be the Ictalurid fish, in which

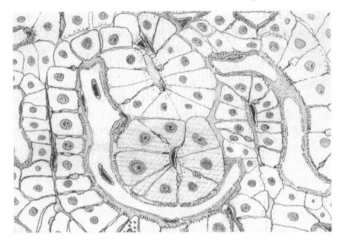

Figure 4.6 Drawing depicting the architecture of hepatic tubules and sinusoids of the liver. In this arrangement, basal aspects of some hepatocytes face adjacent sinusoids across the space of Disse. However, when portions of hepatic tubules are immediately adjacent to other tubules, no sinusoids separate the hepatocytes of neighboring tubules and no direct proximity to the sinusoids is seen in these sites.

morphologic features of sinusoidal macrophages resemble Kupffer cells (Hinton and Pool, 1976; Hampton *et al.*, 1987). Perisinusoidal, interhepatocytic macrophages are occasionally present in mammals, but are apparently much more common in livers of fish (Table 4.1). Rocha *et al.* (1996) presented evidence in brown trout (*Salmo trutta*) that these cells leave sinusoids to take up residence in hepatic tubules. In addition, they are found in close proximity to cholangioles of tubules, suggesting some shared function which may influence clearance. Macrophage aggregates are considered stromal components of the liver of fish. In some instances, macrophage aggregates can be used as indicators of contaminant exposure (Wolke *et al.*, 1985; Blazer *et al.*, 1987).

Important aspects of fish liver physiology related to interpretation of toxicity

The liver of fish has a central function in maintenance of organismic homeostasis by regulating blood composition through interconversion and (re)distribution of nutrients, by metabolism of hormones to avoid their build-up in the plasma, by removal of metabolic wastes, and by excretion of bile fluid. The important metabolic role of the liver for processing and routing of nutrients is related to its unique location receiving the venous drainage from the hepatic portal vein before its entry into the general circulation. After meals, in the absorptive stage, the liver takes up nutrients from the plasma, and it releases nutrients in catabolic states such as starvation or stress. In this way, the liver smoothes out potentially large fluctuations of blood composition and nutrient supply of peripheral tissues. The liver further influences blood composition by secreting major plasma proteins such as fibrinogen and albumin. The fibrinogens are required for blood clotting, whereas the albumins are important for maintaining blood osmolality, blood pH buffering, providing an amino acid source and as a transport carrier for hormones as well as exogenous chemicals such as metals and organics. These functions are potential targets for the action of chemicals.

Research during the last two decades has established that piscine liver has many of the same metabolic pathways and enzymes as known for mammalian liver (Cowey and Walton, 1989; Moon and Foster, 1995). The overall importance of individual pathways as well as their nutritional and endocrine regulation, however, are less well understood. In the following, those aspects which appear to distinguish fish liver metabolism from that of mammals will be emphasized.

The importance of fish liver for organismic homeostasis and environmental acclimation is already evident with the relationship between weight of liver and that of body weight, i.e. the liver somatic index (LSI). Numerous studies have demonstrated that alterations of the nutritional state, season, temperature, reproductive status, as well as exposure to toxicants readily change the LSI of fish (Yarbrough et al., 1976; Adams and McLean, 1985; Segner and Braunbeck, 1988, 1990; Moon et al., 1989; Blasco et al., 1992; Braunbeck and Segner, 1992; Everaarts et al., 1993; Böhm et al., 1994). Short-term stress usually decreases the LSI either because of depressed feeding or because of an enhanced energy drain. Under conditions of chronic stress, however, the liver cells may undergo an adaptive hyperplasia and/or hypertrophy, resulting in an increase of the LSI. Nevertheless, changes in cell number are usually not the cause of fluctuating LSI (Böhm et al., 1994).

The liver is a glucose-utilizing, glucose-producing and glucose-storing organ. As such, it acts as a glucostat in the vertebrate organism, regulating glucose levels of blood. When glycemia is challenged, the liver elicits adaptive metabolic responses, thereby maintaining a level of blood glucose which is optimal for animal function. Glycemic levels and glucose turnover rates vary among fish species. Glucose concentrations mostly range between 2 and 10 mM, and glucose turnover rates are between 0.6 and 5.7 μmol kg^{-1} body weight min^{-1} (Moon, 1988; Weber and Zwingelstein, 1995). These values are 20–100 times lower than those for resting mammals of equivalent size. The lower body temperature and lower metabolic rate of fish may account for this difference.

Hepatic removal of excess glucose from the blood is comparatively inefficient in teleost fish. In glucose tolerance tests, fish were found to respond like diabetic man (Cowey and Walton, 1989). The evidence available to date suggests that the diabetic-like behavior of fish is related to a low hepatic capacity for glucose uptake, rather than a lack of insulin or insulin receptors. In addition, regulatory mechanisms appear to operate more slowly, i.e. requiring hours instead of minutes as is the case for mammals (Moon and Foster, 1995; Weber and Zwingelstein, 1995). Hexokinase, the enzyme responsible for glucose phosphorylation and the maintenance of the membrane gradient for glucose, is present at low activities in fish liver (Knox et al., 1980; Sundby et al., 1991), whereas glucokinase (a high K_m hexokinase) that is responsible for increased hepatic glucose phosphorylation during high levels of plasma glucose appears to be absent from fish liver (Cowey *et al.*, 1977; Sundby *et al.*, 1991). Recently, however, an adaptable glucokinase-like enzyme in the liver of Atlantic salmon, *Salmo salar*, has been described (Tranulis *et al.*, 1996).

Despite the poor response of fish to exogenous glucose load as described above, hepatic glycogen stores increase when fish are fed high carbohydrate diets (Hilton and Atkinson, 1982; Shimeno *et al.*, 1985; Hutchins *et al.*, 1998). A major part of intracellular glucose in teleost hepatocytes is channeled into glycogen. Glycogen repletion occurs mainly by way of the direct pathway through glucose-6-phosphate and not by the indirect pathway through lactate/pyruvate (Pereira *et al.*, 1995). However, this situation may change with the nutritional conditions of the fish. The pancreatic hormone insulin supports glycogen synthesis, whereas glucagon and glucagon-like peptide strongly stimulate glycogenolysis (Foster and Moon, 1989; Mommsen and Moon, 1989; Pereira *et al.*, 1995).

Glycogen levels in fish liver vary over a wide range of concentrations, from 20 to 2000 μmol g^{-1} liver, and in cyprinids glycogen can make up more than 4 percent of body weight (Moon and Foster, 1995). After white muscle, liver glycogen is the major carbohydrate store in the fish body. Hepatic glycogen is rapidly mobilized during stress situations by catecholamines (Chavin, 1973; Ottolenghi *et al.*, 1984; Janssens and Lowrey, 1987). The existence of α-adrenoceptors to mediate catecholamine effects in fish liver are well established, and the presence of α_1-adrenoceptors linked to Ca^{2+} mobilization has been demonstrated by Fabbri *et al.* (1998).

Another metabolic situation associated with mobilization of hepatic glycogen is food deprivation. In man, liver glycogen reserves are completely spent within the first 24 h of fasting to maintain blood glucose levels (Cahill, 1970). By contrast, starving teleosts partially protect their liver glycogen (Gas and Serfaty, 1972; Love, 1980; Navarro and Gutierrez, 1995). Despite the relative constancy of hepatic glycogen, teleosts generally do not suffer severe hypoglycemia (Moon *et al.*, 1989; Navarro and Gutierrez, 1995; Moon, 1998). The conservation of plasma glucose during starvation of fish appears to be achieved by strongly decreased glucose turnover rates (Weber and Zwingelstein, 1995) and enhanced gluconeogenesis in the liver (French *et al.*, 1983; Moon, 1988; Navarro and Gutierrez, 1995). Glucose synthesis (gluconeogenesis) in the liver of food-deprived fish utilizes amino acids from protein breakdown or glycerol from lipid catabolism (Sheridan and Mommsen, 1991). In mammals, during prolonged starvation, glucose is partly replaced as energy substrate by the ketone bodies acetoacetate and 3-hydroxybutyrate, which are generated from hepatic fatty acid oxidation (Cahill, 1970). Ketogenesis reduces muscle protein catabolism by decreasing tissue glucose dependence and thus the need for increases in hepatic gluconeogenesis. This strategy, however, is apparently not utilized by fish; hepatic generation of ketone bodies is not detectable in fasting teleosts (Zammit and Newsholme, 1979; Segner *et al.*, 1997).

During exercise in mammals, the liver has important functions supporting muscle metabolism by supplying glucose fuel; also, during recovery after exercise, it converts significant amounts of the lactate released by the working muscle into glucose (Cori cycle). In fish, the role of the liver during exercise seems to differ from the mammalian model. Milligan and Girard (1993) demonstrated that white muscle of exercising teleost fish is not dependent on bloodborne glucose to fuel its increased energy demand, rather it behaves as a self-supporting unit relying mainly on its endogenous glycogen reserves. Likewise, the Cori cycle, which converts muscle-derived lactate by liver gluconeogenesis into glucose, appears to be of minor importance in fish. Studies of American eel, *Anguilla rostrata*, and skipjack tuna, *Katsowonus pelanus*, demonstrated that < 1 percent of glucose carbon was contributed by lactate, which is minor compared with Cori cycle values in mammals of 10–40 percent (Cornish and Moon, 1985; Weber *et al.*, 1986; Moon, 1988). The dominant mechanism for glycogen restoration in muscles of exercising teleost fish appears to be *in situ* glycogenesis from lactate, directly in the muscle (Milligan and Girard, 1993).

The liver is a major storage site of lipids in many species of fish. Some species such as salmonids rarely display hepatic lipid storage (or only under pathologic conditions), other species such as many cyprinids store both lipids and glycogen, and still other species such as cod store almost exclusively lipid. A number of factors including, but not limited to, nutrition, sexual cycle, and temperature modify liver lipid metabolism (Peute *et al.*, 1978; Cossins, 1983). Lipids reach the liver from the intestine in the form of chylomicrons (Sargent *et al.*, 1989; Sheridan, 1998). The liver repackages dietary lipids, complexing that with lipid synthesized *de novo* and with protein to deliver those lipoproteins to the peripheral tissues (Sheridan, 1988; Sargent *et al.*, 1989; Tocher, 1995). A peculiar lipoprotein which is synthesized by the liver of reproducing female fish is the egg yolk precursor vitellogenin. During catabolic states, fish liver oxidizes lipids by both the mitochondrial and the peroxisomal systems (Sargent *et al.*, 1989). Also, a number of hormones have been shown to exert a lipolytic action on lipids stored in fish liver (Sheridan and Bern, 1986; Sheridan, 1987).

The principal site of *de novo* synthesis of fatty acids in fish is the liver (Sargent *et al.*, 1989; Segner and Böhm, 1994). Although knowledge on the lipogenetic pathway and its regulation in fish remains fragmentary, it appears that cytoplasmic acetyl coenzyme A (acetyl-CoA), the precursor of fatty acid synthesis, is derived by way of citrate cleavage pathway from mitochondrial acetyl-CoA, and that dietary lipid levels suppress hepatic lipogenesis in fish. However, the response of teleost lipogenesis to carbohydrates and to hormones is not understood (Segner and Böhm, 1994).

Amino acid homeostasis in fish is governed by the liver. Fish liver is the main site of transdeamination and gluconeogenesis (Juerss and Bastrop, 1995). Blood amino acid levels rise considerably in hepatectomized fish (Kenyon, 1967). The main pathway for hepatic amino acid catabolism and ammoniogenesis appears to be transdeamination with little importance of direct deamination or the purine nucleotide cycle (Cowey and Walton, 1989; Juerss and Bastrop, 1995). Some fish species such as toadfish, *Opsanus tau*, are known to generate urea instead of ammonia from amino acid breakdown (Mommsen and Walsh, 1991).

In summary, liver metabolism is a potential target for the toxic action of chemicals. It is important to differentiate between effects which (1) are toxic for the liver cells *per se*, i.e. enzyme inhibition (Heath, 1995), or altered gene expression [TCDD and phosphoenol pyruvate (PEPCK)], lindane-induced steatosis as a consequence of a block in lipid secretion, and effects which (2) do not harm the liver cells but disturb their support of peripheral tissues, for example the inhibition of vitellogenin or choriogenin synthesis by antiestrogens, or exaggerated hepatic hormone catabolism. In cases when liver cells are damaged by toxicants, this can be detected by clinical analysis of blood enzymes (Heath, 1995). Toxicity arising from the interaction of xenobiotics or metals with pathways of hepatic intermediary metabolism or with specific liver receptors is one of the least documented aspects of hepatic target organ toxicity in fish.

Liver xenobiotic metabolism

The liver is the major site of biotransformation in fish. It performs both phase I and phase II reactions of xenobiotic metabolism. It is particularly important to understand how metabolism is related to target organ toxicity. One of the earliest examples of mammals in which hepatotoxicity was shown to be due to the tissue-specific formation of a reactive intermediate able to bind covalently to hepatic macromolecules was bromobenzene (Caldwell *et al.*, 1986). Although a number of studies have analyzed the liver conversion of xenobiotics in fish, the relation between hepatic biotransformation and hepatotoxicity or the importance of distribution of xenobiotic biotransformation enzymes in the different cell types of fish liver for development of liver toxic lesions have been hardly addressed. From field studies, evidence is available on the association of enhanced expression of hepatic biotransformation enzymes, particularly monooxygenases (cytochromes P450), and the occurrence of toxicopathic liver lesions, including neoplastic alterations (Myers *et al.*, 1987, 1994). Stegeman and Hahn (1994) as well as Buhler and Wang-Buhler (1998) have reviewed the various cytochrome P450s in fish, including their induction and role in environmental monitoring. In this brief coverage, we restrict our attention to cytochrome P4501A (CYP1A).

The best-characterized biotransformation enzyme in fish is CYP1A, which shows highest specific activities in the liver (Stegeman and Hahn, 1994). The enzyme is localized in hepatocytes, biliary epithelial cells, and vascular endothelial cells (Hinton, 1993; Goksoyr and Husoy, 1998; Sarasquete and Segner, 2000). In contrast to mammals in which CYP1A shows a heterogeneous distribution throughout the liver parenchyma, no zonation can be observed in teleost liver (Lorenzana *et al.*, 1989; Smolowitz *et al.*, 1991). Subcellularly, CYP1A is localized at the membranes of the rough endoplasmic reticulum and the nuclear envelope (Lester *et al.*, 1993).

The various liver cell types appear to differ in CYP1A induction profiles after xenobiotic exposure. Environmental exposure led to a particularly strong increase of CYP1A immunoreactivity in the biliary and endothelial cells of cod and flounder, whereas hepatocyte staining remained weak (Husoy *et al.*, 1996). In contrast, in lemon sole (*Microstomus kitt*), CYP1A induction was stronger in hepatocytes than in vascular cells (Husoy *et al.*, 1996). In scup, 3,3′,4,4′-tetrachlorobiphenyl treatment strongly increased CYP1A expression in both hepatocytes and non-hepatocytes (Smolowitz *et al.*, 1991). Possibly, the cell-specific induction response varies with the type of inducer (readily versus slowly metabolizable compounds), dose of inducer, or with the fish species (Anulacion *et al.*, 1998; Goksoyr and Husoy, 1998). Another relevant factor influencing hepatic CYP1A expression could be the exposure route; Van Veld *et al.* (1997) demonstrated that a stronger increase of CYP1A immunostaining occurred in hepatocytes of mummichog (*Fundulus heteroclitus*) when exposure took place via the water rather than via the food.

Studies with reference hepatotoxicants in fish

Background

Analysis of toxicity in livers of fish following exposure to reference hepatotoxicants is a particularly useful way to produce baseline toxicologic data in these species. Because the reference hepatotoxicants may lack environmental relevance, financial support for work to compare qualitatively and quantitatively piscine with better-described rodent models has been limited. As much is known relative to mechanisms of toxicity of these compounds in mammalian species, this information is useful in predicting and interpreting responses of fish liver to xenobiotics. However, relatively few classic hepatotoxicants – acetaminophen, allyl formate, bromobenzene, carbon tetrachloride, and chloroform – have been studied for their toxic effects on livers of fish; also, the few studies are largely restricted to rainbow trout (*Oncorhynchus mykiss*). Unfortunately, the information generated has – to varying extent – been confusing, conflicting, and largely inconclusive (see detailed coverage below), leaving appreciable gaps in our basic understanding of hepatic toxicology in these vertebrates.

Three reference hepatotoxicants – acetaminophen (AP), allyl formate (AF), and carbon tetrachloride (CCl_4) – have received the most attention in studies with fish. Although the first and last are classic centrolobular mammalian hepatotoxicants, the second, AF, is unique in that its toxicity is to the periportal hepatocytes of mammals.

AP at high doses and in an acute time frame produces centrolobular necrosis in rodent liver (Mitchell *et al.*, 1973a,b). The hepatotoxic response of trout to AP was investigated by Droy (1988) and was less than that seen when rats, mice, or hamsters were given similar or lower levels of the drug (Mitchell *et al.*, 1973a; Ioannides *et al.*, 1983). Droy (1988) reported an increase in serum glutamic pyruvic transaminase (SGPT) in a dose-dependent manner following exposure of trout to AP. However, maximal enzyme elevation in serum was at least an order of magnitude lower than those reported for hamsters (Lupo *et al.*, 1987) and rats (Smith and Mitchell, 1985). Of the parameters investigated, only liver sodium showed a dose–response effect. Serum bilirubin, liver potassium, water content, and percent water failed to elicit a response relationship in AP-exposed trout.

In trout, focal hepatocyte necrosis, as evidenced by nuclear lysis, pyknosis, increased cytoplasmic eosinophilia (decreased basophilia), and vascular disruption, was seen in hepatic parenchyma (Droy, 1988). However, the magnitude of the toxicity was much less in trout than in mammals. For instance, doses of 750 mg kg^{-1} AP induced necrotic lesions in 99 percent of mouse livers 48 h after treatment (Mitchell *et al.*, 1973a). Likewise, Price *et al.* (1987) showed 100 percent incidence of necrosis in rats following 48-h exposures to 1000 mg kg^{-1} AP. In contrast, two of three trout exhibited hepatotoxic lesions 48 h after exposure to 800 mg kg^{-1} AP. Necrotic lesions were rare and no zonation of effect (perivenous necrosis), as reported in rodent liver (Mitchell *et al.*, 1973a), was seen.

The results of the trout study cited above correlate with an absence of AP-induced biochemical changes in mullet liver (Thomas and Wofford, 1984). Glutathione protects mammalian liver against acute AP-induced hepatotoxicity by inactivating a reactive electrophilic intermediate produced via cytochrome P450 activation (Hinson, 1980). Also, mouse liver glutathione decreases in a dose-dependent manner to a maximum 75 percent depletion following 1-h exposures of up to 375 mg kg^{-1} acetaminophen (Mitchell *et al.*, 1973b). Exposures of mullet to 400 mg kg^{-1} AP did not cause significant depletion of glutathione stores (Thomas and Wofford, 1984). Absence of glutathione depletion in mullet was explained by a lack of bioactivation of AP to an electrophilic intermediate. Inability to produce this toxic metabolite is consistent with the low toxicity seen in trout liver following AP exposure. Based upon these results from mullet (Thomas and Wofford, 1984) and trout (Droy, 1988) following AP exposure, it is likely either that constitutive levels of teleost liver cytochrome P450 cannot metabolize this compound in an analogous manner to most mammalian species or that AP is converted to a non-toxic form. The fact that isolated trout hepatocytes produced lower levels of AP metabolites than mammals (Parker *et al.*, 1981) supports decreased bioactivation of AP by trout.

Analysis of metabolite profiles following preferential induction of specific cytochrome P450 (CYP) isoforms has determined mechanisms of AP-induced toxicity in mammals. For instance, in rats, CYP1A catalyzes formation of higher amounts of reactive AP metabolite *N*-acetyl-P-benzoquinone imine (NAPQI) than inert metabolite 3-hydroxy-AP (Harvison *et al.*, 1988). Phenobarbital (PB) induces the CYP2B proteins that result in enhanced 3-hydroxy-AP formation and potentiation of AP – induced liver necrosis in rats and mice (Mitchell *et al.*, 1974a), but that protect against liver toxicity in Syrian hamsters. Whereas fish lack CYP2B-like or PB-inducible forms, some species may have CYP2E-like proteins (Kaplan *et al.*, 1991) that have been shown in mammals to be one of the primary activators of AP to NAPQI. The demonstration of the capacity to form reactive intermediates of AP indicates that differences in metabolism alone do not account for lack of liver toxicity in fish. Quantitative structural differences in the hepatocyte surface area in proximity to sinusoidal lumens may account for observed differences. However, at present, these ideas are merely speculative but deserve future attention.

The reference hepatoxicant AF was of considerable interest for studies in fish because it exerts selective periportal zonal toxicity in rodent liver (Rees and Tarlow,

1967). AF requires alcohol dehydrogenase (Schar *et al.*, 1985), catalyzing conversion of allyl alcohol to the highly reactive compound, acrolein, which binds to tissue macromolecules and induces liver necrosis (Rees and Tarlow, 1967). Droy *et al.* (1989) showed morphologic, biochemical and functional changes in trout liver following *per os* administration of AF, and toxicity was also demonstrated through elevation of specific enzymes in serum. Trout histopathology after AF (Droy *et al.*, 1989) included severe necrosis and hemorrhage, quantitatively similar to the magnitude of toxic response reported in portal areas of rat livers (Rees and Tarlow, 1967; Reid, 1972). Some areas of trout parenchyma appeared less altered than others, but no preferential zonation of liver effect could be discerned following exposure of trout to 100 μL kg^{-1} AF. However, when a lower dose of AF (30 μL kg^{-1}) was administered and liver analyzed after 48 h, necrosis and hemorrhage in hepatic parenchyma were located near venular profiles in some sections. This may be similar to the portal necrosis reported in rat liver following exposure to AF (Rees and Tarlow, 1967). The induction of massive necrosis after 100 μL kg^{-1} AF could have masked the induction potential for zonal effects. No hepatotoxicity was seen following 10 μL kg^{-1} AF.

Trout light microscopic studies of AF toxicity were followed by ultrastructural analyses (Droy *et al.*, 1989). The increased resolution of these preparations suggested that the structural defect inducing the severe hemorrhage was endothelial cell destruction, with subsequent disruption of sinusoidal walls permitting entry of serum and erythrocytes to interhepatocytic space. Droy *et al.* (1989) showed that most quantitative indices examined after AF exposure correlated with morphologic alterations. SGPT levels peaked at 48 h after exposure to 100 μL kg^{-1} AF and returned to control levels by 72 h. As a single dose was administered, this pattern seems reasonable. Serum bilirubin levels were not elevated at 24 h after a toxic dose. Presumably, this response would take longer and perhaps require a more sustained loss of hepatic function. However, once elevated, this parameter remained high, suggesting that pre-exposure hepatic function had not been achieved by 72 h. Tissue K$^+$ was significantly decreased from control values by 24 h, showed partial recovery by 48 h, and returned to control levels by 72 h. Droy *et al.* (1989) interpreted this as an initial loss or leakage of K$^+$ through altered plasma membranes of hepatocytes, followed by restabilization of tissue with viable hepatocytes having the ability to retain K$^+$. Na$^+$ influx was observed, and high levels of this ion were maintained throughout the test period.

AF hepatotoxicity is thought to be mediated through alcohol dehydrogenase (Reid, 1972; Belinsky *et al.*, 1984). Pyrazole inhibits alcohol dehydrogenase and, when administered prior to AF, significantly attenuates hepatotoxicity (Reid, 1972; Serafini-Cessi, 1972; Ohno *et al.*, 1985). A pattern of heterogeneous, portal venous distribution of alcohol dehydrogenase would account for regional differences in toxicity seen in trout after the 30 μL kg^{-1} dose of AF. However, histochemical studies on trout liver have indicated no preferential localization of this enzyme (Schar *et al.*, 1985).

Thiol compounds, including glutathione, are known to protect liver from AF-mediated toxicity. Glutathione was administered with AF in rat liver post-mitochondrial fractions, and this significantly reduced AF inhibition of protein synthesis (Rees and Tarlow, 1967). Glutathione content is important in AF toxicity of rat liver where prior use of depleting agents significantly potentiates toxicity, as assessed by lethality (Hanson and Anders, 1978). However, studies localizing glutathione in trout liver showed homogeneous distribution (Droy, 1988).

The role of oxygen tension may be an important determinant of zone-specific hepatotoxicity after AF. Using isolated perfused rat liver, Belinsky *et al.* (1984) reported that AF is metabolized in both pericentral and periportal regions of the liver lobules in a pattern unlikely to give rise to zone-specific hepatotoxicity. These same workers in a later study (Belinsky *et al.*, 1986) concluded that mechanisms other than thiol distribution within the lobule account for periportal toxicity. Isolated perfused rat liver studies (Badr *et al.*, 1986) showed that AF caused liver toxicity when introduced to the liver via either retrograde or anterograde flow (i.e. preferentially exposing centrolobular or periportal hepatocytes respectively). Hepatocytes were only damaged in areas where oxygen tension was high. Decreasing oxygen tension by lowering the rate of flow of perfusate significantly reduced toxicity. Interestingly, perfusion rates of trout (5.2 mL min^{-1} kg^{-1} body weight; Schmidt and Weber, 1973) are significantly less than those of mammals (66 mL min^{-1} kg^{-1} body weight; Rabinovici and Wiener, 1963). The role of oxygen and possible oxygen tension gradients deserve further attention.

CCl_4, the last of the reference hepatotoxicants for detailed consideration, has been used by several investigators (Racicot *et al.*, 1975; Gingerich *et al.*, 1978; Statham *et al.*, 1978; Hiraoka *et al.*, 1979; Weber *et al.*, 1979; Pfeifer *et al.*, 1980) in controlled laboratory exposures of fish, primarily trout. As with AP, a mosaic of altered hepatocytes in centrolobular zones contrasts with unaltered periportal hepatocytes (Diaz Gomez *et al.*, 1975) in rodents exposed to CCl_4. However, this pattern does not occur in trout and English sole after exposure to CCl_4. Rather, only occasional hepatocytes, in no apparent pattern, show alteration. For example, Pfeifer *et al.* (1980) reported that characteristic centrolobular necrosis was not a prominent histopathologic finding. Statham *et al.* (1978) found subcapsular necrosis in two out of three of their CCl_4-treated trout, apparently caused by direct contact (solvent effect) of the compound with the liver. Gingerich *et al.* (1978) described liver pericentral necrosis in only 25 percent of trout, and Racicot *et al.* (1975) found necrosis in 20 percent of trout livers 18 h after exposure. In contrast, nearly 100 percent of rodents receiving a similar dose of CCl_4 exhibited pericentral hepatocyte necrosis (Weber *et al.*, 1979).

As was described above, it is difficult, if not impossible, to differentiate pericentral and periportal areas of liver parenchyma. Earlier interpretations of alterations as 'pericentral responses' in trout studies were likely in error. For example, Gingerich *et al.* (1978) did not describe operational procedures for differentiation of periportal versus pericentral areas in their preparations. Careful examination of photomicrographs of the Gingerich *et al.* (1978) study indicates that perivenous necrosis was found only in areas adjacent to subcapsular necrosis, i.e. that necrosis attributed to solvent effect (Statham *et al.*, 1978). In a manner characteristic of conclusions of other investigators, Weber *et al.* (1979) suggested that trout are less sensitive to the toxic effects of CCl_4.

Oral dosing studies with CCl_4 were used by Droy (1988) as this method of delivery is related to a more natural exposure route and obviated the solvent effect seen after intraperitoneal injections. Necrotic hepatocytes were more abundant in the high-dose group (3 mL kg^{-1} CCl_4), and loss of cytoplasmic basophilia with swelling characterized affected hepatocytes. In rats used as biologic controls for the experiment, Droy (1988) found a quantity of necrosis not approximated by trout even after greater doses of the toxicant. Why are trout livers more resistant to CCl_4-induced injury? One hypothesis for this lack of response is that trout liver does not have the metabolic machinery capable of bioactivating CCl_4. Evidence against this includes the report of increased lipid peroxidation and ^{14}C-labeled CCl_4 residues in trout liver 6 h after exposure to a single intraperitoneal injection of 1 mL kg^{-1} CCl_4 (Statham *et al.*, 1978). Also from mullet, *Mugil cephalus*, and croaker, *Micropogonias undulatus*, increased lipid peroxidation as a result of CCl_4 treatment has been reported (Wofford and Thomas, 1988). Further, *in vitro* studies provide strong evidence that fish liver cells are able to metabolize CCl_4 (Råbergh and Lipsky, 1997). However, it should be remembered that this threefold elevation of lipid peroxidation could have been due to CCl_4-induced direct solvent injury which does not require metabolic activation (Berger *et al.*, 1986). In addition, presence of ^{14}C-labeled CCl_4 residues in trout liver may represent distributed but non-metabolized compound; acid/methanol extractions for protein binding analysis were not performed (Statham *et al.*, 1978). Indirect evidence in support of trout liver possessing the metabolic capability to bioactivate CCl_4 was provided by Kleinow *et al.* (1990), who showed that non-inducible winter flounder P450 was preferentially decreased after exposure to this compound.

A second hypothesis that could explain why CCl_4 is less hepatotoxic in trout would be the presence of higher amounts of antioxidants or free radical scavengers such as glutathione. However, when trout and rat were compared after exposure to monochlorobenzene (Dalich and Larson, 1985), rat hepatic glutathione was about three times that of trout. Therefore, it seems unlikely that lack of susceptibility of trout to CCl_4 can be explained by excessive amounts of antioxidant. In addition, Råbergh and Lipsky (1997) showed that CCl_4 treatment of trout hepatocytes resulted in a serious depletion of cellular glutathione levels. Droy (1988) also considered the possibility that interspecies differences in oxygen tension and glutathione distribution patterns might explain the trout response to CCl_4. In rodents, CCl_4 is metabolized to trichloromethyl radical which rapidly combines with oxygen to form the very reactive trichloromethyl peroxyl radical (Packer *et al.*, 1978). Therefore, different free radical species of CCl_4 are present under aerobic and anaerobic conditions in rat liver. Hepatic injury is maximal in centrolobular areas of mammalian liver where oxygen tension is lowest (Burk *et al.*, 1983, 1984). Also, hypoxia potentiates covalent binding of metabolites to rat liver microsomal lipids and proteins (Shen *et al.*, 1982). Glutathione is effective against CCl_4-induced lipid peroxidation in rat liver only in the presence of oxygen where trichloromethyl peroxyl radical is readily formed (Burk *et al.*, 1983). The relative abundance of glutathione in portal – relative to centrolobular – areas also contributes to the protection seen in portal hepatocytes of rodent liver (Smith *et al.*, 1979). Therefore, glutathione appears to protect rat liver portal hepatocytes from CCl_4 because of preferential distribution of glutathione and the likely ability of this compound to scavenge trichloromethyl peroxyl radical but not trichloromethyl radical. In trout, Droy (1988) reported a refractiveness to the hepatotoxic effects of CCl_4, with scattered individual hepatocyte necrosis being the typical response. Perhaps the inability of CCl_4 to produce appreciable toxicity in trout liver could be due to comparatively high oxygen tension. Under these conditions, the trichloromethyl radical would be converted to the trichloromethyl peroxyl radical, which would combine with glutathione and be rendered inert. Alternatively, the high oxygen tension in fish liver may prevent reductive dehalogenation of CCl_4 to an activated metabolite. Lastly, one other possibility may be the absence of Kupffer cells in fish liver which have been shown to be an important factor in CCl_4 toxicity in mammals (Moslen, 1995).

Mechanistic understanding of piscine hepatic responses to reference hepatotoxicants is hampered by our lack of a structural model of the organ in fish coupled with deficiency in understanding of oxygen tension differentials within specific regions of the liver microvasculature, if they exist. Although the oxygen differential in part explains some rodent responses to reference hepatotoxicants, we currently lack understanding in fish of this and of vascular parenchymal relationships in general. Perhaps it is erroneous to expect the tubular liver of fish with its differences in tissue and cellular architecture to respond like rodent liver. Biochemical properties of fish liver are poorly understood as well. Often, for fish we have inadequate knowledge concerning the metabolism of reference hepatotoxicants. A lack of information on organ-specific isoforms of serum enzymes in fish hampers our ability to interpret clinical enzyme tests. We cannot say with certainty that serum enzyme alterations are due solely to hepatic responses and not to toxic injury in cells of other organs and tissues as well. Finally, patterns and magnitude of responses following different routes of administration are needed to sort out effects of toxicant from those of carrier vehicle. What is needed is development of a field of fish-specific hepatotoxicity, one encompassing unique histologic, cellular biologic, molecular, biochemical and physiologic features of this organ in health and disease.

Hepatocellular adaptation

Cytopathologic (ultrastructural) alterations in hepatocytes of trout have been studied by Braunbeck with his colleagues and students, and have been reviewed (Braunbeck, 1994, 1998). This approach has been important in demonstrating that cells of teleost liver respond with specific organellar alterations associated with exposure to various agents. Correlation between these alterations and the conventional liver toxic manifestations of necrosis and bile stasis in fish are needed.

In these investigations, both intact fingerling trout and isolated hepatocytes were used, although most studies were with the former. Compounds investigated included: the intermediate in dye manufacture 4-chloroaniline; the cyclodiene insecticide endosulfan; the herbicide atrazine; the organophosphate pesticides diazinon and disulfoton; linuron; and the mycotoxin ochratoxin. After exposure durations of 4–5 weeks (Braunbeck, 1994), fixation was administered by vascular infiltration using cardiac perfusion. Morphologic alterations were quantitatively assessed using stereologic analysis (Segner and Braunbeck, 1990). For this brief review, the changes have been arranged by organelle.

Nuclei of hepatocytes were frequent sites of alteration. Stimulation of mitosis and augmentation of the number of binucleated hepatocytes followed exposure. Compounds not producing nuclear change were endosulfan and disulfoton.

Rough endoplasmic reticulum of control trout was arranged as extensive stacks of parallel cisternae around the centrally placed nuclei. With exposure, alteration was apparent. Dilation of cisternae, fragmentation of membrane stacks, vesiculation, and reduction were common changes in this organelle for all compounds except diazinon. Ochratoxin stimulated additional change including fenestration, fusion of opposing cisternal membranes, and transformation of rough endoplasmic reticulum into huge myelin bodies.

Smooth endoplasmic reticulum of control fish, a network of undulating, anastomosing tubular and vesicular profiles near Golgi apparatus, was affected by exposure to all compounds except 4-chloroaniline. All other compounds induced smooth endoplasmic reticulum augmentation. Degree of proliferation varied from slight changes after linuron to extensive change with diazinon. Ochratoxin led to degranulation of rough endoplasmic reticulum, making differentiation into smooth or rough forms difficult.

Whereas Golgi apparatus of control trout displayed up to five fenestrated cisternae with budding of many vesicles of varying dimensions and containing numerous very low-density lipoprotein (VLDL) granules, those of exposed trout showed extensive variability. 4-Chloroaniline and endosulfan reduced VLDL secretion. In contrast, linuron caused hypertrophy of Golgi and stimulated VLDL production, whereas diazinon provoked an increase of Golgi vesicles without increase of VLDL production. Ochratoxin caused complete morphologic disintegration of Golgi (Braunbeck, 1994).

Mitochondrial diversity of form was the common pattern after exposure, but individual compounds differed in the alterations produced in this organelle. Control trout showed spherical to elongated profiles in close association with rough endoplasmic reticulum lamellae and peroxisomes close to nuclei. Both linuron and 4-chloroaniline produced a conspicuous proliferation of mitochondria. Endosulfan caused dilation of intermembranous space, formation of membrane whorls and lysis of the organelle. The atrazine pattern was unique with elongated longitudinal cristae and myelin formation in the matrix and intermembranous space. Diazinon exposure was associated with megamitochondria. Ochratoxin caused collapse of mitochondria with matrix volume reduced to zero.

Peroxisomes were also studied, with visual verification based on presence of catalase after cytochemical localization using alkaline diaminobenzidine

(Braunbeck *et al.*, 1990). Normally, these organelles lack a matrix core and are small spherical particles. Proliferation of peroxisomes was weak after linuron and diazinon but was conspicuous after endosulfan exposure. Proliferation of peroxisomes was accompanied by formation of slender, tail-like projections. With exposure to endosulfan, peroxisomes formed aggregates of up to twenty profiles and displayed intensely staining matrices reminiscent of mammalian peroxisomes (Braunbeck, 1994). Interestingly, exposure to 4-chloroaniline was associated with a reduction in peroxisomes.

Lysosomes, rare in control hepatocytes, were increased after exposure to all but 4-chloroaniline and ochratoxin. After diazinon or linuron, lysosomes contained stacks of membrane fragments. Glycogenosomes, i.e. organelles degrading glycogen within lysosomes, were seen after diazinon and, especially, after 4-chloroaniline.

Whereas control hepatocytes were occupied by extensive glycogen depots, lipid droplets were sparse. Linuron or ochratoxin stimulated cytosolic lipid deposits and steatosis. 4-Chloroaniline was associated with accumulation of cholesterol-like crystals and condensation of glycogen into dense masses which eventually ended up in glycogenosome-like particles.

Although not as impressive as the alteration of hepatocytes, other cell types were different after exposure. Macrophage immigration and formation of aggregates was seen after toxicant exposure. Elevated phagocytic activity of glycogen, debris, and entire hepatic cells (apoptosis) was evident. Atrazine exposure was associated with immigration of granulocytes. Ito cells were proliferated after exposure to diazinon.

The above reactions reveal that trout liver responds differently to a variety of compounds, including a herbicide, a cyclodiene pesticide, organophosphate pesticides, and a specific toxin. Integration of these types of studies with biochemical assays could lead to a better appreciation of liver toxicity and has been initiated using marker enzymes for the various organelles (Braunbeck, 1994).

Hepatobiliary system

Bile formation is an essential function of liver cells. The biliary excretion of substances includes three major steps: (1) the uptake of substances at the sinusoidal surface of the hepatocytes, (2) their intracellular transport and possibly metabolism, and (3) the excretion into the bile fluid at the bile canaliculus (Groothuis and Meijer, 1996). This fluid is then conducted through a hierarchy of bile passageways to the bile duct. In most fish, the gall bladder, which stores and concentrates bile, is present and serves as a means to collect bile for analysis. More recently, as advances have been made in other vertebrates, studies have focused on understanding how fish hepatocytes function in the formation of bile and in xenobiotic clearance.

The basolateral uptake of substances from the blood at the sinusoidal surface of the rodent hepatocytes includes bile acids, fatty acids, and other choleophilic anions such as bilirubin or various xenobiotic and pharmaceutical compounds

(Groothuis and Meijer, 1996). The uptake across the cell membrane is effected by multiple Na^+-dependent and Na^+-independent carrier systems. In fish, Råbergh et al. (1994) studied the uptake of taurocholic acid and cholic acid in hepatocytes from rainbow trout, and they provided evidence for the existence of a Na^+-independent carrier system with a saturable component and a non-saturable component. With these characteristics, hepatocellular uptake of bile acids in trout resembles the corresponding systems described for rodent liver cells, but the fish carriers are distinguished by their high efficiency at low temperatures.

The bile canaliculus contains about 10 percent of the plasma membrane of the mammalian hepatocyte. For fish, no data are available, but pronounced species differences may be expected considering, for instance, the difference in bile canalicular organization between cyprinid fish, with the presence of a so-called unicellular canaliculus (Vogt and Segner, 1997), and the non-cyprinids, with the intercellular canaliculus typical for the vertebrate liver. In the latter case, canalicular lumina are formed by the apposing membranes of neighboring hepatocytes and are delimited by tight junction complexes that are functionally leaky and permit paracellular exchange between plasma and the canaliculus (Arias et al., 1993). The bile canaliculus is a contractile structure because of the presence of cytoskeletal elements in the pericanalicular cytoplasm. At the surface of the canalicular microvilli, ectoenzymes are present which can be visualized histochemically as lead phosphate precipitates in both rodent and piscine liver when incubating cryostat sections with lead and ATP (Hampton et al., 1988). Finally, the canalicular membrane contains several carriers which transport non-permeable molecules into the canalicular lumen. Luminal accumulation of osmotically active, impermeant solutes promotes the movement of water and electrolytes through tight junctions and canalicular membranes (Ballatori and Truong, 1992).

Several ATP-dependent transporters have been differentiated in rodent liver, including carriers for taurine- or glycine-conjugated as well as non-conjugated bile acids, for organic anions other than bile acids (e.g. glutathione conjugates), and for organic cations. For the last group of carriers, the P-glycoproteins, three classes can be differentiated in the mammalian liver, with classes I and II being responsible for the transport of amphipathic cationic and neutral compounds and class III transporting phospholipids (Elferink et al., 1996). In mammals, P-glycoproteins are encoded by a multigene family consisting of two multidrug resistance (mdr) genes in humans designated MDRI and MDRIII, and three mdr genes in mice and hamsters (Ng et al., 1989). All P-glycoproteins demonstrate high levels of homology, but particularly MDRI has been implicated in xenobiotic transport, and, therefore, has attracted much attention among toxicologists. The 3-ATP binding site and the 3'-terminal end of P-glycoprotein were sequenced for one fish species, the flounder Pleuronectes americanus, and were found to be homologous with mammalian species (Chan et al., 1992). Immunohistochemical studies using antibodies against mammalian P-glycoprotein demonstrated the presence of an immunoreactive product at the bile canaliculi of fish liver (Hemmer et al., 1995). Further studies will have to explore the functional role of these transporters for hepatic xenobiotic excretion in fish.

Cornelius (1991) reviewed the literature on bile pigments and their identification in cyclostomes, elasmobranchs and teleosts. Several aspects of that review are of direct relevance to our consideration. (1) Most studies have identified bile pigments by more primitive techniques and only in gall bladder (GB) bile. (2) During the past decades (1980s at the time of writing) more modern high-resolution, high-performance liquid chromatography (HPLC) techniques have been used. (3) Little information is available in fish concerning: identity of bile pigments in newly secreted hepatic duct bile; metabolism of these tetrapyrrole pigments and their enterohepatic transport; and any chemical alterations that occur in their structure during storage in the gall bladder. Cornelius (1991) concluded that GB bile of most fish studied to date contains unconjugated biliverdin (BV) and/or conjugated bilirubin and he regarded the potential for heterogeneity in the composition of bile pigments in GB bile as real, due mainly to the existence or quantity of BV reductase.

In vivo imaging studies of trout liver provided information on the intrahepatic biliary system in this fish. These were performed using epifluorescence illumination and water-submersible objectives (McCuskey *et al.*, 1986). Anesthetized trout were cannulated via the dorsal aorta and the portal vein, and a lateral incision was made to sever the liver surface. Sodium fluorescein was introduced to analyze epithelial and vascular relationships of the organ. With time, following introduction of the fluorescein, fluorescence was accentuated in biliary structures of hepatic tubules, with the result that canalicular and ductular lumina in central portions of hepatic tubules were imaged (see Figures 4.6 and 4.7; Hinton *et al.*, 1987). This pattern was the result of passage of label from sinusoids through fenestrae of lining endothelium to space of Disse, followed by uptake across basal membranes of hepatocytes and transcytosis into the biliary space. As reviewed above, CCl_4 causes hepatotoxicity resulting in increases of certain serum enzymes, especially glutamic pyruvic transaminase. When Gingerich *et al.* (1978) used biliary clearance of sulfobromophthalein (BSP) in plasma as an indicator of toxicity, decreased biliary clearance was seen after exposure to the reference hepatoxicant CCl_4. Although the injury could have been limited to hepatocytes, recent studies have shown that biliary epithelial cells also contribute to bile formation and that these cells may have been targeted as well.

Studies with a cartilaginous fish, for example the little skate (*Raja erinacea*), were proceeding at the same time (Smith *et al.*, 1987) as those imaging teleost liver. Hepatocyte plates in the little skate were demonstrated to be several cells thick and these relationships were retained in culture, representing a unique model for the study of cellular and subcellular organization of hepatocytes (Figure 4.8). Later, Henson *et al.* (1995) studied the cytoskeletal organization in these cell aggregates and, using confocal imaging and conventional epifluorescence microscopy in conjunction with fluorescent markers and immunocytochemistry, examined the structure and function of the cytoskeleton in these cells. Their results indicated that the microtubule cytoskeleton plays a fundamental role in the mediation of transcytosis, endocytosis, and bile excretory function in these hepatocytes.

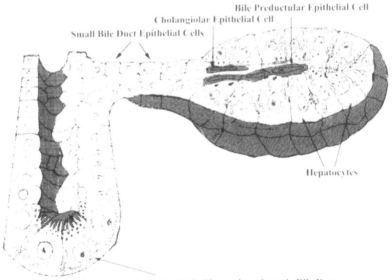

Figure 4.7 Proposed structural model of the hepatic tubule with associated binary passageways. Lateral hepatocyte plasma membranes of adjacent cells and hepatic tubules form bile canaliculi. After a short course, canaliculi opened into the biliary space at the center of the tubule, i.e. the lumen of the hepatic tubule. Here, transitional biliary epithelial cells, the bile preductular epithelial cells (Hinton and Pool, 1976) (hatched cells at the center of the hepatic tubule), form junctional complexes with hepatocytes, providing a wall of biliary passageway (Hampton *et al.*, 1988) in a manner analogous to exocrine pancreatic acini and centroacinar cells (Banks, 1993). The next portion of the biliary passageway is the cholangiole. Mural elements of the cholangiole are exclusively cuboidal biliary epithelial cells. The subsequent, downstream, intrahepatic biliary passageways (bile ducts) are lined by increasingly taller columnar epithelial cells and the passageway develops an increasingly greater luminal diameter. This model is centered on the excretory, i.e. biliary, function of the liver. Secretory products of the hepatocytes are released over lateral and basal plasma membranes (hatched areas of hepatic tubule on the left of the drawing) into intercellular space, including space of Disse, and into sinusoids for transport to other sites within the animal. Nutrients and related substances which are stored in hepatocytes reach these cells by passage from sinusoids through endothelial fenestrae and into the space of Disse, where they are presented to the finger-like processes of the hepatocytes for uptake.

Bile stasis is recognized as a hallmark feature of hepatic toxicity; however, we could find no indications that this has been investigated in fish species. Boyer and associates have conducted extensive investigations on biliary secretion in elasmobranchs (Boyer *et al.*, 1976a,b, 1993). Ballatori and Boyer (1992a,b) examined mechanisms of taurine transport in hepatocytes of the little skate, *Raja erinacea*, and recent investigations have examined effects of mercuric chloride on these cells and showed that taurine efflux was inhibited at mercury concentrations (20–40 μM) that had no effect on intracellular ATP levels or ATP/adenosine diphosphate (ADP) ratios, consistent with a direct interaction with the channel (Ballatori and Boyer, 1996).

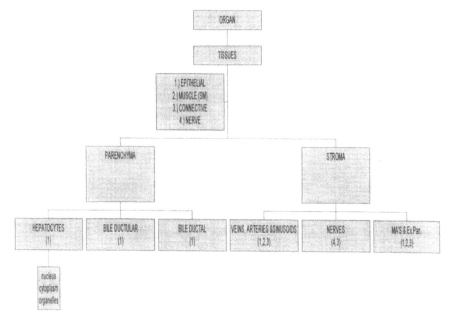

Figure 4.8 Hierarchy of organization in fish liver. The liver is composed of the four tissue types (1, epithelial; 2, smooth muscle; 3, connective tissue; and 4, neural) distributed within two major divisions, the parenchyma and the stroma. Parenchyma represents the epithelial cells largely responsible for the function of the organ. Stroma includes: nerves; hepatic vascular elements such as arteries, arterioles, capillary-like vessels termed sinusoids, venules, and veins; and the connective tissue compartments which contain collagen and macrophages, sometimes occurring as aggregates (MAs) and – depending on the species – islets of exocrine pancreas (ExPan). Each of the cell types contains a nucleus and cytoplasm with organelles, which for simplicity are shown only for hepatocytes.

In vitro model systems

Isolated primary cells or established cell lines offer a number of technical, ethical, and scientific advantages for studies on target organ toxicity. The use of isolated liver cells of fish started in the late 1970s, but it was only during the 1990s that the application of this methodologic approach for fish toxicologic studies became more widespread (Baksi and Frazier, 1990; Pesonen and Andersson, 1997; Monod *et al.*, 1998; Segner, 1998a; Braunbeck and Segner, 1999). Among the various cell types being present in fish liver, *in vitro* studies to date considered only hepatocytes except for the study of Blair *et al.* (1995), who described the isolation and culture of biliary epithelial cells from trout liver.

In vitro systems of fish liver cells have been mainly utilized to investigate xenobiotic metabolism, chemical cytotoxicity, genotoxicity, and the induction of specific cellular toxicant responses such as induction of metallothioneins or CYP1A (a few selected examples of *in vitro* toxicologic studies using isolated trout liver cells are given in Table 4.3).

Table 4.3 Selected toxicologic studies with isolated liver cells from bony fish; only studies with rainbow trout (*Oncorhynchus mykiss*) have been considered.

Subject of the study	References
Xenobiotic metabolism	Bailey *et al.* (1982), Loveland *et al.* (1987), Cravedi and Baradat (1991)
Induction of biotransformation enzymes	Andersson and Forlin (1985), Pesonen and Andersson (1991), Devaux *et al.* (1992), Masfaraud *et al.* (1992), Pesonen *et al.* (1992), Miller *et al.* (1993a,b), Sadar *et al.* (1995, 1996), Scholz *et al.* (1997), Scholz and Segner (1999)
Thermal acclimation of biotransformation enzymes	Andersson and Koivusaari (1986), Jensen *et al.* (1996)
Toxicity testing of environmental samples	Pesonen and Andersson (1992), Gagné and Blaise (1995, 1997), Zahn *et al.* (1995)
Mutagenesis and genotoxicity studies	Walton *et al.* (1984), Masfaraud *et al.* (1992), Devaux *et al.* (1997)
Xenobiotic induction of vitellogenesis	Jobling and Sumpter (1993), White *et al.* (1994), Petit *et al.* (1995), Schrag *et al.* (1998)
Cytotoxicity of heavy metals	Denizeau and Marion (1990)
Metallothionein induction	Hyllner *et al.* (1989), Gagné *et al.* (1990), Olsson *et al.* (1990)
Cytotoxicity of organic toxicants	Reader *et al.* (1996), Råbergh and Lipsky (1997)
Ultrastructural effects by various model xenobiotics	Zahn and Braunbeck (1993, 1995), Zahn *et al.* (1996), Braunbeck (1998)

The standard technique for isolation of fish liver cells is perfusion of the liver by collagenase-containing solutions (Mommsen *et al.*, 1994; Segner, 1998b; Braunbeck and Segner, 1999). Following isolation, a number of different culture techniques can be applied to the hepatocytes. Five cellular model systems from fish liver have been used for *in vitro* toxicologic studies:

1   Freshly isolated liver cells. Freshly isolated cells can be maintained at good viability in suspension for short periods only (Klaunig *et al.*, 1985). Incubation periods typically used for suspension cultures of fish liver cells range from 2 to 8 h. During this period, the cells are maintained in either buffered salines or in complete cell culture media. Suspension cultures offer the possibility of using high cell densities, which is an advantage for xenobiotic metabolism studies. In fact, this culture system has been frequently applied in biotransformation studies (Baksi and Frazier, 1990; Monod *et al.*, 1998). Hepatic metabolism of xenobiotics *in vivo* and *in vitro* differs quantitatively, i.e. the two approaches differ in the relative amount of metabolites produced; however, it appears to agree qualitatively, i.e. the two approaches generate

the same metabolites (Monod *et al.*, 1998). For instance, the major route of *in vivo* biotransformation of benzo(a)pyrene in fish leads to the formation of 3-hydroxy-benzo(a)pyrene, benzo(a)pyrene-7,8-diol and benzo(a)pyrene-9,10-diol. The same metabolites were detected in the medium after incubation of benzo(a)pyrene with suspensions of brown bullhead, *Ameiurus nebulosus* (Steward *et al.*, 1990), and English sole (*Parophrys vetulus*; Nishimoto *et al.*, 1992) hepatocytes. Thus, hepatocyte suspensions are well suited to reveal species differences in xenobiotic metabolism (Coulombe *et al.*, 1984; Sikka *et al.*, 1993).

2   Monolayer cultures. For incubation periods longer than 1 day, the monolayer technique is most frequently used for cultivation of fish hepatocytes. In this system, the cells are seeded as one layer in Petri dishes or in multiwell plates. Attachment of teleost hepatocytes to the surface of the culture dishes can be a problem, particularly with trout hepatocytes (Segner, 1998a). As culture media, most laboratories use serum-free standard formulations, as developed for rodent cells, such as minimum essential medium or Leibovitz medium. Monolayer culture of piscine hepatocytes supports survival of the isolated cells at good viability for periods of 5–8 days (Maitre *et al.*, 1986; Blair *et al.*, 1990; Braunbeck and Storch, 1992; Segner *et al.*, 1995; Segner, 1998b).

   Monolayer systems of fish liver cells are a good experimental model to investigate the cellular responses to toxicants which require activation of gene transcription and translation. Examples include the xenobiotic induction of CYP1A synthesis via the arylhydrocarbon receptor pathway, the induction of metallothionein synthesis by certain metal species, or the induction of vitellogenin synthesis through the estrogen receptor pathway (Segner and Braunbeck, 1998).

   Particular emphasis in studies using hepatocyte cultures have been given to the analysis of CYP1A regulation. The expression of this enzyme which is induced by dioxins, dibenzofurans, polychlorinated biphenyls, polyaromatic hydrocarbons, and other environmental toxicants experiences a rapid decline in monolayer cultures of rodent hepatocytes (Skett, 1994). This decrease in CYP1A levels is not due to a general loss of cell viability but rather appears to be the result of a dedifferentiation process, with the reversion of adult hepatocytes to a fetal-like metabolism (Guguen-Guillouzo and Guillouzo, 1983). In cultured fish hepatocytes, however, CYP1A levels appear to be rather stable and retain the ability to be induced by xenobiotics (Vaillant *et al.*, 1989; Pesonen and Andersson, 1991; Jensen *et al.*, 1996; Segner, 1998b). The physiologic basis for the conservation of cytochrome P4501A in piscine liver cells, contrary to the rodent model, is not understood.

   Induction of cytochrome P4501A activity in monolayer cultures of rainbow trout hepatocytes exposed to the prototype inducer β-naphthoflavone was first demonstrated by Vaillant *et al.* (1989). CYP1A activity has been mostly assessed through measurement of the 7-ethoxyresorufin-*O*-deethylase (EROD) activity. Particularly the development of methods allowing the analysis of EROD in the intact cell, without need for laborious preparation of microsomes (Hahn *et al.*, 1996; Behrens *et al.*, 1998), fostered the application of *in vitro* models for CYP1A induction studies. During recent years, measurement of hepatocellular CYP1A *in vitro* at the protein and mRNA level became possible through introduction of homologous immunochemical and molecular probes (Pesonen *et al.*, 1992; Scholz *et al.*, 1997). The time-course of CYP1A induction *in vitro* appears to be similar to that *in vivo*, although the *in vitro* response shows a lower magnitude of induction (Pesonen and Andersson, 1997; Monod *et al.*, 1998).

3   Aggregate cultures and co-culture systems. *In vitro* studies with mammalian hepatocytes have demonstrated that the cellular microenvironment has an important influence on viability and functional performance of isolated hepatocytes (Guguen-Guillouzo and Guillouzo, 1983; Skett, 1994). A major factor that changes with the transfer of liver cells from the *in vivo* to the *in vitro* situation is the cellular microenvironment, i.e. the hetero- and homotypic cell contacts and the extracellular matrix. Thus, mammalian toxicologists have developed various approaches to overcome this problem, for instance 'sandwich' cultures, coating of culture plates with collagen-like matrices, co-cultures or spheroid cultures (Guguen-Guillouzo and Corlu, 1993; Skett, 1994). With fish liver cells, two approaches have been explored (Segner, 1998b): the aggregate system (Flouriot *et al.*, 1993; Cravedi *et al.*, 1996), which cultivates the hepatocytes in a three-dimensional spheroid, and the co-culture system, which incubates the hepatocytes together with established fish cell lines (Scholz *et al.*, 1998). Both systems support the *in vitro* conservation of liver-specific functions. Furthermore, the aggregate system extends the possible culture period of fish hepatocytes from 5–8 days, as obtained with monolayers, up to 30 days. During this period, the cells remain responsive to estrogen induction of vitellogenin synthesis as well as to xenobiotic induction of CYP1A (Flouriot *et al.*, 1993; Cravedi *et al.*, 1996). Unfortunately, the potential of these systems for liver toxicity studies remains largely unused.

4   Liver slices. An alternative approach to conserve the *in vivo* microenvironment of liver cells is the use of tissue slices. Hepatic tissue slice technology allows investigation of metabolic activity while maintaining cells intact within their three-dimensional tissue architecture. An important improvement of this method was achieved when microtomes were developed which were capable of rapidly generating thin (< 250 µm) and reproducible slices (Kane and Thohan, 1996). A disadvantage of the method is the fact that the slices seem to remain metabolically competent only for relatively short culture periods (Singh *et al.*, 1996). As already mentioned for the aggregate cultures, the application of liver slices in fish toxicology is still in its infancy (Kane and Thohan, 1996; Singh *et al.*, 1996).

5   Liver cell lines. Although a number of liver-derived continuous cell lines from fish are available, most of these lines obviously do not express liver-specific functions (Segner, 1998a). Recently, two liver cell lines were established which possess at least one facet of hepatic metabolism, i.e. inducible CYP1A activity; these are the PLHC-1 cell line originating from a

hepatoma of *Poecilopsis lucida* (Hightower and Renfro, 1988; Hahn *et al.*, 1993) and the RTL-W1 cell line from a normal trout liver (Lee *et al.*, 1993). Regulation and xenobiotic induction of CYP1A in these cell lines appears to be similar to primary hepatocytes (Celander *et al.*, 1997). These cell lines may be promising practical tools for induction studies, for instance to establish toxicity equivalency factors, however their usefulness for (mechanistic) studies on liver toxicity remains to be established.

Chronic toxicity – liver carcinogenesis studies

Our consideration to this point has been primarily with acute toxic responses in this target organ. Although our intention is not to review the field of fish liver carcinogenesis, the responses of hepatic cells to carcinogens and the multistep process leading from normal morphology to overt neoplasia in this organ teach us much about toxicity and mechanisms used by subpopulations of hepatocytes to withstand toxic injury and to respond to growth stimuli. This section is therefore of relevance to our consideration of the liver as a target organ for chronic toxicity.

Fish models for environmental carcinogenesis have been reviewed, and the interested reader is referred to Bailey *et al.* (1996) and Bunton (1996). The liver is the most common organ in which experimentally induced tumors in fish develop, and this organ is also the site of tumors linked to pollutant exposure in wild species. In 1964, following detection of liver neoplasms in white sucker (*Catostomus commersonii*) and brown bullhead (*Ictalurus nebulosus*) from a reservoir in Deep Creek, Maryland, USA, a contaminant etiology was proposed (Dawe *et al.*, 1964). Since then, hepatic neoplasms have been found with high or repeated prevalences in populations of additional species from North America (Harshbarger and Clark, 1990), from the North Sea and from central European rivers (Peters *et al.*, 1987; Vethaak and Wester, 1996). Comparisons of tumor prevalences in age-matched groups from multiple sampling sites and comprehensive analyses of contaminants in sediment and in fish tissue have clearly established a link between sediment contamination and liver neoplasia for two species, English sole (*Pleuronectus vetulus*) (Malins *et al.*, 1987) and brown bullhead (Baumann and Harshbarger, 1995). Controlled laboratory administration of contaminated sediment or sediment extracts from various sources produced liver tumors in rainbow trout and medaka (Maccubbin *et al.*, 1987; Metcalfe *et al.*, 1988; Fabacher *et al.*, 1991).

Various aspects of hepatic carcinogenesis in fish are of importance for our consideration because this is a form of chronic toxicity and the progression from exposure to endpoint has been well defined (Bailey *et al.*, 1996). Except for rare direct-acting chemical carcinogens, most potentially carcinogenic substances must be metabolized to ultimate carcinogenic forms. Various events which are known, or are considered to be involved, between exposure and endstage include the following. (1) The ultimate carcinogenic form of the genotoxic carcinogen binds to target cellular macromolecules (DNA) and an adduct is formed. (2) Once DNA adducts are formed, they may be excised and repaired correctly or (3) the resultant toxicity may cause the death of the affected cell. In these instances, the process may be halted. (4) If the DNA is misrepaired and/or if the cell divides before repair, daughter cells with DNA mutations will result. (5) In a clonal expansion, the initial morphologic stage associated with the neoplastic process, a focus of cellular alteration, occurs. (6) The further promotion and progression of foci to the more advanced 'bridging lesion', the adenoma, and to the endstage, hepatocellular carcinoma, occurs over a chronic time frame and is regarded as a multistep process (Farber, 1980), with a striking similarity exemplified between the process as seen in certain fish and that of the more conventional and better characterized rodent models (Bannasch *et al.*, 1996). How these altered fish cells survive and preferentially grow, in a toxic milieu, provides us with valuable information on the fish-relevant cell biology of survival and growth.

Because foci of cellular alteration occur after exposure to known carcinogens and before development of adenomas and carcinomas, the lesions have received appreciable attention. Three basic phenotypes are seen with conventional hematoxylin and eosin procedures. These are basophilic, eosinophilic and clear cell. Teh and Hinton (1993) have shown that the same battery of enzyme histochemical procedures which mark foci of rodent liver mark these lesions in medaka liver. Moore and Myers (1993) reviewed enzyme characteristics of foci and included a review of immunohistochemical and biochemical observations on cytochrome P4501A and glutathione-S-transferase (GST) of foci and neoplasms. Foci generally are deficient in phase I but enriched in certain phase II enzymes, quinone oxidoreductase and UDP-glucose dehydrogenase (UDPGdH). However, they concluded that the pattern associated with GST needs additional work in fish.

Köhler and Van Noorden (1998) and Köhler *et al.* (1998) established metabolic changes and cell proliferation indices for extrafocal liver, foci and neoplasms of European flounder. Small foci with increased activity of G6PDH and elevated expression of proliferating cell nuclear antigen (PCNA) appeared prior to development of tumors. High initial velocities of glucose-6-phosphate dehydrogenase (G6PDH) and high PCNA labeling index of early foci persisted in hepatocellular adenomas and carcinomas (Köhler and Van Noorden, 1998).

That fish mature hepatocytes are targets of both the toxic injurious and carcinogenic effects of known carcinogens is well supported by published studies to date. Following the hepatotoxic changes dictated by the type of initiating compound, its concentration and/or the duration of exposure, the earliest possible preneoplastic or neoplastic change appears to involve fully differentiated, mature hepatocytes. Foci of altered hepatocytes are detected by tinctorial (i.e. staining) differences between the focus and the surrounding extrafocal hepatocytes. Teh and Hinton (1993), using freeze-dried, vacuum-embedded fresh tissue, have studied enzyme histochemistry of foci in medaka induced by brief exposure to diethylnitrosamine (DEN). Foci detected by conventional staining also showed enzyme alteration, and foci appeared first and preceding the other lesions.

Although a number of phenotypes can be demonstrated using conventional stains (Okihiro, 1996) and although it is important to distinguish each in thoroughly addressing the histogenesis of hepatic neoplasia (Bannasch *et al.*, 1996), three are more commonly described (Couch and Courtney, 1987; Boorman *et al.*, 1997). These are basophilic, eosinophilic, and clear cell foci (Couch and Courtney, 1987; Hinton, 1998). Biochemical, cytochemical and ultrastructural properties of foci in rat hepatocarcinogenesis (Bannasch *et al.*, 1985, 1996) and, to a lesser extent, in mouse (Ruebner *et al.*, 1996) have been reported previously.

Adenomas appear to be bridging lesions between certain foci and eventual hepatocellular carcinomas (described below). When serial analyses are performed, adenomas typically overlap with the occurrence of foci of altered hepatocytes, but are distinguished from them by more distinct borders, by thickened hepatic tubules, by cellular pleomorphism, and by variable overall shape. Adenomas are typically larger than foci (Couch and Courtney, 1987; Hinton, 1993).

Characteristics of hepatocellular carcinomas in fish have been previously described (Hendricks *et al.*, 1984; Hinton, 1993; Bailey *et al.*, 1996). Overall size varies from focal to extensive, the latter often occupying a major portion of the overall organ. Masses of cells show numerous bizarre mitotic figures. In addition, the masses are often solid and resemble engorged tubules. The margin is less well defined, and invasion into otherwise normal surrounding parenchyma is common. Both poorly- and well-differentiated hepatocellular carcinomas are seen as subsets of hepatocellular carcinomas. It is possible that well-differentiated carcinomas develop from cells in some of the adenomas that, as a group, begin to decrease in numbers coincident with the first appearance of the well-differentiated carcinomas (Couch and Courtney, 1987; Hinton *et al.*, 1992; Okihiro, 1996). Metastasis occurs (Okihiro and Hinton, 2000) by direct extension or by hematogenous route.

The second cell that could putatively give rise to hepatocellular, biliary and, perhaps, mixed cell neoplasms is the biliary epithelial cell. As shown above, the bile preductular cells are positioned within the hepatic tubule. Okihiro and Hinton (2000) provided evidence that bile preductular cells proliferate after partial hepatectomy in both medaka and trout. Further serial analysis of the repair/regeneration showed the appearance of cells with transitional features between biliary epithelium and hepatocytes, and, finally with additional time, the hepatocyte phenotype characterized these cell populations. Taken together, these pieces of evidence suggest that the bile preductular epithelial cell may be a stem cell capable of giving rise to both adult cell types within the hepatic tubule. This is conjecture at this time, but the question is worthy of future investigation. Further similarity between descriptions of fine structure of oval cells in rodents and bile preductular cells in trout were demonstrated by Hampton *et al.* (1988). More detailed serial progression between initiation and first appearance of neoplasms is needed, and this is under investigation in medaka in which the interval of interest is 16 weeks. Based on the sheer number of the bile preductular epithelial cells (Hampton *et al.*, 1989), trout and medaka as well as other teleosts are excellent models for analysis of the role(s) of this cell type. Okihiro (1996) added new cytologic markers, including lectins and cytofilaments, for these cells, thereby increasing the list of those reviewed by Hinton (1993).

Another cell type that may be involved in formation of neoplasms is the perisinusoidal cell of Ito. Bannasch *et al.* (1981) first described a pathologic liver condition involving these cells in rats experimentally exposed to *N*-nitrosomorpholine, spongiosis hepatis (SH). In fish, the SH lesions were first reported by Hinton *et al.* (1984) and studied in altered pathologic states by Couch (1991) and by Bunton (1990). Fish SH lesions appear to be derived from pathologic conditions similarly to those in mammals. Morphogenetic observations (electron microscopy and light microscopy levels) suggest that SH in rats (Bannasch *et al.*, 1981, 1986) and in fish (Couch, 1991), following exposure to certain nitrosamines, has as its cell of origin the perisinusoidal cell. It is considered possible that SH could progress to tumor formation.

Experimentally induced endothelial cell neoplasms in the livers of fish have been reported only once. Grizzle and Thiyagarajah (1988) described cavernous hemangiomas in *Rivulus marmoratus* exposed continuously for 8 months to DEN (beginning as larval fish). These lesions began as peliosis as early as after 4 weeks of exposure in the larvae and progressed (four cases) to cavernous hemangiomas characterized as large blood-filled cavities lined by single layers of swollen endothelial cells. There is a question as to whether these lesions are neoplastic or are representative of advanced peliosis hepatis. It is apparent that the sinusoids and their endothelia, however, are the structures involved in their origin.

Information gaps and future directions

As we reviewed the literature for preparation of this chapter, it became apparent that we have little fish-specific, cellular biologic underpinning from which to interpret responses of fish liver cells and tissues. For example, the development of fatty liver or steatosis in rodents may arise by various mechanisms, but the research to determine mechanisms of fatty liver of fish simply has not been carried out. While the drugs and metabolic inhibitors which were used to tease out mechanisms in rodents lack environmental relevance, our ability to interpret responses of livers in free-ranging aquatic vertebrates is hampered without such background information. Given the ranges of normality between salmonids which apparently store little lipid in liver and that stored in livers of fish such as cod (*Gadus morhua*), the possibility of using a comparative approach to immense advantage seems most attractive.

In recent reviews of toxic responses of the liver (Moslen, 1995; Vandenberghe, 1996), rodent studies have extended to the role(s) played by individual cell types, and coverage includes the analysis of interactions between cell types. We lack that degree of depth in studies of fish liver toxicity. The co-culture systems which use hepatocytes and non-hepatocytic cell types may be used to investigate important interrelationships between these cells. The detailed advances being made in hepatocyte uptake and secretion of bile acids, organic anions, endogenous and exogenous compounds in the little skate, reviewed above, need to be transferred to investigations of teleost fish as well.

Although analysis of serum enzymes has potential as a partially invasive technique which could be applied to individual fish to gain information concerning health and toxicity, additional effort is necessary if the full advantages of such techniques are to be realized. Not only is it essential to note target organs from whose cells enzymes may leak into the serum but it is also important to have confidence concerning the source of such proteins. We need organ-specific isozyme analysis of serum enzymes in fish.

Finally, we call for the formation of a field of fish hepatology which uses integrated structural and biochemical/physiologic approaches to investigate the wealth of responses which are possible in the various aquatic species.

References

Adams, S.M. and McLean, R.B. 1985. Estimation of largemouth bass, *Micropterus salmoides* Lacépède, growth using the liver somatic index and physiological variables. *Journal of Fish Biology* 26: 111–126.

Andersson, T. and Forlin, L. 1985. Spectral properties of substrate-cytochrome P-450 interaction and catalytic activity of xenobiotic metabolizing enzymes in isolated rainbow trout liver cells. *Biochemical Pharmacology* 34: 1407–1413.

Andersson, T. and Koivusaari, U. 1986. Oxidative and conjugative metabolism of xenobiotics in isolated liver cells from thermally acclimated rainbow trout. *Aquatic Toxicology* 8: 85–92.

Anulacion, B.F., Myers, M.S., Willis, M.L. and Collier, T.K. 1998. Quantitation of CYP1A expression in two flatfish species showing different prevalences of contaminant-induced hepatic disease. *Marine Environmental Research* 46: 7–11.

Arias, I.M., Che, M., Gatmaitan, X., Leveille, C., Nishida, T. and St. Pierre, M. 1993. The biology of the bile canaliculus. *Hepatology* 17: 318–329.

Badr, M.Z., Belinksy, S.A., Kauffman, F.C. and Thurman, R.G. 1986. Mechanism of hepatotoxicity to periportal regions of the liver lobule due to allyl alcohol: role of oxygen and lipid peroxidation. *Journal of Pharmacology and Experimental Therapeutics* 238: 1138–1142.

Bailey, G.S., Taylor, M.J. and Selivonchick, D.P. 1982. Aflatoxin B1 metabolism and DNA binding in isolated hepatocytes from rainbow trout (*Salmo gairdneri*). *Carcinogenesis* 3: 511–518.

Bailey, G.S., Williams, D.E. and Hendricks, J.D. 1996. Fish models for environmental carcinogenesis: The rainbow trout. *Environmental Health Perspectives* 104: 5–21.

Baksi, S.M. and Frazier, J.M. 1990. Isolated fish hepatocytes – model systems for toxicology research. *Aquatic Toxicology* 16: 229–256.

Ballatori, N. and Boyer, J.L. 1992a. Taurine transport in skate hepatocytes. 1. Uptake and efflux. *American Journal of Physiology* 262: 445–450.

Ballatori, N. and Boyer, J.L. 1992b. Taurine transport in skate hepatocytes. 2. Volume activation, energy and sulfhydryl dependence. *American Journal of Physiology* 262: 451–460.

Ballatori, N. and Boyer, J.L. 1996. Disruption of cell volume regulation by mercuric chloride is mediated by an increase in sodium permeability and inhibition of an osmolyte channel in skate hepatocytes. *Toxicology and Applied Pharmacology* 140: 404–410.

Ballatori, N., and Truong, A.T. 1992. Glutathione as a primary osmotic driving force in hepatic bile formation. *American Journal of Physiology* 263: 617–624.

Banks, W.J. 1993. Epithelia. In *Applied Veterinary Histology*, 3rd edn. Banks, W.J. (eds), pp. 48–67. Mosby-Year Book, St Louis.

Bannasch, P., Bloch, M. and Zerban, H. 1981. Spongiosis hepatis, specific changes of the perisinusoidal liver cells induced in rats by N-nitrosomorpholine. *Laboratory Investigations* 44: 252–264.

Bannasch, P., Moore, M.A., Hacker, H.J., Klimek, F., Mayer, D., Enzmann, H., and Zerban, H. 1985. Potential significance of phenotypic instability in focal and nodular liver lesions induced by hepatocarcinogens. In *Hepatology*. Brunner, H. and Thaler, H. (eds), pp. 191–209. Raven Press, New York.

Bannasch, P., Griesemer, R.A., Anders, F., Becker, R., Cabral, J.R. and Poreta, G.D. 1986. Early preneoplastic lesions. In *Long-term and Short-term Assays for Carcinogens: A Critical Appraisal*. IARC Scientific Publications No. 83. Montesano, R., Bartsch, H., Vainaio, H., Wilbourn, J. and Yamasaki, H. (eds), pp. 85–101. International Agency for Research on Cancer/International Programme on Chemical Safety Commission of the European Communities, Lyon.

Bannasch, P., Zerban, H. and Hacker, H.J. 1996. Foci of altered hepatocytes, rat. In *Monographs on Pathology of Laboratory Animals. Digestive System*, 2nd edn. Jones, T.C., Popp, J.A. and Mohr, U. (eds), pp. 3–37. Springer-Verlag, Berlin.

Baumann, P.C. and Harshbarger, J.C. 1995. Decline in liver neoplasms in wild brown bullhead catfish after coking plant closes and environmental PAHs plummet. *Environmental Health Perspectives* 103: 168–170.

Behrens, A., Schirmer, K., Bols, H.C. and Segner, H. 1998. Microassay for rapid measurement of 7-ethoxyresorufin-O-deethylase activity in intact fish hepatocytes. *Marine Environmental Research* 46: 369–373.

Belinsky, S.A., Matsumura, T., Kauffman, F.C. and Thurman, R.G. 1984. Rates of allyl alcohol metabolism in periportal and pericentral regions of the liver lobule. *Molecular Pharmacology* 25: 158–164.

Belinsky, S.A., Badr, M.Z., Kauffman, F.C. and Thurman, R.G. 1986. Mechanism of hepatotoxicity in periportal regions of the liver lobule due to allyl alcohol: studies on thiols and energy status. *Journal of Pharmacology and Experimental Therapeutics* 238: 1132–1137.

Berger, M.L., Bhatt, H., Combes, B. and Estabrook, R.W. 1986. CCl_4-induced toxicity in isolated hepatocytes: the importance of direct solvent injury. *Hepatology* 6: 36–45.

Blair, J.B., Miller, M.R., Pack, D., Barnes, R., Teh, S.J. and Hinton, D.E. 1990. Isolated trout liver cells: establishing short-term primary cultures exhibiting cell–cell interactions *in vitro*. *Cellular and Developmental Biology* 26: 237–249.

Blair, J.B., Ostrander, G.K., Miller, M.R. and Hinton, D.E. 1995. Isolation and characterization of biliary epithelial cells from rainbow trout liver. *In Vitro Cellular and Developmental Biology* 31: 780–789.

Blasco, J., Fernandez, J. and Gutierrez, J. 1992. Fasting and refeeding in carp, *Cyprinus carpio* L.: the mobilization of reserves and plasma metabolite and hormone variations. *Journal of Comparative Physiology* B162: 539–546.

Blazer, V.S., Wolke, R.E., Brown, J. and Powell, C.A. 1987. Piscine macrophage aggregate parameters as health monitors: effect of age, sex, relative weight, season and site quality in largemouth bass (*Micropterus salmoides*). *Aquatic Toxicology* 10: 199–215.

Blouin A, Bolender R.P. and Weibel E.R. 1977. Distribution of organelles and membranes between hepatocytes and nonhepatocytes in the rat liver parenchyma. A stereological study. *Journal of Cellular Biology* 72: 441–55.

Böhm, R., Hanke, W. and Segner, H. 1994. The sequential restoration of plasma metabolite levels, liver composition and liver structure in refed carp, *Cyprinus carpio. Journal of Comparative Physiology* B164: 32–41.

Boorman, G.A., Botts, S., Bunton, T.E., Fournie, J.W., Harshbarger, J.C., Hawkins, W.E., Hinton, D.E., Jokinen, M.P., Okihiro, M.S. and Wolfe, M.J. 1997. Diagnostic criteria for degenerative, inflammatory, proliferative nonneoplasia and neoplastic liver lesions in medaka (*Oryzias latipes*): Consensus of a National Toxicology Program Pathology Working Group. *Toxicologic Pathology* 25: 202–210.

Boyer, J.L., Schwarz, J. and Smith, N. 1976a. Biliary secretion in elasmobranchs. I. Bile collection and composition. *American Journal of Physiology* 230: 970–973.

Boyer, J.L., Schwarz, J. and Smith, N. 1976b. Biliary secretion in elasmobranchs. II. Hepatic uptake and biliary excretion of organic anions. *American Journal of Physiology* 230: 974–981.

Boyer, J.L., Hagenbuch, B., Ananthanarayanan, M., Suchy, F., Stieger, B. and Meier, P.J. 1993. Phylogenic and ontogenic expression of hepatocellular bile acid transport. *Proceedings of the National Academy of Sciences of the United States of America* 90: 435–438.

Braunbeck, T. 1994. Detection of environmentally relevant concentrations of toxic organic compounds using histological and cytological parameters: substance-specificity in the reaction of rainbow trout liver? In *Sublethal and Chronic Effects of Pollutants on Freshwater Fish*. Muller, R. and Lloyd, R. (eds), pp. 15–29. Blackwell Science, Oxford.

Braunbeck, T. 1998. Cytological alterations in fish hepatocytes following *in vivo* and *in vitro* sublethal exposure to xenobiotics – structural biomarkers of environmental contamination. In *Fish Ecotoxicology*. Braunbeck, T., Hinton, D.E. and Streit, B. (eds), pp. 61–140. Birkhäuser Verlag, Basel.

Braunbeck, T. and Segner, H. 1992. Preexposure temperature acclimation and diet as modifying factors for the tolerance of golden ide (*Leuciscus idus melanotus*) to short-term exposure to 4-chloroaniline. *Ecotoxicology and Environmental Safety* 24: 72–94.

Braunbeck, T. and Segner, H. 1999. Isolation and culture of teleost hepatocytes. In *Isolated Hepatocytes – Preparation, Properties and Applications*, 6th edn. Berry, J.M. (ed.). Elsevier, Amsterdam.

Braunbeck, T. and Storch, V. 1992. Senescence of hepatocytes isolated from rainbow trout (*Oncorhynchus mykiss*) in primary culture. *Protoplasma* 170: 138–159.

Braunbeck, T., Burkhardt-Holm, P. and Storch, V. 1990. Liver pathology in eels (*Anguilla anguilla* L.) from the Rhine river exposed to the chemical spill at Basle in November 1986. *Limnologie aktuell* 1: 371–392.

Buhler, D.R. and Wang-Buhler, J.L. 1998. Rainbow trout cytochrome P450s: purification molecular aspects, metabolic activity, induction and role in environmental monitoring. *Comparative Biochemistry and Physiology* 121C: 107–138.

Bunton, T.E. 1990. Hepatopathology of diethylnitrosamine in the medaka (*Oryzias latipes*) following short-term exposure. *Toxicologic Pathology* 18: 313–323.

Bunton, T.E. 1996. Experimental chemical carcinogenesis in fish. *Toxicologic Pathology* 24: 603–618.

Burk, R.F., Patel, K. and Lane, J.M. 1983. Reduced glutathione protection against rat liver microsomal injury by carbon tetrachloride. Dependence on O_2. *Biochemistry Journal* 215: 441–445.

Burk, R.F., Lane, J. and Patel, K. 1984. Relationship of oxygen and glutathione in protection against carbon tetrachloride-induced hepatic microsomal lipid peroxidation and covalent binding in the rat: rationale for the use of hyperbaric oxygen to treat carbon tetrachloride ingestion. *Journal of Clinical Investigations* 74: 1996–2001.

Cahil, G.F. 1970. Starvation in man. *New England Journal of Medicine* 282: 668–675.

Caldwell, J., Sangster, S.A. and Sutton, J.D. 1986. The role of metabolic activation in target organ toxicity. In *Target Organ Toxicity*, Vol. 1. Cohen, G.M. (ed.), pp.37–54. CRC Press, Boca Raton.

Celander, M., Bremer, J., Hahn, M.E. and Stegeman, J.J. 1997. Glucocorticoid–xenobiotic interactions: dexamethasone-mediated potentiation of cytochrome P4501A induction by β-naphthoflavone in fish hepatoma cell line (PLHC-1). *Environmental Toxicology and Chemistry* 16: 900–907.

Chan, K.M., Davies, P.L., Childs, L., Veinot, L. and Ling, V. 1992. P-glycoprotein genes in the winter flounder, *Pleuronectes americanus*: isolation of two types of genomic clones carrying 3′-terminal exons. *Biochimica et Biophysica Acta* 1171: 65–72.

Chavin, W. 1973. Teleostean endocrine and para-endocrine alterations of utility in environmental studies. In *Responses of Fish to Environmental Change*. Chavin, W. (ed.), pp. 199–238. Springfield, Illinois.

Cornelius, C.E. 1991. Bile pigments in fishes: a review. *Veterinary and Clinical Pathology* 20: 106–115.

Cornish, I., and Moon, T.W. 1985. Glucose and lactate kinetics in American eels, *Anguilla rostrata*. *American Journal of Physiology* 249: R67–R72.

Cossins, A.R. 1983. Adaptive responses of fish membranes to altered environmental temperature. *Biochemical Society Transactions* 11: 322–333.

Couch, J.A. 1991. Spongiosis hepatis: Chemical induction, pathogenesis, and possible neoplastic fate in a teleost fish model. *Toxicologic Pathology* 19: 237–250.

Couch, J.A. and Courtney, L.A. 1987. N-Nitrosodiethylamine-induced hepato-carcinogenesis in estuarine sheepshead minnow (*Cyprinodon variegatus*): neoplasms and related lesions compared with mammalian lesions. *Journal of the National Cancer Institute* 79: 297–321.

Coulombe, R.A., Bailey, G.S. and Nixon, J. 1984. Comparative activation of aflatoxin B1 to mutagens by isolated hepatocytes from rainbow trout (*Salmo gairdneri*) and coho salmon (*Oncorhynchus kisutch*). *Carcinogenesis* 5: 29–33.

Cowey, C.B. and Walton, M.J. 1989. Intermediary metabolism. In *Fish Nutrition*, 2nd edn. Halver, J.E. (ed.), pp. 259–329. Academic Press, San Diego.

Cowey, C.B., Knox, D., Walton, M.J. and Adron, J.W. 1977. The regulation of gluconeogenesis by diet and insulin in rainbow trout (*Salmo gairdneri*). *The British Journal of Nutrition* 38: 463–470.

Cravedi, J.P. and Baradat, M. 1991. Comparative metabolic profiling of chloramphenicol by isolated hepatocytes from rat and trout (*Onorhynchus mykiss*). *Comparative Biochemistry and Physiology* 100C: 649–652.

Cravedi, J.P., Paris, A., Monod, G., Devaux, A., Flouriot, G. and Valotair, Y. 1996. Maintenance of cytochrome P450 content and phase I and phase II enzyme activities in trout hepatocytes cultured as spheroideal aggregates. *Comparative Biochemistry and Physiology* 113C: 241–246.

Dalich, G.M. and Larson, R.E. 1985. A comparative study of the hepatotoxicity of monochlorobenzene in the rainbow trout (*Salmo gairdneri*) and the sprague-dawley rat. *Comparative Biochemistry and Physiology* 80C: 115–122.

Dawe, C.J., Stanton, M.F. and Schwartz, F.J. 1964. Hepatic neoplasms in native bottom-feeding fish of Deep Creek Lake, Maryland. *Cancer Research* 24: 1194–1201.

Denizeau, F. and Marion, M. 1990. Toxicity of cadmium, copper, and mercury to isolated trout hepatocytes. *Canadian Journal of Fisheries and Aquatic Science* 47: 1038–1052.

Devaux, A., Pesonen, M., Monod, G. and Andersson, T. 1992. Glucocorticoid-mediated potentiation of P-450 induction in primary culture of rainbow trout hepatocytes. *Biochemistry and Pharmacology* 43: 898–901.

Devaux, A., Pesonen, M. and Monod, G. 1997. Alkaline comet assay in rainbow trout hepatocytes. *Toxicology In Vitro* 11: 71–79.

Diaz Gomez, M.I., De Castro, C.R., D'Acasta, N., DeFenos, O.M., De Ferreyra, E.C. and Castro, J.A. 1975. Species differences in carbon tetrachloride-induced hepatotoxicity: the role of CCl_4 activation and of lipid peroxidation. *Toxicology and Applied Pharmacology* 34: 102–114.

Droy, B.F. 1998. Effect of reference hepatotoxicants on rainbow trout liver. Ph.D. Thesis. University of Morgantown, West Virginia.

Droy, B.F., Davis, M.E. and Hinton, D.E. 1989. Mechanism of allyl formate-induced hepatotoxicity in rainbow trout. *Toxicology and Applied Pharmacology* 98: 313–324.

Elferink, R.P.J.O., Frijters, C.M.G. and Groen, A.K. 1996. Regulation of canalicular transport activities. *Journal of Hepatology* 24: 94–99.

Elias, H. and Sherrick, J.C. 1969. *Morphology of the Liver*, pp. 6–9. Academic Press, New York.

Everaarts, J., Shugart, L., Gustin, M., Hawkins, W. and Walker, W. 1993. Biological markers in fish: DNA integrity, biological parameters and liver somatic index. *Marine Environmental Research* 35: 101–108.

Fabacher, D.L., Besser, J.M., Schmitt, C.J., Harshbarger, J.C., Peterman, P.H. and Lebo, J.A. 1991. Contaminated sediments from tributaries of the Great Lakes: chemical characterization and carcinogenic effects in medaka (*Oryzias latipes*). *Archives of Environmental Contamination and Toxicology* 21: 17–34.

Fabbri, E., Capuzzo, A. and Moon, T.W. 1998. The role of circulating catecholamines in the regulation of fish metabolism: an overview. *Comparative Biochemistry and Physiology* 120: 177–192.

Farber, E. 1980. The sequential analysis of liver cancer induction. *Biochimica et Biophysica Acta* 605: 149–166.

Flouriot, G., Vaillant, C., Salbert, G., Pelissero, C., Guiraud, J.M. and Valotaire, Y. 1993. Monolayer and aggregate cultures of rainbow trout hepatocytes – long-term and stable liver-specific expression in aggregates. *Journal of Cellular Science* 105: 407–416.

Foster, G.D. and Moon, T.W. 1989. Insulin and the regulation of glycogen metabolism and gluconeogenesis in American eel hepatocytes. *General and Comparative Endocrinology* 73: 374–381.

French, C.J., Hochachka, P.W. and Mommsen, T.P. 1983. Metabolic organization of liver during spawning migration of sockeye salmon. *American Journal of Physiology* 245: R827–R830.

Gagné, F. and Blaise, C. 1995. Evaluation of the genotoxicity of environmental contaminants in sediments to rainbow trout hepatocytes. *Environmental Toxicology and Water Quality* 10: 217–229.

Gagné, F. and Blaise, C. 1997. Evaluation of cell viability, mixed function oxidase activity, metallothionein induction, and genotoxicity in rainbow trout hepatocytes exposed to industrial effluents. II. Validation of the rainbow trout hepatocyte model for ecotoxicity testing of industrial wastewater. *Environmental Toxicology and Water Quality* 12: 305–314.

Gagné, F., Marion, M. and Denizeau, F. 1990. Metal homeostasis and metallothionein induction in rainbow trout hepatocytes exposed to cadmium. *Fundamental and Applied Toxicology* 14: 429–437.

Gas, N. and Serfaty, A. 1972. Cytophysiology of the carp liver (*Cyprinus carpio* L). Ultrastructural modifications following winter fasting. *Journal of Physiology* 64: 57–67.

Gingerich, W.H., Weber, L.J. and Larson, R.E. 1978. Carbon tetrachloride-induced retention of sulfobromophthalein in the plasma of rainbow trout. *Toxicology and Applied Pharmacology* 43: 147–158.

Goksoyr, A. and Husoy, A.M. 1998. Immunochemical approaches to studies of CYP1A localization and induction by xenobiotics. In *Fish Ecotoxicology*. Braunbeck, T., Hinton, D.E. and Streit, B. (eds), pp. 165–202. Birkhauser, Basel.

Grizzle, J.M. and Thiyagarajah, A. 1988. Diethylnitrosamine-induced hepatic neoplasms in the fish *Rivulus ocellatus marmoratus*. *Diseases of Aquatic Organisms* 5: 39–50.

Groothuis, G.M. and Meijer, D.K.F. 1996. Drug traffic in the hepatobiliary system. *Journal of Hepatology* 24 (Suppl. 1): 3–28.

Guguen-Guillouzo, C. and Corlu, A. 1993. Recent progress on long-term primary hepatocyte culture: importance of cell microenvironment. *Cytotechnology* 11: 3–5.

Guguen-Guillouzo, C. and Guillouzo, A. 1983. Modulation of functional activities in cultured rat hepatocytes. *Molecular and Cellular Biochemistry* 53/54: 35–56.

Hahn, M.E., Lamb, T.M., Schultz, M.E., Smolowitz, R.M. and Stegeman, J.J. 1993. Cytochrome P4501A induction and inhibition by 3,3′,4′-tetrachlorobiphenyl in an Ah receptor-containing fish hepatoma cell line, PLHC-1. *Aquatic Toxicology* 26: 185–208.

Hahn, M.E., Woodward, B.K., Stegeman, J.J. and Kennedy, S.W. 1996. Rapid assessment of induced cytochrome P4501A protein and catalytic activity in fish hepatoma cells grown in multiwell plates: response to TCDD, TCDF, and two planar PCBs. *Environmental Toxicology and Chemistry* 15: 582–591.

Hampton, J.A., McCuskey, P.A., McCuskey, R.S. and Hinton, D.E. 1985. Functional units in rainbow trout (*Salmo gairdneri*, Richardson) liver. I. Histochemical properties and arrangement of hepatocytes. *Anatomical Record* 213: 166–175.

Hampton, J.A., Klaunig, J.E. and Goldblatt, P.J. 1987. Resident sinusoidal macrophages in the liver of the brown bullhead (*Ictalurus nebulosus*): an ultrastructural, functional and cytochemical study. *Anatomical Record* 219: 338–346.

Hampton, J.A., Lantz, R.C. and Hinton, D.E. 1989. Functional units in rainbow trout (*Salmo gairdneri*, Richardson) liver. III. Morphometric analysis of parenchyma, stroma, and component cell types. *American Journal of Anatomy* 185: 58–73.

Hampton, J.A., Lantz, R.C., Goldblatt, P.J., Laurén, D.J. and Hinton, D.E. 1998. Functional units in rainbow trout (*Salmo gairdneri*, Richardson) liver. II. The biliary system. *Anatomical Record* 221: 619–634.

Hanson, S.K. and Anders, M.W. 1978. The effect of diethyl maleate treatment, fasting and time of administration of allyl alcohol hepatotoxicity. *Toxicology Letters* 1: 301–305.

Harshbarger, J.C. and Clark, J.B. 1990. Epizootiology of neoplasms in bony fish of North America. *Science of the Total Environment* 94: 1–32.

Harvison, P.J., Guengerich, F.P., Rashed, M.S. and Nelson, S.D. 1988. Cytochrome P-450 isozyme selectivity in the oxidation of acetaminophen. *Chemistry Research and Toxicology* 1: 47–52.

Heath, A.G. 1995. *Water Pollution and Fish Physiology*, 2nd edn. CRC Lewis Publishers, Boca Raton.

Hemmer, M.J., Courney, L.A. and Ortego, L.S. 1995. Immunohistochemical detection of P-glycoprotein in teleost tissues using mammalian polyclonal and monoclonal antibodies. *Journal of Experimental Zoology* 272: 69–77.

Hendricks, J.D., Meyers, T.R. and Shelton, D.W. 1984. Histological progression of hepatic neoplasia in rainbow trout (*Salmo gairdneri*). *National Cancer Institutes Monograph* 65: 321–336.

Henson, J.H., Capuano, S., Nesbitt, D., Hager, D.N., Nundy, S., Miller, D.S., Ballatori, N. and Boyer, J.L. 1995. Cytoskeletal organization in clusters of isolated polarized skate hepatocytes – structural and functional evidence for microtubule-dependent transcytosis. *Journal of Experimental Zoology* 271: 273–284.

Hess, F.A., Weibel, E.R. and Preisig, R. 1973. Morphometry of dog liver: normal baseline data. *Virchows Archives B Cellular Pathology* 12: 303–317.

Hightower, L.E. and Renfro, J.L. 1988. Recent applications of fish cell culture to biomedical research. *Journal of Experimental Zoology* 248: 290–302.

Hilton, J.W. and Atkinson, J.L. 1982. Response of rainbow trout (*Salmo gairdneri*) to increased levels of available carbohydrates in practical trout diets. *British Journal of Nutrition* 47: 597–607.

Hinson, J.A. 1980. Biochemical toxicology of acetaminophen. *Review of Biochemistry and Toxicology* 2: 103–130.

Hinton, D.E. 1993. Cells, cellular responses, and their markers in chronic toxicity of fishes. *In Aquatic Toxicology: Molecular, Biochemical and Cellular Perspectives*. Malins, D.C. and Ostrander, G.K. (eds), pp. 207–239. Lewis Publishers, Boca Raton.

Hinton, D.E. 1998. Structural considerations in teleost hepatocarcinogenesis: Gross and microscopic features including architecture, specific cell types and focal lesions. In *Pathobiology of Spontaneous and Induced Neoplasms in Fishes: Comparative Characterization, Nomenclature and Literature*. Dawe, C.J., Harshbarger, J.C., Wellings, S.R. and Strandberg, J.D. (eds). Academic Press, New York.

Hinton, D.E. and Pool, C.R. 1976. Ultrastructure of the liver in channel catfish *Ictalurus punctatus* (Rafinesque). *Journal of Fish Biology* 8: 209–219.

Hinton D.E., Lantz, R.C. and Hampton, J.A. 1984. Effect of age and exposure to a carcinogen on the structure of the medaka liver: A morphometric study. *National Cancer Institute Monographs* 65: 239–249.

Hinton, D.E., Lantz, R.C., Hampton, J.A., McCuskey, P.R. and McCuskey, R.S. 1987. Normal versus abnormal structure: considerations in morphologic responses of teleosts to pollutants. *Environmental Health Perspectives* 71: 139–146.

Hinton D.E., Teh, S.J., Okihiro, M.S., Cooke, J.B. and Parker, L.M. 1992. Phenotypically altered hepatocyte populations in diethylnitrosamine-induced medaka liver carcinogenesis: resistance, growth and fate. *Marine Environmental Research* 34: 1–5.

Hiraoka, Y., Nakagawa, H. and Murachi, S. 1979. Blood properties of rainbow trout in acute hepatotoxicity by carbon tetrachloride. *Bulletin of the Japanese Society of Scientific Fisheries* 45: 527–532.

Husoy, A.M., Myers, M.S. and Goksoyr, A. 1996. Cellular localization of cytochrome P4501A (CYPIA) induction and histology in Atlantic cod (*Gadus morhua*) and European flounder (*Platichthys flesus*) after environmental exposure to contaminants by caging in Sorfoden, Norway. *Aquatic Toxicology* 36: 53–74.

Hutchins, C.G., Rawles, S.D. and Gatlin, D.M. 1998. Effects of dietary carbohydrate kind and level on growth, body composition and glycemic response of juvenile sunshine bass (*Morone chrysops* female x *M. saxatilis* male). *Aquaculture* 161: 187–199.

Hyllner, S.J., Andersson, T., Haux, C. and Olsson, P.E. 1989. Cortisol induction of metallothionein in primary culture of rainbow trout hepatocytes. *Journal of Cellular Physiology* 139: 24–28.

Ioannides, C., Steele, C.M. and Parke, D.V. 1983. Species variation in the metabolic activation of paracetamol to toxic intermediates: role of cytochromes P-450 and P-448. *Toxicology Letters* 16: 55–61.

Janssens, P.A. and Lowrey, P. 1987. Hormonal regulation of hepatic glycogenolysis in the carp, *Cyprinus carpio*. *American Journal of Physiology* 252: R653–R660.

Jensen, E.G., Thauland, R. and Søli, N.E. 1996. Measurement of xenobiotic metabolising enzyme activities in primary monolayer cultures of immature rainbow trout hepatocytes at two acclimatisation temperatures. *Alternatives to Laboratory Animals* 24: 727–740.

Jobling, S. and Sumpter, J.P. 1993. Detergent components in sewage effluent are weakly estrogenic to fish: an *in vitro* study using rainbow trout (*Oncorhynchus mykiss*) hepatocytes. *Aquatic Toxicology* 27: 361–372.

Juerss, K. and Bastrop, R. 1995. Amino acid metabolism in fish. In *Biochemistry and Molecular Biology of Fishes*, Vol. 4. Hochachka, P.W. and Mommsen, T.P. (eds), pp. 159–189. Elsevier, Amsterdam.

Jungermann, K. and Sasse, D. 1978. Heterogeneity of liver parenchymal cells. *Trends in Biochemical Science* 3: 198–202.

Kane, A.S. and Thohan, S. 1996. Dynamic culture of fish hepatic tissue slices to assess phase I and phase II biotransformation. In *Techniques in Aquatic Toxicology*. Ostrander, G.K. (ed.), pp. 371–391. CRC Press, Boca Raton.

Kaplan, L.A.E., Schultz, M.E., Schultz, R.J. and Crivello, J.F. 1991. Nitrosodiethylamine metabolism in the viviparous fish *Poeciliopsis*: evidence for the existence of liver P450pj activity and expression. *Carcinogenesis* 12: 647–652.

Kenyon, A.J. 1967. The role of the liver in the maintenance of plasma proteins and amino acids in the eel, *Anguilla anguilla* L., with reference to amino acid deamination. *Comparative Biochemistry and Physiology* 22: 169–175.

Klaunig, J.E., Ruch, R.J. and Goldblatt, P.J. 1985. Trout hepatocyte culture: isolation and primary culture. *In Vitro Cellular and Developmental Biology* 21: 221–228.

Kleinow, K.M., Droy, B.F., Buhler, D.R. and Williams, D.E. 1990. Interaction of carbon tetrachloride with beta-naphthoflavone-mediated cytochrome P450 induction in winter flounder (*Pseudopleuronectes americanus*). *Toxicology and Applied Pharmacology* 104: 367–374.

Knox, D., Walton, M.J. and Cowey, C.B. 1980. Distribution of enzymes of glycolysis and gluconeogenesis in fish tissues. *Marine Biology* 56: 7–10.

Köhler, A. and Van Noorden, C.J.F. 1998. Initial velocities *in situ* of G6PDH and PGDH and expression of proliferating cell nuclear antigen (PCNA): sensitive diagnostic markers of environmentally induced hepatocellular carcinogenesis in a marine flatfish (*Platichthys flesus* L.). *Aquatic Toxicology* 40: 233–252.

Köhler, A., Bahns, S. and Van Noorden, C.J.F. 1998. Determination of kinetic properties of G6PDH and PGDH and the expression of PCNA during liver carcinogenesis in coastal flounder. *Marine Environmental Research* 46: 179–183.

Lee, L.E.J., Clemons, J.H., Bechtel, D.G., Caldwell, S.J., Han, K.B., Pasitschniak-Arts, M., Mosser, D. and Bols, N.C. 1993. Development and characterization of a rainbow trout liver cell line expressing cytochrome P450-dependent monooxygenase activity. *Cell Biology and Toxicology* 9: 279–294.

Lester, S.M., Braunbeck, T.A., Teh, S.J., Stegeman, J.J., Miller, M.R. and Hinton, D.E. 1993. Hepatic cellular distribution of cytochrome CYP1A1 in rainbow trout (*Oncorhynchus mykiss*): an immunohisto- and cytochemical study. *Cancer Research* 53: 3700–3706.

Lorenzana, R.M., Hedstrom, O.R., Gallagher, J.A. and Buhler, D.R. 1989. Cytochrome P450 isozyme distribution in normal and tumor-bearing hepatic tissue from rainbow trout (*Salmo gairdneri*). *Experimental and Molecular Pathology* 50: 348–361.

Love, R.M. 1980. *The Chemical Biology of Fishes*. Academic Press, London.

Loveland, P.M., Wilcox, J.S., Pawlowski, N.E. and Bailey, G.S. 1987. Metabolism and DNA binding of aflatoxicol and aflatoxin B1 *in vivo* and in isolated hepatocytes from rainbow trout (*Salmo gairdneri*). *Carcinogenesis* 8: 1065–1070.

Lupo, S., Yodis, L.A., Mico, B.A. and Rush, G.F. 1987. *In vivo* and *in vitro* hepatotoxicity and metabolism of acetaminphen in Syrian hamsters. *Toxicology* 44: 229–239.

Maccubbin, A.E., Ersing, N., Weinar, J. and Black, J.J. 1987. In vivo carcinogen bioassays using rainbow trout and medaka fish embryos. In *Short-term Bioassays in the Analysis of Complex Environmental Mixtures*, Vol. V, 36, 36. Sandhu, S.S., DeMarini, D.M., Mass, M.J., Moore, M.M. and Mumford, J.K. (eds), pp. 209–223. Plenum Press, New York.

McCuskey, P.A., McCuskey, R.S. and Hinton, D.E. 1986. Electron microscopy of cells of the hepatic sinusoids in rainbow trout (*Salmo gairdneri*). In *Cells of the Hepatic Sinusoid*, Vol. 1. Kirn, A., Knook, D.L. and Wisse, E. (eds), pp. 489–494. Kupffer Cell Foundation, Leiden, The Netherlands.

Maitre, J.K., Valotaire, Y. and Guguen-Guillozo, C. 1986. Estradiol-17 stimulation of vitellogenin synthesis in primary cultures of rainbow trout hepatocytes. *In Vitro Cellular and Developmental Biology* 22: 337–343.

Malins, D.C., McCain, B.B., Myers, M.S., Brown, D.W., Krahn, M.M., Roubal, W.T., Schiewe, M.H., Landahl, J.T. and Chan, S.-L. 1987. Field and laboratory studies of the etiology of liver neoplasms in marine fish from Puget Sound. *Environmental Health Perspectives* 71: 5–16.

Masfaraud, J.F., Devaux, A., Pfohlleskowicz, A., Malaveille, C. and Monod, G. 1992. DNA adduct formation and 7-ethoxyresorufin O-deethylase induction in primary culture of rainbow trout hepatocytes exposed to benzo[a]pyrene. *Toxicology in Vitro* 6: 523–531.

Metcalfe, C.D., Cairns, V.W. and Fitzsimons, J.D. 1988. Experimental induction of liver tumours in rainbow trout (*Salmo gairdneri*) by contaminated sediment from Hamilton Harbour, Ontario. *Canadian Journal of Fisheries and Aquatic Science* 45: 2161–2167.

Miller, M.R., Saito, N., Blair, J.B. and Hinton, D.E. 1993a. Acetaminophen toxicity in cultured trout liver cells. II. Maintenance of cytochrome P450 1A1. *Experimental and Molecular Pathology* 58: 127–138.

Miller, M.R., Wentz, E., Blair, J.B., Pack, D. and Hinton, D.E. 1993b. Acetaminophen toxicity in cultured trout liver cells. I. Morphological alterations and effects on cytochrome P450 1A1. *Experimental and Molecular Pathology* 58: 114–126.

Milligan, C.L. and Girard, S.S. 1993. Lactate metabolism in rainbow trout. *Journal of Experimental Biology* 180: 275–193.

Mitchell, J.R., Jollow, D.J., Potter, W.Z., Davis, D.C., Gillette, J.R. and Brodie, B.B. 1973a. Acetaminophen-induced hepatic necrosis. I. Role of drug metabolism. *Journal of Pharmacology and Experimental Therapeutics* 187: 185–194.

Mitchell, J.R., Jollow, D.J., Potter, W.Z., Gillette, J.R. and Brodie, B.B. 1973b. Acetaminophen-induced hepatic necrosis. IV. Protective role of glutathione. *Journal of Pharmacology and Experimental Therapeutics* 187: 211–217.

Mommsen, T.P. and Moon, T.W. 1989. Metabolic actions of glucagon-family hormones in liver. *Fish Physiology and Biochemistry* 7: 279–288.

Mommsen, T.P. and Walsh, P.J. 1991. Urea synthesis in fishes: evolutionary and biochemical perspectives. In *Biochemistry and Molecular Biology of Fishes*. Hochachka, P.W. and Mommsen, T.P. (eds). Elsevier, Amsterdam.

Mommsen, T.P., Moon, T.W. and Walsh, P.J. 1994. Hepatocytes: isolation, maintenance and utilization. In *Biochemistry and Molecular Biology of Fishes*. Vol. 3. *Analytical Techniques*. Hochachka, P.W. and Mommsen, T.P. (eds), pp. 355–372. Elsevier, Amsterdam.

Monod, G., Devaux, A., Valotaire, Y. and Cravedi, J.-P. 1998. Primary cell cultures from fish in ecotoxicology. In *Fish Ecotoxicology*. Braunbeck, T., Hinton, D.E. and Streit, B. (eds), pp. 39–60. Birkhäuser Verlag, Basel.

Moon, T.W. 1988. Adaptation, constraint and the function of the gluconeogenetic pathway. *Canadian Journal of Zoology* 66: 1059–1068.

Moon, T.W. 1998. Glucagon: From hepatic binding to metabolism in teleost fish. *Comparative Biochemistry and Physiology* B121: 27–34.

Moon, T.W. and Foster, G.D. 1995. Tissue carbohydrate metabolism, gluconeogenesis and hormonal and environmental influences. In *Biochemistry and Molecular Biology of Fishes*. Vol. 4. *Metabolic Biochemistry*. Hochachka, P.W. and Mommsen, T.P. (eds), pp. 65–100. Elsevier Sciences Publishers B.V., New York.

Moon, T.W., Foster, G.D. and Plisetskaya, E.M. 1989. Changes in peptide hormones and liver enzymes in the rainbow trout deprived of food for 6 weeks. *Canadian Journal of Zoology* 67: 2189–2193.

Moore, M.J. and Myers, M.S. 1993. Pathobiology of chemical-associated neoplasia in fish. In *Aquatic Toxicology – Molecular Biochemical and Cellular Perspectives*. Malins, D.C. and Ostrander, G.K. (eds), pp. 327–386. CRC Press, Boca Raton.

Moslen, M.T. 1995. Toxic responses of the liver. In *Toxicology – The Basic Science of Poisons*, 5th edn. Klaassen, C.D., Amdur, M.O. and Doull, J. (eds), pp. 403–416. McGraw Hill, New York.

Myers, M.S., Rhodes, L.D. and McCain, B.B. 1987. Pathologic anatomy and patterns of occurrence of hepatic neoplasms, putative preneoplastic lesions, and other idiopathic hepatic conditions in English sole (*Parophrys vetulus*) from Puget Sound, Washington. *Journal of the National Cancer Institute* 78: 333–363.

Myers, M.S., Stehr, C.M., Olson, O.P., Johnson, E., McCain, B.B., Chan, S.-L. and Varanasi, U. 1994. Relationship between toxicopathic hepatic lesions and exposure to chemical contaminants in English sole (*Parophrys vetulus*), starry flounder (*Platichthys stellatus*), and white croaker (*Genyonemus lineatus*) from selected marine sites on the Pacific Coast, USA. *Environmental Health Perspectives* 102: 200–215.

Navarro, I. and Gutierrez, J. 1995. Fasting and starvation. In *Biochemistry and Molecular Biology of Fishes*, Vol. 4. Hochachka, P.W. and Mommsen, T.P. (eds), pp. 393–434. Elsevier, Amsterdam.

Ng, W.F., Sarangi, R.L., Zastawny, L., Veinot-Drebot, L. and Ling, T. 1989. Identification of members of the P-glycoprotein multigene family. *Molecular and Cellular Biology* 9: 1224–1232.

Nishimoto, M., Yanagida, G.K., Stein, J.E., Baird, W.M. and Varanasi, U. 1992. The metabolism of benzo(a)pyrene by English sole (*Parophrys vetulus*): comparison between isolated hepatocytes *in vitro* and *in vivo*. *Xenobiotika* 22: 949–961.

Ohno, Y., Ormstad, K., Ross, D. and Orrenius, S. 1985. Mechanism of allyl alcohol toxicity and protective effects of low-molecular-weight thiols studied with isolated rat hepatocytes. *Toxicology and Applied Pharmacology* 78: 169–179.

Okihiro, M. and Hinton, D.E. 2000. Partial hepatectomy and bile duct ligation in rainbow trout (*Oncorhynchus mykiss*): Histologic, immunohistochemical and enzyme histochemcial characterization of hepatic regeneration and biliary hyperplasia. *Toxicologic Pathology* 28: 342–356.

Okihiro, M.S. 1996. Regenerative, hyperplastic, and neoplastic-hepatic growth in medaka (*Oryzias latipes*) and rainbow trout (*Oncorhynchus mykiss*): an investigation of the role of the bipolar hepatic stem cell. Ph.D. thesis. Comparative Pathology. Department of Anatomy, Physiology & Cell Biology, School of Veterinary Medicine, University of California, Davis.

Olsson, P.E., Hyllner, S.J., Zafarullah, M., Andersson, T. and Gedamu, L. 1990. Differences in metallothionein gene expression in primary cultures of rainbow trout hepatocytes and the RTH-149 cell line. *Biochimica et Biophysica Acta* 1049: 78–82.

Ottolenghi, C., Puviani, A.C., Baruffaldi, A. and Brighenti, L. 1984. Effect of insulin on glycogen metabolism in isolated catfish hepatocytes. *Comparative Biochemistry and Physiology* A78: 705–710.

Packer, J.E., Slater, T.F. and Willson, R.L. 1978. Reactions of the carbon tetrachloride-related peroxy free radical with amino acids: pulse radiolysis evidence. *Life Sciences* 23: 2617–2620.

Parker, R.S., Morrissey, M.T., Moldens, P. and Selivonchick, D.P. 1981. The use of isolated hepatocytes from rainbow trout (*Salmo gairdneri*) in the metabolism of acetaminophen. *Comparative Biochemistry and Physiology* 70B: 631–633.

Pereira, C., Vijayan, M.M., Storey, K.B., Jones, R.A. and Moon, T.W. 1995. Role of glucose and insulin in regulating glycogen synthase and phosphorylase activities in rainbow trout hepatocytes. *Journal of Comparative Physiology* 165: 62–70.

Pesonen, M. and Andersson, T. 1991. Characterization and induction of xenobiotic metabolizing enzyme activities in a primary culture of rainbow trout hepatocytes. *Xenobiotica* 21: 461–471.

Pesonen, M. and Andersson, T. 1992. Toxic effects of bleached and unbleached paper mill effluents in primary cultures of rainbow trout hepatocytes. *Ecotoxicology and Environmental Safety* 24: 63–71.

Pesonen, M. and Andersson, T.B. 1997. Fish primary hepatocyte culture: An important model for xenobiotic metabolism and toxicity studies. *Aquatic Toxicology* 37: 253–267.

Pesonen, M., Goksoyr, A. and Andersson, T. 1992. Expression of P4501a1 in a primary culture of rainbow trout hepatocytes exposed to beta-naphthoflavone or 2,3,7,8-tetrachlorodibenzo-para-dioxin. *Archives of Biochemistry and Biophysics* 292: 228–233.

Peters, N., Köhler, A. and Kranz, H. 1987. Liver pathology in fishes from the Lower Elbe as a consequence of pollution. *Diseases of Aquatic Organisms* 2: 87–97.

Petit, F., Valotaire, Y. and Pakdel, F. 1995. Differential functional activities of rainbow trout and human estrogen receptors expressed in the yeast *Saccharomyces cerevisiae*. *European Journal of Biochemistry* 233: 584–592.

Peute, J., van der Gaag, M.A. and Lambert, J.G.D. 1978. Ultrastructure and lipid content of the liver of the zebrafish *Brachydanio rerio*, related to vitellogenin synthesis. *Cell Tissue Research* 186: 297–308.

Pfeifer, K.F., Weber, L.J. and Larson, RE. 1980. Carbon tetrachloride-induced hepatotoxic response in rainbow trout, *Salmo gairdneri*, as influenced by two commercial fish diets. *Comparative Biochemistry and Physiology* 67C: 91–96.

Price, V.F., Miller, M.G. and Jollow, D.J. 1987. Mechanisms of fasting-induced potentiation of acetaminophen hepatotoxicity in the rat. *Biochemistry and Pharmacology* 36: 427–433.

Råbergh, C.M.I. and Lipsky, M.M. 1997. Toxicity of chloroform and carbon tetrachloride in primary cultures of rainbow trout hepatocytes. *Aquatic Toxicity* 37: 169–182.

Råbergh, C.M.I., Ziegler, K., Isomaa, B., Lipsky, M.M. and Ericksson, J.E. 1994. Uptake of taurocholic acid in isolated hepatocytes from rainbow trout. *American Journal of Physiology* 267: G380–G386.

Rabinovici, N. and Wiener, E. 1963. Hemodynamic changes in the hepatectomized liver of the rat and their relationships to regeneration. *Journal of Surgical Research* 3: 3–8.

Racicot, J.G., Gaudet, M. and Leray, C. 1975. Blood and liver enzymes in rainbow trout (*Salmo gairdneri* Rich.) with emphasis on their diagnostic use: Study of CCl$_4$ toxicity and a case of *Aeromonas* infection. *Journal of Fish Biology* 7: 825–835.

Reader, S., Saint Louis, R.M., Pelletier, E. and Denizeau, F. 1996. Accumulation and biotransformation of tri-*n*-butyltin by isolated rainbow trout hepatocytes. *Environmental Toxicology and Chemistry* 15: 2049–2052.

Rees, K.R. and Tarlow, M.J. 1967. The hepatotoxic action of allyl formate. *Biochemical Journal* 104: 757–761.

Reid, W.D. 1972. Mechanism of allyl alcohol-induced hepatic necrosis. *Experientia* 11: 1058–1061.

Rocha, E., Monteiro, R.A.F. and Pereira, C.A. 1996. The pale-grey interhepatocytic cells of brown trout (*Salmo trutta*) are a subpopulation of liver resident macrophages or do they establish a different cellular type? *Journal of Submicroscopic Cytology and Pathology* 28: 357–368.

Rocha, E., Monteiro, R.F. and Pereira, C.A. 1997. Liver of the brown trout, *Salmo trutta* (Teleostei, Salmonidae): a stereological study at light and electron microscopic levels. *Anatomical Record* 247: 317–328.

Ruebner, B.H., Bannasch, P., Hinton, D.E., Cullen, J.M. and Ward, J.M. 1996. Foci of altered hepatocytes, mouse. In *Monographs of Pathology of Laboratory Animals, Digestive System*, 2nd edn. Jones, T.C., Popp, J.A. and Mohr, U. (eds), pp. 38–49. Springer-Verlag, Berlin.

Sadar, M.D., Ash, R. and Andersson, T.B. 1995. Picrotoxin is a cyp1a1 inducer in rainbow trout hepatocytes. *Biochemical and Biophysical Research Communications* 214: 1060–1066.

Sadar, M.F., Ash, R., Sundqvist, J., Olsson, P.E. and Andersson, T.B. 1996. Phenobarbital induction of cyp1a1 gene expression in a primary culture of rainbow trout hepatocytes. *Journal of Biological Chemistry* 271: 17635–17643.

Sarasquete, C. and Segner, H. 2000. Cytochrome P4501A in teleostean fishes: immunochemical studies. *Science of the Total Environment* 247: 313–332.

Sargent, J., Henderson, R.J. and Tocher, D.R. 1989. The lipids. In *Fish Nutrition*, 2nd edn. Halver, J.E. (ed.), pp. 153–218. Academic Press, San Diego.

Schar, M., Maly, I.P. and Sasse, D. 1985. Histochemical studies on metabolic zonation of the liver in the trout (*Salmo gairdneri*). *Histochemistry* 83: 147–151.

Schmidt, D.C. and Weber, L.J. 1973. Metabolism and biliary excretion of sulfo-bromophthalein by rainbow trout (*Salmo gairdneri*). *The Journal of the Fisheries Research Board of Canada* 30: 1301–1308.

Scholz, S. and Segner, H. 1999. Induction of CYP1A in primary culture of rainbow trout (*Oncorhynchus mykiss*) liver cells: concentration–response relationships for our model substances. *Ecotoxicology and Environmental Safety* 43: 252–260.

Scholz, S., Behn, I., Honeck, H., Hauck, C., Braunbeck, T. and Segner, H. 1997. Development of a monoclonal antibody for ELISA of CYP1A in primary cultures of rainbow trout (*Oncorhynchus mykiss*) hepatocytes. *Biomarkers* 2: 287–294.

Scholz, S., Braunbeck T. and Segner, H. 1998. Viability and differential function of rainbow trout liver cells in primary culture: Coculture with two permanent fish cells. *In Vitro Cellular and Developmental Biology of the Animal* 34: 762–771.

Schrag, B., Ensenbach, U., Navas, J.M. and Segner, H. 1998. Evaluation of xenoestrogenic effects in fish on different organization levels. In *Reproductive Toxicology*. del Mazo, J. (ed.), pp. 207–214. Plenum Press, New York.

Segner, H. 1998a. Isolation and primary culture of teleost hepatocytes. *Comparative Biochemistry and Physiology* A120: 71–81.

Segner, H. 1998b. Fish cell lines as a tool in aquatic toxicology. In *Fish Ecotoxicology*. Braunbeck, T., Hinton, D.E. and Streit, B. (eds), pp. 1–38. Birkhäuser Verlag, Basel.

Segner, H. and Böhm, R. 1994. Enzymes of lipogenesis. In *Biochemistry and Molecular Biology of Fishes*, Vol. 3. Hochachka, P.W. and Mommsen, T.P. (eds), pp. 313–326. Elsevier, Amsterdam.

266 *David E. Hinton, Helmut Segner, and Thomas Braunbeck*

Segner, H. and Braunbeck, T. 1988. Hepatocellular adaptation to extreme nutritional conditions in ide, *Leuciscus idus melanotus* L. (Cyprinidae). A morphofunctional analysis. *Fish Physiology and Biochemistry* 5: 79–97.

Segner, H. and Braunbeck, T. 1990. Adaptive changes of liver composition and structure in golden ide during winter acclimatization. *Journal of Experimental Zoology* 255: 171–185.

Segner, H. and Braunbeck, T. 1998. Cellular response profile to chemical stress. In *Ecotoxicology*. Schürmann, G. and Markert, B. (eds), pp. 521–569. John Wiley/ Spektrum, New York/Heidelberg.

Segner, H., Scholz, S. and Böhm, R. 1995. Carp (*Cyprinus carpio*) hepatocytes in primary culture: morphology and metabolism *Actes du Colloque* 18: 77–82.

Segner, H., Dölle, A. and Böhm, R. 1997. Ketone body metabolism in the carp *Carpinus carpio*: biochemical and 1H-HMR spectroscopical analysis. *Comparative Biochemistry and Physiology* 116B: 257–262.

Serafini-Cessi, F. 1972. Conversion of allyl alcohol into acrolein by rat liver. *Biochemical Journal* 128: 1103–1107.

Shen, E.S., Garry, V.F. and Anders, M.W. 1982. Effect of hypoxia on carbon tetrachloride hepatotoxicity. *Biochemistry and Pharmacology* 31: 3787–3793.

Sheridan, M.A. 1987. Effects of epinephrine and norepinephrine on lipid mobilization from coho salmon liver incubated *in vitro*. *Endocrinology* 120: 2234–2239.

Sheridan, M.A. 1988. Lipid dynamics in fish: aspects of absorption, transportation, deposition and mobilization. *Comparative Biochemistry and Physiology* B90: 679–690.

Sheridan, M.A. 1998. Structure–functional relationships of the somatostatin peptide family. *American Zoologist* 38: 142A.

Sheridan, M.A. and Bern, H.A. 1986. Both somatostatin and the caudal neuropeptide, urotensin II, stimulate lipid mobilization from coho salmon liver incubated *in vitro*. *Regulatory Peptides* 14: 333–344.

Sheridan, M.A. and Mommsen, T.P. 1991. Effects of nutritional state on *in vivo* lipid and carbohydrate metabolism of coho salmon, *Oncorhynchus kisutch*. *General and Comparative Endocrinology* 81: 473–483.

Shimeno, S., Hosokawa, H., Takeda, M., Kayijama, H. and Kaisho, T. 1985. Effect of dietary lipid and carbohydrate on growth, feed conversion and body composition in young yellowtail. *Bulletin of the Japanese Society of Scientific Fisheries* 51: 1893–1989.

Sikka, H.C., Steward, A.R., Zaleski, J., Kandaswami, C., Rutkowski, J.P., Kumar, S. and Gupta, R.C. 1993. Comparative metabolism of benzo(a)pyrene by the liver microsomes and freshly isolated hepatocytes of brown bullhead and carp. *Polycyclic and Aromatic Hydrocarbons* 3: 1087–1094.

Singh, Y., Cooke, J.B., Hinton, D.E. and Miller, M.G. 1996. Trout liver slices for metabolism and toxicity studies. *Drug Metabolic Disposition* 24: 7–14.

Skett, P. 1994. Problems in using isolated and cultured hepatocytes for xenobiotic metabolism/metabolism-based toxicity testing – solutions? *Toxicology In Vitro* 8: 491–504.

Smith, C.V. and Mitchell, J.R. 1985. Acetaminophen hepatotoxicity *in vivo* is not accompanied by oxidant stress. *Biochemical and Biophysical Research Communications* 133: 329–336.

Smith, D.J., Grossbard, M., Gordon, E.R. and Boyer, J.L. 1987. Isolation and characterization of a polarized isolated hepatocyte preparation in the skate *Raja erinacea*. *Journal of Experimental Zoology* 241: 291–296.

Smith, M.T., Loveridge, N., Wills, E.D. and Chayen, J. 1979. The distribution of glutathione in the rat liver lobule. *Biochemical Journal* 182: 103–108.

Smolowitz, R.M., Hahn, M.E. and Stegeman, J.J. 1991. Immunohistochemical localization of cytochrome P-4501A1 induced by 3,3',4,4'-tetrachlorobiphenyl and by 2,3,7,8-tetrachlorodibenzo furan in liver and extrahepatic tissues of the teleost *Stenotomus chrysops* (scup). *Drug Metabolic Disposition* 19: 113–123.

Statham, C.N., Croft, W.A. and Lech, J.J. 1978. Uptake, distribution, and effects of carbon tetrachloride in rainbow trout (*Salmo gairdneri*). *Toxicology and Applied Pharmacology* 45: 13140.

Stegeman, J.J. and Hahn, M.E. 1994. Biochemistry and molecular biology of monooxygenases: current perspectives on forms, functions and regulation of cytochrome P450 in aquatic species. In *Aquatic Toxicology*. Malins, D.C. and Ostrander, G.K. (eds), pp. 87–203. Lewis Publishers, Boca Raton.

Steward, A.R., Zaleski, J. and Sikka, H.C. 1990. Metabolism of benzo[a]pyrene and (–)-trans-benzo[a]pyrene-7,8-dihydrodiol by freshly isolated hepatocytes of brown bullheads. *Chemico-Biological Interactions* 74: 119–138.

Sundby, A., Hemre, G.I., Borrebaek, B., Christophersen, B. and Blom, A.K. 1991. Insulin and glucagon family peptides in relation to activities of hepatic hexokinase and other enzymes in fed and starved Atlantic salmon, *Salmo salar*, and cod, *Gadus morhua*. *Comparative Biochemistry and Physiology* 100B: 467–470.

Teh, S.J. and Hinton, D.E. 1993. Detection of enzyme histochemical markers of hepatic preneoplasia and neoplasia in medaka (*Oryzias latipes*). *Aquatic Toxicology* 24: 163–182.

Thomas, P. and Wofford, H.W. 1984. Effects of metals and organic compounds on hepatic glutathione cysteine, and acid-soluble thiol levels in mullet (*Mugil cephalus* L.) *Toxicology and Applied Pharmacology* 76: 172–182.

Tocher, D.R. 1995. Glycerophospholipid metabolism. In *Biochemistry and Molecular Biology of Fishes*, Vol. 4. Hochachka, P.W. and Mommsen, T.P. (eds), pp. 119–157. Elsevier, Amsterdam.

Tranulis, M.A., Dregni, O., Christophersen, B., Kroghdal, A. and Borrebaek, B. 1996. A glucokinase-like enzyme in the liver of Atlantic salmon (*Salmo salar*). *Comparative Biochemistry and Physiology* 114B: 35–39.

Vaillant, C., Monod, G., Valotaire, Y. and Riviere, J.L. 1989. Measurement and induction of cytochrome P450 and monooxygenase activities in a primary culture of rainbow trout (*Salmo gairdneri*) hepatocytes. *Comptes Rendues de l'Académie des Sciences Paris, Serie III* 308: 83–88.

Vandenberghe, J. 1996. Hepatotoxicology: structure, function and toxicological pathology. In *Toxicology, Principles and Applications*. Niesink, R.J.M., De Vries, J. and Hollinger, M.A. (eds), pp. 668–701. CRC Press, Boca Raton.

Van Veld, P.A., Vogelbein, W.K., Cochran, M.K., Goksoyr, A. and Stegeman, J.J. 1997. Route specific cellular expression of cytochrome P4501A (CPY1A) in fish (*Fundulus heteroclitus*) following exposure to aqueous and dietary benzo(a)pyrene. *Toxicology and Applied Pharmacology* 142: 348–359.

Vethaak, A.D. and Wester, P.W. 1996. Diseases of flounder *Platichthys flesus* in Dutch coastal and estuarine waters, with particular reference to environmental stress factors. II. Liver Histopathology. *Diseases of Aquatic Organisms* 26: 99–116.

Vogt, G. and Segner, H. 1997. Spontaneous formation of intercellular bile canaliculi and hybrid biliary-pancreatic canaliculi in co-culture of hepatocytes and exocrine pancreas cells from carp. *Cell Tissue Research* 289: 191–194.

Walton, D.G., Acton, A.B. and Stich, H.F. 1984. DNA repair synthesis following exposure to chemical mutagens in primary liver. *Cancer Research* 44: 1120.

Weber, J.M. and Zwingelstein, G. 1995. Circulatory substrate fluxes and their regulation. In *Biochemistry and Molecular Biology of Fishes*, Vol. 4. Hochachka, P.W. and Mommsen, T.P. (eds), pp. 15–32. Elsevier, Amsterdam.

Weber, J.M., Brill, R.W. and Hochachka, P.W. 1986. Mammalian metabolite flux rates in a teleost fish: lactate and glucose turnover in tuna. *American Journal of Physiology* 250: 452–458.

Weber, L.J., Gingerich, W.H. and Pfeifer, K.F. 1979. Alterations in rainbow trout liver function and body fluids following treatment with carbon tetrachloride or monochlorobenzene. In *Pesticide and Xenobiotic Metabolism in Aquatic Organisms*. Khan, M.A.Q., Lech, J.J. and Menn, J.J. (eds), pp. 401–413. ACS Symposium Series 99, Washington, DC.

White, R., Jobling, S., Hoare, S.A., Sumpter, J.P. and Parker, M.G. 1994. Environmentally persistent alkylphenolic compounds are estrogenic. *Endocrinology* 36: 175–182.

Wofford, H.W. and Thomas, P. 1988. Peroxidation of mullet and rat liver lipids in vitro: effects of pyridine nucleotides, iron, incubation buffer, and xenobiotics. *Comparative Biochemistry and Physiology* C89: 201–205.

Wolke, R.E., Murchelano, R.A., Dickstein, C.D. and George, C.J. 1985. Preliminary evaluation of the use of macrophage aggregates (MA) as fish health monitors. *Bulletin of Environmental Contamination and Toxicology* 35: 222–227.

Yarbrough, J.D., Heitz, J.R. and Chambers, J.E. 1976. Physiological effects of crude oil exposure in the striped mullet, *Mugi cephalus*. *Life Sciences* 19: 755–760.

Zahn, T. and Braunbeck, T. 1993. Isolated fish hepatocytes as a tool in aquatic toxicology – sublethal effects of dinitro-*o*-cresol and 2,4-dichlorophenol. *Science of the Total Environment* Suppl.: 721–734.

Zahn, T. and Braunbeck, T. 1995. Cytotoxic effects of sublethal concentrations of malachite green in isolated hepatocytes from rainbow trout (*Oncorhynchus mykiss*). *Toxicology In Vitro* 9: 729–741.

Zahn, T., Hauck, C., Holzschuh, J. and Braunbeck, T. 1995. Acute and sublethal toxicity of seepage waters from garbage dumps to permanent cell lines and primary cultures of hepatocytes from rainbow trout (*Oncorhynchus mykiss*) – a novel approach to environmental risk assessment for chemicals and chemical mixtures. *Zentralblatt für Hygiene und Umweltmedizin* 196: 455–479.

Zahn, T., Arnold, H. and Braunbeck, T. 1996. Cytological and biochemical response of R1 cells and isolated hepatocytes from rainbow trout (*Oncorhynchus mykiss*) to subacute *in vitro* exposure to disulfoton. *Experimental Toxicology and Pathology* 48: 47–64.

Zammit, V.A. and Newsholme, E.A. 1979. Activities of enzymes of fat and ketone body metabolism and effects of starvation on blood concentrations of glucose and fat fuels in teleost and elasmobranch fish. *Biochemical Journal* 184: 313–322.

5 Response of the teleost gastrointestinal system to xenobiotics

Kevin M. Kleinow and Margaret O. James

Introduction

The primary function of the fish gastrointestinal tract (GIT) is digestion and nutrient absorption. This complex process requires a unique integration of biochemical and physical events which allow for the breakdown of foodstuffs, transport of the resulting nutrients, and elimination of waste products. The complexity of function is reflected in the complexity of organization with multiple cell types and tissue layers. Cell and tissue types qualitatively and quantitatively change along the length and across the wall of the GIT. These changes provide segregation of not only structure but also function and the resulting luminal conditions.

Along with digestive and absorptive functions, the GIT plays a variety of other important roles in fish. Adaptations which facilitate digestion and absorption, such as a large surface area, a thin lining, vigorous mixing, presence of proteolytic enzymes, and low pH values, also potentially serve to make the GIT a portal to the outside world. Fortunately, the GIT performs an important barrier function through processes such as mucus secretion, cell proliferation, biotransformation, efflux transport, and immune-related capacities. Resident microflora which limit the growth of less desirable organisms also provide essential factors and alternate biotransformation pathways. Excretion and elimination functions serve as a necessary conduit for digestive waste as well as xenobiotic and physiologic by-products. Less intuitive, but critical in nature, the GIT in select groups of fish serves in other capacities, including osmoregulation, buoyancy, ion regulation, and placental function.

As might be expected, the GIT is a fertile arena for the absorption, biotransformation, elimination, and toxicity of xenobiotics. Although enclosed within the fish, the GIT provides a unique interface between the external environment and internal milieu. The extent of this interface is expanded significantly by the relatively large size of the GIT (in most fish, the GIT is the largest organ system) and by its large surface area, as promoted by mucosal folds. The environment of the GIT is extremely dynamic. Mucus and cells of the luminal lining are continually in flux and change in conjunction with food passage. Likewise, the food bolus is continually changing in character along its passage, with mucus, epithelial cells, secretory and elimination products being added, and nutrients absorbed. Blood flow, neurologic function, muscle activity, and secretory function show large variations with the presence or absence of a meal, or even with changes in the size or character of a meal. The diversity of functions necessary for the GIT to carry out its physiologic role as well as structural differences along its length provide not only a large number of mechanistic modalities for the interaction with toxicants but also a variety of localized conditions for these interactions to occur.

The importance of the GIT as a portal of entry varies with the nature of the chemical. Non-polar compounds tend to accumulate in food and sediments rather than in the water, and thus are often considered in relation to the GIT. Ionic components such as metals may also make it in sufficient quantities into the diet. In species of fish which osmoregulate by the ingestion of water, the drinking of water serves as a vehicle of toxicant delivery. Xenobiotics may be contained or absorbed or may elicit a toxic effect in the GIT, dependent on local concentrations, residence time, and toxicologic character. Toxicant effects on the GIT may also occur following systemic delivery. Important processes relative to contaminants and aquatic species such as xenobiotic bioavailability, bioaccumulation, and biomagnification rely heavily in principle upon the action of the GIT and the interaction of toxicants with it.

Toxicant effects upon the GIT of fish may range from mild changes in motility, secretion, and absorptive functions to more severe effects associated with mucosal integrity, blood flow or neuromuscular control. These effects may in turn influence mucosal barrier function, biochemical reactions, microflora, nutrition, and, ultimately, the ability of the organism to thrive. Severity of clinical signs is based, as with other systems, upon the degree of effect. Because of the redundancy incorporated in the GIT for many functions, localized toxicity is often subtle in nature. Impairment of the nutritional or barrier roles of the organ may result in apparently independent effects at distal sites.

In the course of this chapter, we hope to achieve several goals. The first is to describe the fish GIT in terms of its contribution and complexity by briefly describing its anatomy and physiology. The second is to describe how toxicants may interact with the GIT through the processes of absorption, biotransformation, and elimination. Finally, to review what is known regarding chemical toxicity to the fish GIT.

Structure of the gastrointestinal tract

Esophagus

Relatively short and extremely distensible in most teleosts, the esophagus facilitates movement of food from the pharyngeal region to the stomach. In most teleosts, the esophagus consists of an innermost mucosa, a fibrous submucosa, one or two layers of striated muscle in the muscularis, and a serosa composed of fibrous connective tissue and a simple squamous epithelium (Figure 5.1). Several notable features of the teleost esophagus are associated with the mucosa. To facilitate movement of food, the stratified epithelial lining of the mucosa is abundantly

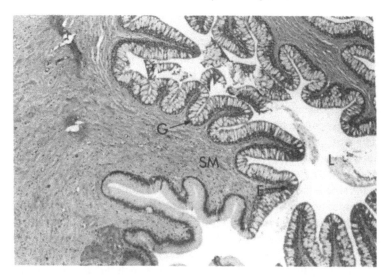

Figure 5.1 Esophagus of the channel catfish, *Ictalurus punctatus*. Lining the lumen (L) is a stratified squamous epithelium (E) and numerous large goblet cells (G). Submucosal muscle (SM) projects into the submucosa and contributes to the composition of primary, secondary, and tertiary longitudinal folds. These folds provide for the distensible nature of the esophagus. The muscularis and serosa (not shown) complete the structure (63×).

endowed with goblet cells and presents numerous primary, secondary and even tertiary folds which allow distension during the swallowing of food. In some species, the mucosal folds are spirally arranged to aid in the rapid movement of food (Chandy, 1956). For numerous species, taste buds are situated in the mucosa along these longitudinal folds (Mehrotra and Khanna, 1969). In other species, the esophageal epithelium is keratinized, protecting against abrasion from hard foods such as diatoms (Ghazzawi, 1935; Khalilov *et al.*, 1963). A variety of glandular structures secreting serous, mucous or enzymatic secretions have been identified within the esophagus on a regional basis (Sarbahi, 1939; Weinreb and Bilstad, 1955; Iwai, 1962). Arising from the submucosa in select species, small esophageal teeth penetrate the mucosal epithelium into the esophageal lumen (Isokawa *et al.*, 1965).

Beyond its obvious transit role, a number of additional functions have been described for the esophagus. The posterior esophagus contains gastric-type glands in select species, indicating function in accessory digestion (Ghazzawi, 1935; Western, 1969), whereas esophageal sacs aid in food storage, trituration, and mucus production in other species (Isokawa *et al.*, 1965; Khanna and Mehrotra, 1970). Accessory respiratory abilities have been ascribed to esophageal intraepithelial capillaries (Liem, 1967), and concentrations of nerve cells perform gustatory functions (Srivastava, 1958; Swarup, 1959). Physostome fish can fill or relieve gas from their swimbladder via the pneumatic duct which connects to the esophagus (Pelster, 1997).

Stomach

Of the 21 000 plus species of living fish, approximately 15 percent do not contain an acid-secreting stomach. Included in this group are the Cyprinidae (minnows and carps), Catostomidae (suckers), Labridae (wrasses), Scaridae (parrotfish), Odacidae, and some members of the Blennidae (comtooth blennies). In a number of stomachless species, the anterior portion of the intestine is enlarged to varying degrees. These enlargements have been referred to as an intestinal bulb, duodenum, swollen part of the intestine, large arm of the intestine, or, in some cases erroneously, the stomach (Al-Hussaini, 1947; Kapoor, 1958). For those species possessing a true stomach (85 percent), a large degree of variability exists in the shape and size of the stomach (Figure 5.2). There is little gross distinction between the esophagus and the stomach in most fish. Longitudinal folds of the esophagus may extend into the stomach for some distance. The stomachs of species such as the Dipneusti (lungfish) and Chimaeridae (chimaeras) are simple straight tubes. In other species, the esophageal end (cardiac) and pyloric ends of the stomach have fused along the line of the lesser curvature such that the stomach appears like a blind pouch. In elasmobranchs, the pyloric limb of the stomach is smaller than the cardiac portion, presenting a J-shaped stomach. Similarly, many teleosts have J- or U-shaped stomachs.

For many teleosts, the stomach may be divided into two or three regions: the cardiac stomach, the transitional fundic stomach (variable between species), and the pyloric stomach. The cardiac stomach is usually characterized by the presence of the serous cardiac glands, whereas the pyloric region lacking these glands may be further distinguished in some species by a greater development of the muscular layer. Four major layers are distinguishable in the stomach of most fish: the mucosa, submucosa, muscular coat, and serosa. The mucosa is more or less distinguished into additional layers, dependent on the stomach region and species. These layers include the lamina epithelialis (luminal surface columnar epithelium), lamina propria, stratum compactum, stratum granulosum, and muscularis mucosa. In addition, the mucosa in the cardiac region of the stomach contains cardiac glands (much like chief cells of other species) which lie between the lamina propria and the columnar epithelial cells lining the stomach lumen. These cardiac glands converge to form gastric pits (Figure 5.3). The gastric mucosa varies in thickness in different parts of the stomach because of the degree of development of these gastric glands. The delineation of the mucosa into the multitude of layers is variable between species. The stratum compactum is a variable layer often composed of collagen which aids in delineating the surrounding lamina propria and the stratum granulosum. The stratum compactum is an adaptive characteristic in many carnivorous fish, such as trout, which acts as a strengthening and protective layer to prevent overdistention of the stomach during food intake (Burnstock, 1959; Bucke, 1971). In species such as the channel catfish which does not have a stratum compactum, a thick muscularis serves the same purpose (Burnstock, 1959).

The pyloric region of the stomach is modified in members of the Clupeodei (herrings and anchovies), Channoidei (snakeheads), Characinoidei and Mugiloidei (mullets) to act as a gizzard for trituration and mixing (Schmitz and Baker, 1969).

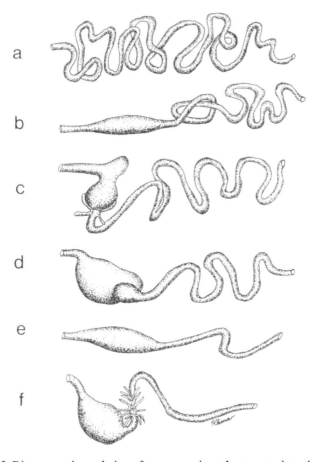

Figure 5.2 Diagrammatic rendering of representative teleost gastrointestinal tracts. (A) Agastric herbivore (some cyprinids). (B) Agastric with intestinal bulb (some cyprinids). (C) Microphagous species with pyloric region of the stomach modified into a muscular gizzard for trituration and mixing (herring and mullets). (D) Omnivorous gastric species without pyloric ceca and with a relatively long intestine (catfish). (E) Gastric carnivorous species with no pyloric ceca but extended intestine (pike). (F) Gastric carnivorous species with pyloric ceca and short intestine (trout).

Often, these fish demonstrate a protective inner coating of the stomach consisting of a cuticle, a horny keratinous layer, a thick mucous sheet, a layer of non-cellular material or a stratified non-cellular or keratinized layer (Zambriborsch, 1953; Mahadevan, 1954; Lopez and DeCarlo, 1959; Swarup, 1959; Chandy and George, 1960; Castro *et al.*, 1961). In addition, the submucosa is often reduced in these fish whereas the muscularis is well developed, especially the circular muscle layer. The presence of a gizzard as well as lengthening of the intestine, loss of teeth, proliferation of gill rakers, and development of epibranchial organs are often associated with fish with microphagous habits (Nelson, 1967; Schmitz and Baker, 1969).

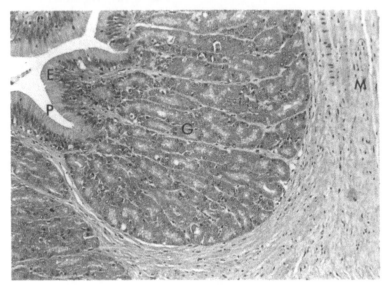

Figure 5.3 Mucosal fold of the fundic stomach of the channel catfish (*Ictalurus punctatus*). Simple columnar epithelium (E) with basal nuclei and light apical cytoplasm line the stomach lumen. Gastric glands (G) project through the lamina propria and empty into gastric pits (P) which are spaced over the surface of the mucosal fold. A submucosa supports the mucosal folds. Muscularis (M) and serosal layers complete the structure (not shown) (160×).

Pyloric ceca

Adjacent to the pyloric sphincter in the proximal intestine of many fish are structures referred to as pyloric ceca. Ceca are found in most fish groups, but are generally lacking in Chondrichthyes (sharks and rays), Cyprinidae and Catostomidae (stomachless carp and suckers), Esocidae (pikes) (Jacobshagen, 1915; Pernkopf and Lehner, 1937) and some species of Blennidae (blennies) (Buddington and Diamond, 1987). These structures, blind sacs of narrow diameter and varying number (1–1000) (Suyehiro, 1942), tend to be better developed in carnivores with short alimentary tracts than in either herbivores or carnivores with relatively long tracts. Evidence suggests that, for ceca-containing species, ceca may represent a large portion of the post-gastric surface area of the GIT (Marcotte and De La Noue, 1984). Values of the relative contribution of ceca to total areas range from 16 percent in striped bass to 70 percent in cod and trout and up to 90 percent in tuna (Buddington and Diamond, 1987). Histologically, ceca are similar to the adjacent intestine with columnar absorptive cells lining the cecal lumen (Blake, 1930; Groman, 1982; Buddington, 1983; Yasutake and Wales, 1983). Functional and histologic studies of the ceca, in a variety of species, have demonstrated uptake of amino acids and glucose (Collie, 1985; Buddington and Diamond, 1987) as well as digestion and absorption of lipids and waxy esters (Greene, 1914; Patton *et al.*, 1975; Bauermeister *et al.*, 1979). Enzymatically, the brush border of ceca contains membrane-bound dipeptidases (Ash, 1980) and disaccharidases, whereas in the lumen secretory products of the liver and pancreas (Frange and Grove, 1979) are present, as found in the proximal intestine. Only finely fragmented or solubilized food along with digestive secretions enter the ceca from the intestine. Studies with glass beads indicate size sieving by the ceca to particles of less than 150 μm (Buddington and Diamond, 1987). Ceca maintain a small resident volume and a modest food residence time of similar duration to the intestine (Buddington and Diamond, 1987). Although microflora are present, ceca apparently do not maintain a resident microflora, as would be necessary for fermentative activities.

Intestine

The intestine of fish varies significantly in length from less than the body length in some carnivores to as many as twenty times the body length in certain herbivores or detritivores (Horn, 1989; Gerking, 1994; Opuszynski and Shireman, 1995). Although herbivores generally have longer intestines than carnivores, there are significant exceptions in both directions from that generalization (Robertson and Klumpp, 1983; Clements and Bellwood, 1988). Intestinal length is also highly dependent on nutritional status, with the intestine becoming shorter in herbivores and detritivores when they are fed a high-quality diet (Horn, 1989). The plasticity of intestinal length can be considerable with changes in diet in some species of omnivores as well (Lange, 1962; Vickers, 1962; Aganovic and Vukovic, 1966; Vukovic, 1966; Shuljak, 1968; Gas and Noaillac-Depeyre, 1974). Fasting as either an experimental exercise or as a consequence of natural seasonal food availability has been shown to reduce mucosal mass of the intestine (Boge *et al.*, 1981; McLeese and Moon, 1989). Corresponding declines in intestinal surface area have been noted with changes in mucosal mass and folding patterns (McLeese and Moon, 1989), although relative mucosal gradients along the intestine appear to be maintained.

The intestine may be topographically laid out more or less in a straight line, as exemplified by the rainbow trout, arranged in a convoluted fashion filling all the available space, as with the goldfish, or ordered in tightly opposing circular arrays, as with the winter flounder (Ferraris and Ahearn, 1983). For most teleosts, the intestine is not clearly delineated grossly into definitive regions, as is the case with mammals. Although this is the case, the proximal intestine (pyloric intestine) closest to the stomach or esophagus generally is larger and has a larger lumen than either the middle or distal intestine (except rectal). Likewise, for many species of teleosts, the posterior intestine contains the highest concentration of goblet cells (Al-Hussaini, 1949; Grizzle and Rogers, 1979). In general, the intestine of both gastric and agastric fish shows a variety of longitudinal mucosal ridges which project into the intestinal lumen (Figure 5.4A and B). Large differences exist between species in the degree of this epithelial folding because in part of the amount of roughage passing through the intestine (Kapoor *et al.*, 1975) or feeding habits (Ferraris, 1982). In carnivorous species with a short gut, extensive folding

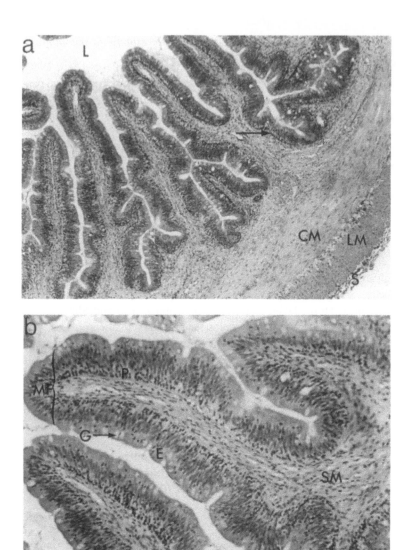

Figure 5.4 (A and B) The proximal or pyloric intestine of the channel catfish (*Ictalurus punctatus*) is composed of branched mucosal folds (MF) which project into the intestinal lumen (L). Each fold or branch is covered on the luminal surface by a simple columnar epithelium (E) with an occasional goblet cell (G). A thin lamina propria (P) composed of fibrous connective tissue and often containing lymphocytes (arrow) is directly below the epithelium and above the submucosa (SM). The muscularis consists of an inner circular (CM) and outer longitudinal (LM) layer of smooth muscle. Covering the outer muscle layer is a thin serosa (S). (A, 63×; B, 160×).

is evident in the proximal intestine, whereas in herbivores less epithelial folding can be observed. The folding of the posterior intestine is much less developed than the proximal intestine in carnivores, while in herbivores the proximal and distal intestine are more similar. Gut length and mucosal folding may be alternative strategies for maximizing absorption. Although there are exceptions, the intestinal lumen of most fish is lined by a simple columnar epithelium which has a brush border (Kapoor *et al.*, 1975; Krementz and Chapman, 1975). In contrast to mammals, true villi and crypt regions are not evident in many fish intestinal mucosae, but rather there are longitudinal folds (Kapoor *et al.*, 1975: Field *et al.*, 1978). In some primitive species of fish, the columnar epithelia are ciliated (Mohsin, 1962). Often, these species have a poorly developed muscular layer, relying on the cilia to aid in the movement of food (Patt and Patt, 1969). The absorptive epithelia of the distal intestine are often characterized by deep intermicrovillous invaginations, cytoplasmic tubules and a variety of vacuoles and vesicles (reviewed by McLean and Donaldson, 1990). These supranuclear cytoplasmic structures have been associated with endocytosis of proteins in adult goldfish, carp, tench, barbel, perch, clarias, catfish, cod, and rainbow trout as well as in the larvae of a variety of saltwater and freshwater species (reviewed by Sire and Vernier, 1992). Unlike the stomach, the trout intestine does not contain a clearly defined submucosa or a muscularis mucosa, but still maintains a lamina epithelialis, lamina propria, stratum compactum, and stratum granulosum in the mucosa. Both the circularis and longititudinalis muscles are present, but are much thinner than those found in the stomach. In the catfish, the muscularis mucosae is similarly absent (Grizzle and Rogers, 1976). It appears that in general the teleostean intestine lacks a muscularis mucosae (Bucke, 1971; Magid, 1975). A variety of other structures is found periodically as distinct species characteristics. These include structures such as crypts (Jacobshagen, 1937), ileorectal valves (Al-Hussainei, 1947; Maggese, 1967), rectal ceca (Singh, 1966), anal sphincters (Dawes, 1929), and annulospiral septa in the rectum (Burnstock, 1959; Bullock, 1963).

Rectum

The rectum, similar to the teleost intestine, is composed of mucosa, submucosa, muscularis, and serosa. It is demarcated in some species from the intestine by an intestinal sphincter composed largely of smooth muscle. Such a valve, however, is not present in a great many species (Mohsin, 1962). Rectum is often distinguished from the rest of the teleost GIT by a rich endowment of goblet cells and a wide lumen. Often, the rectal intestine has shorter intestinal folds (Grizzle and Rogers, 1976) and may contain structures such as annulospiral septa which protrude into the rectal lumen (Burnstock, 1959; Yasutake and Wales, 1983).

Trophotaenial placenta

Several species of viviparous fish have evolved unique hindgut-derived tissues which serve as the fetal component of a placenta. These structures are present in nearly all members of goodeid fish (Schindler and DeVries, 1987a) and are represented by two species of Ophidiiform family Bythitidae (Lombardi and Wourms, 1978; Lombardi, 1983) and one species within the gadiform family Parabrotulidae (Turner, 1936; Wourms and Lombardi, 1979). Trophotaeniae are rosette or ribbon-like extraembryonic expansions of the embryonic posterior gut. These processes, containing a vascularized connective tissue core covered by a surface epithelium, extend into the ovarian cavity during gestation (Lombardi and Wourms, 1985; Sire and Vernier, 1992). Nutrients synthesized and secreted by the ovarian epithelium into the ovarian lumen are taken up by a tubulovesicular system of the apical cytoplasm of the trophotenia (Schindler and DeVries, 1987b). These structures are lost at parturition.

Gastrointestinal innervation

The teleost gastrointestinal tract is controlled by extrinsic as well as intrinsic autonomic innervation. The activity of the intrinsic or enteric system, although functionally autonomous, can be modified by the extrinsic pathways. Extrinsic control is derived through nerve fibers of the vagi and splanchnic nerves. Anterior splanchnic nerves originating from the right sympathetic chain via the celiac ganglion run along the celiac and mesenteric arteries to the anterior gastrointestinal tract. A posterior splanchnic nerve finds origin in a single strand fused sympathetic chain in the posterior abdomen (Chevrel, 1889; Young, 1931) and innervates the posterior portion of the GIT (Young, 1931; Nilsson, 1976, 1983; Uematsu, 1985, 1986; Lundin and Holmgren, 1986; Uematsu *et al.*, 1989). In most teleosts, the splanchnic nerves are composed of post-ganglionic fibers (Nilsson, 1983), whereas in some species preganglionic neurons also innervate ganglionic cells (Nilsson, 1983; Gibbins, 1993). The vagus nerve, composed of cranial autonomic parasympathetic fibers, branches and supplies the gut. Innervation of the teleost GIT by the vagus is largely confined to the esophagus, stomach, and proximal intestine (Nilsson, 1983). The intrinsic or enteric nervous system contains two major plexi. Between the longitudinal and circular muscle layers of the GIT lie the ganglionated myenteric plexus, and within the submucosa the submucosa plexus (Figure 5.5). These plexi supply nerve fibers which terminate in all layers including the muscles, vasculature, and mucosa (Kirtisinghe, 1940; Holmgren and Nilsson, 1983; Holmgren, 1985).

Gastrointestinal tract vasculature

In most teleost fish, the celiacomesenteric artery supplies the majority of blood to the gastrointestinal tract (Godsil, 1954; Nawar, 1955, Singh, 1960; Peukat, 1965; Thorarensen *et al.*, 1991). This vessel originates from the dorsal aorta near the cranial end of the head kidney and courses via a variety of branches to most organs of the visceral cavity. The pattern of vascular branching from the celiacomesenteric artery varies somewhat between species.

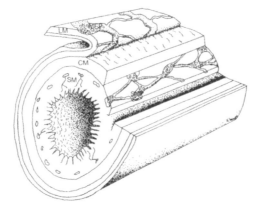

Figure 5.5 Two major nerve plexi contribute to the intrinsic or enteric nervous system of the teleost gastrointestinal tract. The ganglionated myenteric plexus (MEP) lies between the longitudinal (LM) and circular muscle (CM) layers. A submucosal plexus (SMP) is found within the submucosa (SM).

Through the use of corrosion casts, a comprehensive view of the arterial and venous vasculature has been obtained for the chinook and coho salmon gastrointestinal tract (Thorarensen *et al.*, 1991) (Figure 5.6A and B). As has been reported in other *Oncorhynchus* species (Smith and Bell, 1975; Olson and Meisheri, 1989), the hindgut receives blood flow via the celiacomesenteric arcade and interconnects directly with the dorsal aorta. These latter interconnections supply blood to the artery intestinalis dorsalis (supplying the dorsal mid- to distal intestine), which undergoes further anastomoses with the artery intestinalis ventralis supplying the ventral intestine. Similar interconnenctions have been described for other teleost species (Nawar, 1955; Petukat, 1965). Numerous other arterial vessels and anastomoses exist providing a vascular supply network of collateral circulation. Converging in the spleen, anastomoses occur between the artery gastrosplenica (supplies stomach and spleen), the artery gastrica (supplies dorsal surface of stomach) and the artery duodenosplenica (supplying the pyloric cecae and the spleen). In salmonids, the most extensive capillary networks were centered around the pyloric ceca, correlating to the absorptive function and high percentage of post-gastric surface area represented by these organs. The venous drainage, although quite variable, forms several loops. One such loop includes the venous gastrica lateralis (stomach wall), the venous gastrosplenica (to and in the spleen), venous splenica (spleen), the venous duodenosplenica (ceca) and the venous portae hepatis, thus completing a gastroduodensplenic loop. A second loop comprising the venous intestinalis ventralis (ventral intestine) and the venous intestinalis dorsalis (dorsal intestine) connects at the rectum and drains into the venous portae hepatis at the other end. Along the intestine between these two vessels, a complex of interdigitating feather-like venuoles supplies the intestinal wall. As with other previous studies with chinook salmon (Green, 1913) and the rainbow trout (Robinson and Mead, 1973), lymphatics and lacteals were not identified in the Thorarensen *et al.* (1991) studies. Although earlier studies have suggested the

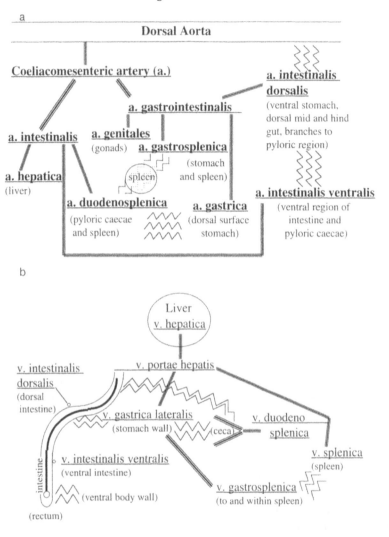

Figure 5.6 Schematic representation of the arterial (A) and venous (B) vasculature of the salmonid gastrointestinal tract.

existence of lymphatics in teleost fish (Kampmeier, 1970; Wardle, 1971), current studies have questioned their existence. Alternatively, a secondary circulation with different characteristics and alternate distribution and anatomic links has been suggested for teleost fish (Vogel and Claviez, 1981; Vogel, 1985; Steffensen *et al.*, 1986; Satchell, 1991; Steffensen and Lomholt, 1992). Given this apparent controversy, should these structures be absent from the gastrointestinal tract then all materials absorbed by the GIT would enter via the hepatic portal vein.

Other vascular patterns have been more generally described, such as for the catfish (Grizzle and Rogers, 1976). In the catfish, the celiacomesenteric artery gives rise to the pneumatic artery supplying the swim bladder before sending a branch to the esophagus. The celiacomesenteric artery further divides into the gastric artery, which supplies most of the stomach, while the vessel continues as the mesenteric artery. The first artery of the mesenteric branch supplies the gall bladder and liver while the mesenteric artery continues branching to supply the pyloric intestine via the anterior intestinal arteries, the spleen via the splenic artery and, finally, right and left branches of the mesenteric artery supplying the middle and posterior intestine. Blood primarily collects from the gastrointestinal tract via the hepatic portal vein. Just anterior to the spleen, intestinal veins enter the hepatic portal vein, whereas the gastric vein from the stomach enters the hepatic portal vein just prior to its entering the liver.

Gastrointestinal tract physiology

Gut motility

As with their mammalian counterparts, the gastrointestinal tract of teleosts is endowed with muscular layer(s) which are capable of imparting a variety of muscular movements, including swallowing, mixing, peristalsis, and segmental and pendular contractions. These movements are controlled via higher centers by adrenergic and cholinergic routes as well as by local paracrine and neurologic responses. Spinal adrenergic innervation results in a variety of smooth muscle effects in teleosts, including excitation, inhibition or a combination thereof dependent upon the fish species (Nilsson, 1983). Vagal stimulation may result in both excitatory and inhibitory effects upon GIT smooth muscle. Excitatory effects appear to be cholinergically and, in some cases, non-adrenergic non-cholinergic (NANC) (Venugopalan *et al.*, 1994, 1995) mediated, whereas vagal inhibitory effects involved in the control of gastrointestinal motility appear in numerous species to be non-adrenergic non-cholinergic in origin (Campbell, 1975; Stevenson and Groves, 1977; Holmgren and Nilsson, 1981). The neurotransmission status of putative neurotransmitters in the enteric system of teleost fish has been reviewed by Donald (1997). Included on a list of putative neurotransmitters in the enteric system of the GIT of teleost fish are catecholamines (A/NA), acetylcholine (ACh), 5-hydroxytryptamine (5-HT), nitric oxide (NO), vasoactive intestinal polypeptide (VIP), bombesin (BOM) tachykinins (SP, NK), somatostatin (SOM), galanin (GAL), gastrin/cholecystokinin (gastrin/CKK), neuropeptide Y (NPY), calcitonin gene-related peptide (CGRP), and opioid peptides Met/Leu enkephalin and neurotensin (NT). Stimulation of enteric nerves may result in the release of neurotransmitters in various combinations both qualitatively and quantitatively. Often, the responses are species specific (Nilsson, 1983; Jensen and Holmgren, 1994), may differ between the stomach and intestine (Holmgren and Jonsson, 1988), and may exhibit interactions between neurotransmitters (Thorndyke and Holmgren, 1990).

Many parameters influence gastrointestinal transit times as related to elicitation of local paracrine and neurologic responses. Meal size relative to the size of the organism in terms of weight, volume and fullness of the stomach is one such example. Large meal size often increases the time required for digestion (Jobling *et al.*, 1977) and decreases absorption efficiency (Pandian, 1967, Solomon and Brafield, 1972). The effect of meal size on gastric evacuation times appears to follow a proportional linear relationship for a variety of fish and diets (Grove and Crawford, 1980), while in others this relationship was best described in non-linear fashion (Garber, 1983; Brodeur, 1984). Feeding frequency, length of premeal fast, type of meal, variations in water temperature and thermal history may also influence GIT transit times. In general, increased feeding frequency decreases GIT transit times. Diet composition appears to have a complicated relationship to gastric evacuation times. Dietary lipid has been shown to delay gastric emptying (Kitchell and Windell, 1968; Windell *et al.*, 1969) under certain dietary circumstances, whereas little effect has been demonstrated elsewhere even when the same species has been examined (Windell *et al.*, 1972; Persson, 1982). Other dietary constituents such as crude starch have been shown to accelerate food passage (Spannhof and Plantikow, 1983), and the rate of food passage has even been related to the dietary invertebrate species (Kionka and Windell, 1972; Gannon, 1976). Temperature effects upon gastric evacuation rates can be quite pronounced. Rainbow trout evacuate the stomach of dietary oligochetes in 16.4 h at 20°C and 58.5 h at 5°C (Windell *et al.*, 1976). Likewise, gastric emptying times for pelleted food in sockeye salmon decreased from 147 h at 3°C to 18 h at 23°C (Brett and Higgs, 1970). Catfish, when examined at 10.0°C, 15.5°C, 21.1°C, 23.9°C, 26.6°C, and 29.4°C exhibited the fastest gastric evacuation rate at 26.6°C and the slowest at 10°C. Catfish stomachs were empty or nearly empty at all temperatures after 24 h. In general, evacuation rates decrease as the temperature decreases and increase with rising temperatures up to the species tolerance temperature (Shrable *et al.*, 1969; Brett and Higgs, 1970; Elliott, 1972; Windell *et al.*, 1976). The largest change in evacuation rates on a per unit temperature basis occurs at lower temperatures. Food passage through the intestine also responds to temperature and the amount of food ingested. Intestinal passage times appear to be highly variable, perhaps related to the method of dietary administration and handling stress. Total food passage times of 15–16 h and 24–26 h have been reported for juvenile rainbow trout held at 18°C and 9–10°C respectively (Fauconneau *et al.*, 1983). Armstrong and Blackett (1966) demonstrated nearly complete digestion of salmon fry by Dolly Varden trout at 24 h post-ingestion when held at 15°C. In catfish, the intestinal content of a meal peaked at 12 h after feeding and the intestine was nearly empty by 24 h (Shrable *et al.*, 1969).

Blood flow

One of the characteristic features of the gastrointestinal tract is the large variance seen in blood flow to the region. Blood flows are altered dramatically to facilitate transport of nutritive substances during the digestive process. Blood flow to the gut of the sea raven by the celiac artery, for example, doubles upon feeding from 2.9 mL min^{-1} kg^{-1} in the unfed animal (Axelsson *et al.*, 1989). Similarly, blood flow in the Atlantic cod gastrointestinal tract increases from 40 percent to approximately 52 percent of the total cardiac output upon feeding (Axelsson and Fritsche, 1991). The increased flows, besides facilitating absorption and transport of nutrients, may facilitate uptake of xenobiotics associated with nutrients (e.g. lipids) or as independent entities (water soluble). The relative importance of blood flow on the absorptive process is dependent upon the blood flow becoming a rate-limiting step in the uptake process. In mammals, this is rarely the case, except when a general circulatory failure is evident (Houston and Wood, 1980; Nimmo, 1980). For a variety of fish species under unfed conditions blood flows can be quite low, with 0.7–3.8 percent and 1.5–5.5 percent of the cardiac output being reported for the stomach and intestine respectively (Barron *et al.*, 1987; White *et al.*, 1988; Kolok *et al.*, 1993). Under conditions of low blood flow, the GI tract is usually minimally involved in absorptive processes as it is this process which increases blood flow to the region. Situations in which flow may be limited in the face of active digestion is under exercise or under temperature change. A redistribution of blood flow occurs during exercise in the Atlantic cod. In this case, vascular resistance in the celiacomesenteric vessels almost doubles, decreasing blood flows by nearly 33 percent (Axelsson *et al.*, 1989). In *Catostomus macrocheilus*, flow rates were approximately 0.6 percent of the cardiac output with exercise compared with 1.5 percent at rest (Kolok *et al.*, 1993). Temperature has also been shown to influence the blood flow to the intestine and the stomach (Barron *et al.*, 1987). Elevated temperature increases the cardiac output of rainbow trout; however, blood flow to the GI as a percent of cardiac output was relatively stable at 6°C and 12°C (3.7–3.8 percent). When temperatures reached 18°C, fractional flows dropped two- and fivefold in the intestine and stomach respectively.

Gastrointestinal pH

pH has been characterized throughout the GIT in relatively few teleost species. In those species examined, large differences are apparent. Approximately 15 percent of teleosts lack a true stomach and acid digestion (Grondel *et al.*, 1987). Several hypotheses have been forwarded regarding the loss or lack of a stomach in agastric species (Jacobshagen, 1937; Hirsch, 1950; Verighina, 1969). One hypothesis for freshwater species suggests that production of hydrochloric acid is difficult as chlorine levels in freshwater are low. Alternatively, it has been hypothesized that species with no stomach have developed specialized masticatory structures or consume a diet with a high proportion of indigestible ballast. The latter requires great quantities of food with no real need for either the HCl digestion or the storage ability of the conventional stomach. It is thought that it may be advantageous on a dietary basis for specific species not to have a stomach. For example, the coral-eating agastric parrot fish with an upper GIT pH of 8.4 would require production of large amounts of acid to acidify the large amounts of ingested calcium carbonate from coral. This would be difficult to maintain even under marine conditions (Lobel, 1980, 1981). Likewise, it has been suggested that alkaline conditions in agastric species may detoxify dietary toxins such as those from blue–green algae (Lobel, 1981). Surveys of gastric herbivorous fish have correlated gastric pH with digestive strategy (Lobel, 1980, 1981). Those fish with strong thick-walled stomachs (gizzard-like grinding structures) maintained pH values closer to neutrality, whereas fish with thin-walled non-grinding stomachs had gastric pH values ranging from 2.4 to 4.2, apparently relying on chemical breakdown for the digestive process. Even within a family of fish, gastric pH values vary substantially. Although the mullets *Liza dumerili*, *Mugil cephalus* and *Mugil curema* have alkaline pH values throughout their GIT (stomach, ceca and intestine), *Liza falcipinnis* has a gastric pH ranging between 3.4 and 4.5 (Payne, 1976, 1978). Investigations with various species of tilapia have demonstrated gastric pH values ranging from 1.4 to nearly 2.0 (Moriarty, 1973; Payne, 1976, 1978; Bowen, 1981) and intestinal pH values ranging from 6.8 to 8.8 (Fish, 1960; Nagase, 1964). In general, acidic gastric pH values were found to be capable of lyzing and digesting algae. Interestingly, in several species, including tilapia, secretion of gastric acid stops at the end of daylight feeding hours with an ensuing rise in gastric pH (Bowen, 1982). Consistent with this, stomach pH decreases upon ingestion of food for a variety of elasmobranchs and teleosts, with values changing from near neutrality before a meal to a pH of around 2 post-ingestion (Western, 1971; Moriarty, 1973).

Mucus

Functionally, mucus serves a protective and lubricative role throughout the teleost gastrointestinal tract. In most teleosts, large numbers of secretory goblet cells are present in the esophagus, much fewer in the stomach (although gastric columnar epithelium is mucoid in nature in many species), a relatively moderate number in the proximal intestine, and a variable but usually significant representation in the distal intestine (Al-Hussaini, 1949; Kapoor *et al.*, 1975; Grizzle and Rogers, 1976; Yasutake and Wales, 1983). The relative amounts of mucus produced appears to correlate well with the numbers of goblet cells in the respective sections. There are indications that the chemical composition of mucus may change with trophic status as well as along the length of the GIT within a species. In a comparison of acid and neutral mucopolysaccharides in intestinal mucus, herbivores were found to have greater amounts of sialic acids (sialomucins) than omnivores and carnivores (Jirge, 1970). Grass carp exhibited less acidic mucus in the posterior than in the anterior intestine (Trevisan, 1979). Similar changes in mucus composition along the length of the GIT have been widely described for mammals (Marshall and Allen, 1978; La Mont, 1985). Among the few characteristics definitively attributed to teleost mucus are an osmotic barrier (Marshall, 1978) and a contributor to the ion gradient across epithelial tissues of the esophagus (Shephard, 1982).

Epithelial cell renewal

As has been described for their mammal counterparts, teleost fish maintain an epithelial cell renewal system in the mucosa of the gastrointestinal tract (Vickers,

1962; Hyodo, 1965a; Gas and Noaillac-Depeyre, 1974; Stroband and Debets, 1978). Several important differences exist between fish and mammals which may be influential in both xenobiotic disposition and toxicity. The first of these relates to epithelial renewal times. Whereas rodents have intestinal epithelial renewal times of approximately 2–3 days (Lipkin, 1973), turnover rates are generally much slower for fish and vary widely with the fish species, intestinal segment, and acclimation temperature. Strong regional differences in cell turnover have been demonstrated in carp held at 12°C, with renewal times of 8, 37 and 52 days for the proximal, middle, and distal intestine (Gas and Noaillac-Depeyre, 1974). Under similar conditions, carp acclimated to 7°C resulted in slower renewal times of 54 and 82 days for the proximal and distal intestine respectively (Gas and Noaillac-Depeyre, 1974). Goldfish held at higher temperatures have generally demonstrated more rapid cell turnover rates. At 17–20°C, renewal rates were 6 days for the proximal intestine and 9 days for the distal intestine (Vickers, 1962), while goldfish held at 20°C and 37°C in other studies demonstrated a transition time for labeled cells of 4 and 2 days respectively (Garcia and Johnson, 1972). Adding variation to the mix, Hyodo-Taguchi (1970) demonstrated turnover values of 20 days for goldfish held at 25°C. Significant variations in proliferative activity between mucosal folds of an individual and between individuals have been reported for the grass Carp (Stroband and Debets, 1978)

Epithelial cell proliferation varies in location between mammals and fish. The proliferative zone of the mammalian small intestine is confined to the lower reaches of the crypts of Lieberkuhn (Galjaard *et al.*, 1972; Lipkin, 1973; Cheng and Leblond, 1974a,b). Although a distinct proliferative zone has been described for goldfish in the lower region of the mucosa fold (Vickers, 1962; Hyodo, 1965b), [³H]-thymidine labeling studies with other species have described an ill-defined proliferative zone. In these studies, DNA synthesis was evident in both mid- and basal portions of the mucosal folds (Gas and Noaillac-Depeyre, 1974; Stroband and Debets, 1978; Kiss *et al.*, 1988; Kleinow *et al.*, 1996a). Kiss *et al.* (1988) and Stroband and Debets (1978) have suggested that cell proliferation in the fish mucosal epithelium may occur both via undifferentiated stem cells and via mitotic division of mature functional absorptive cells. This observation may also be attributed to a prolongation in the intestinal cells' ability to proliferate (Kiss *et al.*, 1988). Whatever the situation, the fish appears to possess an extended zone and/or interval of proliferation compared with the mammalian small intestine (Lipkin, 1973).

An additional finding relates to regional [³H]-thymidine labeling of DNA synthesis activity. A variety of investigations suggest that a proximal to distal gradient exists in the labeling index in the fish GIT (Wurth and Musacchia, 1964; Gas and Noaillac-Depeyre, 1974; Kiss *et al.*, 1988; Kleinow *et al.*, 1996). For most studies, the highest proliferative indices occur in the proximal and middle intestine followed by the distal intestine. Where it was examined, the proliferative indices of the esophagus and stomach were intermediate between the middle and distal intestine. Fasted catfish have demonstrated greater [³H]-thymidine cell labeling in the distal intestine than in either the proximal or middle intestine when examined over an extended time-course on a per mucosal fold basis (Kleinow *et al.*, 1996). Upon subsequent feeding, the distal intestine exhibited lower numbers of labeled cells, whereas no significant differences were noted with other regions of the intestine. Results from mammals suggest that intraluminal factors contained in the fecal bulk play a major role in the control of colonic cell renewal (Stragand and Hagemann, 1977). Also, starvation has been demonstrated to decrease the rate of cell proliferation in the intestine of mammals (Hooper and Blair, 1958; Brown *et al.*, 1963). In contrast to what is seen with mammals, the catfish data suggest that feeding stimulates marginal increases in epithelial cell turnover in the distal intestine. Feeding behavior of poikilotherms, the effect of temperature on cell turnover and general differences in cell proliferation may account for these findings.

Digestion

On a qualitative basis, the digestive process in teleost species is quite similar to that of other vertebrates. With this being said, differences do exist. Among the most commonly recognized are quantitative differences in digestive enzyme concentrations and activities. These differences are largely an adaptation to the diet and facilitate nutrient digestibility. Beyond these differences a number of unique features place a signature appearance upon the digestive and absorptive processes in teleosts. For the most part, the importance of these differences as related to digestion are unknown. The most outstanding among these distinctions are the extensive vacuolar network and extraluminal protein digestion in the posterior intestine, unique pancreatic lipase activity and distinctive differences in transport lipids, the apparent lack of lipid-carrying lymphatics, and the fermentative activities of the posterior intestine.

Protein digestion and absorption

Digestion of dietary proteins in fish occurs primarily through the action of gastric and pancreatic secretions. Pancreatic enzymes such as trypsins, chymotrypsins A and B as well as elastases, through hydrolysis of protein in the gut and pyloric ceca, act to supply free amino acids, dipeptides, and tripeptides. Enzymes associated with intestinal cells, dipeptidases and tripeptidases of the microvilli or the cytoplasm, provide additional production of free amino acids. Protein can be absorbed in the GIT of fish as free amino acids and peptides, as for their mammalian counterparts, or uniquely as whole proteins (Smith, 1989). Absorption of free amino acids occurs throughout the intestine using stereospecific carrier-mediated sodium-linked active transport. The selectivity of these transport systems is based on the structure of the amino acid side-chain. Di- and tripeptides are also absorbed using transport systems in the fish (Ash, 1985). These systems appear to be dependent upon proton-linked active transport. Perhaps the most interesting aspect in the absorption of proteinaceous materials in teleosts in the endocytotic uptake in protein macromolecules by a well-developed vacuolar system in the posterior intestine. Endocytosis of macromolecules by this highly developed vacuolar system in the epithelial cells has been demonstrated in early life stages of various teleost species (Watanabe, 1981, 1982a,b, 1984a,b,c), in adult agastric species such as goldfish (Gauthier and Landis, 1972; Iida and Yamamoto, 1985; Iida *et al.*, 1986), carp (Noaillac-Depeyre and Gas, 1973; Rombout *et al.*, 1985), tench (Noaillac-Depeyre and Gas, 1976), barbel (Rombout, 1977), and the grass carp (Stroband and Van ver Veen, 1981), and also in adult teleost fish with a stomach such as the perch (Noaillac-Depeyre and Gas, 1979), clarias (Stroband and Kroon, 1981), catfish (Noaillac-Depeyre and Gas, 1983), and cod (Lied and Solbakken, 1984). Localization of the endopeptidase cathepsin D, cathepsin B, lysosomal proteolytic enzymes and detection of acid phosphatase activity within vacuoles containing exogenous protein in the posterior intestine suggest that these structures are phagolysosomes operative in extraluminal protein digestion (Georgopoulou *et al.*, 1985, 1986a; Iida *et al.*, 1986). Several studies have correlated elevations in posterior intestinal cathepsin D activity with feeding (Georgopoulou *et al.*, 1986a,b). This activity remains throughout dietary bolus transit. In addition to these pathways, protein tracer studies also suggest that the posterior intestine is capable of paracellular protein absorption. Such a pathway bypassing protein degradation by the vacuolar system has been histochemically demonstrated in rainbow trout (Georgopoulou *et al.*, 1988), carp (Rombout *et al.*, 1985) and in trophoteniae of Goodeidae fish (Schindler and De Vries, 1987b). These routes of protein absorption appear to support a double function: nutritional and immune. Once in the enterocyte, amino acids may undergo deamination, transamination or protein synthesis before entering the bloodstream.

Lipid digestion and absorption

Triglycerides and wax esters are the primary neutral lipids available to fish through their diet (Sargent, 1976; Cowey and Sargent, 1977). Dietary lipids are absorbed by the anterior intestine and, in those species possessing them, the pyloric ceca (Gauthier and Landis, 1972; Noaillac-Depeyre and Gas, 1974, 1976, 1979, 1983; Stroband and Debets, 1978; Bauermeister *et al.*, 1979; Sire *et al.*, 1981; Lie *et al.*, 1987). Intestinal absorption and mobilization of dietary fat is generally much slower in teleost fish than in their mammalian counterparts (Turner and Barrowman, 1978; Bauermeister *et al.*, 1979; Sire *et al.*, 1981; Honkanen *et al.*, 1985). Studies with rainbow trout suggest that maximum fat absorption occurs 18–24 h after ingestion (Bauermeister *et al.*, 1979; Sire *et al.*, 1981). Pancreatic lipases, along with bile salts, are primarily responsible for the digestion of neutral lipids. Because of the non-specificity of fish pancreatic lipase (Patton *et al.*, 1975; Lie and Lambertsen, 1985) and the action of β–monoglyceride lipase, the luminal hydrolysis of dietary triglycerides results in the production of free fatty acids and glycerol. These products with the action of bile form lipid micelles which move through the aqueous milieu to the enterocyte apical plasma membrane. At this point, micelles dissociate and diffuse through the membrane assisted by re-esterification of fatty acids within the enterocytes. The luminal production of free fatty acids and glycerol in the fish intestine is in contrast to mammals, in which monoacylglycerol and fatty acids are the primary hydrolysis products. The significance of this difference is that in fish the major route of intracellular esterification occurs via glycerol-3-phosphate, which leads to the synthesis of both glycerophospholipids and triglycerides (Leger and Bauchard, 1972; Lie and Lambertsen, 1985; Lie *et al.*, 1987). Conversely, 2-monoacylglycerol, the major hydrolysis product in mammals, leads only to the synthesis of triglycerides. Fish, as a result, generally have much higher levels of phospholipids per unit triglycerides consumed than mammals. Fatty acids, once absorbed into intestinal epithelial cells, are re-esterified primarily as triglycerides (Bauermeister *et al.*, 1979; Sire *et al.*, 1981; Vetter *et al.*, 1985; Van Veld *et al.*, 1987). Intracellular dietary triglyceride is then incorporated into fat vacuoles by the smooth endoplasmic reticulum (Bauermeister *et al.*, 1979; Vetter *et al.*, 1985; Sire *et al.*, 1981), which may or may not be further processed by the Golgi apparatus (Vernier and Sire, 1986). These vacuolar structures are transient in intestinal epithelial cells until release. Dietary lipids present in the enterocytes as stored lipid droplets are exported as lipoproteins to interstitial spaces of the lamina propria and to the lumen of lymphatics (cautionary note) (Sire *et al.*, 1981). Mammals release these products as chylomicrons and very low-density lipoproteins (VLDLs). In fish, several contrasting stories have emerged. Studies in rainbow trout have shown that dietary lipids were released into lymphatic routes largely as triglycerides associated with VLDL-like particles, much like their mammalian counterparts (Sire *et al.*, 1981; Babin and Vernier, 1989). (There is controversy regarding the existence of lymphatic systems in fish.) Likewise, in all species studied with a standard diet, the size of lipoproteins was closer to VLDLs than to the chylomicrons of mammals (Noaillac-Depeyre and Gas, 1974, 1976; Stroband and Debets, 1978; Bauermeister *et al.*, 1979; Sire *et al.*, 1981). Other investigations have demonstrated direct release of lipid hydrolysis products into the circulation (Robinson and Mead, 1973; Kayama and Iijima, 1976; Iijima *et al.*, 1985, 1990a; Mankura *et al.*, 1987), including small amounts of free fatty acids and free fatty acids associated with circulatory high-density lipoproteins (HDLs) (Iijima *et al.*, 1990a). The bulk of dietary lipid, in these studies, appears to be resynthesized to triglycerides and secreted into blood in the form of HDLs (Kayama and Iijima, 1976; Mankura *et al.*, 1987; Iijima *et al.*, 1990a,b). Export of lipoproteins appears to differ between species, with export only via the portal route in carp (Noaillac-Depeyre and Gas, 1974), lymph in trout (Sire *et al.*, 1981), or via both in the tench and the perch (Noaillac-Depeyre and Gas, 1976, 1979).

Wax esters, common in dietary zooplankton, slowly undergo enzymatic hydrolysis in the proximal GIT (Patton and Benson, 1975; Patton *et al.*, 1975; Tocher and Sargent, 1984; Lie and Lambertson, 1985) to produce fatty alcohols which are further oxidized by the intestine to form fatty acids (Thyagarajan *et al.*, 1979). Products of waxy esters are also re-esterified into mainly triglycerides and phospholipids (Bauermeister and Sargent, 1979; Thyagarajan *et al.*, 1979; Mankura *et al.*, 1987).

Carbohydrate digestion and absorption

The primary carbohydrates in the diet of fish include storage products such as starch and glycogen and structural components such as cellulose and chitin. Amylase produced largely in the pancreas converts starch and glycogen to oligosaccharides which in turn are acted upon by pancreatic and intestinal carbohydrases to form monosaccharides. Monosaccharides are actively absorbed by mucosal enterocytes of the intestine (Jobling, 1995). This process appears to be coupled by a carrier molecule to the passive transport of sodium ions. The sodium gradient driving this process is maintained by active transport of sodium in the opposing direction. The carrier molecules exhibit monosaccharide specificity and saturability. Intestinal transport of glucose varies more than 200-fold among the species examined, with an apparent correlation with diet. In general, the transport rate for glucose is lowest in the carnivorous trout, is of intermediate values in the catfish, and is greatest for the herbivorous carp (Buddington *et al.*, 1985).

The digestion of structural carbohydrates such as chitin and cellulose appears to enlist at least in part the aid of gut microflora (Jobling, 1995; Seeto *et al.*, 1996). Chitin-digesting enzymes produced by the pancreas and gut microflora produce *N*-acetyl-glucosamine as the end product. Many species of marine herbivorous fish have in their gut, especially in posterior regions, microorganisms capable of fermentative digestion of polysaccharides to short-chain fatty acids (SCFAs), much like terrestrial vertebrate herbivores (Horn, 1989, 1992). Like mammals, these SCFAs appear to supply energy to meet basal metabolic energy requirements. Anaerobic bacteria and SCFAs in the gut of fish have also been found in several freshwater species, including the omnivorous *Cyprinus carpio* and *Dorosoma cepedianum* and the carnivorous *Micropterus salmoides* (Smith *et al.*, 1996).

Intestine and osmoregulation

Marine teleost fish live in a hyperosmotic environment presenting osmotic gradients as large as 600–800 mosmol kg^{-1} (Karnaky, 1980). The net result of these gradients is a loss of water to the environment, primarily across the gills. To compensate for this water loss, drinking rates are three to ten times higher in marine species than in freshwater-adapted animals. Rates of oral ingestion of water vary, correlated with osmotic water permeability, but generally rates in marine teleosts are approximately 1 mL 100 g^{-1} h^{-1}. The bulk of the water and salt uptake occurs in the small intestine (Evans, 1979). Sodium chloride is taken up in the intestine via a sodium chloride co-transport, with water following osmotically (Groot and Bakker, 1988; Musch *et al.*, 1990). Basolateral (serosal) Na^+,K^+-ATPase generates an electrochemical gradient for sodium. This gradient provides the driving force for Na^+K^+2Cl transport across the mucosal (apical) membrane. Net sodium uptake is via basolateral sodium/potassium exchange, as well as paracellular diffusion down an electrochemical gradient generated by

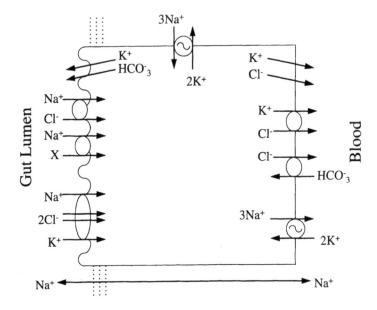

Figure 5.7 Intestinal ion fluxes involved in osmoregulation of marine teleost fish. Basolateral Na⁺,K⁺-ATPase generates an electrochemical gradient for Na⁺. The Na⁺ gradient provides the driving force for Na⁺K⁺2Cl⁻ co-transport across the apical (mucosal) membrane. Na⁺ uptake occurs via Na⁺K⁺ exchange as well as via paracellular diffusion. Water follows Na⁺ into the systemic circulation (adapted from Loretz, 1995).

basolateral chloride channels (Figure 5.7). Basolateral potassium chloride secretion also provides for net uptake of chloride. Apical (mucosal) potassium channels provide for net secretion of potassium into the gut lumen. Hickman (1968) demonstrated that in the southern flounder (*Paralichthyes lethostigma*) over 95 percent of the ingested sodium, potassium and chloride, 69 percent of the calcium, but less than 15 percent of the magnesium and sulfate are absorbed in the intestine. The bulk of the magnesium and sulfate are excreted rectally. Excess sodium and chloride are eliminated by the gill because urine and fecal matter are generally hypotonic or isotonic compared with the blood (Evans, 1993).

Xenobiotic absorption and disposition

Bioavailability of dietary xenobiotics

The GIT is a major route of xenobiotic uptake for teleost fish species (Pizza and O'Connor, 1983; Spigarelli *et al.*, 1983; Bruggeman *et al.*, 1984; Thomann and Connolly, 1984; Maccubbin *et al.*, 1985; Walker, 1985; Fisher *et al.*, 1986; Connolly and Pedersen, 1988; Gobas *et al.*, 1988). Compounds which bioaccumulate in lower trophic levels are of prime consideration for GIT absorption in fish. In addition, chemicals in the sediments or water may also become available to the GIT depending on the feeding habits of the fish or methods of osmoregulation. The relative importance of dietary absorption is dependent upon the chemical under consideration and upon the conditions of exposure. For example, studies with polychlorinated biphenyls (PCBs) in brown trout indicate a ninefold higher PCB level from the diet than from the water (Spigarelli *et al.*, 1983), whereas studies with goldfish exposed to DDT in the water and diet indicate that the importance of each route was dependent on the relative compartmental concentrations over the exposure period (Rhead and Perkins, 1984). Accumulation of Kepone in spot from food and water was shown to be additive; however, the extent of dietary uptake was related to feeding rate (Fisher *et al.*, 1986). Likewise, metal uptake in fish appears to occur primarily across the gill and the intestine (Dallinger *et al.*, 1987), although the relative importance of each pathway under varying circumstances has not been clearly delineated.

Among the many topics regarding dietary exposure still lacking clear definition is the issue of biomagnification. While most investigators agree the dietary exposure is an important component of the process, most of the evidence gathered is indirect in nature. Investigations examining PCBs in Lake Ontario salmonids, for example, have demonstrated bioconcentration factors (BCFs) which far exceed those expected from direct uptake from the water (MacKay, 1982; Oliver and Niimi, 1988). Through the pelagic food chain, a greater divergence from the BCF–K_{ow} line has been observed with each trophic level, independent of variations in lipid content. Lake trout from PCB-contaminated lakes, with varying trophic complexity, have demonstrated that PCB concentrations increased with food chain length and lipid content (Rasmussen *et al.*, 1990). A 3.5-fold biomagnification factor was correlated for each successive trophic level. Lake trout collected from lakes containing intermediate trophic levels such as *Mysis* and pelagic forage fish (alewife, ciscoes, smelt, whitefish) contained significantly higher lipid-normalized levels of PCBs than those taken from lakes lacking these organisms. A positive correlation has also been reported between trophic position and mercury levels in biota from Ontario lakes (Cabana and Rasmussen, 1994) as well as polychlorinated dibenzo-*p*-dioxins and dibenzofurans in littoral and pelagic food chains in the northern Baltic Sea (Broman *et al.*, 1992). Mechanistic models by Thomann (1989) and Thomann *et al.* (1986) have suggested that 90–99 percent of contaminants in Lake Michigan lake trout result from concentration via the food chain.

The bioavailability of xenobiotics from the fish GIT has been quantitatively assessed using either pharmacologic modeling-based techniques or assimilation-type approaches. Pharmacologic approaches which integrate systemic uptake throughout the exposure period are calculated from the ratio of the areas under the blood concentration–time curves of orally and intravascularly (100 percent) administered doses of a drug or toxicant. The validity of this method depends upon equivalent clearance mechanisms for both routes of administration. Toxicants which alter their own distribution or elimination may not adhere to this premise (Walsh, 1997). This approach has found use with select toxicants in fish (Kleinow and Brooks, 1986; Kleinow *et al.*, 1989; Barron *et al.*, 1991; McCloskey *et al.*, 1998), whereas it has found wide application with aquaculture drugs (Droy *et al.*, 1990; Bowser *et al.*, 1992; Kleinow *et al.*, 1992a,b, 1994; Jarboe *et al.*, 1993;

Martinsen *et al.*, 1993; Horsberg *et al.*, 1996). Assimilation/absorption efficiency approaches, based on chemical residues in the fish at a given sampling time, are an adequate means for the estimation of bioavailability of compounds which are not significantly biotransformed or eliminated (Gobas *et al.*, 1988; Fisk *et al.*, 1996). Bioavailability values determined by either technique may range widely depending upon the chemical characteristics, character of the dietary vehicle, intestinal transit times, and conditions within the gastrointestinal tract. Using assimilation approaches, Niimi and Oliver (1988) demonstrated absorption efficiencies of halogenated organic chemicals in rainbow trout between 50 percent and 80 percent, with molecular size as a determinant factor. Gobas *et al.* (1988) and Opperhuizen and Sijm (1990) demonstrated similar absorption efficiencies with organochlorine data from other fish species. Non-chlorinated compounds and abundant data from aquaculture drugs demonstrate a greater range in bioavailability, with some compounds nearly completely absorbed (Cravedi *et al.*, 1987; Dauble and Curtis, 1989; Jarboe *et al.*, 1993) from the diet while others are partially (Cravedi *et al.*, 1987; Kleinow *et al.*, 1992b) or poorly (Grondel *et al.*, 1987; Plakas *et al.*, 1988) absorbed.

Assimilation-based work with chlorinated paraffins (CPs), a class of polychlorinated alkanes, illustrates a number of features operative in GIT bioavailability. Bengtsson *et al.* (1979) demonstrated that CPs of short carbon chain length and low chlorination had the highest uptake rate in fish. Consistent with these findings, CPs with high molecular weights (MW > 600) were found to have low or non-existent accumulation in fish, presumably due to permeability considerations (Zitko, 1974; Lombardo *et al.*, 1975; Bengtsson *et al.*, 1979). Subsequent work with rainbow trout has demonstrated that high chlorination of short-chain CPs results in an increase in bioaccumulation, whereas medium and long-chain CPs, when highly chlorinated, exhibit a reduction in assimilation efficiency (Fisk *et al.*, 1996). These findings appear to be multifactorial in their etiology, with part of the bioaccumulation differences related to the action of the gastrointestinal tract. Increased chlorination increases molecular volume, hindering either absorption across the mucosal membrane or perhaps incorporation into micelles. This process reduces bioavailability, especially for longer chain length compounds. In contrast, chlorination of shorter chain length CPs results in greater bioaccumulation because of reduced biotransformation. In general, lipophilic contaminants with high triglyceride solubility (Van Veld, 1990), but not excessively high log K_{ow} values (Clark *et al.*, 1990; Opperhuizen and Sijm, 1990), MW < 600 (Bruggeman *et al.*, 1984; Niimi and Oliver, 1988), low degrees of chlorine substitution (Tanabe *et al.*, 1982) and small molecular volumes (< 0.25 nm³) are more readily absorbed (Niimi and Oliver, 1988).

Metabolites of polyaromatic hydrocarbons (PAHs) are known to be formed and persist for protracted periods of time in invertebrate species (Little *et al.*, 1985; McElroy, 1985). *In vivo* food chain studies with PAH carcinogens also suggest that fish are capable of absorbing these lower trophic level biotransformation products (McElroy and Sisson, 1989; James *et al.*, 1991). A variety of studies have examined the dietary bioavailability of individual PAH metabolites; some of which are potentially hazardous to the fish. Benzo(a)pyrene (BaP) metabolites BaP-9-glucuronide and BaP-9-sulfate, as presented in the *in situ* isolated perfused intestine, were transported intact from the intestinal lumen to the systemic circulation with little breakdown or conversion to other products (James *et al.*, 1996). 3-OH-BaP on the other hand was extensively conjugated to BaP-3-sulfate and BaP-3-glucuronide, while a 26- to 50-fold greater amount on a relative basis was evident in the blood (James *et al.*, 1996). Besides extensive first pass intestinal metabolism, these data demonstrate some selectivity in metabolite transport to or retention/presentation in the blood. Similar radiolabel equivalent concentrations and biotransformation data for post-infusate, mucosa and blood suggest that 3-OH-BaP retention in mucosa and the appearance of 3-OH-BaP-derived molar equivalents in blood may be dependent upon biotransformation. This does not appear to be the case for BaP-9-sulfate or BaP-9-glucuronide, where the major forms in the blood and mucosa were the compounds of administration. *In vivo* studies with both natural foodstuffs and formulated carriers have demonstrated the gastrointestinal absorption of BaP-7,8-dihydrodiol, the proximate carcinogenic metabolite of BaP (Kleinow *et al.*, 1989; McElroy and Sisson, 1989; James *et al.*, 1991). Using integrative bioavailability techniques over the time-course of exposure, winter flounder demonstrated nearly equivalent BaP-7,8-dihydrodiol bioavailability as evident with BaP (Kleinow *et al.*, 1989). Tissue residue data with southern flounder indicated greater absorption with BaP-7,8-dihydrodiol than BaP when viewed by static sampling (James *et al.*, 1991). Compositely, these results demonstrate that metabolites of highly lipophilic compounds such as BaP may be readily absorbed from the GIT. Transfer of inactive or active metabolites along the food chain to subsequent trophic levels may have toxicologic significance for select compounds.

A few studies have examined the effect of dosage form and dosage on the gastrointestinal bioavailability of xenobiotics in fish. Studies with the aquaculture drug oxolinic acid in yellowtail have indicated that reduction in particle size enhanced bioavailability (Endo *et al.*, 1987). The bioavailability of xenobiotics has also been shown to vary with dose. Oxolinic acid bioavailability in rainbow trout was reported to range from 13.6 percent to 91 percent over the dosages of 5–100 mg kg^{-1} (Cravedi *et al.*, 1987; Bjorklund and Bylund, 1991; Kleinow *et al.*, 1994). These studies indicated that the bioavailability of oxolinic acid was higher on a percentage basis at lower dosages. Drug ionization and varying carrier vehicles were also likely modifiers to bioavailability in the papers cited. Other studies with oxolinic acid in Atlantic salmon indicated no difference in bioavailability when dosed in the feed at 9 and 26 mg kg^{-1} (Hustvedt *et al.*, 1991). Similarly, studies with sulfadimethoxine showed marginally lower bioavailabilities when doses were raised threefold from 42 to 126 mg kg^{-1} (Kleinow *et al.*, 1992b). The effect of dose appears to be compound and species dependent.

The effect of dose and beta-naphthoflavone (BNF) induction upon biotransformation and bioavailability of [^{3}H]-BaP have been examined using an isolated *in situ* perfused catfish intestine (Kleinow *et al.*, 1998). Total [^{3}H]-BaP molar equivalents (Meq) entering the systemic circulation in the *in situ* preparation varied with dose and with BNF treatment of the fish. A 10-fold increase in BaP dose administered resulted in a significant 3.0- to 6.3-fold increase in systemically available (blood) BaP Meq. BNF pretreatment resulted in significant 5.5- and 2.6-fold (depending upon dose) increases in systemically available BaP Meq for the 2 and 20 µM doses respectively. Metabolites accounted for 10.2% and 23.1% of the increased BaP Meq entering blood of BNF-treated animals for the 2 and 20 µM doses respectively. Although the authors suggested that metabolism may influence bioavailability by providing substrates for specific transporters (Lindwall and Boyer, 1987; Kobayashi *et al.*, 1991) or by metabolizing the lipophilic BaP parent compound to metabolites with polarities more favorably suited for absorption, it appears that additional features may be operative. BNF has been shown to alter structural components and the biochemistry in fish systems (Takahashi *et al.*, 1995; James *et al.*, 1997). Independent studies have noted histologic changes in the catfish GIT at the dosages used in the *in situ* studies, suggesting that intestinal integrity may be an operative feature with these findings (K. M. Kleinow, unpublished). Similar investigations examining the bioavailability of the BaP metabolites 3-OH-BaP, BaP-9-glucuronide and BaP-9 sulfate as administered in the intestinal lumen did not report any significant differences with BNF pre-exposure (James *et al.*, 1996).

Gastric and intestinal evacuation rates are influential determinants of gastrointestinal bioavailability in aquatic species. Gastric emptying rates are especially important as the major site of absorption is the proximal intestine owing to the high surface area and associated lipid absorption. In mammals, longer retention times in the intestine results in greater absorption, whereas shorter times result in lower absorption. It has been demonstrated in a variety of fish species that GIT passage rates decrease substantially with lower temperatures (Shrable *et al.*, 1969; Windell *et al.*, 1976). This effect could increase absorptive contact time for those drugs and matrices in which bioavailability is limited primarily by transit time rather than chemical form (Winter, 1988). As an illustration, catfish administered oxolinic acid in the diet demonstrated temperature-related differences in bioavailability ranging between 56.0 percent at 24°C and 91.8 percent at 14°C. Digestive physiology suggests that these generalizations may not be as unequivocal under all circumstances which influence gastrointestinal transit times in fish. For example, lower temperatures increase GIT retention times, and decrease digestibility. Studies with methylmercury indicate a large interanimal variation in bioavailability which was related to the amount of food gavaged to the fish (catfish) (McCloskey *et al.*, 1998). Absorption efficiency of methylmercury from the gut decreased as the quantity of food increased. The absorption of xenobiotics from the diet is directly related to the digestibility of the matrix containing the xenobiotic and the contact time of the xenobiotic with the absorptive surface. Both of these factors are directly related to the transit times of the xenobiotic-containing matrix through the GIT.

Absorptive process

Xenobiotics presented in the diet are liberated during the digestive process. For teleosts, this process varies substantially with the digestive strategy of the fish. In gastric carnivorous species, this is a gradual process as the outer layers of the prey are eroded off in the stomach. For other species, mechanical mastication and grinding may accelerate the process. Whatever the mechanism, xenobiotics, when exposed, partition from the various constituents of the chyme into either the surrounding aqueous fluid or lipid-soluble constituents of the digesta, depending upon the chemical's physical characteristics. As the fluid of the digesta is a polar environment, lipid constituents liberated in the intestine with aid of pancreatic lipase form lipid droplets and by the action of bile salts form micelles. This action allows lipophilic xenobiotics to become associated with and distributed throughout the aqueous phase as an incorporated component in the dispersed micelles (Vetter *et al.*, 1985). The presence of fat in the diet appears to facilitate absorption of lipophilic contaminants in fish (Van Veld, 1990) as well as in mammals through this association (Bloedow and Hayton, 1976; Kuksis, 1984; Yoshitomi *et al.*, 1987). The xenobiotic-carrying capacity of the micelle may be influenced by micelle fatty acid composition. Studies in mammals indicate that PAH association with micelles is favored by long-chain monounsaturated fatty acids and their acyl glycerol components, while the aqueous solubilization of polychlorinated biphenyls has been shown to be favored by polyunsaturated components of the micelle (Laher and Barrowman, 1983). Micelle fatty acid chain length appears to influence solubilization of both hydrocarbons (Laher and Barrowman, 1983) and vitamins (Takahashi and Underwood, 1974), with long-chain mixtures resulting in greater carrying capacity than shorter chain mixtures. *In vitro* studies with 20 μM [^{14}C]- 3,4,3′,4′-tetrachlorobiphenyl (TCB) have demonstrated increasing solubilization with micelle chain length, progressing from 17 percent of the dose for micelles composed of lauric acid (12:0) to 23 percent for myristic acid (14:0), and 43 percent for linoleic acid (18:2) (Doi *et al.*, 2000). Micelles composed of a mixture of acyl chain lengths exhibited a considerably reduced TCB solubility (5 percent). Evaluation of the systemic bioavailability of TCB as delivered by micelles of different composition in a catfish *in situ* intestinal preparation reflected the TCB micelle solubility as demonstrated in the *in vitro* studies (Doi *et al.*, 2000). Systemic bioavailability was nearly twofold greater for linoleic (18:2) micelles than the micelle mixture when administered at the same total dose into the intestinal preparation. Other studies with fat-soluble compounds in mammals have demonstrated the influence of dietary lipid saturation status upon the mucosal transport of exogenous agents (Hollander *et al.*, 1977; Hollander, 1980, 1981). Unsaturated fatty acids of similar chain length can reduce gut transport by as much as 50 percent compared with their saturated counterpart. Excessive dietary fat may also reduce uptake of fat-soluble contaminants by restricting the digestibility of available lipids (Andrews *et al.*, 1978). This latter phenomenon appears to result from a competition for toxicants by dietary triglyceride and digestive transport products.

Absorption of chemicals across the wall of the GIT requires passage through a number of real or functional layers with varying physicochemical properties. These layers include the unstirred water layer, the cell-surface charge/acid microclimate

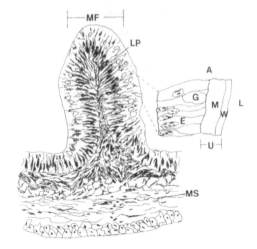

Figure 5.8 Diagrammatic representation of the absorptive surface of the teleost intestinal mucosal fold. MF, mucosal fold; LP, lamina propria; MS, muscle; L, lumen; M, mucous layer; W, water layer; U, unstirred water layer; A, acid microclimate on mucosal cell membrane; G, goblet cells; E, epithelial cells.

and the mucosal cell membrane (Figure 5.8). For xenobiotics to be absorbed from the GIT, they must transverse the environment that each layer imposes. The character, often the size, of a number of these layers is variable, depending upon prevailing conditions.

The unstirred water layer in practicality includes the layer of mucus which acts as a hydrated, but permeable, gel at the mucosal cell surface and the overlying static water layer. Mucus as a grossly visible substance is composed of water, electrolytes, mucopolysaccarhides, mucoproteins, bacteria, mucosal cells, and transported substances. It is continually laid down at the surface of the GIT mucosa and moves towards the lumen to become part of the digesta. The rates of formation and loss, as well as the character of the unstirred layer, are dependent in part upon the nature of the ingesta and physical events such as abrasion, and are influenced by processes including inflammation or disease. The character and production of mucin changes with region. Mucus is evident throughout the entire length of the fish GIT, although larger numbers of goblet cells occur in the esophagus and aboral intestinal segments. The chemical character of the mucin layer changes regionally, with alterations of mucin components such as sialic acid and sulfate (Trevisan, 1979). These features, as demonstrated in mammals, alter the xenobiotic-binding characteristics of the layer.

In a toxicologic framework, the unstirred layer often acts as a barrier to the absorption of the xenobiotics, largely as a result of the static nature of the fluid space and the polarity of the associated fluid. Hydrophobic chemicals are most susceptible to the transport limitations imposed by this layer as both solubility and mixing restrictions apply. Restrictions upon polar molecules occur primarily by the static characteristics of the layer. Within the confines of this layer, a number of important processes occur. Mucosal lipases liberate fatty acids and xenobiotics from micelles, xenobiotics may be bound to chemical constituents of the layer, and xenobiotics may be modified by select reactions of microflora trapped in this layer, and local xenobiotic concentration gradients to and from the mucosal cells may be set up. Mucus may modulate xenobiotic transport as the mucin and mucosal cell turnover rate is in opposition to xenobiotic uptake. This layer is known in mammals to alter intestinal absorption of drugs (Braybrooks *et al.*, 1975; Kellaway and Marriot, 1975), nutritive agents (Vahouny *et al.*, 1980; Smithson *et al.*, 1981; Leeds, 1982) and water permeability (Gordon, 1974). Autoradiographic evidence in the catfish indicates that this layer is interactive with xenobiotics such as BaP. High BaP concentrations and gradients are associated with the mucus even well beyond clearance of digesta (Kleinow *et al.*, 1996). This suggests that the mucus may act as a barrier to absorption or as a source of xenobiotic for absorption, especially in the early and later intervals of the time-course respectively. It is likely that the mucous layer may also modify the process of xenobiotic elimination in the GIT, although there have been no studies to support this statement for fish.

The luminal intestinal membrane has one of the highest melting points and is one of the most viscous in the vertebrate body (Brasitus and Schachter, 1980). Common to most membranes, the luminal mucosal membrane as a barrier is defined in large measure by its lipophilicity. This property modulates movement of polar xenobiotics. Lipid-soluble compounds may pass through the luminal membrane. The possible exceptions are the compounds with excessively high K_{ow} values (Gobas *et al.*, 1988). Movement of xenobiotics across the lipoid mucosal membrane may occur by a variety of processes such as passive diffusion, facilitated diffusion, active transport, and pinocytosis. In addition, passage may occur by paracellular pathways at the 'tight junctions' of adjoining enterocytes. All of these processes are known to be operative in mammals. Less information is available for fish. It is likely that diffusion will have applicability to the greatest number of lipophilic xenobiotics (Schanker *et al.*, 1958; Thomson and Dietschy, 1981; Kuksis, 1984). Two types of diffusion appear to be operative, with both following a bidirectional downhill concentration gradient. Passive diffusion, a non-saturable process which is dependent upon the concentration gradient and partition coefficient of the contaminant, involves integration of xenobiotics into and through the membrane lipids. This process may be influenced by the lipid composition of the intestinal mucosal membrane as altered nutritionally or by temperature in teleost fish (Hazel, 1984). The second diffusive process is carrier mediated or facilitated, but because a carrier is required the movement is saturable. A number of agents carried by facilitated systems have been identified in mammals, including short-chain fatty acids (Stremmel *et al.*, 1985), lead (Aungst and Fung, 1981), cadmium (Foulkes, 1980), and fat-soluble vitamins (Kuksis, 1984).

The rate of diffusion of a xenobiotic across the mucosal membrane is proportional to the concentration gradient, the membrane surface area and the permeability coefficient of the compound. Xenobiotic permeability depends on the diffusivity of the molecule through the membrane, the membrane/aqueous medium partition coefficient, and the thickness of the membrane. Membrane surface area is a major determinant of xenobiotic absorption. Intestinal length, besides being variable between species especially of differing trophic levels, is also variable within a species. Diet, nutritional status, nutritional status at development, and sex have all been documented to influence absorptive surface area/gut length in fish. In part, the plasticity is related to diet quality. Fish placed on higher quality diets over time will shorten intestinal length (Horn, 1989) as well as increase gut length with low protein/high roughage diets (Sibly, 1981). Starvation will significantly alter the size of the gut, diminishing over time the composite size of the GIT. It is likely that if the nutritional status of the fish is influential in modulating the efficiency of nutrient absorption by altering the surface area available then it is likely that these changes will as a by-product influence xenobiotic uptake as well.

The actual mechanism for movement of lipophilic contaminants across the intestinal wall to the systemic circulation of the fish is a matter of debate. A model championed by Vetter *et al.* (1985) and supported by fluorescence and radiotracer studies of BaP absorption in the killifish intestine indicates that BaP and dietary fat were co-digested, co-dispersed and co-transported across the luminal membrane (Vetter *et al.*, 1985). BaP uptake into the enterocyte is facilitated by the continuous synthesis of triglyceride from absorbed fatty acids and the mobilization of the resulting lipid products to the lamina propria and systemic circulation. An alternative conceptual model suggests that the uptake of fat-soluble contaminants across the intestine occurs primarily as a result of passive diffusion, independent of a fat co-assimilation process (Gobas *et al.*, 1993). In this case, the driving force of the absorption process is meal digestion and absorption (Figure 5.9). The volume of the ingesta decreases and the composition of the ingesta changes with digestion. This results in a greater tendency of the chemical to leave the intestinal lumen for the systemic circulation. Both lines of thought involve the movement of dietary lipid in the process. The importance of these apparently modest differences is most evident when considering the dietary aspects of biomagnification in which fish attain contaminant concentrations greater than the diet they ingest. Dietary xenobiotic assimilation occurs against an apparent concentration gradient during a considerable portion of the biomagnification process. According to the first line of thought, higher body burdens could result as systemic lipids transported to the tissues are transformed into energy (an increased fugacity). This is accomplished at the tissue level as xenobiotics are co-assimilated with lipid in the intestine. Alternatively, for the second process, the active process of GIT digestion would produce a gradient into the organism. Recent studies have examined the effect of a TCB body burden upon the subsequent systemic bioavailability of [14C]-TCB (carried in micelles) using an *in situ* isolated perfused catfish intestinal preparation (Doi *et al.*, 2000). In these studies, a 10-day dietary pre-exposure to either 0.5 or 5 mg kg^{-1} of unradiolabeled TCB reduced the subsequent systemic bioavailability of [14C]-TCB to the blood of the isolated preparation (perfusing the animal's own blood to the isolated segment). At the same time, the mucosal [14C]-TCB concentrations were nearly equivalent for all groups and the post-infusate (what was left in the intestinal segment lumen) concentrations were inversely related to

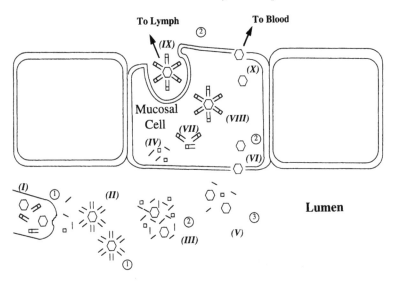

Figure 5.9 Gobas's fugacity concept of dietary chemical absorption and magnification. Arabic numbers represent changes in chemical fugacities which may result with the digestive process. Roman numerals represent the digestive and absorptive process: chemical in food (I); chemical in micelle (II); fatty acids are liberated from micelles and absorbed into enterocytes providing a greater leaving tendency for the chemical (III, IV); chemical in the lumen is absorbed (V, VI); chemical is repackaged and absorbed or absorbed directly (VII, VIII, IX, X) (adapted from Gobas *et al.*, 1993).

the blood concentrations, with the lowest levels in the controls. Because of the experimental design (micelle delivery with no progressive digestive process involved, and no access to peripheral tissue depots other than vasculature), little can be said regarding the validity of either hypothesis; however, in this system, TCB pre-exposure appears to be operative at the level of the systemic circulation. Other studies have described a decreased uptake of PCBs with increased body burden in fish (Clark *et al.*, 1990; Barber *et al.*, 1991). Long-term accumulation studies in catfish have demonstrated that the PCB body burden increases exponentially with exposure time after dietary administration (Hansen *et al.*, 1976). However, even though the body burden of PCB increases, the concentrations in the tissues equilibrate and may even decrease with changes in growth rates and food ingestion. These results suggest that once steady-state levels are reached in the animal the rate of uptake decreases, even though body burden may increase. Investigations with hexachlorobiphenyl (2,2′,3,3′,5,5′-HCB) in catfish (Dabrowska *et al.*, 1996) showed that HCB uptake was independent of prior contaminant exposure. In this study, the accumulation curve of dietary HCB was much steeper for animals not exposed than those exposed to contaminated sediments. These results were attributed to a difference in depuration rate between the two groups. It is plausible that rates of uptake may change with an existing body burden, which may or may not affect the total chemical assimilated, depending upon the residence time of the ingesta. Animals without any prior exposure may have an increased rate of uptake until they reach steady state, when there is a decrease in uptake.

Luminal pH and membrane charge characteristics also appear to play a part in the absorption of weak organic acids and bases. For weak acids and bases, the pH of the GIT luminal medium relative to the dissociation constant (pK) of the chemical is a major determinant of the lipid/membrane versus aqueous partition coefficient. For these compounds, the non-ionized form of the compound associates with lipid components and more readily transfers through membranes, whereas ionized forms are more aqueously soluble. The ratio of ionized to non-ionized forms of the chemical is dependent on the compound's character (whether it is a weak acid or base), the compound's pK and the GIT pH. Although pK is a characteristic of the compound (it can change with temperature, which is important in poikilotherms), the pH is a function of the fish species and of the specific location in the GI tract. Gastrointestinal pH may be biologically significant in regard to absorption of food-related xenobiotics (Guarino *et al.*, 1988). Thus, in some fish species the membrane/aqueous partition coefficient changes with pH alterations along the GI tract, while in others it is constant as is the pH. The ratio of ionized to non-ionized chemical is thus on a finite continuum for each specific weak organic acid or base as determined by pH in specific regions of the GIT. It is clear from pharmacologic studies in mammals that for many weak organic acids and weak organic bases the absorption from the GIT does not strictly follow expected profiles according to their ionization. This appears to be related to a number of confounding features: (1) for small molecules movement may occur through paracellular channels rather than through the lipoid mucosal membrane; (2) facilitated transport by carrier molecules is utilized by some chemicals, and (3) ion trapping may establish equilibrium-based restrictions.

Opposed to the fixed negative charges of the mucosal membrane surface is a proton-rich sheet referred to as the acid microclimate. This functional layer may also influence the permeability of weak acids and bases. Weak acids that are primarily anionic at intestinal pH become protonated at the acid microclimate, whereas weak bases remain protonated. The former become more lipid soluble, facilitating movement through the membrane, while the latter remain protonated, favoring the approach to the negatively charged membrane surface.

Active transport, carrier-mediated transport, and paracellular mechanisms are known to be important in ion and nutrient processes in the teleost GIT. Active transport in the fish intestine is an important mechanism for the normal absorptive process of many dietary solutes, such as amino acids, sugars, electrolytes, and bile salts. Toxicants are known to be transported by these mechanisms in the mammalian GIT. Such mechanisms tend to be more compound specific and limited in scope than diffusive processes. Competitive interactions for such carriers have been shown to occur between the normally transported substrates and structurally similar toxicants. For example, 5-fluorouracil competes for the pyrimidine absorptive pathway (Schanker and Jeffrey, 1961). Cobalt and manganese are known to compete for the iron transport system (Schade *et al.*, 1970; Thomson *et al.*, 1971), and a variety of drugs are known to be carried by nutrient-based transport mechanisms in mammals (Watkins and Klaassen, 1997). Beyond associative correlations with mammals, the operation of selective carriers has been demonstrated in the fish GIT. For example, zinc uptake in the intestine of the winter flounder appears to be dependent upon specific carriers. Absorption of zinc can be saturated by high zinc loads and is inhibited by other metals, including copper, cobalt, chromium, cadmium, nickel, magnesium, and mercury (Shears and Fletcher, 1983).

Pinocytosis is best known as the transport mechanism responsible for the absorption of macromolecules, such as antigens, in the early life stages of mammals. Relatively few compounds such as azo dyes (Benson *et al.*, 1957) have been shown to be transported by the membrane invagination and vesicle formation typical of this process. The very active pinocytotic activity in the distal intestine of teleost fish would suggest that compounds transported by these mechanisms may be readily absorbed. Few studies have yet addressed this question. Likewise, the paracellular transport of toxicants is of unknown importance in teleost fish. Transmembrane movement of toxicants through pores and leaky junctions appears to be dependent upon molecular size and charge. This type of movement may be generated by diffusion or by osmotically driven bulk water flow. The physiology of fish, especially marine species, indicates that such processes may deserve consideration.

All of the factors such as pH, dietary carrier vehicle, digestibility, xenobiotic availability, and uptake modality may dictate a regional character to xenobiotic absorption. The proximal intestine or ceca of fish contain considerably more absorptive surface area than distal regions of the gut. The combination of high surface area, involvement with nutrient-assisted transport (lipid and amino acids) (Gauthier and Landis, 1972; Noaillac-Depeyre and Gas, 1976, 1979) and its location for first contact with the highest xenobiotic concentrations suggests that the proximal intestine is the most important site of xenobiotic absorption in the fish GIT. Other regions of the teleost intestine are capable of xenobiotic absorption, as has been shown in studies with the antibiotic sulfadimethoxine (SDM) (Kleinow *et al.*, 1992b). These *in vivo* SDM studies, using surgically prepared rainbow trout, indicate that although all segments of the GIT are capable of absorption the stomach region exhibited the lowest absorption and the distal intestine the highest. The relative importance of each GIT segment in the absorptive process is modulated by position in the food stream (first come first served) and the regional preference of the absorptive mechanisms utilized. Regionally selective absorption has been demonstrated with both organic chemicals and metals (Pentreath, 1976; Shears and Fletcher, 1983).

Xenobiotic distribution within the GIT

Radiotracer and autoradiographic investigations with the tritiated carcinogen BaP have given some insights into local absorption and regional disposition in the catfish GIT (Kleinow *et al.*, 1996). Following dietary exposure to [³H]-BaP, the whole tissue radioassay of washed serial segments of the catfish GIT revealed both localized and regional differences in BaP Meq concentrations along the length the GIT wall. These findings appear to be a reflection of the transit movements of the food bolus containing the [³H]-BaP as well the character of the gut wall. Large differences in local BaP concentrations were evident over small distances, even within a specific region of the catfish intestine. Such differences appear to be the result of both the contact time between the food bolus with the mucosal wall at a given locale and the time since last contact. Aboral and oral movement of contaminated food occurs in a less than uniform manner in the teleost GIT. Local positioning of mucosal folds and the local strength and duration of propulsive movements provide specific exposure environments within the GIT. Likewise, food movement stimulates localized mucus secretion or loss and physical abrasion of exfoliating epithelial cells. These responses are among the many features which may customize the local exposure.

Strong regional differences in [³H]-BaP accumulation and time-course of depuration were also noted in the GIT of the channel catfish (Kleinow *et al.*, 1996). BaP Meq concentrations exhibited logarithmic declines over the first 96 h after gavage for the proximal and middle intestine when the fish was subsequently fasted. Over the same interval, the distal intestine exhibited higher GIT BaP Meq concentrations and a much slower decline out to the last sampling at 192 h after administration. Autoradiographs of cryosectioned intestinal tissues from each segment demonstrated that [³H]-BaP Meq values were heavily concentrated in the mucus/unstirred layer which presented a concentration gradient through the epithelial cells and lamina propria of the mucosal folds (Figure 5.10). In addition to the persistence of [³H]-BaP in the unstirred layer, image analysis of the [³H]-BaP signal demonstrated a long-term persistence of BaP Meq values within the mucosal folds of the distal intestine, much like the whole tissue counts had indicated. Kleinow *et al.* (1996) suggested that either the unstirred layer retained luminal [³H]-BaP and presented a persistent [³H]-BaP source resulting in the continued concentration gradient to the underlying intestinal tissue (well beyond gut clearance of the food) or the distal intestinal tissue was excreting [³H]-BaP out of the GIT cells which was later trapped by the overlying mucus. In either case, mucus appeared to facilitate BaP persistence and provided an opportunity for continued absorption or, in the later scenario, reabsorption. To define this phenomenon further, studies examined the influence of subsequent ingestion of dietary bulk on [³H]-BaP dynamics in the catfish. These studies indicated that subsequent feeding after [³H]-BaP ingestion dramatically reduced BaP Meq concentrations in all segments and greatly accelerated its elimination from the distal GIT. Autoradiographic examination demonstrated that the large losses of [³H]-BaP with feeding were evident in both the mucus and the tissue of the mucosal fold. [³H]-Thymidine-based cell turnover studies indicated that the [³H]-BaP losses from the mucosal fold itself were not accounted for by a stimulation of cell turnover by the additional feeding. Taken together, these results suggest that the mucous layer may play a role in setting up local xenobiotic concentration gradients which

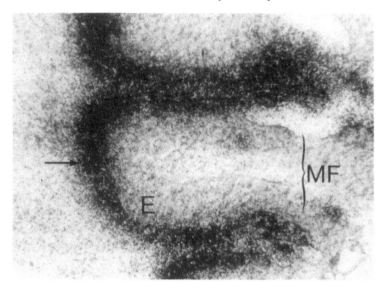

Figure 5.10 Autoradiographs of cryosectioned mucosal folds from the distal intestine of fasted catfish 192 h after exposure to tritiated benzo(a)pyrene ([³H]-BaP) in the diet. Black dots (arrow) represents distribution of [³H]-BaP molar equivalents. MF, mucosal fold; M, mucous layer; E, epithelial cells (40×) (Kleinow *et al.*, 1996).

may be operative over greater time intervals and equilibrium conditions (existing body burdens) than dictated by food passage and luminal concentrations. Additionally, strong local and regional differences in xenobiotic concentrations may exist in the GIT, paving the way for site-specific or regional effects.

Biotransformation

As well as providing the means of absorption of xenobiotics into the body, the gastrointestinal tract can also modify the structure and properties of xenobiotics (James and Kleinow, 1994). While some xenobiotics taken in orally are absorbed or excreted from the GIT as parent compound, other xenobiotics may be wholly or partially biotransformed in the stomach or intestine before systemic absorption or elimination in feces. Upon entering the GIT, xenobiotics may interact with the intestinal mucosa and the microflora. Either or both of these interactions may result in biotransformation, depending on the compound and the pathway of biotransformation. Intestinal biotransformation of environmental chemicals, which alters their lipid and water solubility, is likely to affect the chemicals' intestinal bioavailability and may affect the toxicity.

Biotransformation reactions are conveniently considered as phase 1 (introduction or uncovering of a functional group) or phase 2 (combination of a functional group with a small, usually polar, endogenous molecule) reactions. Phase 1 reactions, including hydrolysis of conjugates formed by phase 2 reactions (which are sometimes referred to as phase 3 reactions), are generally more likely to increase the toxicity of an environmental toxicant than phase 2 reactions.

Studies of xenobiotic biotransformation in fish have mainly focussed on the liver, the organ that has quantitatively the highest expression of most biotransformation enzymes and is normally considered the major site of xenobiotic biotransformation. It is fair to generalize that, for most pathways, biotransformation enzymes are present in the intestine in much lower total amounts than in liver. In fish, intestine mucosal tissue containing the biotransformation enzymes is a considerably smaller percentage of body mass than liver, and for many enzymes of xenobiotic biotransformation the level of expression in intestine is low compared with liver. If large amounts of a chemical are ingested, as is often the case for therapeutic drugs, the capacity of the intestine for biotransformation is likely to be overwhelmed. Chemicals that are present in the diet of fish at low concentrations, such as most environmental chemicals, are less likely to saturate the metabolic capacity of the intestine, so for these chemicals the intestine has the potential to play an important role in their biotransformation.

Phase 1 reactions in intestinal mucosa

Those reactions that have been demonstrated in fish intestine include cytochrome P450 (CYP)-dependent monooxygenations, conjugate hydrolyzes, epoxide hydrolysis and ester hydrolysis but not (to date) flavin monooxygenase-dependent reactions. In considering toxicity, CYP-dependent monooxygenase reactions are particularly important.

OXIDATIVE ENZYMES

Cytochromes P450   Cytochrome P450 concentrations in fish intestinal mucosa are normally lower than are found in liver, and the total amount of CYP in intestinal mucosal cells is substantially less because there is less intestinal mucosa than liver. Recent studies with selective substrates, selective inhibitors, and molecular biological probes have shown that there are several forms of cytochrome P450 in fish intestine.

Lee *et al.* (1998) have shown that the major form in uninduced intestine of trout was in the CYP3 family (CYP3A27), with smaller amounts of a CYP2 family isozyme, CYP2K1. A CYP3A-like cytochrome P450 was identified in the intestine of the Atlantic cod (Husoy *et al.*, 1994). CYP2K1 (previously termed LM2) has been shown to activate aflatoxin in trout liver to its carcinogenic metabolite aflatoxin B_1-8,9-oxide (Williams and Buhler, 1983). The role of CYP2K1 in intestinal activation of aflatoxin B_1 is not yet established, but may contribute to activation of dietary aflotoxin B_1. The substrate selectivity of CYP3A27 is not known, but related enzymes in mammalian liver and intestine contribute to the metabolism of many xenobiotics and steroids (Rendic and Di Carlo, 1997). Hydroxylation of the PAH BaP at position 3 is catalyzed by CYP1A2 and -3 forms (Bauer *et al.*, 1995). While 3-hydroxy-BaP is not considered as toxic as benzo ring metabolites of BaP, such as BaP-7,8-dihydrodiol-9,10-oxide, 3-hydroxy-BaP may be activated by further metabolism (Glatt *et al.*, 1987). As yet, the role of CYP3A27 in xenobiotic toxicity is not known. CYP2K1 and CYP3A27 are constitutive enzymes; their inducibility in fish intestine has not been demonstrated, although there are indications that these forms are under gender-specific control.

CYP1A expression is associated with toxicity for several reasons. CYP1A catalyzes the formation of reactive metabolites from PAH, heteroaromatic compounds, planar polyhalogenated biphenyls and related large planar molecules. It is known from studies in mammals that benzo ring metabolism of BaP (i.e. metabolism at the 7,8,9 and 10 positions) is primarily carried out by CYP1A (Bauer *et al.*, 1995). Furthermore, the presence of CYP1A implies the presence of the aromatic or aryl hydrocarbon (Ah) receptor as CYP1A expression is regulated through the Ah receptor. Compounds that bind to the Ah receptor are often implicated in toxicity by several mechanisms (Whitlock, 1993). Several studies have shown that CYP1A concentrations, measured by Western blotting, were very low in uninduced spot, mummichog and Atlantic cod intestine (Van Veld *et al.*, 1988a, 1997; Husoy *et al.*, 1994). Although CYP1A concentrations in intestines from other fish species were not measured by Western blotting, studies of the metabolism of BaP and ethoxyresorufin in the intestine also suggested that CYP1A expression was low in uninduced fish. Ethoxyresorufin *O*-deethylase (EROD) activity is primarily due to CYP1A (Stegeman and Hahn, 1994), whereas benzo(a)pyrene-3-hydroxylase (AHH) activity results from several isozymes, including CYP1A as discussed above. For example, in uninduced catfish, EROD activity approached the limits of detection in intestine but was easily measured in liver, whereas AHH activities were similar in intestine and liver (Table 5.1). Addition of BNF to catfish microsomes resulted in different inhibitory profiles for EROD and AHH, in that the IC_{50} for EROD was 0.08 μm but was 2.2 μm for AHH (James *et al.*, 1997). These findings suggested that there was very low expression of CYP1A in uninduced catfish intestine and that CYPs other than CYP1A support intestinal AHH activity (James *et al.*, 1997).

Studies in marine and freshwater fish have shown that CYP1A and CYP1A-dependent monooxygenase activities, such as EROD and AHH, are readily induced in the intestine by dietary exposure to Ah agonists (Table 5.1). Van Veld *et al.* (1987) showed that oral administration of BaP, 16 mg kg⁻¹ food, to the marine fish spot resulted in increased biotransformation of BaP in the intestine, but not in the liver. Similarly, incorporation of 3-methylcholanthrene (3MC), 10 mg kg⁻¹ food, in the diet of spot resulted in an 18-fold induction of intestinal CYP1A expression and a 36-fold induction of intestinal EROD activities to undetectable levels (Van Veld *et al.*, 1988a). The spot that were fed 3MC, 10 mg kg⁻¹ food, did not show hepatic induction of CYP1A, AHH or EROD. The composition of the diet was shown to be an important determinant of intestinal monooxygenase activity in a freshwater fish, the channel catfish (James *et al.*, 1997). Intestinal, but not hepatic, EROD and AHH activities were higher in catfish maintained on a commercial fish chow than in catfish maintained on a purified diet, and were

Table 5.1 Intestinal microsomal monooxygenase activities in control and treated fish.

Species	Treatment	EROD (pmol min⁻¹ mg⁻¹ protein)	AHH (pmol min⁻¹ mg⁻¹)	Total CYP (nmol mg⁻¹ protein)	CYP1A (nmol mg⁻¹ protein)	Reference
Spot	Control (3 groups)[a]	56 ± 7	42 ± 5	0.188 ± 00.17	0.013, 0.010	Van Veld et al. (1988a)
	BaP, 10 mg kg⁻¹ (4 groups)[a]	1000 ± 180	600 ± 100	0.40 ± 1.00		
	3MC, 10 mg kg⁻¹ (3 groups)[a]	2018 ± 136	720 ± 102	0.36 ± 0.042	0.193, 0.243	
	Starved (3 groups)[a]	Not detectable	Not detectable	0.11 ± 0.02		
Spot	Control (3 groups)[a]		40 ± 20	0.24 ± 0.04		Van Veld et al. (1987)
	BaP, 16 mg kg⁻¹ (3 groups)[a]		340 ± 100	0.47 ± 0.11		
Toadfish	Control (3)		33 ± 32	0.19 ± 0.04		Van Veld et al. (1988b)
	BaP, 10 mg kg⁻¹ (3)		320 ± 60	0.37 ± 0.04		
Mummichog	Control (16)	65 ± 9				Van Veld et al. (1991)
	BNF, 250 mg kg⁻¹ (18)	667 ± 71				
Mummichog	Control (5)				Not detectable	Van Veld et al. (1997)
	BaP, 10 mg kg⁻¹ (5)				0.009 ± 0.004	
Catfish	Purified diet (8)	1.2 ± 0.4	3.9 ± 0.7	0.11 ± 0.02		James et al. (1997)
	Commercial chow (8)	3.6 ± 0.8	23.1 ± 5.4	0.13 ± 0.02		
	Chow + BNF, 10 mg kg⁻¹ (4)	11.5 ± 2.0	70 ± 13	0.14 ± 0.02		
	Chow + BNF, 100 mg kg⁻¹ (4)	3.0 ± 1.0	40 ± 15	0.10 ± 0.02		

Notes
Values are means ± S.E.
Gaps in the table indicate the parameter was not studied.
a The groups were three to five fish.

further induced by low dietary concentration of BNF (Table 5.1). It is not known which components of the commercial chow induced catfish EROD and AHH activity, and, presumably, CYP1A. In several species, induction of CYP1A was specifically studied and was shown to increase as expected in intestinal microsomes; however, total CYP content was not increased dramatically. Indeed, in catfish intestinal microsomes, no increase in total CYP was observed following BNF induction (Table 5.1). This is perhaps because, even after induction, CYP1A was a very small fraction of the total intestinal microsomal CYP content. Another possibility was that, although CYP1A was up-regulated, one or more constitutive isozymes were down-regulated by the dietary BNF.

In comparing activities in catfish, mummichog, spot, and toadfish, it is apparent that control and induced intestinal EROD and AHH activities were much lower in catfish than in the marine fish. Another striking difference between catfish and mummichog was the non-responsiveness of the catfish intestinal AHH and EROD activities to high BNF doses. These differences in induction may be due to differences in the clearance of BNF from intestine, which may relate to the lower basal EROD and AHH activities in the catfish intestine. BNF was a potent inhibitor of both EROD and AHH activities (Slaga *et al.*, 1977; Takahashi *et al.*, 1996; James *et al.*, 1997). If it is not cleared quickly from the intestine, residual BNF could mask the induction of CYP1A. Overall, the results summarized in Table 5.1 suggest that the intestine is likely to play a more important role in the initial metabolism of dietary PAH in species with high basal activity.

HYDROLYTIC ENZYMES

Of particular interest are epoxide hydrolase, ester hydrolases, β-glucuronidase, β-glucosidase, and sulfatase. Although hydrolysis of conjugates is, chemically, a similar reaction to hydrolysis of esters, these reactions are not always considered to be phase 1 reactions. In the intestine, however, conjugate hydrolysis has been shown in mucosa and microflora, and may be of importance in enterohepatic cycling of xenobiotic compounds.

Epoxide hydrolase   This enzyme catalyzes the addition of water to potentially toxic epoxide and arene oxide metabolites. While the resulting dihydrodiols are less toxic than the epoxide or arene oxide, the dihydrodiol may be further metabolized by CYP to ultimate dihydrodiol-epoxide carcinogens. A well-known example is benzo(a)pyrene-7,8-dihydrodiol-9,10-oxide, formed by the consecutive actions of CYP1A, expoxide hydrolase, and CYP1A (Sims *et al.*, 1974). Epoxide hydrolase activity with styrene oxide and benzo(a)pyrene-4,5-oxide as substrates has been found in intestinal as well as in hepatic microsomes from several fish species (James *et al.*, 1979, 1997) (Table 5.2). Fish with high epoxide hydrolase activity may be less susceptible to the toxicity of epoxides and arene oxides formed in the intestine by CYP1A and other CYPs, although this has not been documented.

Ester and amide hydrolases   Xenobiotics with ester linkages may commonly be

Table 5.2 Styrene oxide hydrolase activity in hepatic and intestinal microsomes from fish.

Species	Liver	Intestine	Reference
Splake	0.125 ± 0.021	0.051 ± 0.007[a]	Laitinen *et al.* (1981)
Sheepshead	5.6 ± 2.4	0.7 ± 0.3	James *et al.* (1979)
Redfish	2.0	1.0	James *et al.* (1979)
Large skate	1.8 ± 0.2	0.1 ± 0.0	James *et al.* (1979)
Small skate	0.5 ± 0.4	0.05	James *et al.* (1979)
Catfish	2.1 ± 0.6	2.9 ± 1.6	James *et al.* (1997)

Notes
Values are nmol min^{-1} mg^{-1} protein.
a   The values for splake intestine are nmol min^{-1} mg^{-1} of post-mitochondrial supernatant, not microsomes.

hydrolyzed in a variety of tissues. Amide linkages are also susceptible to hydrolysis, often by the same non-specific esterases that catalyze hydrolysis of esters; however, amides are commonly more stable than esters. While these hydrolytic reactions may influence toxicity, for example by detoxifying pesticides such as parathion and related organophosphates, there appears to have been few studies of these enzymes in intestinal tissue.

β-Glucuronidase and sulfatase   Glucuronide and sulfate conjugates of xenobiotics that are excreted in bile and secreted into the intestine may be hydrolyzed in the intestinal cells or by hydrolases in the intestinal bacteria, and the unconjugated xenobiotic may then be reabsorbed in a process known as enterohepatic cycling. Enterohepatic cycling of the phenol metabolite phenyl-β-D-glucuronide was observed in goldfish. Phenol was excreted into tank water as phenylsulfate, but when bile was collected from phenol-dosed goldfish phenyl-β-D-glucuronide was found (Layiwola and Linnecar, 1981; Layiwola *et al.*, 1983). These authors suggested that phenyl-β-D-glucuronide excreted in the bile was hydrolyzed in the intestine to phenol, reabsorbed and conjugated to form phenylsulfate, which was excreted into tank water (Figure 5.11). The importance of this process in fish has not been extensively studied, but it may contribute to prolonged retention of xenobiotics that are substrates for UDP-glucuronosyltransferase (UGT). Studies in catfish have shown that β-glucuronidase, β-glucosidase, and sulfatase activity were present in catfish intestine, but the rates of hydrolysis of conjugates were lower than rates of formation of conjugates (James *et al.*, 1996, 1997).

Phase 2 reactions in intestinal mucosa

There have been few studies of intestinal phase 2 enzymes in fish or of the effects of inducing agents on these activities. Phase 2 enzymes that have been found in fish intestine include UGT, PAPS-sulfotransferase (ST) and glutathione-S-transferase (GST). As yet, there have been no reports of *N*-acetyltransferase or amino acid transferase activities in intestine. The products of phase 2 reactions

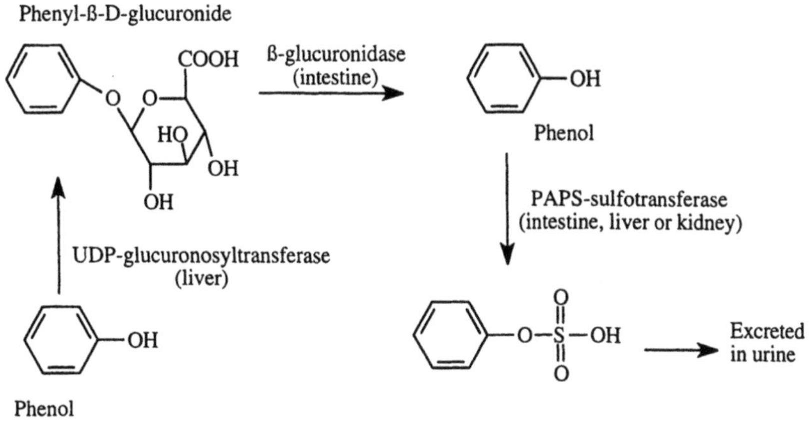

Phenyl-ß-D-glucuronide

Figure 5.11 Enterohepatic recycling of phenol in the goldfish after exposure to phenol (Layiwola *et al.*, 1983).

are usually water soluble and readily excreted. Most phase 2 metabolites are non-toxic, although there are some notable exceptions. The factors that control the expression of phase 2 enzymes are not clearly understood in any fish species. In mammals, the expression of certain forms of UGT and GST has been reported to be Ah receptor linked, so that expression of these proteins increases after exposure to Ah receptor agonists. Induction of UGT and GST following exposure to Ah receptor agonists has been variable in fish, and generally much lower than induction of P450 (James, 1989; George, 1994). Those studies that have been conducted in fish have mainly focussed on the liver, and much less is known about responses of phase 2 enzymes to inducing agents in intestine.

GLUTATHIONE-S-TRANSFERASE (GST)

GST has been widely studied in fish liver because of its importance in detoxifying electrophilic xenobiotics and electrophilic phase 1 metabolites of xenobiotics. Examples of xenobiotics that are themselves substrates for GST include halogenated organics in which the halogen can be readily eliminated, such as atrazine and dibromoethane, and alpha-beta-unsaturated ketones. Although most glutathione conjugates are detoxication products, some chemicals, notably dihaloethanes, form reactive glutathione conjugates (Rannug *et al.*, 1978). Xenobiotic phase 1 metabolites that are substrates for GST include epoxide metabolites of PAH and PCB. Because GSH conjugation of PAH epoxides prevents the epoxide from reacting with tissue macromolecules, it has been suggested that high constitutive expression of GST may protect an organ from damage by these reactive metabolites. Studies with flatfish exposed to PAH and related compounds in Puget Sound showed that species with high GST activity in liver had a lower incidence of hepatic tumors (Collier *et al.*, 1992). Although similar studies have not been conducted with intestine, it is logical to assume that tissues with high

GST and GSH will be protected from the deleterious effects of electrophiles that are detoxified by GSH conjugation.

GST was found in the intestine as well as liver of several marine fish, with 1-chloro-2,4-dinitrobenzene, 1,2-dichloro-4-nitrobenzene, benzo(a)pyrene-4,5-oxide, styrene oxide, and octene-1,2-oxide as substrates (James *et al.*, 1979). The freshwater fish rainbow trout was shown to have GST activity with 1-chloro-2,4-dinitrobenzene as substrate in cytosol prepared from intestinal ceca as well as liver, gills and kidney (Bauermeister *et al.*, 1983). A recent study of catfish intestine showed that a single form of GST expressed in the proximal intestine was highly related to mammalian pi class GST proteins (James *et al.*, 1998). This GST enzyme had high activity with benzo(a)pyrene-4,5-oxide, benzo(a)pyrene-7,8-dihydrodiol-9,10-oxide and ethacrynic acid, as well as good activity with 1-chloro-2,4-dinitrobenzene (B.K.-M. Gadagbui, M.O. James and L. Rowland-Faux, unpublished). Studies of the marine fish plaice showed that intestinal cytosol contained at least two forms of GST, namely GSTA and GSTA1 (Leaver *et al.*, 1997). The sequence of the cDNA for GSTA was related to plant and insect GSTs and to mammalian theta forms of GST (Leaver *et al.*, 1993). While the plaice intestinal GST did not cross-react with antibodies to known mammalian GST forms (Leaver *et al.*, 1997), the catfish proximal intestinal GST cross-reacted with antibodies to human pi antibodies (Gadagbui *et al.*, 1998). Cytosol from the distal intestine of the catfish cross-reacted with human and rat alpha and mu antibodies (Gadagbui *et al.*, 1998).

There are conflicting reports in the literature on the inducibility of GST in liver or intestine of marine or freshwater fish exposed to Ah agonists. The extent of induction of GST in fish, if induction occurs, is rarely more than twofold, whereas the extent of induction of CYP1A-dependent activities (discussed above) is often greater than fivefold. Some studies have shown no effect of Ah agonists on hepatic GST activity (James and Bend, 1980; James *et al.*, 1988; Collier and Varanasi, 1991; Boon *et al.*, 1992), whereas others have shown slight induction, especially if the fish were sampled more than 10 days after the dose of inducing agent (Andersson *et al.*, 1985; Boon *et al.*, 1992). Intestinal and hepatic GST in mummichog sampled from sites heavily polluted with creosote were higher than in mummichog taken from a reference site (Van Veld *et al.*, 1991). Mummichog treated with dietary BNF, 250 mg kg^{-1} food, for 2 weeks had 2.5-fold higher GST in intestine, but no induction of GST in liver, than mummichog maintained on the same diet without BNF (Van Veld *et al.*, 1991). Studies in the marine fish plaice showed that mRNA coding for a form of GST related to mammalian theta class GST enzymes was increased in intestine from male fish following intraperitoneal injection of 50 mg kg^{-1} BNF 72 h before sacrifice (Leaver *et al.*, 1997). A small induction of GST activity was found in the intestinal cytosol of catfish fed a 10-mg kg^{-1} BNF diet for 2–3 weeks relative to control chow-fed fish (James *et al.*, 1997). Intestinal cytosol fractions from catfish fed a 100-mg kg^{-1} BNF diet showed no further increase in GST activity (James *et al.*, 1997). As noted above, high dietary doses of BNF appear to be toxic to catfish intestine (K.M. Kleinow, unpublished).

Many xenobiotics containing alcoholic or phenolic hydroxy groups, thiols, carboxylate groups or amino groups may be conjugated with glucuronic acid in a reaction catalyzed by UDP-glucuronosyltransferase. The UGT family of enzymes are membrane-bound proteins, located on the inside of microsomal particles (Mulder, 1992). In studying this enzyme family, it is important to recognize that optimal activity requires disruption of the microsomal particle (Dutton, 1980). Glucuronidation is usually a detoxication reaction. Glucuronide conjugates are anions at physiologic pH, more water soluble than the parent xenobiotic and normally readily excreted from the animal. There are, however, some examples of glucuronides that are chemically reactive and may be implicated in toxic reactions. For example, glucuronide conjugates of carboxylic acids, or the hydroxy group of *N*-hydroxyarylamines, are somewhat unstable and may decompose, leaving electrophilic metabolites that can bind to cellular macromolecules (Bock, 1994).

Glucuronidation may take place in liver, intestine or other organs. If glucuronidation of dietary xenobiotics occurs in the intestine, the rate and extent of absorption of the conjugate is likely to be different from the parent xenobiotic because of the difference in water and lipid solubility between the parent compound and its glucuronide. Although it was thought that intestinal glucuronidation of a drug or other xenobiotic would lead to elimination of the glucuronide in feces without absorption (unless the glucuronide was hydrolyzed by intestinal glucuronidases as discussed above), recent studies have shown that glucuronides of very lipophilic xenobiotics, such as 9-hydroxy-benzo(a)pyrene, are readily absorbed intact across the intestine (James *et al.*, 1996). This has important implications for the toxicity of lipophilic compounds that can be glucuronidated. The glucuronide may be absorbed and then subsequently hydrolyzed to the parent compound. Alternatively, in the case of unstable glucuronides, the glucuronide may be eliminated or rearranged to a reactive metabolite that can cause cellular damage. In scenarios where the glucuronide formed in the intestine is absorbed and subsequently broken down, the intestine is less likely to be the target organ.

There is good evidence that fish intestine contains one or more of the UGT enzymes. UGT activity with *p*-nitrophenol as substrate has been found in the intestine of several freshwater fish species, including rainbow trout, vendace, perch, and roach (Lindstrom-Seppa *et al.*, 1981). In vendace, activities with *p*-nitrophenol in liver and intestine were similar, whereas in trout and roach activities in intestine were lower than in liver. UGT activity with 1-naphthol and testosterone, but not bilirubin, was found in plaice intestine, although intestinal activities were much lower than hepatic activities (Clarke *et al.*, 1992). UGT activity with 3-, 7- and 9-hydroxybenzo(a)pyrene (9-OH-BaP) as substrates was found in intestine from catfish and brown bullhead (James *et al.*, 1997; M. O. James, unpublished). In the catfish, intestinal activity was similar to hepatic activity for the benzo(a)pyrene phenols, pointing to an important role of the intestine in the first- pass conjugation of these compounds. In the southern flounder, BaP-7,8-dihydrodiol administered by gavage was readily absorbed from the intestine and excreted in bile as a mixture of parent BaP-7,8-dihydrodiol and the glucuronide conjugate (James *et al.*, 1991). DNA adducts of BaP-7,8-dihydrodiol-epoxide were also found in intestine and liver, although the extent of adduction in the intestine was not higher than that found for BaP alone. Although the contribution of intestinal UGT to overall formation of this glucuronide was not directly investigated *in vivo*, it was shown that intestine possessed UGT activity with BaP-7,8-dihydrodiol and it was postulated that intestinal conjugation limited the availability of the BaP-7,8-dihydrodiol for further activation by CYP1A and, thereby, limited the extent of adduction to DNA.

The inducibility of UGT by xenobiotics in fish intestine appears to be less sensitive than CYP1A. Although induction of UGT activity with phenolic substrates has been observed in liver and kidney of some marine fish following treatment with BNF, Aroclors or 3MC, the extent of induction is usually less than twofold (Andersson *et al.*, 1985; Ankley *et al.*, 1986; Clarke *et al.*, 1992; Leaver *et al.*, 1992). UGT activities with 3-, 7- and 9-hydroxy-BaP in channel catfish intestine were not induced by treatment with dietary BNF at diets of 10 or 100 mg kg^{-1} (James *et al.*, 1997). In catfish receiving the higher BNF dose, activities with 7- and 9-OHBaP were lower than in fish fed other diets. *In vitro* studies showed that BNF could inhibit UGT activity with 1 µM 9-OHBaP as substrate, with an IC$_{50}$ of about 50 µM (James *et al.*, 1997). The lower activities in the high-dose fish may have been due to inhibition of UGT activity by residual BNF in the microsomes or to intestinal cell toxicity. Further studies are needed to determine the sensitivity of intestinal UGT to inducing agents.

SULFOTRANSFERASE (ST)

Xenobiotics containing alcoholic or phenolic hydroxy groups or containing amino groups may be conjugated with sulfate, by the action of one or more members of the sulfotransferase enzyme family, and with the co-substrate PAPs. Like glucuronides, sulfate conjugates are anions at physiologic pH, are water soluble, are normally readily excreted from the animal, and can be considered detoxication products. For some chemicals, however, the sulfate group is a relatively good leaving group and may be eliminated from the xenobiotic sulfate conjugate, leaving a reactive, positively charged, and therefore electrophilic, metabolite that can bind cellular macromolecules. This has been demonstrated for sulfate conjugates of *N*-hydroxylated arylamines such as *N*-hydroxy-acetylaminofluorene, hydroxymethylated polycyclic aromatic hydrocarbons such as 7-hydroxymethyl-12-methylbenzanthracene, and hydroxylated terpenes such as 1′-hydroxysafrole (reviewed by Michejda and Kroeger Koepke, 1994). Thus, sulfation of some xenobiotics may be an activation reaction. Although the sulfate group increases the water solubility of a xenobiotic, sulfate conjugates of otherwise lipophilic molecules, such as hydroxylated metabolites of polycyclic aromatic hydrocarbons and other large non-polar molecules, may be absorbed from the intestine. Studies with the channel catfish showed that sulfate conjugates of 3-hydroxybenzo(a)pyrene formed in the intestine were absorbed into the blood (Tong *et al.*, 1995) and that the sulfate conjugate of 9-hydroxybenzo(a)pyrene was absorbed intact from the intestine (James *et al.*, 1996).

There have been relatively few studies of sulfation in fish, and most studies examined only the liver. In studies of the conjugation of hydroxylated benzo(a)pyrene metabolites in channel catfish intestine, it was shown that these substrates were readily sulfated in both liver and intestine (James *et al.*, 1997). Other hydroxylated compounds, such as *p*-nitrophenol, 2-naphthol, 7-hydroxymethyl-12methyl-benzanthracene and 6-hydroxymethylbenzo(a)pyrene were also good substrates for intestinal ST in the catfish intestine (Tong, 1996).

ST activity is not generally inducible by Ah agonists, and there are no reports in the literature of induction of ST activity after exposure to polycyclic aromatic hydrocarbons or related compounds. There is some evidence that ST may be down-regulated by exposure to PAH (Runge-Morris, 1998). ST activity with 9-OH-BaP, but not 3- or 7-OH-BaP, was lower in intestinal cytosol from catfish fed a diet containing 100 mg BNF kg^{-1} chow than in catfish fed 10 mg BNF kg^{-1} chow (James *et al.*, 1997). As was found for UGT, BNF was a weak inhibitor of catfish intestinal 9-OH-BaP ST activity, with an IC$_{50}$ of about 50 μm. It was unclear whether the lower activity of ST with 9-OH-BaP in intestinal cytosol from catfish treated with the high-dose BNF diet was due to residual BNF, down-regulation of ST or to other factors such as overall cellular toxicity from the BNF treatment.

Both UGT and ST activities with BaP phenols were as high in intestine as liver for the catfish. If this proves to be true for other fish, then hydroxylated compounds in the diet are likely to be conjugated in the intestine and the amount absorbed may be influenced by the first-pass intestinal metabolism.

Further metabolism of glutathione conjugates

Glutathione conjugates of xenobiotics must be further metabolized before they are excreted from the body (Chasseaud, 1976). In some cases, there is evidence that glutathione conjugates formed in the intestine may be partially metabolized in the intestine to cysteinylglycine or cysteine conjugates. Although these conjugates are not generally toxic, the action of cysteine conjugate β-lyase on cysteine conjugates leads to thiol compounds, which may be toxic (Dekant *et al.*, 1994). Although this pathway has not been extensively studied in fish intestine, there is evidence that γ-glutamyltranspeptidase and cysteinylglycinase activities are present in intestine (Bauermeister *et al.*, 1983). In mammals, cysteine conjugate β-lyase in intestinal cells and intestinal bacteria can convert cysteine conjugates to thiols (Bakke *et al.*, 1990). Thus, although conjugation with glutathione is normally a detoxication reaction, the conjugates may be further metabolized to thiol metabolites. The thiols could form mixed disulfides with cysteine groups in protein, or react with other groups, yielding adducts that may interfere with normal cellular function. The importance of the cysteine conjugate β-lyase pathway in fish intestine in not yet known.

Integrated metabolism

Several *in vitro* and *in situ* studies have demonstrated the appreciable and varied metabolism of the GIT. Van Veld *et al.* (1988b) using an *in situ* isolated perfused preparation of the toadfish intestine demonstrated extensive first- pass metabolism of BaP (62–92 percent) as well as a significant response to inducers. Blood samples from prepared wild-caught control animals exhibited upon differential extraction 37.5–46.5 percent BaP parent, 25.8–26.7 percent organic soluble metabolites and 26.8–36.7 percent water-soluble metabolites. Animals induced with dietary BaP (10 mg kg⁻¹ food at 5 percent body weight every 3 days) displayed greater intestinal BaP metabolism, with 7.1–9.6 percent parent, 33.8–40.4 percent organic soluble metabolites and 52.5–56.6 percent water-soluble metabolites. *In vitro* studies with intestinal homogenates of BNF-induced (in the diet) winter flounder also exhibited considerable phase I and II BaP metabolism upon induction (McElroy and Kleinow, 1992). These intestinal homogenates demonstrated total BaP metabolite production (demonstrated by differential extraction) which was nearly equivalent to that of the liver from the same animals. Such results suggest that intestinal biotransformational activity may take on greater importance relative to the liver, depending on the route of exposure to the inducer and the dose. As concentrations of BaP in these intestinal *in vitro* preparations were varied from 0.5 to 250 µM, the metabolite profile changed. At the lowest concentrations (0.5 and 5.0 µM), conjugates represented the predominant extraction category on a percentage basis (of BaP, polar, conjugated, and bound), whereas at 50 and 250 µM the parent and bound forms predominated. Studies with the catfish *in situ* intestinal preparation have added further definition to the understanding of GIT BaP metabolism with the identification of specific metabolites (Kleinow *et al.*, 1998). A wide variety of primary and secondary metabolites were formed including unconjugated oxidative products such as hydroxy-BaPs, BaP-diones, BaP-7,8-dihydrodiol, BaP-9,10-dihydrodiol, BaP-7,8,9,10-tetrols and other less abundant unconjugated metabolites. Conjugate concentrations were consistently the most abundant category of metabolite in both the mucosa and the blood for the control and BNF-induced animals. The primary conjugates were glutathione-based products (~ 80 percent) with the balance being sulfate and glucuronide conjugates. The percentage of total recovered radioactivity as metabolites ranged between 8.3 percent and 17.3 percent for the blood and between 5.3 percent and 10.2 percent for the mucosa, although the mucosa contained much higher absolute BaP and metabolite concentrations than the blood. These percentages of metabolized BaP for the blood were considerably less than the toadfish, as described by Van Veld *et al.* (1988b). Comparative AHH assays suggest that this response is in part due to innately lower activities for the catfish. These studies examined the influence of dose (2 and 20 µM based on K_m) and BNF induction status upon the profile of BaP metabolites entering the systemic circulation in the intestinal mucosa and in the intestinal lumen. Quantitatively, the proportions and absolute concentrations of BaP-7,8-dihydrodiol and BaP-9,10-dihydrodiol in blood and mucosa were greater for BNF pretreated animals than for the controls. While BNF increased all metabolites, the increases evident reflected some selectivity in BaP-7,8-dihydrodiol and BaP-9,10-dihydrodiol formation. These data suggest that the catfish intestine responded to dietary BNF by the production of greater proportions of potentially hazardous metabolites which may be transported to distant sites. Likewise, conjugate concentrations responded to these parameters. Control blood and mucosa at a low dose (2 µM) exhibited greater composite BaP conjugate concentrations than unconjugated metabolites. Conversely, at a higher concentration (20 µM) the unconjugated metabolites attained a 1.9- to 2.0-fold higher concentration than the conjugates. Unconjugated metabolites predominated at both concentrations following BNF administration. BNF pretreatment and a higher dose both increased the amounts of intestinal metabolites in absolute terms. Although conjugate concentrations appeared to be sensitive to these mechanisms, they were disproportionate to the larger response of the unconjugated metabolites.

Summary of intestinal metabolism

The fish intestine possesses high activities of many xenobiotic-metabolizing enzymes, especially phase 2 activities, and is clearly important in the first-pass biotransformation of low levels of xenobiotics. Intestinal biotransformation of some xenobiotics can result in increased toxicity to the intestinal cells, although there are few examples where this has been studied in detail in fish.

Elimination

Elimination of xenobiotics by the teleost GIT may occur as a result of several processes. Xenobiotics incompletely absorbed from the diet may present in excretory products. As many compounds are incompletely absorbed, this is a significant pathway of apparent elimination. Other xenobiotics in the feces may be a result of elimination of systemically absorbed xenobiotics. In general, teleosts are extremely good biliary concentrators of chemicals (Statham *et al.*, 1976). Hence, xenobiotics and their metabolites are commonly elaborated into the fecal stream via the bile. Once in the fecal stream, conjugated metabolites are available to hydrolytic enzyme activity, which upon cleavage presents the parent compound for reabsorption. Enterohepatic recirculation of both endogenous and exogenous ligands has been demonstrated in teleost fish (Collicutt and Eales, 1974; Layiwola *et al.*, 1983). This process modulates the amount of metabolites eliminated via the feces. In addition, as defined in earlier sections of this chapter, metabolites of lipophilic environmental contaminants, such as those of BaP, may be absorbed intact from the gastrointestinal tract (James *et al.*, 1996).

As demonstrated in mammals, toxicants may be excreted in the feces by non-biliary pathways (Rozman, 1986). Non-polar lipophilic xenobiotics which are poorly biotransformed and cleared via renal or biliary routes can be exsorbed into the intestinal lumen. Although passive diffusion is an important mechanism for this process, rapid exfoliation of intestinal cells, active secretion (Lauterbach, 1977), and efflux transport (Leu and Huang, 1995) have been demonstrated to be operative for a variety of xenobiotics.

P-Glycoprotein transporter

A number of xenobiotic transporters have been identified in the mammalian gastrointestinal tract. Recent findings suggest that teleost fish may also have a complement of transporters which may not only facilitate elimination of xenobiotics via the GIT but may also modulate xenobiotic bioavailability from the diet. A P-glycoprotein (PGP)-'like' transporter has received some attention regarding these functions. PGP-'like' transporters are plasma membrane-bound efflux pumps found in a medley of organisms ranging from yeast to humans (Van der Blick *et al.*, 1988; Kurelec *et al.*, 1992). They are known to transport a variety of bulky xenobiotics (Yeh *et al.*, 1992; Zimniak and Awasthi, 1993). In teleost fish, conserved forms of this transporter have been immunologically or functionally identified in the gill, proximal renal tubule, liver bile canaliculi, and the gastrointestinal tract (Hemmer *et al.*, 1995; Kleinow *et al.*, 1996). Excluding the gill, the subcellular location and tissue distribution in fish are similar although not identical to that characterized in mammals (Fojo *et al.*, 1987; Thiebaut *et al.*, 1987; Sugawara *et al.*, 1988). Studies with PGP-'like' transporters in the channel catfish have begun to characterize the potential role of PGP in the GIT. As with mammals, GIT PGP appears in the apical plasma membrane of epithelial cells lining the intestinal lumen (Figure 5.12) (Kleinow *et al.*, 2000). Catfish PGP concentrations in the GIT appear to exhibit variable response to prior exposure of xenobiotic agents. Agents, such as the prototypic mammalian PGP inducer vincristine and the P4501A inducer BNF, induced a PGP-like protein in catfish

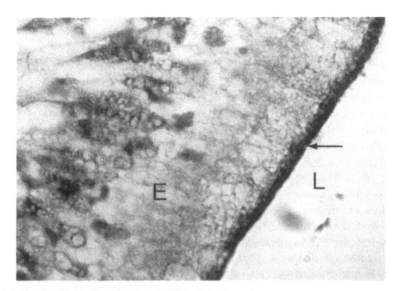

Figure 5.12 Distribution of the P-glycoprotein-like transporter (arrow) in the epithelial cells (E) lining the intestinal lumen (L) of the distal intestine. Immunohistochemistry was performed with C219 monoclonal antibody with alkaline phosphatase staining (269×) (Kleinow *et al.*, 2001).

GIT as observed by immmunohistochemistry (Kleinow *et al.*, 2000) while BNF, TCB, and BaP failed to elicit a response as measured by immunoblotting (Doi *et al.*, in press). Immunohistochemical and immunoblotting experiments indicate low levels of PGP in control animals. Induction of PGP with vincristine and BNF exposure was much more significant in the distal intestine than the proximal intestine in the catfish (Kleinow *et al.*, 2000). In contrast, mammals demonstrate significant levels in both the small intestine and colon (Thiebaut *et al.*, 1987). As with PGP in mammals, the fish PGP-like transporter is at least partially ATP dependent and is inhibited by the prototypic inhibitor verapamil (Doi *et al.*, in press). Studies in mammals have demonstrated co-induction of PGP and CYP1A1 with exposure to isosafrole and 3,4,7,8-tetrachloro-dibenzodioxin (TCDD) (Burt and Thorgeirsson, 1988). Studies with Ah receptor-competent and -deficient mice led to a postulated relationship between the Ah locus and PGP induction (Burt and Thorgeirsson, 1988). There are data in mammals which suggest that this transporter, besides transporting anti-cancer agents, antibiotics and the like, is capable of effluxing environmental agents such as BaP (Yeh *et al.*, 1992). Although not conclusive, there are regional, temporal and *in vitro* data which support the suggestion that such an efflux may also be occurring in the catfish distal intestine (Kleinow *et al.*, 1996, 2000).

Gastrointestinal tract toxicity

Histopathology

A wide variety of chemicals is capable of inducing histopathology in the gastrointestinal tract of fish (Tables 5.3–5.5). Dietary or waterborne exposures to toxic agents have resulted in alimentary pathology, suggesting that appreciable toxicity to the GIT may result following both luminal and systemic exposure. In general, toxic effects occur at much lower concentrations when the toxicant is administered in the diet than in waterborne exposures. Presumably, this is a feature related to the absorptive nature of the GIT and the proximity of susceptible tissues to a concentrated source of toxicant, as may be found in the diet. The GIT, as with most tissues, demonstrates increased damage with both higher dosages and/or increased exposure times to the toxicants. While toxicant-induced gastrointestinal pathology is rather narrowly expressed in fish in regard to the pathologic changes evident, the distribution and extent of lesions appear to vary widely. One region or even mucosal fold may exhibit profound changes while adjacent areas may appear largely normal. Conversely, in another situation, the effect may be much more generalized. It is likely that these types of findings may be related to factors such as residence time of the toxicant in a particular locale (peristalsis of food bolus, speed of absorption, blood flow) accessibility (release from dietary matrix), availability (the protective nature of the mucous coat, permeability) and the extent of regenerative changes over time from exposure to sampling (re-epithelization). In addition to these variable effects, the multilayer complexity of the gastrointestinal tract wall, the positional dynamics of the toxicant within the lumen

Table 5.3 Toxicant-induced histopathology in pharynx, esophagus, pyloric ceca and rectum of fishes.

Chemical	Dose	Species	Pathology	References
Mercuric chloride	1.8 mg L^{-1} and 0.3 mg L^{-1} at 96 h and 5, 10, 20 and 30 days	*Channa punctatus*	Pyloric ceca Most affected Ruptured villi Cell boundaries disintegrated Cells syncytial appearance Inflammatory cells Large amount of mucus	Sastry and Gupta (1978b)
Cadmium chloride	75.5 mg L^{-1} for 3 days	*Notopterus notopterus* (Pallas)	Pharynx Shrinkage of stratified epithelial cells Shriveling of microridges Loss of lateral contacts between epithelial cells Pronounced mucin secretion	Ghosh and Chakrabarti (1992)

Dimecron	0.32 mg L^{-1} for 20 days, renewed every 48 h	*Colisa fasciatus*	Esophagus Damage to stratified epithelial cells Formation of even sheet of microridges Pyloric ceca Disrupted mucosal folds Irregular positioning of microridges Rectum Disintegration of columnar epithelial cells Release of mucus Pyloric ceca Damage more severe than in intestine Degenerated villi and connective tissue core ruptured Mucosal debris in lumen	Sastry and Malik (1979)

Table 5.4 Toxicant-induced histopathology in the stomach of fishes.

Chemical	Dose	Species	Pathology	References
Dimecron	0.32 mg L^{-1} in water, renewed every 48 h for 20 days	Colisa fasciatus	Erosion of mucosa Degeneration of gastric glands Rupture of lateral and basal columnar cell membrane Glandular cells vacuolated Nucleus reduced in volume Cell debris in stomach lumen	Sastry and Malik (1979)
Cadmium chloride	75.5 mg L^{-1} for 3 days	Notopterus notopterus (Pallas)	EM Necrosis of columnar epithelial cells Fragmentation of microridges Vigorous secretion of mucus	Ghosh and Chakrabarti (1992)
Cadmium chloride	6.8 mg L^{-1} 30-day exposure	Heteropneustes fossilis	Spotty erosion of mucosal epithelium	Sastry and Gupta (1979)
Cadmium chloride	20 mg L^{-1} for 30 days, water and toxicant changes on alternate days	Mystus vittatus	Profound vacuolation in the submucosa Elongated gastric glands Wrinkled mucosal epithelium	Datta and Sinha (1989)
Cadmium chloride	20 mg L^{-1} for 30 days, water and toxicant changes on alternate days	Labeo rohita	Intestinal bulb Elongation of columnar epithelial cells Necrosis of columnar cells Disruption and vacuolation of tunica propria and submucosa	Datta and Sinha (1989)

Mercuric chloride	Channa punctatus	10 days: pycnotic nuclei of gastric gland cells, reduced number of pepsinogen granules 15–20 days: mucosal cells degenerated, mucosa detached from submucosa, pepsinogen granules absent from chief cells 30 days: glands atrophied and distorted	Sastry and Gupta (1978a)
	1.8 mg L^{-1} and 0.3 mg L^{-1} for 96 h and 4-, 10-, 15-, 20- and 30-day exposures		
Lead nitrate	Mystus vittatus	Degeneration of columnar epithelial cells Sloughing of columnar cells off basement membrane Vacuolation in tunica propria and submucosa Profound mucus secretion	Datta and Sinha (1989)
	30 mg L^{-1} for 30 days, water and toxicant changes on alternate days		
Lead nitrate	Labeo rohita	Intestinal bulb Erosion of top plate Necrosis of columnar epithelial cells Migration of nuclei towards basement membrane Vacuolation	Datta and Sinha (1989)
	30 mg L^{-1} for 30 days, water and toxicant changes on alternate days		

Table 5.5 Toxicant-induced intestinal histopathology in fish.

Chemical	Dose	Species	Pathology	References
Organophosphate pesticides				
Dimecron	0.32 mg L^{-1} in water, renewed every 48 h for 20 days	Colisa fasciatus	Villi degenerated Villi ruptured Synctial appearing mucosa Mucosa extending in lumen Increase in number of goblet cells Large number mucin granules Lumen filled with mucus	Sastry and Malik (1979)
Sumithion	1 mg L^{-1} in water sampled at 45, 60, 75 and 90 days	Clarias batrachus	Progressive vacuolation with time Ruptured villi Cellular exudate in lumen Mucous cells enlarged	Mandal and Kulshrestha (1980)
Diazinon	0.15-0.37 mg L^{-1} in water; 24-, 96-h and 14-day exposure	Channa punctatus	24 h: slight vacuolation and cytoplasmic granulation in lamina propria 96 h: extensive damage to mucosal folds Vacuolation and granular inclusions in mucosa and submucosa 14 days: submucosal necrosis Muscle degeneration	Anees (1976)
Dimethoate	5.0-10.0 mg L^{-1} in water; 24-, 96-h and 14-day exposure	Channa punctatus	24 h: cytoplasmic vacuolation 96 h: necrotic villi 14 days: dilated submucosal blood vessels Necrotic mucosal folds, cytoplasmic vacuolation	Anees (1976)

Compound	Dose/exposure	Species	Effects	Reference
Methyl parathion	0.90–1.15 mg L^{-1} in water; 24-h, 96-h and 14-day exposure	*Channa punctatus*	24 h: slight damage to longitudinal and circular muscle / 96 h: necrotic villi / 14 days: dilated submucosal blood vessels / Vacuolated and necrotic mucosal folds	Anees (1976)
Elsan	211 ppb, water changed daily and compound added; 7-, 28-, 63- and 90-day exposure	*Channa punctatus*	7 and 28 days: destruction of villus and other layers / 63 days: collapsed villi merged tips flattened appearance / 90 days: severe damage to longitudinal muscle layer	Banerjee and Bhattacharya (1995)
Polyaromatic hydrocarbons				
Benzo(a)pyrene	200 ng g^{-1} via the diet daily	*Centropristis striata*	EM: subcellular inclusions in mucosal cells with flocculent granular material	Fair and Fortner (1987)
Benzo(a)pyrene/cadmium	200 ng BaP 10 µg^{-1} Cd g^{-1} fish via diet daily	*Centropristis striata*	EM: columnar cells with irregular shaped vesicles with material of various presentations	Fair and Fortner (1987)
Benzo(a)pyrene	20 mg kg^{-1} body weight in the diet for 1, 2, 3, 5, 7 and 17 days of treatment	*Dicentrarchus labrax*	EM and light / Day 1: pinocytotic vesicles, modified shape of mitochondria / Day 3–7: increased lysosomes and dense bodies / Day 7: enterocytes crowded with vacuoles / Day 7–17: organelle swelling, cristae regression, mitochondria spherical transformation, nuclei and nucleoli altered, hypertrophy of rough and smooth endoplasmic reticulum, reticulum, cisternae dilated and divided into vesicles, cytolytic degeneration of columnar cells	Lemaire *et al.* (1992)

Table 5.5 Continued.

Chemical	Dose	Species	Pathology	References
β-Naphthoflavone	50 mg kg⁻¹ diet for 10 days	*Ictalurus punctatus*	Erosion of epithelial surface Atrophy of submucosa Separation of submucosa Abundance of lymphocytes	Kleinow *et al.* (1998; unpublished)
Petroleum hydrocarbons (mixture of 2,3-benzothiophene, 2-methylnaphthalene, 2,6-dimethyl-naphthalene, 2,3,6-trimethyl-naphthalene, fluorene, phenanthrene, 1-phenyldodecane, and hepta decylcyclohexane)	125 µg kg⁻¹ in the diet for 28 days	*Oncorhynchus tshawytscha*	EM and light Cellular inclusions in columnar cells; vesiculation of columnar cell cytoplasm	Hawkes *et al.* (1980)

Contaminant	Dose	Species	Effects	Reference
Halogenated hydrocarbons				
Chlorinated biphenyl (mixture of 2-chlorobiphenyl, 2,2'-dichloro-biphenyl, 2,5,2'-trichloro-biphenyl, and 2,5,2'-5'-tetrachlorobiphenyl)	125 μg kg^{-1} fish daily in the diet for 28 days	*Oncorhynchus tshawytscha*	Increased exfoliation of mucosal cells Brush border was reduced or absent in some areas Abnormal cytoplasmic inclusions in columnar cells (irregular shape, fibrillar consistency, electron-dense vesicles and inclusions with clear material)	Hawkes *et al.* (1980)
Petroleum hydrocarbons and chlorinated biphenyl	125 μg kg^{-1} fish for each daily in the diet for 28 days	*Oncorhynchus tshawytscha*	Increased exfoliation of mucosal cells, vesiculation near brush border inclusions, increased endoplasmic reticulum	Hawkes *et al.* (1980)
Heavy metals				
Cadmium	6.8 mg L^{-1}; 30-day exposure	*Heteropneustes fossilis*	Scattered degeneration of columnar epithelia of few villi, cells form syncytical mass Hyperactivity in mucus secretion Lumen filled with mucus	Sastry and Gupta (1979)
Cadmium chloride	75.5 mg L^{-1} for 3 days	*Notopterus notopterus* (Pallas)	Columnar epithelial cells disrupted Microvilli heavily damaged Accelerated mucus production	Ghosh and Chakrabarti (1992)
Cadmium	5 mg kg^{-1} day^{-1} in diet for 15 or 30 days	Rainbow trout	EM and light Increased mucous cell activity Disruption of intestinal brush border Increased renewal rate of absorptive cells	Crespo *et al.* (1986)

Table 5.5 Continued.

Chemical	Dose	Species	Pathology	References
Mercuric chloride	1.8 mg L⁻¹ and 0.3 mg L⁻¹ 96-h, 4-, 10-, 15-, 20- and 30-day exposure	*Channa punctatus*	Ruptured villi Disintegrated lateral cell boundaries Syncytial appearance of cells Hyperactivity of mucus-secreting goblet cells Large amount of mucus in intestinal lumen	Sastry and Gupta (1978a)
Mercuric chloride	16.7 ppb, water changed daily and compound added 7, 28, 63 and 90 days	*Channa punctatus*	7 days: damaged villi with burst tips and anastomoses, necrosis of submucosa 28 days: necrotic, mucosal and goblet cells, collapsed villi 63 days: mucosal folds highly necrotic and disarray in all layers	Banerjee and Bhattacharya (1995)
Lead nitrate	30 mg L⁻¹ for 30 days	*Mystus vittatus*	Erosion of epithelium Dissociation of epithelium from basement membrane Clumped nuclei Vacuolation and degeneration of submucosa Large amount of mucus secreted	Datta and Sinha (1989)
	30 mg L⁻¹ for 30 days	*Labeo rohita*	Necrosis of columnar epithelial cells Clumped nuclei Disruption of basement membrane Vacuolation of submucosa	Datta and Sinha (1989)

Others

| Ammonia as NH_4OH | Channa punctatus | 15.6 mg L^{-1} water changed daily and compound added; 7, 28, 63 and 90 days' exposure | 7 days: villi ruptured and fused, necrotic submucosa, muscles fused
28 days: mucosal cells disintegrated, indistinct blood capillaries, lesions in submucosa
63 days: collapsed villi merged tips, flattened appearance, destruction of all layers
90 days: villi fused, mucosal cells filled with secretory granules, extensive damage to mucosal folds | Banerjee and Bhattacharya (1995) |

and the physiologic/structural differences represented along the gastrointestinal tract provide regional localization of toxic effects transmurally in select layers as well as along the length of the GIT. For example, there is evidence of regional binding and retention of BaP in the posterior intestine of catfish (Kleinow *et al.*, 1996), select regional toxicity of cadmium and lead along the length of the trout intestine (Crespo *et al.*, 1986), and layer select toxicity with agents such as diazinon, ammonia, methyl parathion (gastrointestinal muscle) and diazinon, dimethoate, mercuric chloride, Elsan, lead nitrate and ammonia (submucosa) (Tables 5.3–5.5).

Pathologic changes which occur in the stomach, intestine and other regions of the GIT with toxicant exposure follow a very similar format. Apparent differences are often related to regional structural differences in terms of the cell types or layers involved. For the stomach, degeneration of columnar epithelial cells, elongated and degenerative glandular cells, vacuolation in one or more layers, and elevated mucus secretion are common consequences of toxicant exposure. The most common pathologic changes which occur in the intestine are vacuolation, ruptured necrotic and collapsed villi (folds), syncytical formation, excessive mucus secretion, and vesicle formation with subcellular inclusions (Figure 5.13). Degenerative changes in other structures of the GIT have received much less attention experimentally. However, in studies in which pyloric ceca were examined (Sastry and Gupta, 1978b; Sastry and Malik, 1979; Ghosh and Chakrabarti, 1992), the most common lesions were degenerated or ruptured villi (folds) and excessive mucus production. Ghosh and Chakrabarti (1992) noted degenerative changes in the pharynx and esophagus following exposure to cadmium chloride. Shrinkage of stratified epithelial cells, loss or reduction in microridges and excessive mucin production were noted for both tissues.

Damage to the columnar epithelial cells is a commonly encountered degenerative change in the GIT resulting from toxicant exposure. This layer appears to be at risk for a variety of reasons. Foremost, the GIT is exposed directly to toxicants which may have been accumulated to high concentrations by dietary foodstuffs/organisms and/or are concentrated by the digestive process and nutrient absorption. In addition, the GIT may act as a route of elimination via the bile, or by retrograde diffusive processes across the enterocytes. The net effect is that exposure of the enterocytes occurs not only in the absorptive phase but also during elimination. Such an effect can be amplified by the action of efflux transporters and reabsorptive processes via enterohepatic recirculation. The columnar epithelial cells are also subject to the ancillary chemical and mechanical stessors of the digestive process. Structurally, cells weakened by toxicants are at increased risk to the rigors of what is normally a relatively harsh environment. Columnar epithelial cells are also in dynamic flux during the cell proliferation and maturation cycle. This process, while continually renewing cells and providing a rapid regeneration of the epithelium, also provides a portal of attack for toxicants. Proliferating cells are vulnerable to select toxicants. The maturation process also provides an array of cells which in a biotransformational sense range from incompetent to competent depending on the stage in maturation and the induction or dietary history of the cells. This provides a toxicant with the opportunity to be inactivated or activated, dependent upon the chemical and its relation to biotransformation.

Another common response in the GIT with toxicant exposure are changes in mucus production. Most of the changes examined, and therefore noted, are linked with the increased quantity of mucus elaborated. Increased quantities of mucus have been noted following exposure to a number of agents (lead, cadmium, Dimecron, Sumithion, DDT) (Sastry and Malik, 1979; Hawkes *et al.*, 1980; Mandal and Kulshrestha, 1980; Crespo *et al.*, 1986; Datta and Sinha, 1989). There are some indications that qualitative changes in the mucus composition may also occur with toxicant exposure (Hawkes *et al.*, 1980).

The GIT is unusual among organ systems as it is composed of many different layers and tissue types. Longitudinal and circular muscle layers are prominent features within the GIT, providing both defining muscle tone and gastrointestinal motility. Several studies have examined the effect of toxicants upon these muscle layers. *Channa punctatus* when exposed to the organophosphates diazinon, dimethoate and methyl parathion on an individual basis for 30 days resulted in varying degrees of muscular degeneration (Anees, 1976). These changes occurred while normal feeding behavior was maintained, suggesting that the changes evident upon exposure were not due to toxicant-induced starvation. Banerjee and Bhattacharya (1995) examined the thickness of intestinal muscle layers in *Channa punctatus* following exposure to Elsan® (an organophosphate) [211 parts per

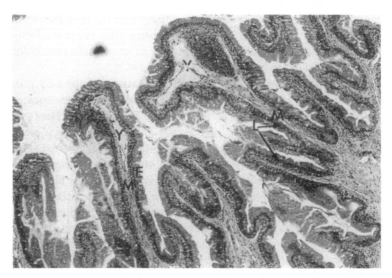

Figure 5.13 The proximal intestine of the channel catfish following exposure to b-naphthoflavone in the diet (50 mg kg⁻¹ diet for 10 days). Note erosion of the epithelial surface (E) on the mucosal fold; atrophy of submucosa (M); abundance of lymphocytes (L); and separation of submucosa (Y) (63×) (see Figure 5.4 for control comparisons; K.M. Kleinow, unpublished).

billion (ppb)], mercuric chloride (16.7 ppb) or ammonia [15.6 parts per million (ppm)] for 7, 28, 63, or 90 days. For each compound, there were trends towards increasing thickness in the circular muscles, whereas the longitudinal muscle layers were either decreased or at near normal levels for most times and treatments.

A variety of ultrastructural changes to the GIT of fish have been noted upon toxicant exposure. As would be expected, PAHs such as BaP are capable of inducing qualitative and quantitative changes in the endoplasmic reticulum. Sea bass exposed to BaP in the diet demonstrated hypertrophy of rough and smooth endoplasmic reticulum (ER) with corresponding dilation of cisternae and division into vesicles (Lemaire *et al.*, 1992). Hawkes *et al.* (1980) also demonstrated proliferation of granular ER with petroleum-exposed salmon. Chlorinated biphenyl exposure on the other hand offered granular ER which were well defined and normal. For a combined petroleum and PCB exposure, increases in ER were evident, as were changes in ER location. In this case, agranular ER was distributed near the luminal surface and extended deeply into the cell.

Subcellular vesicles or inclusions considered as non-living accumulations of metabolites and cell products (Leeson *et al.*, 1985) are prominent features of toxicant-induced ultrastructural change in the fish GIT. Inclusions may be highly variable in size, shape, consistency, and location. Studies by Hawkes *et al.* (1980) which exposed juvenile chinook salmon to chlorinated biphenyls in the diet elicited several types of inclusions: small vesicles containing a clear material of intermediate electron density; variable-sized irregular shaped vesicles with electron-transparent fibrillar material; and small membrane-bound vesicles with electron-dense material. In PAH-exposed salmon, membrane-bound vesicles of varying size and electron density containing a finely granular material were evident in the luminal half of columnar cells of the intestine (Hawkes *et al.*, 1980). BaP and Cd were also shown to stimulate proliferation of inclusions containing flocculent material in sea bass (Fair and Fortner, 1987).

Nuclear changes have also been reported in tissues of the GIT following exposure to a widely varied group of toxicants including organophosphates, BaP, and the heavy metals cadmium, lead and mercury. The changes evident vary with compound, dose, and fish species. Reduction in nuclear volume is a common finding following toxicant exposure. Sastry and Malik (1979) reported a reduction in nuclear volume in mucosal cells of the stomach with exposure to the organophosphate Dimecron®. Similarly, intestinal nuclei were also reduced in volume with exposure of *Heteropneustes fossilis* to 6.8 mg L^{-1} cadmium chloride for 30 days (Sastry and Gupta, 1979). The chronic exposure to mercuric chloride at 0.30 mg L^{-1} for 30 days resulted in pyknotic nuclei in gastric gland cells of the stomach. In the pyloric ceca and intestine, nuclei were irregularly scattered on initial observation, while becoming pyknotic upon longer exposure (Sastry and Gupta, 1978b). Cadmium chloride has also been shown to elicit a similar nuclear response (Sastry and Gupta, 1979). *Mystut vittatus* exposed to lead nitrite exhibited the formation of clumped nuclei within the intestine (Datta and Sinha, 1989), whereas a similar exposure of *Labeo rohitia* to lead resulted in necrosis of columnar epithelial cells of the intestinal bulb, intestine, and rectum. This toxic action appeared to facilitate a migration of nuclei towards the basement membrane. Sea bass exposed to 20 mg BaP kg^{-1} body weight in the diet presented with margination of the nucleoli and a less electron-dense nuclear chromatin (Lemaire *et al.*, 1992).

Neoplasia

Cursory examination of the literature would suggest that chemically induced neoplasia is a relatively uncommon event in the teleost fish gastrointestinal tract compared with other organs such as the liver. This assumption may or may not be true as relatively few studies have focused upon the gastrointestinal tract as an organ of more than passing interest. It is clear, however, that neoplasia does occur in the gastrointestinal tract of fish, most often as a result of an unknown etiology but also as an apparent result of exposure to xenobiotics. In the environment, evidence of gastrointestinal neoplasia has been associated with final oxidation waste water treatment (Grizzle, 1983; Grizzle *et al.*, 1984), contaminated sediments from Long Island Sound (Gardner *et al.*, 1987, 1988), and polluted areas of the Buffalo and East Niagara Rivers as well as other sites (Harshbarger and Clark, 1990). Experimental studies with known carcinogens have also resulted in gastrointestinal neoplasia in fish. Esophageal neoplasms have been observed in guppies and/or zebrafish following exposure to nitrosomorpholine (NM) (Khudoley, 1984), *N*-Nitrosodimethylamine (DMN) (Khudoley, 1984), and *N*-nitrosodiethylamine (DEN) (Khudoley, 1984; Simon and Lapis, 1984). Stomach neoplasms have been induced with *N*-methyl-*N'*-nitro-*N*-nitrosoguanidine (MNNG) (Kimura *et al.*, 1984), while other gastrointestinal tract neoplasms have been demonstrated for NM in zebrafish (Khudoley, 1984), *N*,*N'*-dinitrosopiperazine (DNP) and DEN in the guppy (Khudoley, 1984; Simon and Lapis, 1984). Neoplastic lesions identified in the gastrointestinal tract from all causes include: oral papilloma, fibroma, fibrosarcoma, adenocarcinoma; pharyngeal adenoma, adenocarcinoma; esophageal fibroma; stomach polyps and fibroma; gut adenoma, rhabdomyoma, hemangioma; and anal epithelioma (reviewed by Mawdesley-Thomas, 1975).

Functional effects of toxicants

Digestive enzymes

The effect of toxicants upon GIT digestive enzyme activities has been examined under both *in vitro* and *in vivo* conditions. Among the enzymes examined are gastric, intestinal and cecal proteases, carbohydrases and lipases. Following toxicant exposure, enzymatic activities may go up, may go down or may stay the same within a specific interval or over time. The response elicited appears to be loosely correlated with the enzyme type and the duration of exposure. Compositely, the activity of many enzymes following toxicant exposure appears to be the sum total of the time-related toxicant damage to the enzyme or its biosynthesis and what appears to be a corrective response.

PROTEASES

Alteration of protease function upon toxicant exposure has been demonstrated for a variety of fish species and enzymes. In general, these studies, largely with heavy metals, suggest that activities of proteases such as pepsin (secreted from the gastric glands of the stomach) and trypsin (secreted by the pancreas) are increased upon toxicant exposure. *Channa punctatus* exposed to lead nitrate (3.8 mg L^{-1} for 15 or 30 days) (Sastry and Gupta, 1978a), mercuric chloride (1.8 mg L^{-1} for 96 h) (Sastry and Gupta, 1978c), or (0.3 mg L^{-1} for 15 or 30 days) (Sastry and Gupta, 1978d,e) and cadmium chloride (6.8 mg L^{-1} for 30 days) (Sastry and Gupta, 1979) exhibited increases in pepsin activities ranging from 19.2 percent to 46 percent. Likewise, exposure of *Channa punctatus* to lead nitrite (3.8 mg L^{-1}) (Sastry and Gupta, 1978a), and mercuric chloride at 1.8 mg L^{-1} (Sastry and Gupta, 1978c) or 0.3 mg L^{-1} (Sastry and Gupta, 1978d,e) resulted in increases of intestinal trypsin activities of 72 percent, 79 percent and 161 percent respectively. Corresponding changes for cecal trypsin activities were 92 percent, 48 percent and 40 percent. The only available exception to this general trend was a 47 percent reduction in trypsin activity for *Heteropneustes fossilis* exposed to cadmium chloride at 6.8 mg L^{-1} over a 30-day exposure (Sastry and Gupta, 1979). A few of these studies demonstrated declines in protease activity with short duration exposure and most demonstrated increases with chronic exposure. It is plausible that the increases noted upon chronic exposure may have been related to a digestive response to local tissue damage or were the result of a compensatory repair.

In contrast to pepsin and trypsin, the activities of several di- and tripeptidases are negatively impacted following toxicant exposure. Uniformly, declines in aminotripeptidase activity were noted upon mercuric chloride (0.3 mg L^{-1} for 15 and 30 days) (Sastry and Gupta, 1980), cadmium chloride (6.8 mg L^{-1} for 30 days) (Sastry and Gupta, 1979) and lead nitrate (2.8 mg L^{-1} for 15 and 30 days or 6.8 mg L^{-1} for 96 h) exposures for both the intestine and ceca. Compositely, these declines ranged from 16 percent to 23 percent at 15 days of exposure and 31 percent to 55 percent at the 30-day interval. The decreases noted were nearly equivalent for the intestine and ceca following mercuric chloride exposure (Sastry and Gupta, 1980) and were somewhat greater for intestine than ceca for lead nitrate (Sastry and Gupta, 1979). When normalized with an equivalent duration of exposure (30 days) and tissue (intestine), lead nitrate (2.8 mg L^{-1}) demonstrated the largest decline at 55 percent, followed by cadmium chloride (6.8 mg L^{-1}) at 40 percent and mercuric chloride at 39 percent. If the relative dose and formula weights are considered, it appears that mercuric chloride affects aminotripeptidase to a greater degree than lead nitrate, which has a greater effect than cadmium chloride. These enzymes, unlike the gastric pepsin and pancreatic trypsin, are secreted from the intestinal wall and act in the intestine. It is plausible that the effects of these toxicants upon enzymatic activity may be related to the accompanying damage to the intestinal mucosa.

Other intestinal peptidases such as glycl-1-glycine dipeptidase, glycyl-1 leucine dipeptidase and leucyl-1-glycine dipeptidases in *Channa punctatus* and *Heteropneustis fossilis* exhibited declines in activity, most of which were significant, following exposure to mercuric chloride (Sastry and Gupta, 1980), cadmium chloride (Sastry and Gupta, 1979) and lead nitrate (Sastry and Gupta, 1979). Triglycine tripeptidase activity of *Channa punctatus* did not demonstrate any significant changes upon exposure to mercuric chloride at 1.8 mg L^{-1} for 96 h.

CARBOHYDRASES

The effects of toxicants upon enzymes involved in carbohydrate digestion appear to be considerably more variable than the proteases (Sastry and Gupta, 1978a,c,d,e, 1979, 1980). With the large number of varying parameters including toxicants and dose, it is difficult to correlate the results and specific trends between studies with the data available. Interpretation of the results is further complicated by changes associated with adaptive mechanisms of recovery in longer term exposures. The fish pancreatic enzyme amylase, which splits carbohydrates into double sugars, demonstrates this variability of response. *Channa punctatus*, when exposed to lead nitrate (3.8 mg L^{-1}) for 15 or 30 days, demonstrated significant increases in gastric, intestinal and cecal amylase activity at 15 days, and then significant decreases at 30 days (Sastry and Gupta, 1978a). In contrast, *Channa punctatus* exposed to mercuric chloride at 0.3 mg L^{-1} for 7, 15 and 30 days exhibited increases, declines and unaltered amylase activities at 7 and 15 days for stomach, intestine and pyloric ceca. By 30 days of mercuric chloride exposure, significant increases in amylase activities were noted for each of the tissues (Sastry and Gupta, 1978d). For *Channa punctatus* exposed to mercuric chloride at 1.8 mg L^{-1} for 96 h, there were significant increases in the stomach and intestinal amylase activity, while significant decreases were noted for pyloric ceca (Sastry and Gupta, 1978c).

Other carbohydrases such as maltase and lactase which split disaccharides into monosaccharides have also been examined (Sastry and Gupta, 1978a, 1979, 1980). Maltase in the stomach, intestine, and pyloric ceca following 15 days of exposure to lead nitrate (3.8 mg L^{-1}) (Sastry and Gupta, 1978a) or mercuric chloride (0.3 mg L^{-1}) (Sastry and Gupta, 1980) demonstrated elevated activities relative to controls, whereas by 30 days activities were decreased in all three organs (Sastry and Gupta, 1980). Exposure of *Heteropneustes fossilis* to cadmium chloride at 6.8 mg L^{-1} for 30 days resulted in maltase activities which were within 1.4–1.8 percent of controls (Sastry and Gupta, 1979). Interestingly, for these same exposures to lead nitrate and mercuric chloride, lactase activities dropped by the 15-day sampling and approached near control levels by 30 days (Sastry and Gupta, 1978a, 1980). Cadmium chloride on the other hand demonstrated a small decline in lactase activity (Sastry and Gupta, 1979).

LIPASES

Gastric, intestinal, and cecal lipase activities have been examined for *Channa punctatus* following exposure to mercuric chloride at 0.3 mg L^{-1} for 15 and 30 days (Sastry and Gupta, 1980) and lead nitrate at either 2.8 mg L^{-1} for 15 and 30 days or 6.8 mg L^{-1} for 96 h (Sastry and Gupta, 1979). Lipase activity was consistently lower for both compounds in all tissues, for all treatments and exposure times. In general, the longer 30-day exposures resulted in lower lipase activities than that observed at 15 days. Significant reductions of approximately 25 percent were evident for mercuric chloride at 30 days. Declines noted for lead nitrate ranged from 4 percent to 31 percent; however, these differences were not significant. Intestinal and cecal lipase following exposure of either compound suffered nearly equivalent reductions in activity; however, changes of a lesser magnitude were generally observed for the stomach.

ION TRANSPORT ENZYMES

ATPases   As with mammals, various ATPases have been shown to be affected by toxicants in the fish GIT. This important group of enzymes provides the ion transport essential for numerous processes including osmotic pressure regulation, cell volume regulation, ion extrusion mechanisms, membrane permeabilities and nutrient uptake (Skou *et al.*, 1987). Through the use of ATP and the action of these enzymes, cellular ion homeostasis is maintained. Generation of electrochemical gradients by ion transport also provide the driving force for many other processes in the GIT.

Distinct inhibitory alterations in the activity of GIT Na$^+$,K$^+$-ATPase and other ATPases have been noted with chromium VI (Thaker *et al.*, 1996), mercuric chloride (Lakshmi *et al.*, 1991), endosulfan (Sastry *et al.*, 1988), lead (Crespo *et al.*, 1986), and cadmium (Schoenmakers *et al.*, 1992). Studies with chromium mercuric chloride and endosulfan suggest that both increasing dose and duration result in greater inhibition of enzyme activity (Sastry *et al.*, 1988; Lakshmi *et al.*, 1991; Thaker *et al.*, 1996). Of the various ATPases, Na$^+$,K$^+$-ATPase or Ca^{2+}-ATPase activities were generally the most severely affected (Lakshmi *et al.*, 1991; Thaker *et al.*, 1996). Na$^+$,K$^+$-ATPase, Ca^{2+}-ATPase, Mg^{2+},HCO$_3$$^-$-ATPase, total ATPase, Ca^{2+},HCO$_3$$^-$-ATPase and Mg^{2+}-ATPase represent the order from greatest to least affected in multiple dose/multiple day static renewal studies exposing *Periophthalmus dipes* to 5, 10, or 15 mg L^{-1} of chromium VI (Thaker *et al.*, 1996). Select chromium concentrations resulted in initial stimulation of enzyme activities early in the time-course before exhibiting inhibitory effects. In *Boleophthalmus dentatus*, mercuric chloride at 1.50 mg L^{-1} for 3 days elicited reduction in intestinal ATPase function of approximately 50 percent for Na$^+$,K$^+$-ATPase, 45 percent for Ca^{2+},HCO$_3$$^-$-ATPase, 40 percent for Ca^{2+}-ATPase and 30 percent for each of Mg^{2+}-ATPase and Mg^{2+},HCO$_3$$^-$-ATPase (Lakshmi *et al.*, 1991).

In vitro studies with *Oreochromis mossambicus* basolateral plasma membrane preparations examined the effect of changing cadmium concentrations upon Na$^+$,K$^+$-ATPase, Ca^{2+}-ATPase and Na$^+$Ca^{2+} exchange activities (Schoenmakers *et al.*, 1992). In these studies, Ca^{2+}-ATPase was very sensitive to cadmium, with an IC$_{50}$ of 8.2 ± 3.0 pM cadmium. Schoenmakers *et al.* (1992) suggest that the inhibition occurred via the calcium-binding site. The sensitivity of Na$^+$,K$^+$-ATPase was somewhat less with an IC$_{50}$ of 2.6 ± 0.6 μM cadmium. Here, inhibition appears to occur via the Mg^{2+}-dependent site on the Na^+,K^+-ATPase. The Na^+Ca^{2+} exchanger, while the most important quantitatively for calcium transport, was only partially blocked by the binding of cadmium. It was suggested that cadmium acted as a competitive inhibitor of the exchanger via an interaction with the calcium site. This interaction exhibited an IC_{50} of 73 ± 11 nM.

Disruption of Na^+,K^+-ATPase activity by toxicants in the fish GIT has been linked to changes in intestinal ion fluxes. Crespo *et al*. (1986) observed that rainbow trout exposed to lead nitrate in the diet at 10 mg Pb kg^{-1} of fish day^{-1} demonstrated significant declines in middle intestine mucosal Na^+,K^+-ATPase activity from 45.9 nmol min^{-1} mg^{-1} protein to values of 29.7 and 29.3 nmol min^{-1} mg^{-1} protein at 15 and 30 days of exposure. Cadmium (5 mg Cd kg^{-1} fish in the diet) on the other hand resulted in small non-significant declines to values of 36.6, 32.1 and 39.6 nmol min^{-1} mg^{-1} protein at 10, 15 and 30 days of exposure.

Following lead exposure for 30 days, the trout middle intestine transepithelial sodium influx (J_{ms}) and outflux (J_{sm}) were significantly altered (Crespo *et al*., 1986). Sodium J_{ms} and J_{sm} dropped from 3.21 to 1.79 and from 1.71 to 0.91 μmol h^{-1} cm^{-2} respectively. The net Na^+ flux was significantly different at 15 days, but not for 30 days. Transepithelial chlorine fluxes in the middle intestine of trout were also significantly altered by lead. Chlorine J_{ms} declined from 3.32 to 1.86 at 30 days and J_{sm} decreased from 1.77 to 0.99 μmol h^{-1} cm^{-2}. Posterior intestine J_{ms} was significantly decreased by 30 days for both sodium and chlorine, whereas J_{sm} was unmodified. J_{net} was significantly lower as well. No significant alterations were noted in the posterior intestine following cadmium exposure. These studies indicated that cadmium and lead resulted in different effects upon Na^+K^+ activity and sodium chlorine fluxes. The authors attributed the decrease in Jms after lead treatment to declines in Na^+,K^+-ATPase activity in the middle intestine. J_{ms}/J_{sm} ratios changed with duration of treatment, indicating that J_{ms} and J_{sm} varied differently over time. Differences in effect of lead upon the middle and posterior intestine were attributed to differences between the epithelia lining the two segments. Measurements of epithelia resistivity indicated that the posterior intestine was much tighter than the middle intestine. Such findings were reflected by the large change in J_{net} in the posterior intestine compared with the middle intestine. These results suggest that regions differ with regard to their ionic flux response to toxicants and that toxicants may offer widely varying effects. In this case, Crespo *et al*. (1986) suggested that the lack of response by cadmium may be related to binding by metallothionein-like proteins (Noel-Lambot *et al*., 1978).

Electrophysiology

Lionetto *et al*. (1998) examined the effect of cadmium chloride on electro-physiologic parameters in the intestine of *Anguilla anguilla*. Chloride absorption expressed as a short circuit current (lsc, μA cm^{-2}) and the transepithelial potential difference responded in a concentration-dependent manner to cadmium chloride in the mucosal or serosal bathing solution. The maximal inhibition (90 percent) occurred at serosal and mucosal cadmium concentrations of 0.1 mmol L^{-1} and

5.0 mmol L^{-1} respectively. On a cellular level, cadmium chloride appears to block the Na$^+$,K$^+$-ATPase and also the Na$^+$K$^+$2Cl$^-$ co-transporter. These studies also resulted in a significant increase of transepithelial resistance from control values of 20.2 Ω cm^{-2} to a treatment resistance of 36.4 Ω cm^{-2}. This change in resistivity may reflect the effect of cadmium chloride upon paracellular pathways or tight junctions of the epithelium. Cadmium reduced dilution potentials by nearly half (V_t control 12.2 mV to 6.5 mV) and increased permeability ratios from controls (controls –0.09; cadmium chloride –0.38), indicating that indeed permeability of the tight junctions was modified by the metal. Sodium permeability was reduced, impairing the overall process of sodium chloride absorption. This process is critical as it is the main mechanism to recover water via the intestine to replace that which was osmotically loss in marine teleosts.

Nutrient absorption and transport

Alterations in nutrient absorption from the gastrointestinal tract have been demonstrated following exposure of *Channa punctatus* to the organochlorine endosulfan (Sastry and Siddiqui, 1982; Sastry *et al.*, 1988), the organophosphate quinalphos (Sastry and Siddiqui, 1982) and the heavy metal mercuric chloride (Sastry and Rao, 1983). For each of these agents, nutrient absorption was inhibited under both *in vitro* and *in vivo* conditions. The *in vitro* component of these studies demonstrated a strong dose–response-related inhibition. Endosulfan with dosages of 0.1–100 mg L^{-1} inhibited fructose transport by 1.5–15.5 percent, tryptophan transport by 2.8–20.1 percent and glucose transport by 5.8–13.4 percent (Sastry and Siddiqui, 1982; Sastry *et al.*, 1988). Toxicant exposure, ATPase activity and transport function of the GIT were linked in studies with endosulfan (Sastry *et al.*, 1988). *In vitro* intestinal sacs exposed to endosulfan concentrations ranging from 0.1 to 100 mg L^{-1} demonstrated decreases in intestinal brush border Na$^+$,K$^+$-ATPase activities of 3.5–77 percent. The corresponding fructose transport dropped 15.5 percent, while tryptophan transport dropped 20.1 percent with the greatest decline at the maximal 100 mg L^{-1} dose under analogous *in vitro* conditions. These studies indicated that endosulfan entering the fish intestine either through the water or food could reduce Na$^+$,K$^+$-ATPase activity and could result in a corresponding decrease in transport of nutrients. Although changes in *in vitro* Na$^+$,K$^+$-ATPase activity and fructose and tryptophan transport were not stoichiometric, there is correlative evidence elsewhere for their association. Sastry *et al.* (1988) attributed the inhibition of Na$^+$,K$^+$-ATPase as one of the factors responsible for the decline in nutrient transport. Other factors such as damage to the intestinal mucosa were suggested as additional reasons for changes in absorption. Similarly, quinalphos (0.1–100 mg L^{-1}) inhibited glucose transport from 2.1 percent to 14.6 percent (Sastry and Siddiqui, 1982). Mercuric chloride reduced glucose transport from control values of 43.9 μM glucose g^{-1} h^{-1} to 14.2 μM g^{-1} h^{-1} at 0.1 nM and 3.4 μM g^{-1} h^{-1} at 10 mM (Sastry and Rao, 1983). Fructose transport was reduced from 23.5 to 15.6 μM g^{-1} h^{-1} at 0.1 μM mercuric chloride and to 5.3 μM g^{-1} h^{-1} at 10 mM (Sastry and Rao, 1983). Maximally, these changes with mercuric chloride represent decreases of nearly 92 percent and 78 percent for glucose and fructose transport respectively.

The *in vivo* exposures of *Channa punctatus* to these agents also demonstrated significant reductions in nutrient transport. The declines noted were greater both with increased duration of exposure and at elevated concentrations of endosulfan and quinalphos (Sastry and Siddiqui, 1982; Sastry *et al.*, 1988). These trends did not extend to glucose or fructose transport following *in vivo* exposure to 3μg L^{-1} of mercuric chloride for 10 or 30 days. Although mercuric chloride significantly reduced transport at both the 10- and 30-day intervals, the transport evident was only slightly less for the 30-day exposure than for the 10-day exposure (Sastry and Rao, 1983). It is of interest to note that in fish (as for their mammalian counterparts) monosaccharides and amino acid transport across the intestine are energy-dependent processes tied to sodium ion transport. Thus, the effects as noted for transport may be related to the enzymes involved in providing energy to these transport systems.

Alkaline phosphatase

Alkaline phosphatase in the intestine is a brush border enzyme involved in transphosphorylation. The endogenous substrate in unknown, but the enzyme is capable of hydrolyzing many phosphate esters. This enzyme has been associated with nutrient transport across the intestine. Alkaline phosphatase activity demonstrates a variety of responses to toxicants in fish. Exposure to the organophosphate Dimecron® (0.32 mg L^{-1} for 20 days) or the heavy metal cadmium chloride (6.8 mg L^{-1} for 30 days) resulted in declines in gastric and intestinal alkaline phosphatase activity of *Channa punctatus* and *Heteropneustes fossilis* respectively (Sastry and Gupta, 1979; Sastry and Malik, 1979). The declines in alkaline phosphatase activities were approximately 9 percent for the stomach and 49.5 percent for the intestine following Dimecron® exposure. Decreases of 8.5 percent and 10.8 percent were evident for the stomach and intestine following cadmium chloride exposure. Mercuric chloride on the other hand elicited significant increases in intestinal and cecal alkaline phosphatase activities following 1.8 mg HgCl$_2$ L^{-1} for 96 h (Sastry and Gupta, 1978c) and 0.3 mg L^{-1} for 7, 15, and 30 days (Sastry and Gupta, 1978e). The corresponding stomach activities either declined or stayed the same. *Channa punctatus* exposed to lead nitrate at 3.8 mg L^{-1} for 15 or 30 days resulted in alkaline phosphatase activities which were near control levels or only slightly elevated for the stomach, intestine, and pyloric ceca (Sastry and Gupta, 1978a).

Radiation toxicity

Teleost fish are susceptible to radiation toxicity. Investigations with the goldfish gastrointestinal tract suggest that the fish is nearly as sensitive to radiation as mammals (Hyodo, 1964). Radiation-induced intestinal damage and survival time after irradiation are strongly temperature dependent (Hyodo, 1964, 1965a,b). This phenomenon appears to be related to differences in the time-course of radiation-induced destruction or cell death. Hyodo (1965a) indicated that goldfish held at 22°C and irradiated with 8 kRad of X-rays survived on average 8.6–10 days following exposure. At 4°C, all fish survived for more than 110 days; however, approximately 75 percent of fish died between 150 and 200 days after irradiation, with a mean survival time of 170 days. Gastrointestinal lesions, including flattened mucosal folds, increased size of epithelial cells, enlarged nuclei and diminished cell number, were nearly identical for the two temperature groups. Lesions in fish held at 22°C occurred at 8–10 h post-irradiation, whereas for those held at 4°C lesions did not start to appear until 100 days post-exposure. Congruent with these findings, goldfish held at 4°C did not exhibit differences in epithelial cell number by 100 days post-irradiation, whereas by 181 days cell number had markedly diminished. Compositely, these results indicated that the survival time was fifteen to twenty times longer at 4°C than at 22°C. Other studies with fish have also indicated that post-irradiation survival time was much longer if animals were kept at low temperatures (Gros and Bloch, 1957; Etoh and Egami, 1965). These authors correlated the timing of intestinal damage directly to radiation death, thus suggesting that survival times were a reflection of the time necessary to develop histologic lesions at each temperature. Furthermore, it has been suggested that the rate of lesion development was dependent upon rates of cellular metabolism. GIT autoradiographic studies with [^{3}H]-thymidine-treated goldfish indicate that post-irradiation temperature influences both the metabolic rate and generation time of intestinal epithelial cells. The rates of [^{3}H]-thymidine incorporation and the number of mitotic figures labeled at 3.6 and 24 h were either non-existent or much lower at 4°C than at 22°C (Hyodo, 1965b).

Subsequent studies have further defined the influence of radiation dose upon GIT toxicity and recovery (Hyodo-Taguchi, 1970). Fish exposed to 8 kRad X-rays followed by a dose of [^{3}H]-thymidine indicated that the number of labeled cells were similar between control and treated animals at 3 and 6 h. By 24 h at 25°C and 48 h at 15°C, degenerative cells were evident and the number of labeled cells approached zero. When fish were irradiated at a lower dose of 1 or 2 kRad, there was a temporary reduction in the number of labeled cells, followed by cell recovery. Following irradiation at 1, 2 and 8 kRad, intestinal epithelium DNA synthesis and mitotic activity were affected. The 1- and 2-kRad groups exhibited recovery after an initial fall in a similar temperature dependent manner.

In vivo *observations*

Few studies have described the effects of gastrointestinal tract toxicants on a whole organism basis. One such study by Woodward *et al.* (1995) related the deficit of early life stage recruitment of trout to arsenic, cadmium, copper, and lead concentrations in dietary benthic invertebrates. Early life stages of brown and rainbow trout when fed diets of benthic invertebrates from various contaminated sites exhibited a variety of clinical signs related to the GIT. More than 50 percent of fish receiving contaminated diets exhibit some degree of constipation. Brown trout exposed to diet from contaminant sites appear to be predisposed to swollen abdomens, resulting from a stomach and distal intestine excessively filled with food. Gut impaction, while not evident in rainbow trout, did occur in 3–9 percent of contaminated brown trout. Percentages were dependent on site and degree of contamination of dietary food items. Both brown trout and rainbow trout fed contaminated diet deposited fecal material in long ribbons. This was in contrast to reference animals whose fecal material was shorter and narrower in diameter. Other variants relative to digestive function were also evident. Brown trout exhibited vacuolation and sloughing of intestinal mucosal epithelial cells and exhibited depletion of zymogen granules (precursors of digestive enzymes) from the exocrine pancreas. Although fish survival did not appear to be altered by the diet, Woodward *et al.* (1995) point out that experimental animals were not subjected to the same pressures and hazards as in the natural environment. Constipation, impaction, and morphologic alteration have been reported with metals in other species (Scharding and Oehme, 1973).

Summary

It is clear that a variety of important processes take their origin in the GIT. Processes such as digestion, neurotransmission and osmoregulation are fairly well defined for teleost fish; however, we remain relatively uninformed regarding other topics. Likewise, there are numerous shortcomings in the understanding of the interaction of this organ with toxicants. In part, this is due to the lack of investigational attention and, in part, it is due to an incomplete appreciation for the important biochemistry and physiology.

Even with the foregoing limitations, it is evident that the GIT is a fertile crossroads for toxicant interactions. As a portal of access for contaminants, the GIT is often subject to high toxicant concentrations. An array of mechanisms which serve a protective role are also influential in determining chemical bioavailability and form. The status of cell turnover, mucus production, biotransformation, efflux transporter action, and tissue redundancy may have much to do with the toxicity observed. These protective mechanisms may not only define the local impact of dietary insult but, for select toxicants, may also define systemic toxicity. As a result, it is highly likely that features which influence these protective processes, such as diet, will also play a determinant role in the assessment of toxicant risk to the GIT.

A multitude of sites, modalities and conditions exist in the GIT for the interaction with xenobiotics. Even with the paucity of information currently available, it is apparent that toxicants are operative at many different levels in the GIT. Enzymatic activity, nutrient transport, ion transport, neural activity muscle activity, and structural integrity are known to be affected. Structural and functional differences represented along the GIT as well as between teleost species may preclude broad-sweeping generalizations regarding toxicant interactions with the GIT (bioavailability biotransformation, toxicity). Likewise, the unique features of the teleost GIT are likely to generate unique toxicologic responses beyond those generally predictable from mammalian systems.

Finally, in concert with the need for mechanistic understanding is the need to recognize the breadth of chemicals which may have effects upon the GIT as well as the scope of their effects. This is especially true for environmental chemicals at environmentally realistic dosages. Although many effects of GIT toxicants may present as subtle injury in the laboratory, these same effects may be less tolerated in the environment where dietary resource allocation may be less generous.

Acknowledgments

The authors' work as discussed in this chapter has been supported by US Public Health Service grants ES-05781 and P42 ES-07375 as well as by grants FD-R-000158 and FD-0-01466 from the US Food and Drug Administration.

References

Aganovic, M. and Vukovic, T. 1966. Odnos duzine crevnog trakta i duzine tijela kod tri lokalne populacije ostrulja (*Aulopige hugelii* Heck). *Ribarstvo Jugeslavije* 21: 8–11.

Al-Hussaini, A.H. 1949. On the functional morphology of the alimentary tract of some fish in relation to differences in their feeding habits: cytology and physiology. *Quarterly Journal of Microscopical Science* 90: 323–354.

Al-Hussaini, A.H. 1947. The feeding habits and the morphology of the alimentary tract of some teleosts living in the neighbourhood of the Marine Biological Station, Ghardaqa, Red Sea. *Publications of the Marine Biological Station, Ghardaga (Red Sea)* 5: 4–61.

Andersson, T., Pesonen, M. and Johansson, C. 1985. Differential induction of cytochrome P450 monooxygenase, expoxide hydrolase, glutathione transferase and UDP-glucuronosyl-transferase activities in the liver of rainbow trout by β-naphthoflavone or Clophen 150. *Biochemical Pharmacology* 34: 3309–3314.

Andrews, J.W., Murray M.W. and Davis, J.M. 1978. The influence of dietary fat levels and environmental temperature on digestible energy and absorbability of animal fat in catfish diets. *Journal of Nutrition* 108: 749–752.

Anees, M.A. 1976. Intestinal pathology in a fresh water teleost. *Channa punctatus* (Bloch) exposed to sub-lethal and chronic levels of three organophosphorous insecticides. *Acta Physiologica Latinoamericana* 26: 63–67.

Ankley, G.T., Blazer, V.S., Reinert, R.E. and Agosin, M. 1986. Effects of Aroclor 1254 on cytochrome P-450 dependent monooxygenase, glutathione S-transferase and UDP-glucuronosyltransferase activities in channel catfish liver. *Aquatic Toxicology* 9: 91–103.

Armstrong, R.H. and Blackett, R.F. 1966. Digestion rate of the Dolly Varden. *Transactions of the American Fisheries Society* 95: 429–430.

Ash, R. 1980. Hydrolytic capacity of the trout (*Salmo gairdneri*) intestinal mucosa with respect to three specific dipeptides. *Comparative Biochemistry and Physiology* 65B: 173–176.

Ash, R. 1985. Protein digestion and absorption. In *Nutrition and Feeding in Fish*. Cowey, C.B., Mackie, A.M. and Bell, J.G. (eds), pp. 69–93. Academic Press, London.

Aungst, B.J. and Fung, H.L. 1981. Kinetic characterization of in vitro lead transport across the rat small intestine. *Toxicology and Applied Pharmacology* 69: 39–47.

Axelsson, M. and Fritsche, R. 1991. Effects of exercise, hypoxia and feeding on the gastrointestinal blood flow in the Atlantic cod (*Gadus morhua*). *Journal of Experimental Biology* 158: 181–198.

Axelsson, M., Driedzic, W.R., Farrell, A.P. and Nilsson, S. 1989. Regulation of cardiac output and gut blood flow in the sea raven, *Hemitripterus americanus*. *Fish Physiology and Biochemistry* 6: 315–326.

Babin, P.J. and Vernier, J.M. 1989. Plasma lipoproteins in fish. *Journal of Lipid Research* 30: 467–489.

Bakke, J.E., Feil, V.J. and Mulford, D.J. 1990. Biliary excretion and intestinal metabolism in the intermediary metabolism of pentachlorothioanisole. *Xenobiotica* 20: 601–605.

Banerjee, S. and Bhattacharya, S. 1995. Histopathological changes induced by chronic nonlethal levels of elsan, mercury and ammonia in the small intestine of *Channa punctatus* (Bloch). *Ecotoxicology and Environmental Safety* 31: 62–68.

Barber, M.C., Suarez, L.A. and Lassiter, R.R. 1991. Modelling bioaccumulation of organic pollutants in fish with an application to PCBs in Lake Ontario salmonids. *Canadian Journal of Fisheries and Aquatic Science* 48: 318–337.

Barron, M.G., Tarr, B.D. and Hayton, W.L. 1987. Temperature-dependence of cardiac output and regional blood flow in rainbow trout, *Salmo gairdneri* Richardson. *Journal of Fish Biology* 31: 735–744.

Barron, M.G., Plakas, S.M. and Wilga, P.C. 1991. Chlorpyrifos pharmacokinetics and metabolism following intravascular and dietary administration in channel catfish. *Toxicology and Applied Pharmacology* 108: 474–482.

Bauer, E., Guo, Z., Ueng, Y.F., Bell, L.C. Zeldin, D. and Gunergerich, F.P. 1995. Oxidation of benzo(a)pyrene by recombinant human cytochrome P450 enzymes. *Chemistry Research and Toxicology* 8: 136–142.

Bauermeister, A.E.M. and Sargent, J.R. 1979. Biosynthesis of triacylglycerols in the intestines of rainbow trout (*Salmo gairdneri*) fed marine zooplankton rich in wax esters. *Biochimica et Biophysica Acta* 575: 358–364.

Bauermeister, A.E.M., Pirie, B.J.S. and Sargent, J.R. 1979. An electron microscopic study of lipid absorption in the pyloric caeca of rainbow trout (*Salmo gairdneri*) fed wax ester-rich zooplankton. *Cell Tissue Research* 200: 475–486.

Bauermeister, A., Lewendon, A., Ramage, P.I.N. and Nimmo, I.A. 1983. Distribution and some properties of the glutathione S-transferase and g-glutamyl transpeptidase activities of rainbow trout. *Comparative Biochemistry and Physiology* C74: 89–93.

Bengtsson, B., Svenberg, O., Linden, E., Lunde, G. and Ofstad, E.B. 1979. Structure related uptake of chlorinated paraffins in bleaks (*Alburnus alburnus*. L) *Ambio* 8: 121–122.

Benson, J.A., Culver, P.J., Ragland, S., Jones, C.M., Drummey, G.D. and Bougas, E. 1957. The D-xylose absorption test in malabsorption syndromes. *New England Journal of Medicine* 256: 335–339.

Björkland, H.V. and Bylund, G. 1991. Comparative pharmacokinetics and bioavailability of oxolinic acid and oxytetracycline in rainbow trout (*Onchorhynchus mykiss*). *Xenobiotica* 21: 1511–1520.

Blake, I.H. 1930. Studies on the comparative histology of the digestive tube of certain teleost fishes. I. A predaceous fish, the sea bass *Centropristes striatus*. *Journal of Morphology* 50: 39–70.

Bloedow, D.C. and Hayton, W.L. 1976. Effects of lipids on bioavailability of sulfisoxazole acetyl, dicumarol and griseofulvin in rats. *The Journal of Pharmaceutical Sciences* 65: 328–334.

Bock, K.W. 1994. UDP-glucuronosyltransferases and their role in metabolism and disposition of carcinogens. In *Conjugation-dependent Carcinogenicity and Toxicity of Foreign Compounds. Advances in Pharmacology*, Vol. 27. Anders, M.W. and Dekant, W. (eds), pp. 367–383. Academic Press, San Diego.

Boge, G., Rigal, A. and Peres, G. 1981. A study of in vivo glycine absorption by fed and fasted rainbow trout (*Salmo gairdneri*). *Journal of Experimental Biology* 91: 285–295.

Boon, J.P., Everaarts, J.M., Hillebrand, M.T., Eggens, M.L., Pijnenburg, J. and Goksoyr, Å. 1992. Changes in levels of hepatic biotransformation enzymes and haemoglobin levels in female plaice (*Pleuronectes platessa*) after oral administration of a technical polychlorinated biphenyl mixture (Clophen A40). *Science of the Total Environment* 114: 113–133.

Bowen, S.H. 1981. Digestion and assimilation of periphytic detrital aggregate by *Tilapia mossambica*. *Transactions of the American Fisheries Society* 110: 239–245.

Bowen, S.H. 1982. Feeding, digestion and growth-qualitative considerations. In *The Biology and Culture of Tilapias*. Pullin, R.S.V. and Lowe-McConnell, R.H. (eds), p. 141. ICLARM Conf. Proc. 7, Manila: Int. Cent. Living Aquat. Resour. Manage.,

Bowser, P.R., Wooster, G.A., St Leger, J. and Babish, J.G. 1992. Pharmacokinetics of enrofloxacin in fingerling rainbow trout (*Oncorhynchus mykiss*). *Journal of Veterinary Pharmacology and Therapeutics* 15: 62–71.

Brasitus, T.A. and Schachter, D. 1980. Lipid dynamics and lipid protein interactions in rat enterocyte basolateral and microvillus membranes. *Biochemistry* 19: 2763–2769.

Braybrooks, M.P., Barry, B.W. and Abbs, E.T. 1975. The effect of mucin on the bioavailability of tetracycline from the gastrointestinal tract: in vitro correlations. *Journal of Pharmacy and Pharmacology* 27: 508–515.

Brett, J.R. and Higgs, D.A. 1970. Effects of temperature on the rate of gastric digestion in fingerling sockeye salmon (*Oncorhynchus nerka*). *Journal of the Fisheries Research Board of Canada* 27: 1767–1779.

Brodeur, R.D. 1984. Gastric evacuation rates for two foods in the black rockfish, *Sebastes melanops* Girard. *Journal of Fish Biology* 24: 287–298.

Broman, D., Naf, C., Rolff, C., Zebuhr, Y., Fry, B. and Hobbie, J. 1992. Using ratios of stable nitrogen isotopes to estimate bioaccumulation and flux of polychlorinated dibenzo-p-dioxins (PCDDS) and dibenzofurans (PCDFS) in 2 food chains from the Northern Baltic. *Environmental Science and Technology* 11: 331–345.

Brown, H.O., Levine, M. and Lipkin, M. 1963. Inhibition of intestinal epithelial cell renewal and migration induced by starvation. *American Journal of Physiology* 205: 868–872.

Bruggeman, W.A., Opperhuizen, A., Wijbenga, A. and Hutzinger, O. 1984. Bioaccumulation of super-lipophilic chemicals in fish. *Environmental Toxicology and Chemistry* 7: 173–189.

Bucke, D. 1971. The anatomy and histology of the alimentary tract of the carnivorous fish the pike *Esox lucius* L. *Journal of Fish Biology* 3: 421–431.

Buddington, R.K. 1983. Digestion and feeding of the white surgeon, *Acipenser transmontanus*. Ph.D. dissertation. University of California, Davis.

Buddington, R.K. and Diamond, J.M. 1987. Pyloric ceca of fish: a new absorptive organ. *American Journal of Physiology* 252: G65–G76.

Buddington, R.K., Chen, J. and Diamond, R. 1985. Abstract. In *Proceedings of the American Association for Advancement of Science Annual Meeting*, Los Angeles, 26–31 May.

Bullock, W.L. 1963. Intestinal histology of some salmonid fishes with particular reference to the histopathology of acanthocephalan infections. *Journal of Morphology* 112: 23–44.

Burnstock, G. 1959. The morphology of the gut of the brown trout (*Salmo trutta*). *Quarterly Journal of Microscopical Science* 100: 183–198.

Burt, R.K. and Thorgeirsson, S.S. 1988. Co-induction of MDR-1 multidrug resistance and cytochrome P-450 genes in rat liver by xenobiotics. *Journal of the National Cancer Institute* 80: 1383–1386.

Cabana, G. and Rasmussen, J.B. 1994. Modelling food chain structure and contaminant bioaccumulation using stable nitrogen isotopes. *Nature* 372: 255–257.

Campbell, C. 1975. Inhibitory vagal innervation of the stomach in fish. *Comparative Biochemistry and Physiology* 50C: 169–170.

Castro, N.M., Sasso, W.S. and Katchburian, E. 1961. A histological and histochemical study of the gizzard of the *Mugil* sp. *Acta Anatomica* 45: 155–163.

Chandy, M. 1956. On the oesophagus of the milk-fish *Chanos chanos* (Forskal). *Journal of the Zoological Society of India* 8: 79–84.

Chandy, M. and George, M.G. 1960. Further studies on the alimentary tract of the milk fish *Chanos* in relation to its food and feeding habits. *Proceedings of the National Institute of Sciences of India* B26: 126–134.

Chasseaud, L.F. 1976. Conjugation with glutathione and mercapturic acid excretion. In *Glutathione Metabolism and Function*. Aries, I.M. and Jacoby, W.B. (eds), pp. 77–114. Raven Press, New York.

Cheng, H. and Leblond, C.P. 1974a. Origin, differentiation and renewal of the four main epithelial cell types in the mouse small intestine. I. Columnar cells. *American Journal of Anatomy* 141: 461–480.

Cheng, H. and Leblond, C.P. 1974b. Origin, differentiation and renewal of the four main epithelial cell types in the mouse small intestine. V. Unitarian theory of the origin of the four epithelial cell types. *American Journal of Anatomy* 141: 537–562.

Chevrel, R. 1889. Systeme nerveu grand-sympathique des elasmobranchs et des poissons osseux. *Arch Zool Exp Gen* 5: 1–194.

Clark, K.E., Gobas, F.A.P.C. and Mackay, D. 1990. Model of organic chemical uptake and clearance by fish from food and water. *Environmental Science and Technology* 24: 1203–1213.

Clarke, D.J., Burchell, B. and George, S.G. 1992. Differential expression and induction of UDP-Glucuronosyltransferase isoforms in hepatic and extrahepatic tissues of a fish, *Pleuronectes platessa*: immunochemical and functional characterization. *Toxicology and Applied Pharmacology* 115: 130–136.

Clements, K.D. and Bellwood, D.R. 1988. A comparison of the feeding mechanisms of two herbivorous labroid fishes, the temperate *Odax pullus* and the tropical *Scarus rubroviolaceus*. *Australian Journal of Marine and Freshwater Research* 39: 87–107.

Collicutt, J.M. and Eales, J.G. 1974. Excretion and enterohepatic cycling of ^{125}I-L-thyroxine in channel catfish, *Ictalurus punctatus* Rafinesque. *General Comparative Endocrinology* 23: 390–402.

Collie, N.L. 1985. Intestinal nutrient transport and the effects of development, starvation, and seawater adaptation. *Journal of Comparative Physiology* B156: 163–174.

Collier, T.K. and Varanasi, U. 1991. Hepatic activities of xenobiotic metabolizing enzymes and biliary levels of xenobiotics in English sole *(Parophrys vetulus)* exposed to environmental contaminants. *Archives of Environmental Contamination and Toxicology* 20: 462–473.

Collier, T.K., Singh, S.V., Awasthi, Y.C. and Varanasi, U. 1992. Hepatic xenobiotic metabolizing enzymes in two species of benthic fish showing different prevalences of contaminant-associated liver neoplasms. *Toxicology and Applied Pharmacology* 113: 319–324.

Connolly, J.P. and Pedersen, C.J. 1988. A thermodynamic-based evaluation of organic chemical accumulation in aquatic organisms. *Environmental Science and Technology* 22: 99–103.

Cowey, C.B. and Sargent, J.R. 1977. Lipid nutrition in fish. *Comparative Biochemistry and Physiology* 57B: 269–273.

Cravedi, J.P., Choubert, G. and Delous, G. 1987. Digestibility of chloramphenicol, oxolinic acid and oxytetracycline in rainbow trout and influence of these antibiotics on lipid digestibility. *Aquaculture* 60: 133–141.

Crespo, S., Nonnotte, G., Colin, D.A., Leray, C., Nonnotte, L. and Aubree, A. 1986. Morphological and functional alterations induced in trout intestine by dietary cadmium and lead. *Journal of Fish Biology* 28: 69–80.

Dabrowska, H., Fisher, S.W., Dobrowski, K. and Staubus, A.E. 1996. Dietary uptake efficiency of HCBP in channel catfish: The effect of fish contaminant body burden. *Environmental Toxicology and Chemistry* 15: 746–749.

Dallinger, R., Prosi, F., Segner, H. and Back, H. 1987. Contaminated food and uptake of heavy metals by fish. A review and a proposal for further research. *Oceologia* 73: 91–98.

Datta, D. and Sinha, G.M. 1989. Responses induced by long term toxic effects of heavy metals on fish tissues concerned with digestion, absorption, and excretion. *Gegenbaurs Morphological Jahrb Leipzig* 135: 627–657.

Dauble, D.D. and Curtis, L.R. 1989. Rapid branchial excretion of dietary quinoline by rainbow trout *(Salmo gairdneri)*. *Canadian Journal of Fisheries and Aquatic Science* 46: 705–713.

Dawes, B. 1929. The histology of the alimentary tract of the plaice *(Pleuronectes platessa)*. *Quarterly Journal of Microscopical Science* 73: 243–274.

Dekant, W., Vamvakas, S. and Anders, M.W. 1994. Formation and fate of nephrotoxic and cytotoxic glutathione S-conjugates: Cysteine conjugate β-lyase pathway. In *Conjugation-dependent Carcinogenicity and Toxicity of Foreign Compounds. Advances in Pharmacology*, Vol. 27. Anders, M.W. and Dekant, W. (eds), pp. 367–384. Academic Press, San Diego.

Doi, A.M., Lou, Z., Holmes, E., Li, C.-L.J., Venugopal, C. S., James, M.O. and Kleinow, K.M. 2000. Effect of micelle fatty acid composition and 3,4,3',4' tetrachlorobiphenyl (TCB) exposure on intestinal [^{14}C]-TCB bioavailability and biotransformation in channel catfish *in situ* preparations. Toxicological Sciences 55: 85–96.

Doi, A.M., Holmes, E. and Kleinow, K.M. 2001. P-glycoprotein in catfish intestine: inducibility by xenobiotics and functional properties. *Aquatic Toxicology* (in press).

Donald, J.A. 1997. Autonomic nervous system. In *Physiology of Fishes*, 2nd edn. Evans, D.H. (ed.), pp. 407–439. CRC Press, Boca Raton.

Droy, B.F., Goodrich, M.S., Lech, J.J. and Kleinow, K.M. 1990. Bioavailability, disposition and pharmacokinetics of ormetroprim in rainbow trout (*Salmo gairdneri*). *Xenobiotica* 20: 147–157.

Dutton, G.J. 1980. *Glucuronidation of Drugs and Other Compounds*. CRC Press, Boca Raton.

Elliott, J.M. 1972. Rates of gastric evacuation in brown trout (*Salmo trutta* L.). *Freshwater Biology* 2: 1–18.

Endo, T., Onozawa, M., Hamaguchi, M. and Kusuda, R. 1987. Enhanced bioavailability of oxolinic acid by ultra-fine size reduction in yellowtail. *Nippon Suisan Gakkaishi* 53: 1711–1716.

Etoh, H. and Egami, N. 1965. Effect of temperature on survival period of the fish, *Oryzias latipes*, following irradiation with different doses of X-rays. *Annotationes Zoologica Japonica* 38: 113–121.

Evans, D.H. 1979. Fish. In *Comparative Physiology of Osmoregulation in Animals*. Maloiy, G.M.O. (ed.), pp. 305–390. Academic Press, Orlando.

Evans, D.H. 1993. Osmotic and ionic regulation. In *The Physiology of Fishes*. Evans, D.H. (ed.), pp. 315–341. CRC Press, Boca Raton.

Fair, P.H. and Fortner, A.R. 1987. Effect of ingested benzo(a)pyrene and cadmium on tissue accumulation, hydroxylase activity and intestinal morphology of the Black Sea Bass, *Centropristis striata*. *Environmental Research* 42: 185–195.

Fauconneau, B., Choubert, G., Blanc, D., Breque, J. and Luquet, P. 1983. Influence of environmental temperature on flow rate of foodstuffs through the gastrointestinal tract of rainbow trout. *Aquaculture* 34: 27–39.

Ferraris, R.P. 1982. Glucose and alanine transport in herbivorous and carnivorous marine fish intestines. Ph.D. Thesis. Department of Zoology, University of Hawaii, Honolulu.

Ferraris, R.P. and Ahearn, G.A. 1983. Intestinal glucose transport in carnivorous and herbivorous marine fishes. *Journal of Comparative Physiology* 152: 79–90.

Field, M., Karnaky, Jr, K.J., Smith, P.L., Bolton, J.E. and Kinter, W.B. 1978. Ion transport across the isolated intestinal mucosa of the winter flounder *Pseudopleuronectes americanus*. I. Functional and structural properties of cellular and paracellular pathways for Na and Cl. *Journal of Membrane Biology* 41: 265–293.

Fish, G.R. 1960. The comparative activity of some digestive enzymes in the alimentary canal of Tilapia and Perch. *Hydrobiologia* 15: 161–178.

Fisher, D.J., Clark, J.R., Roberts, M.H., Connolly, J.P. and Mueller, L.H. 1986. Bioaccumulation of Kepone by spot (*Leiostomus xanthurus*): importance of dietary accumulation and ingestion rate. *Aquatic Toxicology* 9: 161–178.

Fisk, A.T., Cymbalisty, C.D., Bergman, A. and Muir, D.C.G. 1996. Dietary accumulation of C_{12} and C_{16}-chlorinated alkanes by juvenile rainbow trout (*Oncorhynchus mykiss*). *Environmental Toxicology and Chemistry* 15: 1775–1782.

Fojo, A.T., Ueda, K., Salmon, D.J., Poplack, D.G., Gottesman, M.M. and Pastan, I. 1987. Expression of multidrug-resistance gene in human tumors and tissues. *Proceedings of the National Academy of Sciences of the United States of America* 84: 265–269.

Foulkes, E.C. 1980. Some determinants of intestinal cadmium transport in the rat. *Journal of Environmental Pathology and Toxicology* 3: 471–481.

Frange, R. and Grove, D. 1979. Digestion. In *Fish Physiology*, Vol. 8. Hoar, W.S., Randall, D.J. and Brett, J.R. (eds), pp. 162–260. Academic Press, New York.

Gadagbui, B.K.M., James, M.O., Rowland-Faux, L. and Sikazwe, D.N. 1998. Distribution of glutathione S-transferase (GST)-pi-cross-reactive protein in the intestine of the channel catfish. *Toxicological Sciences* 42 (Suppl.): 274.

Galjaard, V., Van der Meer Fieggen, W. and DeBoth, N.J. 1972. Cell differentiation in gut epithelium. In *Cell Differentiation*. Harris, R., Allin, P. and Viza, D. (eds), pp. 322–328. Munksgaard, Copenhagen.

Gannon, J.E. 1976. The effect of differential digestion rates of zooplankton by Alewife, *Alosa pseudoharengus*, on determinations of selective feeding. *Transactions of the American Fisheries Society* 105: 89–95.

Garber, K.J. 1983. Effect of fish size, meal size and dietary moisture on gastric evacuation of pelleted diets by yellow perch, *Perca flavescens*. *Aquaculture* 34: 41–49.

Garcia, M.N. and Johnson, H.A. 1972. Cell proliferation kinetics in goldfish acclimated to various temperatures. *Cell Tissue Kinetics* 5: 331–339.

Gardner, G.R., Yevich, P.P., Malcolm, A.R. and Pruell, R.J. 1987. Carcinogenic effects of Black Rock Harbor sediment on American oysters and winter flounder. National Cancer Institute and US Environmental Protection Agency Collaborative Program on environmental cancer. Project report to the National Cancer Institute. US EPA, Environmental Research Laboratory, Narragansett, Rhode Island, ERLC Contribution No. 901.

Gardner, G.R., Yevich, P.P., Malcolm, A.R., Rogerson, P.F., Mills, L.J., Senecal, A.G., Lee, T.C., Harshbarger, J.C. and Cameron, T.P. 1988. Tumor development in American oysters and winter flounder exposed to a contaminated marine sediment under laboratory and field conditions. In *Aquatic Toxicology*, Vol. 11. Malins, D.C. and Jensen, A. (eds), pp. 403–404. Elsevier Science Publishers, B.V. Amsterdam.

Gas, N. and Noaillac-Depeyre, J. 1974. Renouvellement de l'epithelium intestinal de la Carpe (*Cyprinus carpio* L.). Influence de la saison. *C.r. hebd. Sèanc. Acad. Sci., Paris* 279D: 1085–1088.

Gauthier, G.F. and Landis, S.C. 1972. The relationship of ultrastructural and cytochemical features to absorptive activity in the goldfish intestine. *Anatomical Record* 172: 675–702.

George, S.G. 1994. Enzymology and molecular biology of phase II xenobiotic-conjugating enzymes in fish. In *Aquatic Toxicology: Molecular, Biochemical and Cellular Perspectives*. Malins, D.C. and Ostander, G.K. (eds), pp. 37–85. Lewis Publishers, CRC Press, Boca Raton.

Georgopoulou, U., Sire, M.F. and Vernier, J.M. 1985. Macromolecular absorption of proteins by epithelia cells of the posterior intestinal segment and their intracellular digestion in the rainbow trout. Ultrastructural and biochemical study. *Biology of the Cell* 53: 269–282.

Georgopoulou, U., Sire, M.F. and Vernier, J.M. 1986a. Absorption intestinale des proteines sous forme macromoleculaire et leur digestion chez la truite arc-en-ciel. Etude ultrastructurale et biochimique en relation avec la permiere prise de nourriture. *Canadian Journal of Zoology* 64: 1231–1240.

Georgopoulou, U., Sire, M.F. and Vernier, J.M. 1986b. Immunological demonstration of intestinal absorption and digestion of protein macromolecules in the trout (*Salmo gairdneri*). *Cell Tissue Research* 245: 387–395.

Georgopoulou, U., Dabrowski, K., Sire, M.F. and Vernier, J.M. 1988. Absorption of intact proteins by the intestinal epithelium of trout, *Salmo gairdneri*. A luminescence enzyme immunoassay and cytochemical study. *Cell Tissue Research* 251: 145–152.

Gerking, S.D. 1994. *Feeding Ecology of Fish*. Academic Press, San Diego.

Ghazzawi, F.M. 1935. The pharynx and intestinal tract of the Egyptian mullets *Mugil cephalus* and *Mugil capito*. Part II. On the morphology and histology of the alimentary canal in *Mugil capito* (Tobar). *Coastguards and Fisheries Service, Fisheries Research Directorate, Notes and Memoirs Cairo*. 6: 1–31.

Ghosh, A.R. and Chakrabarti, P. 1992. A scanning electron microscopic probe into the cellular injury in the alimentary canal of *Notopterus notopterus* (Pallas) after cadmium intoxication. *Ecotoxicology and Environmental Safety* 23: 147–160.

Gibbins, I.L. 1993. Comparative anatomy and evolution of the autonomic nervous system. In *Comparative Physiology and Evolution of the Autonomic Nervous System*. Nilsson, S. and Holmgren, S. (eds). Burnstock, G. (series ed.). Hardwood Academic, Chur.

Glatt, H., Seidel, A., Ribeiro, O., Kirkby, C., Hirom, P. and Oesch, F. 1987. Metabolic activation to a mutagen of 3-hydroxy-trans-7,8-dihydroxy-7,8-dihydrobenzo(a)pyrene, a secondary metabolite of benzo(a)pyrene. *Carcinogenesis* 8: 1621–1627.

Gobas, F.A.P.C., Muir, D.C.G. and Mackay, D. 1988. Dynamics of dietary bioaccumulation and fecal elimination of hydrophobic organic chemicals in fish. *Chemosphere* 17: 943–962.

Gobas, F.A.P.C., McCorquodale, J.R. and Haffner, G.D. 1993. Intestinal absorption and biomagnification of organochlorines. *Environmental Toxicology and Chemistry* 12: 567–576.

Godsil, H.C. 1954. A descriptive study of certain tuna-like fishes. *State of California Department of Fish and Game, Fisheries Bulletin* 97: 190.

Gordon, H.A. 1974. Intestinal water absorption in young and old, germ-free and conventional rats. *Experientia* 15: 214–215.

Green, C.W. 1913. The fat-absorbing function of the alimentary tract of the king salmon. *Bulletin of the United States Fisheries Bureau* 33: 149–175.

Grizzle, J.M. 1985. Black bullhead: An indicator of the presence of chemical carcinogens. In *Water Chlorination: Chemistry, Environmental Impact and Health Effects*, Vol. 5. Jolley, R.L., Condie, L.W., Johnson, J.D., Katz, S., Minear, R.A., Mattice, J.S. and Jacobs, V.A. (eds), pp. 451–462. Lewis Publishers, Chelsea, MI.

Grizzle, J.M. and Rogers, W.A. 1976. *Anatomy and Histology of the Channel Catfish*. Auburn University Agricultural Experiment Station, Auburn, AL.

Grizzle, J.M., Melius, P. and Strength, D.R. 1984. Papillomas on fish exposed to chlorinated wastewater effluent. *Journal of the National Cancer Institute* 73: 1133–1142.

Groman, D.B. 1982. Histology of the striped bass. Monograph no 3. American Fisheries Society Monograph, Bethesda.

Grondel, J.L., Nouws, J.F.M., DeJong, M., Schutte, A.R. and Driessens, F. 1987. Pharmacokinetics and tissue distribution of oxytetracycline in carp, *Cyprinus carpio* L., following different routes of administration. *Journal of Fish Diseases* 10: 153–163.

Groot, J.A. and Bakker, R. 1988. NaCl transport in the vertebrate intestine. In *Comparative and Environmental Physiology, NaCl Transport in Epithelia*. Gregor, R. (ed.), p. 103. Springer-Verlag, Berlin.

Gros, C.M. and Bloch, J. 1957. Influence de la temperature sur la survie d'un poisson *Carassius carassius* (L.) apres irradiation totale. *Compt Rend Soc Biol* 151: 602–604.

Grove, D.J. and Crawford, C.D. 1980. Correlation between digestion rate and feeding frequency in the stomachless teleost, *Blennius pholis* L. *Journal of Fish Biology* 16: 235–247.

Guarino, A.M., Plakas, S.M. Dickey, R.W. and Zeeman, M. 1988. Principles of drug absorption and recent studies of bioavailability in aquatic species. *Veterinary and Human Toxicology* 30 (Suppl. 1): 41–44.

Hansen, L.G., Wiekhorst, W.B. and Simon, J. 1976. Effects of dietary Aroclor 1242 on channel catfish (*Ictalurus punctatus*) and the selective accumulation of PCB components. *Journal of the Fisheries Research Board of Canada* 33: 1343–1352.

Harshbarger, J.C. and Clark, J.B. 1990. Epizootiology of neoplasms in bony fish of North America. *The Science of the Total Environment* 94: 1–32.

Hawkes, J.W., Gruger, Jr, E.H. and Olson, O.P. 1980. Effects of petroleum hydrocarbons and chlorinated biphenyls on the morphology of the intestine of chinook salmon (*Oncorhynchus tshawytscha*). *Environmental Research* 23: 149–161.

Hazel, J.R. 1984. Effects of temperature on the structure and metabolism of cell membranes in fish. *American Journal of Physiology*. 246: R460–R470.

Hemmer, M.J., Courtney, L.A. and Ortego, L.S. 1995. Immunohistochemical detection of P-glycoprotein in teleost tissues using mammalian polyclonal and monoclonal antibodies. *The Journal of Experimental Zoology* 272: 69–77.

Hickman, Jr, C.P. 1968. Ingestion, intestinal absorption and elimination of seawater and salts in the southern flounder, *Paralichthyes lethostigma. Canadian Journal of Zoology* 46: 457–466.

Hirsch, G.C. 1950. Magenlose Fische. Versuch einer typenmassigen Erklarung. In *Neue Ergebnisse and Probleme der Zoologie*, pp. 302–326. Akademische Veragsgesellschaft Geest & Portig, K.G., Leipzig.

Hollander, D. 1980. Retinol lymphatics and portal transport. Influence of pH, bile and fatty acids on transport. *American Journal of Physiology* 239: G210–G214.

Hollander, D. 1981. Intestinal absorption of vitamins, A.E.D. and K. *Journal of Laboratory and Clinical Medicine* 97: 449–462.

Hollander, D., Rim, E. and Ruple, P.E. 1977. Vitamin K-2 colonic and ideal in vivo absorption. Bile, fatty acids and pH effects on transport. *American Journal of Physiology* 233: E124–E129.

Holmgren, S. 1985. Substance P in the gastrointestinal tract of *Squalus acanthias. Molecular Physiology* 8: 119–130.

Holmgren, S. and Jonsson, A.-C. 1988. Occurrence and effects on motility of bombesin-related peptides in the gastrointestinal tract of the Atlantic cod, *Gadus morgua. Comparative Biochemistry and Physiology* 89C: 249–256.

Holmgren, S. and Nilsson, S. 1981. On the non-adrenergic, noncholinergic innervation of the rainbow trout stomach. *Comparative Biochemistry and Physiology* 70C: 65–69.

Holmgren, S. and Nilsson, S. 1983. Bombesin-, gastrin/CCK-, 5-hydoxytryptamine-, neurotensin-, somatostatin-, and VIP-like immunoreactivity and catecholamine fluorescence in the gut of the elasmobranch, *Squalus acanthias. Cell Tissue Research* 234: 595–618.

Honkanen, R.E., Rigler, M.W. and Patton, J.S. 1985. Dietary fat assimilation and bile salt absorption in the killifish intestine. *American Journal of Physiology* 249: G399–G407.

Hooper, C.S. and Blair, M. 1958. The effect of starvation on epithelial renewal in the rat duodenum. *Experimental Cell Research* 14: 175–181.

Horn, M.H. 1989. Biology of marine herbivorous fishes. *Oceanography Marine Biology Annual Review* 27: 167–272.

Horn, H.M. 1992. Herbivorous fishes: feeding and digestive mechanisms. In *Plant–Animal Interactions in the Marine Benthos*. John D.M., Hawkins, S.J. and Price, J.H. (eds), Ch. 15. Clarendon Press, Oxford.

Horsberg, T.E., Hoff, K.A. and Nordmo, R. 1996. Pharmacokinetics of florfenicol and its metabolite florfenicol amine in Atlantic salmon. *Journal of Aquatic Animal Health* 8: 292–301.

Houston, J.B. and Wood, S.G. 1980. Gastrointestinal absorption of drugs and other xenobiotics. In *Progress in Drug Metabolism*, Vol. 4. Bridges, J.W. and Chasseaud, L.F. (eds), pp. 57–103. John Wiley and Sons, Chichester.

Husoy, A.-M., Myers, M.S., Willis, M.L., Collier, T.K., Celander, M. and Goksoyr, A. 1994. Immunochemical localization of CYP1A and CYP3A-like isozymes in hepatic and extrahepatic tissues of Atlantic cod (*Gadus morhus* L.) a marine fish. *Toxicology and Applied Pharmacology* 129: 294–308.

Hustvedt, S.O., Salte, R., Kvendset, O. and Vassvik, V. 1991. Bioavailability of oxolinic acid in Atlantic salmon (*Salmo salar* L.) from medicated feed. *Aquaculture* 97: 305–310.

348   Kevin M. Kleinow and Margaret O. James

Hyodo, Y. 1964. Effect of X-irradiation on the intestinal epithelium of the goldfish, *Carassius auratus*. I. Histological changes in the intestine of irradiated fish. *Annotationes Zoologica Japonica* 37: 104–111.

Hyodo, Y. 1965a. Effect of X-irradiation on the intestinal epithelium of the goldfish, *Carassius auratus*. II. Influence of temperature on the development of histopathological changes in the intestine. *Radiation Research* 24: 133–141.

Hyodo, Y. 1965b. Development of intestinal damage after X-irradiation and ^{3}H-thymidine incorporation into intestinal epithelial cells of irradiated goldfish, *Carassius auratus*, at different temperatures. *Radiation Research* 26: 383–394.

Hyodo-Taguchi, Y. 1970. Effect of X-irradiation on DNA synthesis and cell proliferation in the intestinal epithelial cells of goldfish at different temperature with special reference to recovery process. *Radiation Research* 41: 568–578.

Iida, H. and Yamamoto, T. 1985. Intracellular transport of horseradish peroxidase in the absorptive cells of goldfish hindgut *in vitro*, with special references to the cytoplasmic tubules. *Cell Tissue Research* 240: 553–560.

Iida H., Shibata, Y. and Yamamoto, T. 1986. The endosome–lysosome system in the absorptive cells of goldfish hindgut. *Cell Tissue Research* 243: 449–452.

Iijima, N., Kayama, M., Okazaki, M. and Hara, I. 1985. Time course changes of lipid distribution in carp plasma lipoprotein after force-feeding with soybean oil. *Bulletin of the Japanese Society of Scientific Fisheries* 51: 467–471.

Iijima, N., Aida, S. and Kayama M. 1990a. Intestinal absorption and plasma transport of dietary fatty acids in carp. *Nippon Suisan Gakkaishi* 56: 1829–1837.

Iijima, N., Aida, A., Mankura, M. and Kayama, M. 1990b. Intestinal absorption and plasma transport of dietary triglyceride and phosphatidylcholine in the carp. *Comparative Biochemistry and Physiology* 96A: 45–55.

Isokawa, S., Kubota, K. and Kosakai, T. 1965. Some contributions to the study of esophageal sacs and teeth of fishes. *Journal of Nihon University School of Dentistry, Tokyo* 7: 103–111.

Iwai, T. 1962. Studies on the *Plecoglossus altivelis* problems: embryology and histophysiology of digestive and osmoregulatory organs. *Bulletin of the Misaki Marine Biological Institute, Kyoto University* 2: 1–101.

Jacobshagen, E. 1915. Ueber die appendices pyloricae, nebst bemerkungen zur anatomie und morphologie des rumpfdarmes. *Jena Zeitschrift Naturwissenschaften* 53: 445–556.

Jacobshagen, E. 1937. Darmsystem. IV. Mittel und Enddarm. Rumpfdarm. In *Handbuch der Vergleichenden Anatomie der Wirbeltiere*, Vol. III. Bolk, L., Goppert, E., Kallius, E. and Lubosch, W. (eds), pp. 563–724. Urban und Schwarzenberg, Berlin.

James, M.O. 1989. Conjugation and excretion of xenobiotics by fish and aquatic invertebrates. In *Xenobiotic Metabolism and Disposition*. Kato, R., Estabrook, R.W. and Cayen, M.N. (eds), pp. 283–290. Taylor and Francis, UK.

James, M.O. and Bend, J.R. 1980. Polycyclic aromatic hydrocarbon induction of cytochrome P-450 dependent mixed-function oxidases in marine fish. *Toxicology and Applied Pharmacology* 54: 117–133.

James, M.O. and Kleinow, K.M. 1994. Trophic transfer of chemicals in the aquatic environment. In *Aquatic Toxicology: Molecular, Biochemical, and Cellular Perspectives*. Malins, D.C. and Ostrander, G.K. (eds), pp. 1–35. Lewis Publishers, Boca Raton.

James. M.O., Bowen, E.R., Dansette, P.M. and Bend, J.R. 1979. Epoxide hyudrase and glutathione S-transferase activities with selected alkene and arene oxides in several marine species. *Chemico-Biological Interactions* 25: 321–344.

James, M.O., Heard, C.S. and Hawkins, W.G. 1988. Effect of 3-methylcholanthrene on monooxygenase activities, epoxide hydrolase and glutathione S-transferase activities in small estuarine and freshwater fish. *Aquatic Toxicology* 12: 1–15.

James M.O., Schell, J.D., Boyle, S.M., Altman, A.H. and Cromer, E.A. 1991. Southern flounder hepatic and intestinal metabolism and DNA binding of benzo(a)pyrene (BaP) metabolites following dietary administration of low doses of BaP, BaP-7,9-Dihydrodiol or a BaP metabolite mixture. *Chemico-Biological Interactions* 79: 305–321.

James, M.O., Kleinow, K.M., Tong, Z. and Venugopalan, C. 1996. Bioavailability and biotransformation of [^{3}H]-benzo(a)pyrene metabolites in *in situ* intestinal preparations of uninduced and BNF-induced channel catfish. *Marine Environmental Research* 42: 309–315.

James, M.O., Altman, A.H., Morris, K., Kleinow, K.M. and Tong, Z. 1997. Dietary modulation of phase 1 and phase 2 activities with benzo(a)pyrene and related compounds in intestine but not liver of the channel catfish, *Ictalurus punctatus*. *Drug Metabolism and Disposition* 25: 346–354.

James, M.O., Sikazwe, D.N. and Gadagbui, B.K.-M. 1998. Isolation of a pi class glutathione S-transferase from the intestinal mucosa of channel catfish, *Ictalurus punctatus*. *Marine Environmental Research* 46: 57–60.

Jarboe, H.H., Toth, B.R., Shoemaker, K., Greenlees, K.J. and Kleinow, K.M. 1993. Pharmacokinetics, bioavailability, plasma protein binding and disposition of nalidixic acid in rainbow trout (*Oncorhynchus mykiss*). *Xenobiotica* 23: 961–972.

Jensen, J. and Holmgren, S. 1994. The gastrointestinal canal. In *Comparative Physiology and Evolution of the Autonomic Nervous System*. Nilsson, S. and Holmgren, S. (eds), pp. 119–167. Burnstock, G. (series ed.). Hardwood Academic, London.

Jirge, S.K. 1970. Mucopolysaccharide histochemistry of the stomach of fishes with different food habits. *Folia Histochemistry Cytochemistry* 8: 275–280.

Jobling, M. 1995. Digestion and absorption. In *Environmental Biology of Fishes*, pp. 175–210. Chapman & Hall, New York.

Jobling, M., Gwyther, D. and Grove, D.J. 1977. Some effects of temperature, meal size and body weight on gastric evacuation in the dab (*Limanda limanda* [L.]. *Journal of Fish Biology* 10: 291–298.

Kampmeier, O.F. 1970. Lymphatic system of the bony fishes. In *Evolution and Comparative Morphology of the Lymphatic System*. pp. 232–265. C.C. Thomas, Springfield, IL.

Kapoor, B.G. 1958. The anatomy and histology of the digestive tract of a eyprinoid fish, *Catla catla* (Hamilton). *Annali del Musco Cirico di Storia Natruale di Genova* 70: 100–115.

Kapoor, B.G., Smit, H. and Verighira, I.A. 1975. The alimentary canal and digestion in teleosts. In *Advances in Marine Biology*, Vol. 13. Russell, F.S. and Yonge, C.M. (eds), pp. 102–219. Academic Press, London.

Karnaky, Jr, K.J. 1980. Ion secreting epithelia: chloride cells in the head region of *Fundulus heteroclitus*. *American Journal of Physiology* 238: R185–R198.

Kayama, M. and Iijima, N. 1976. Studies on lipid transport mechanism in the fish. *Bulletin of the Japanese Society of Scientific Fisheries* 42: 987–996.

Kellaway, I.W. and Marriot, C. 1975. The influence of mucin on the bioavailability of tetracycline. *Journal of Pharmacy and Pharmacology* 27: 281–283.

Khalilov, F.K.H., Inyushin, V.M. and Vorobjov, N.A. 1963. On morphology and histochemistry of the gut of fishes. *Izvestia Akademii Nauk Kazakhskoj SSR* 2: Seriya Biologicheskaya 82–89.

Khanna, S.S. and Mehrotra, B.K. 1970. Histomorphology of the buccopharynx in relation to feeding habits in teleosts. *Proceedings of the National Academy of Science, India* B40: 61–80.

Khudoley, V.V. 1984. Use of aquarium fish, *Danio rerio* and *Poecilia reticulata*, as test species for evaluation of nitrosamine carcinogenicity. *National Cancer Institute Monographs* 65: 65–70.

Kimura, I., Taniguchi, N., Kumai, H., Tomita, I., Kinae, N., Yoshizaki, K., Ito, M. and Ishikawa, T. 1984. Correlation of epizootiological observations with experimental data: chemical induction of chromatophoromas in the croaker, *Nibea mitsukurii*. *National Cancer Institute Monographs* 65: 139–154.

Kionka, B.C. and Windell, J.T. 1972. Differential movement of digestible and indigestible food fractions in rainbow trout, *Salmo gairdneri*. *Transactions of the American Fisheries Society* 101: 112–115.

Kirtisinghe, P. 1940. The myenteric nerve-plexus in some lower chordates. *Quarterly Journal of Microscopical Science* 81: 521–539.

Kiss, R., De Launoit, Y., Lenglet, G. and Danguy, A. 1988. Autoradiographic investigation on cell proliferation in the digestive mucosa of fishes. (Laboratoire de Biologie Animale et d'Histologie Comparee, Faculet des Sciences, and Laboratoire d'Histologie Faculte de Medecine, University Libre de Bruxelles, Bruxelles, Belgium.) *Acta Zoologica* (Stockholm) 69: 225–230.

Kitchell, J.F. and Windell, J.T. 1968. Rate of gastric digestion in pumpkinseed sunfish, *Lepomis gibbosus*. *Transactions of the American Fisheries Society*. 97: 489–492.

Kleinow, K.M. and Brooks, A.S. 1986. Selenium compounds in the fathead minnow (*Pimephales promelas*). II. Quantitative approach to gastrointestinal absorption, routes of elimination and influence of dietary pretreatment. *Comparative Biochemistry and Physiology* 83C: 71–76.

Kleinow, K.M., Cahill, J.M. and McElroy, A.E. 1989. Influence of preconsumptive metabolism upon the toxicokinetics and bioavailability of a model carcinogen in the flounder (Pseudopleuronectes americanus). *Mount Desert Island Biological Laboratory Bulletin* 28: 122–123.

Kleinow, K., James, M.O. and Lech, J.J. 1992a. Drug pharmacokinetics and metabolism in food producing fish and crustaceans: Methods and examples. In *Xenobiotics and Food-Producing Animals: Metabolism and Residues*. ACS Symposium Series No. 503. Hutson, D.H., Hawkins, D.R., Paulson, G.D. and Struble, C.B. (eds), pp. 98–130. American Chemical Society, Washington, DC.

Kleinow, K.M., Beilfuss, W.L., Droy, B.F., Jarboe, H.H. and Lech, J.J. 1992b. The pharmacokinetics, bioavailability, distribution and metabolism of sulfadimethoxine in the rainbow trout (*Oncorhynchus mykiss*). *Canadian J. Fisheries Aquatic Sci*. 49: 1070–1077.

Kleinow, K.M., Jarboe, H.H., Shoemaker, K. and Greenlees, K.J. 1994. Comparative pharmacokinetics and bioavailability of oxolinic acid in channel catfish (*Ictalurus punctatus*) and rainbow trout (*Oncorhynchus mykiss*). *Canadian Journal of Fisheries and Aquatic Science* 51: 1205–1211.

Kleinow, K.M., Smith, A.A., McElroy, A.E. and Wiles, J.E. 1996. Role of the mucous surface coat, dietary bulk and mucosal cell turnover in the intestinal disposition of benzo(a)pyrene (BaP). *Marine Environmental Research* 42: 65–73.

Kleinow, K.M., James, M.O., Tong, Z. and Venugopalan, C.S. 1998. Bioavailability and biotransformation of benzo(a)pyrene in an isolated perfused *in situ* catfish intestinal preparation. *Environmental Health Perspectives* 106: 155–166.

Kleinow, K.M., Doi, A.M., and Smith, A.A. 2000. Distribution and inducibility of P-glycoprotein in the catfish: immunohistochemical detection using the mammalian C-219 monoclonal. *Marine Environmental Research* 50: 311–317.

Kobayashi, K., Komatsu, S., Nishi, S., Hara, H. and Hayashi, K. 1991. ATP-dependent transport for glucuronides in canalicular plasma membrane vesicles. *Biochemical and Biophysical Research Communications* 176: 622–626.

Kolok, A.S., Spooner, R.M. and Farrell, A.P. 1993. The effects of exercise on the cardiac output and blood flow distribution of the largescale sucker, *Catostomus macrocheilus*. *Journal of Experimental Biology* 183: 301–321.

Krementz, A.B. and Chapman, G.B. 1975. Ultrastructure of the posterior half of the intestine of the channel catfish, *Ictalurus punctatus*. *Journal of Morphology* 145: 441–482.

Kuksis, A. 1984. Intestinal digestion and absorption of fat-soluble environmental agents. In *Intestinal Toxicology*. Schiller, C.M. (ed.), p. 69. Raven Press, New York.

Kurelec, B., Krca, S., Pivcevic, B., Ugarkovic, D., Bachmann, M., Imsiecke, G. and Muller, W.E. 1992. Expression of P-glycoprotein gene in marine sponges. Identification and characterization of the 12 kDa drug-binding glycoprotein. *Carcinogenesis* 13: 69–76.

Laher, J.M. and Barrowman, J.A. 1983. Polycyclic hydrocarbon and polychlorinated biphenyl solubilization in aqueous solutions of mixed micelles. *Lipids* 18: 216–222.

Laitinen, M., Neiminen, M. Pasanen, P. and Hietanen, E. 1981. Tricaine (MS-222) induced modification on the metabolism of foreign compounds in the liver and duodenal mucosal of the splake. *Acta Pharmacologica et Toxicologica* 49: 92–97.

Lakshmi, R., Kundu, R., Thomas, E. and Mansuri, A.P. 1991. Mercuric chloride-induced inhibition of different ATPases in the intestine of mudskipper, *Boleophthalmus dentatus*. *Ecotoxicology and Environmental Safety* 21: 18–24.

La Mont, J.T. 1985. Structure and function of gastrointestinal mucus. *Viewpoints Digestive Disease* 17: 1–4.

Lange, N.O. 1962. Development of the intestine in *Rutilus rutilus* (L.). *Voprosy Ikhtiologii* 2: 336–349.

Lauterbach, F. 1977. *Intestinal Permeation*. Kramer, M. and Lauterbach, F. (eds), pp. 173–195. Excerpta Medica, Amsterdam.

Layiwola, P.J. and Linnecar, D.F.C. 1981. The biotransformation of [^{14}C]phenol in some freshwater fish. *Xenobiotica* 11: 167–171.

Layiwola, P.J., Linnecar, D.F.C. and Knights, B. 1983. Hydrolysis of the biliary glucuronic acid conjugate of phenol by the intestinal mucus/flora of goldfish (*Carassius auratus*). *Xenobiotica* 13: 27–29.

Leaver, M.J., Clarke, D.J. and George, S.G. 1992. Molecular studies of the phase II xenobiotic conjugative enzymes of marine *Pleuronectid* flatfish. *Aquatic Toxicology* 22: 265–278.

Leaver, M.J., Scott, K. and George, S.G. 1993. Cloning and characterization of the major hepatic glutathione S-transferase from a marine teleost flatfish, the plaice (*Pleuronectes platessa*), with structural similarities to plant, insect and mammalian Theta class isoenzymes. *Biochemical Journal* 292: 189–195.

Leaver, M.J., Wright, J. and George, S.G. 1997. Structure and expression of a cluster of glutathione S-transferase genes from a marine fish, the plaice (*Pleuronectes platessa*). *Biochemical Journal* 321: 405–412.

Lee, S.J., Wang-Buhler, J.L., Cok, I., Yu, T.S., Yang, Y.H., Miranda, C.L., Lech, J. and Buhler, D.R. 1998. Cloning, sequencing and tissue expression of CYP3A27, a new member of the CYP3A subfamily from embryonic and adult rainbow trout livers. *Archives of Biochemistry and Biophysics* 360: 53–61.

Leeds, A.R. 1982. Modification of intestinal absorption by dietary fiber and fiber components. In *Dietary Fiber in Health and Disease*. Vahouny, G.V. and Kritchevsky, D. (eds), pp. 53–71. Plenum Press, New York.

Leeson, C.R., Lesson, T.S. and Paparo, A.A. 1985. *Textbook of Histology*, pp. 19, 48–50. Saunders, Philadelphia.

Leger, C. and Bauchard, D. 1972. Hydrolyse des triglycerides par le systeme lipasique du pancreas de truite (*Salmo gairdneri* Rich). Mise en evidence d'un nouveau type de specificite d'action. *C.R. Hebd. Seances Acad. Sci., Paris*. 270D: 2813–2816.

Lemaire, P., Berhaut, J., Lemaire-Gony, S. and LaFaurie, M. 1992. Ultrastructural changes induced by benzo(a)pyrene in sea bass (*Dicentrarchus labrax*) liver and intestine: Importance of the intoxication route. *Environmental Research* 57: 59–72.

Leu, B.L. and Huang, J.D. 1995. Inhibition of intestinal P-glycoprotein and effects on etoposide absorption. *Cancer Chemotherapy and Pharmacology* 35: 432–436.

Lie, O. and Lambertsen, G. 1985. Digestive lipolytic enzymes in cod (*Gadus morrhua*): fatty acid specificity. *Comparative Biochemistry and Physiology* 80B: 447–450.

Lie, O., Lied, E. and Lambertsen, G. 1987. Lipid digestion in cod (*Gadus morrhua*). *Comparative Biochemistry and Physiology* 88B: 697–700.

Lied, E. and Solbakken, R. 1984. The course of protein digestion in Atlantic cod (*Gadus morrhua*). *Comparative Biochemistry and Physiology* 77A: 503–506.

Liem, K.F. 1967. Functional morphology of the integumentary, respiratory and digestive systems of the synbranchoid fish *Monopterus albus*. *Copeia* 2: 375–388.

Lindstrom-Seppa, P., Koivusaari, U. and Hanninen, P. 1981. Extrahepatic xenobiotic metabolism in north-european freshwater fish. *Comparative Biochemistry and Physiology* 69C: 259–263.

Lindwall, G. and Boyer, T.D. 1987. Excretion of glutathione conjugates by primary cultured rat hepatocytes. *Journal of Biological Chemistry* 262: 5151–5158.

Lionetto, M.G., Vilella, S, Trischitta, F., Cappello, M.S., Giordano, M.E. and Schettino, T. 1998. Effects of CdCl$_2$ on electrophysiological parameters in the intestine of the teleost fish, *Anguilla anguilla*. *Aquatic Toxicology* 41: 251–264.

Lipkin, M. 1973. Proliferation and differentiation of gastrointestinal cells. *Physiological Reviews* 53: 891–915.

Little, P.J., James, M.O., Pritchard, J.B. and Bend, J.R. 1985. Temperature dependent disposition of ^{14}C-benzo(a)pyrene in the spiny lobster, *Panulirus argus*. *Toxicology and Applied Pharmacology* 77: 325–333.

Lobel, P.S. 1980. Herbivory by damselfishes and their role in coral reef community ecology. *Bulletin of Marine Science* 30: 273–289.

Lobel, P.S. 1981. Trophic biology of herbivorous reef fishes: alimentary pH and digestive capabilities. *Journal of Fish Biology*. 19: 365–397.

Lombardi, J. 1983. Structure, function and evolution of trophotaeniae: Placental analogues of viviparous fish embryos. Ph.D. dissertation. Clemson University, Clemson, SC.

Lombardi, J. and Wourms, J. 1978. SEM of trophotaeniae, embryonic adaptations in ophidioid, *Microbsptula randalli* and other viviparous fishes. *American Zoologist* 18: 607.

Lombardi, J. and Wourms, J. 1985. The trophotaenial placenta of a viviparous goodeid fish. II. Ultrastructure of trophotaeniae, the embryonic component. *Journal of Morphology* 184: 293–309.

Lombardo, P., Dennison, J.L. and Johnson, W.W. 1975. Bioaccumulation of chlorinated paraffin residues in fish fed Chlorowax 500C. *Journal of the Association of Official Analytical Chemistry* 58: 707–710.

Lopez, R.B. and De Carlo, J.M. 1959. Descripeion anatomica e histologica del aparato digestivo del puyen (*Galaxias attenuuatus*) (Teleostomi, Galaxiidae). *Actas y Trabajos del Primer Congreso Sudamericano de Zoologia. (La Plata)* 56: 8–23.

Loretz, C.A. 1995. Electrophysiology of ion transport in teleost intestinal cells. In *Cellular and Molecular Approaches to Fish Ionic Regulation*. Wood, C.M. and Shuttleworth, T.J. (eds), pp. 25–56. Academic Press, San Diego.

Lundin, K. and Holmgren, S. 1986. Non-adrenergic, non-chlorinergic innervation of the urinary bladder of the Atlantic cod, *Gadus morhua*. *Comparative Biochemistry and Physiology* 84C: 315–323.

McCloskey, J.T., Schultz, I.R. and Newman, M.C. 1998. Estimating the oral bioavailability of methylmercury to channel catfish (*Ictalurus punctatus*). *Environmental Toxicology and Chemistry* 17: 1524–2529.

Maccubbin, A.E., Black, P., Trzeciak, L. and Black, J.J. 1985. Evidence for polynuclear aromatic hydrocarbons in the diet of bottom-feeding fish. *Bulletin of Environmental Contamination and Toxicology* 34: 876–882.

McElroy, A.E. 1985. *In vivo* metabolism of benzo(a)anthracene by the polychaete *Nereis virens*. *Marine Environmental Research* 17: 133–136.

McElroy, A.E. and Kleinow, K.M. 1992. *In vitro* metabolism of benzo(a)pyrene and benzo(a)pyrene-7,8-dihydrodiol by liver and intestinal mucosa homogenates from the winter flounder (*Pseudopleuronectes americanus*). *Marine Environmental Research* 34: 279–285.

McElroy, A.E. and Sisson, J.D. 1989. Trophic transfer of benzo(a)pyrene metabolites between benthic marine organisms. *Marine Environmental Research* 28: 265–269.

MacKay, D. 1982. Correlation of bioconcentration factors. *Environmental Science and Technology* 16: 274–278.

McLean, E. and Donaldson, E.M. 1990. Absorption of bioactive proteins by the gastrointestinal tract of fish. A review. *Journal of Aquatic Animal Health* 2: 1–11.

McLeese, J.M. and Moon, T.W., 1989. Seasonal changes in the intestinal mucosa of winter flounder, *Pseudopeuronectes americanus* (Walbaum), from Passamaquoddy Bay, New Brunswick. *Journal of Fish Biology* 35: 381–393.

Maggese, M.C.I. 1967. Consideraciones anatomo-histogicas sobre el tract digestivo del papamoscas *Cheilodactylus bergi* (*Teleostomi, Cheilodactylidae*). *Physis. Buenos Aires.* 27: 111–124.

Magid, A.M.A. 1975. The epithelium of the gastrointestinal tract of *Polypterus senegalus* (Pisces: Brachiopterygii). *Journal of Morphology* 146: 447–456.

Mahadevan, S. 1954. The digestive system of *Mugil creniabis* (Forsk.) – a plankton feeder. *Journal of the Madras University* B24: 143–160.

Mandal, P.K. and Kulshrestha, A.K. 1980. Histopathological changes induced by the sublethal sumithion in *Clarias batrachus* (Linn). *Indian Journal of Experimental Biology* 18: 547–552.

Mankura, M., Iijima, N., Kayama, M. and Aida, S. 1987. Plasma transport form and metabolism of dietary fatty alcohol and wax ester in carp. *Nippon Suisan Gakkaishi* 53: 1221–1230.

Marcotte, G. and De La Noue, J. 1984. *In vitro* intestinal absorption of glycine and L-alanine by rainbow trout *Salmo gairdneri* Rich. *Comparative Biochemistry and Physiology* 79A: 209–213.

Marshall, T. and Allen, A. 1978. Isolation and characterization of the high molecular weight glycoproteins from pig colonic mucus. *Biochemical Journal* 173: 569–578.

Marshall, W.S. 1978. On the involvement of mucous secretion in teleost osmoregulation. *Canadian Journal of Zoology* 56: 1088–1091.

Martinsen, B., Sohlberg, S., Horsberg, T.E. and Burke, M. 1993. Single dose kinetic study of sarafloxacin after intravascular and oral administration to cannulated Atlantic salmon (*Salmo solar*) held in seawater at 12°C. *Aquaculture* 118: 49–52.

Mawdesley-Thomas, L.E. 1975. Neoplasia in fish. In *Pathology of Fishes*. Ribelin, W.E. and Migaki, G. (eds), pp. 805–870. University of Wisconsin Press, Madison, WI.

Mehrotra, B.K. and Khanna, S.S. 1969. Histomorphology of the oesophagus and the stomach in some Indian teleosts with inference on their adaptational features. *Zoologische Beitrage* 15: 375–391.

Michejda, C.J. and Kroeger Koepke, M.B. 1994. Carcinogen activation by sulfate conjugate formation. In *Conjugation-dependent Carcinogenicity and Toxicity of Foreign Compounds. Advances in Pharmacology*. Vol. 27. Anders, M.W. and Dekant, W. (eds), pp. 331–363. Academic Press, San Diego.

Mohsin, S.M. 1962. Comparative morphology and histology of the alimentary canal in certain groups of Indian teleosts. *Acta Zoologica* (Stockholm) 43: 79–133.

Moriarty, D.J.W. 1973. The physiology of digestion of blue-green algae in the cichlid fish (*Tilapia nilotica*). *Journal of Zoology (London)* 171: 25–39.

Mulder, G.J. 1992. Glucuronidation and its role in regulation of biological activity of drugs. *Annual Review of Pharmacology and Toxicology* 32: 25–49.

Musch, M.W., O'Grady, S.M. and Field, M. 1990. Ion transport of marine teleost intestine. *Methods of Enzymology* 192: 746–752.

Nagase, G. 1964. Contribution to the physiology of digestion in *Tilapia mossambica* Peters; digestive enzymes and the effect of diets on their activity. *Zeitschrift für Vergleichende Physiologie* 49: 270–284.

Nawar, G. 1955. On the anatomy of *Clarias lazera*. III. The vascular system. *Journal of Morphology* 97: 179–214.

Nelson, G.J. 1967. Epibranchial organs in lower teleostean fishes. *Journal of Zoology, London* 153: 71–89.

Niimi, A.J. and Oliver, B.G. 1988. Influence of molecular weight and molecular volume on dietary absorption efficiency of chemicals by fishes. *Canadian Journal of Fisheries and Aquatic Science* 45: 222–227.

Nilsson, S. 1976. Fluorescent histochemistry and cholinesterase staining of sympathetic ganglia in a teleost. *Gadus morhua. Acta Zoologica* (Stockholm) 57: 69–77.

Nilsson, S. 1983. *Autonomic Nerve Function in the Vertebrates*. Springer-Verlag, Berlin.

Nimmo, J. 1980. Drugs: absorption and action. In *Scientific Foundation of Gastroenterology*. Sircus, W. and Smith, A.N. (eds), pp. 141–147. William Heinemann Medical Books, London.

Noaillac-Depeyre, J. and Gas, N. 1973. Absorption of protein macromolecules by the enterocytes of the carp (*Cyprinus carpio* L.). Ultrastructural and cytochemical study. *Zeitschrift Zellforsch* 146: 525–541.

Noaillac-Depeyre, J. and Gas, N. 1974. Fat absorption by the enterocytes of the carp (*Cyprinus carpio* L.). *Cell Tissue Research* 155: 353–365.

Noaillac-Depeyre, J. and Gas, N. 1976. Electron microscopic study on gut epithelium of tench (*Tinca tinca* L.) with respect to its absorptive functions. *Tissue Cell* 8: 511–513.

Noaillac-Depeyre, J. and Gas, N. 1979. Structure and function of the intestinal epithelial cells in the perch (*Perca fluviatilis*). *Anatomical Record* 195: 621–640.

Noaillac-Depeyre, J. and Gas, N. 1983. Etude cytophysiologiqsue de l'epithelium intestinal du poisson-chat (*Ameiurus nebulosus* L.). *Canadian Journal of Zoology* 61: 2556–2563.

Noel-Lambot, F., Gerday, C. and Disteche, A. 1978. Distribution of Cd, Zn and Cu in the liver and gills of the eel *Anguilla anguilla* with special reference to metallothioneins. *Comparative Biochemistry and Physiology* 61C: 177–187.

Oliver, B.G. and Niimi, A.J. 1988. Trophodynamic analysis of polychlorinated biphenyl congeners and other chlorinated hydrocarbons in the Lake Ontario Ecosystem. *Environmental Science and Technology* 22: 388–397.

Olson, K.R. and Meisheri, K.D. 1989. Effects of atrial natriuretic factor on isolated arteries and perfused organs of the trout. *American Journal of Physiology* 256: R10–R18.

Opperhuizen, A. and Sijm, D.T.H.M. 1990. Bioaccumulation and biotransformation of polychlorinated dibenzo-p-dioxins and dibenzofurans in fish. *Environmental Toxicology and Chemistry* 9: 175–186.

Opuszynski, K. and Shireman, J.V. 1995. *Herbivorous Fishes: Culture and Use for Weed Management*. CRC Press, Boca Raton.

Pandian, T.G. 1967. Intake, digestion, absorption and conversion of food in the fishes (*Megalops cyprinoides* and *Ophiocephalus striatus*). *Marine Biology* 1: 16–32.

Patt, D.I. and Patt, G.R. 1969. *Comparative Vertebrate Histology*. Harper and Row, New York.

Patton, J.S. and Benson, A.A. 1975. A comparative study of wax ester digestion in fish. *Comparative Biochemistry and Physiology* 52B: 111–116.

Patton, J.S., Nevenzel, J.C. and Benson, A.A. 1975. Specificity of digestive lipases in hydrolysis of wax esters and triglycerides studies in anchovy and other selected fish. *Lipids*. 10: 575–583.

Payne, A.I. 1976. The relative abundance and feeding habits of the grey mullet species occurring in an estuary in Sierra Leone, West Africa. *Marine Biology* 35: 277–286.

Payne, A.I. 1978. Gut pH and digestive strategies in estuarine grey mullet (*Mugilidae*) and tilapia (*Cichlidae*). *Journal of Fish Biology* 13: 627–629.

Pelster, B. 1997. Buoyancy. In *The Physiology of Fishes*, 2nd edn. Evans, D.H. (ed.), pp. 25–42. CRC Press, Boca Raton.

Pentreath, R.J. 1976. Some further studies on the accumulation and retention of [65]Zn and [54]Mn by the plaice, *Pleuronectes platessa* L. *The Journal of Experimental Marine Biology and Ecology* 21: 179–189.

Pernkopf, E. and Lehner, J. 1937. Vergleichende beschreibung des vorderdarmes bei den einzelnen klassen der kranioten. In *Handbuch der Vergleichende Anatomie der Wirbeltiere*, Vol. 3. Bolk, L., Goppert, E., Kallius, E. and Lubosch, W. (eds), pp. 349–398. Urban Schwarzenberg, Berlin.

Persson, L. 1982. Rate of food evacuation in roach (*Rutilus rutilus*) in relation to temperature, and the application of evacuation rate estimates for studies on the rate of food consumption. *Freshwater Biology* 12: 203–210.

Petukat, S. 1965. Uber die Arteriellan gefaßstamme bei den Teleostiern. *Zoologische Beitrage* NF 11: 437–515.

Pizza, J.C. and O'Connor, J.M. 1983. PCB dynamics in Hudson River striped bass. II. Accumulation from dietary sources. *Aquatic Toxicology* 3: 313–327.

Plakas, S.M., McPhearson, R.M. and Guarino, A.M. 1988. Disposition and bioavailability of ^{3}H-tetracycline in the channel catfish (*Ictalurus punctatus*). *Xenobiotica* 18: 83–93.

Rannug, U., Sundvall, A. and Ramel, C. 1978. The mutagenic effect of 1,2-dichloroethane on *Salmonella typhimurium*. I. Activation through conjugation with glutathione *in vitro*. *Chemico-Biological Interactions* 20, 1–16.

Rasmussen, J.B., Rowan. D.J., Lean, D.R.S. and Carey, J.H. 1990. Food chain structure in Ontario lakes determines PCB levels in lake trout (*Salvelinus namaychus*) and other pelagic fish. *Canadian Journal of Fisheries and Aquatic Science* 47: 2030–2038.

Rendic, S. and Di Carlo, F.J. 1997. Human cytochrome P450 enzymes: a status report summarizing their reactions, substrates, inducers and inhibitors. *Drug Metabolism Reviews* 29: 413–580.

Rhead, M.M. and Perkins, J.M. 1984. An evaluation of the relative importance of food and water as sources of p,p'-DDT to goldfish, *Carassius auratus*. *Water Research* 18: 719–725.

Robertson, D.R. and Klumpp, D.W. 1983. Feeding habits of the southern Australian garfish *Hyporhamphus melanochir*, A diurnal herbivore and nocturnal carnivore. *Marine Ecology Progress Series* 10: 197–201.

Robinson, J.S. and Mead, J.F. 1973. Lipid absorption and deposition in rainbow trout (*Salmo gairdneri*). *Canadian Journal of Biochemistry* 51: 1050–1058.

Rombout, J.H.W.M. 1977. Enteroendocrine cells in the digestive tract of *Barbus conconius* (Cyprinidae). *Cell Tissue Research* 185: 435–450.

Rombout, J.H.W.M., Lamers, C.M.J., Helfrich, M.H., Dekker, A. and Taverne-Thiele, J.J. 1985. Uptake and transport of intact macromolecules in the intestinal epithelium of carp (*Cyprinus carpio* L.) and the possible immunological implications. *Cell Tissue Research* 239: 519–530.

Rozman, K. 1986. Fecal excretion of toxic substances. In *Gastrointestinal Toxicology*. Rozman, K. and Hanninen, O. (eds), pp. 119–145. Elsevier, New York.

Runge-Morris, M. 1998. Regulation of sulfotransferase gene expression by glucocorticoid hormones and xenobiotics in primary rat hepatocyte culture. *Chemico-Biological Interactions* 109: 315–327.

Sarbahi, D.W. 1939. The alimentary canal of *Labeo rohita* (Hamilton). *Journal of the Royal Asiatic Society of Bengal, Science* 5: 87–116.

Sargent, J.R. 1976. The structure, metabolism and function of lipids in marine organisms. *Biochemical and Biophysical Perspectives in Marine Biology* 3: 149–212.

Sastry, K.V. and Gupta, P.K. 1978a. Alterations in the activity of some digestive enzymes of *Channa punctatus*, exposed to lead nitrate. *Bulletin of Environmental Contamination and Toxicology* 19: 549–555.

Sastry, K.V. and Gupta, P.K. 1978b. Effect of mercuric chloride on the digestive system of *Channa punctatus*: A histopathological study. *Environmental Research* 16: 270–278.

Sastry, K.V. and Gupta, P.K. 1978c. Effect of mercuric chloride on the digestive system of a teleost fish, *Channa punctatus*. *Bulletin of Environmental Contamination and Toxicology* 20: 353–360.

Sastry, K.V. and Gupta, P.K. 1978d. Chronic mercuric chloride intoxication in digestive system of *Channa punctatus*. *Journal of Toxicology and Environmental Health*. 4: 777–783.

Sastry, K.V. and Gupta, P.K. 1978e. Chronic mercuric chloride intoxication in the digestive system of *Channa punctatus*. *Journal of Environmental Pathology and Toxicology* 2: 443–446.

Sastry, K.V. and Gupta, P.K. 1979. The effect of cadmium on the digestive system of the teleost fish, *Heteropneustes fossilis*. *Environmental Research* 19: 221–230.

Sastry, K.V. and Gupta, P.K. 1980. Changes in the activities of some digestive enzymes of *Channa punctatus*, exposed chronically to mercuric chloride. *Journal of Environmental Science Health* B15: 109–119.

Sastry, K.V. and Malik, P.V. 1979. Studies on the effect of dimecron on the digestive system of a fresh water fish, *Channa punctatus*. *Archives of Environmental Contamination and Toxicology* 8: 397–407.

Sastry, K.V. and Rao, D.R. 1983. Mercuric chloride induced alterations in the intestinal transport of glucose and fructose in the fresh water murrel, *Channa punctatus*. *Water, Air and Soil Pollution* 19: 143–147.

Sastry, K.V. and Siddiqui, A.A. 1982. Effect of endosulfan and quinalphos on intestinal absorption of glucose in the freshwater murrel, *Channa punctatus*. *Toxicology Letters* 12: 289–293.

Sastry, K.V., Siddiqui, A.A. and Samuel, M. 1988. Alterations in the intestinal absorption of fructose and tryptophan produced by endosulfan in the freshwater teleost fish, *Channa punctatus*. *Journal of Environmental Biology* 9: 295–301.

Satchell, G.H. 1991. *Physiology and Form of fish Circulation*, p. 1. Cambridge University Press, Cambridge.

Schade, S.G., Felsher, B.F., Glader, B.E. and Conrad, M.E. 1970. Effect of cobalt upon iron absorption. *Society of Experimental Biology and Medicine* 134: 741–743.

Schanker, L.S. and Jeffery, J.J. 1961. Active transport of foreign pyrimidines across the intestinal epithelium. *Nature* 190: 727–728.

Schanker, L.S., Tocco, D.J., Brodie, B.B. and Hogben, A.M. 1958. Absorption of drugs from the rat small intestine. *Journal of Pharmacology and Experimental Therapeutics* 123: 81–88.

Scharding, N.N. and Oehme, F.W. 1973. The use of animal models for comparative studies of lead poisoning. *Clinical Toxicology* 6: 419–424.

Schindler, J.F. and De Vries, U. 1987a. Maternal embryonic relationships in the goodeid teleost *Xenoophorus captivus*. Embryonic structural adaptations of viviparity. *Cell Tissue Research* 247: 325–338.

Schindler, J.F. and DeVries, U. 1987b. Protein uptake and transport by trophotaenial cells in two species of goodeid embryos. *Journal of Experimental Zoology* 241: 17–29.

Schmitz, E.H. and Baker, C.D. 1969. Digestive anatomy of the gizzard shad, *Dorsoma cepedianum*, and the threadfin shad, *D. petenense*. *Transactions of the American Microscopical Society* 88: 525–546.

Schoenmakers, T.J.M., Klaren, P.H.M., Flik, G., Lock, R.A.C., Pang, P.K.T. and Wendelaar Bonga, S.E. 1992. Actions of cadmium on basolateral plasma membrane proteins involved in calcium uptake by fish intestine. *Journal of Membrane Biology* 127: 161–172.

Seeto, G.S., Veivers, P.C., Clements, K.D. and Slaytor, M. 1996. Carbohydrate utilisation by microbial symbionts in the marine herbivorous fishes *Odax cyanomelas* and *Crinodus lophodon*. *Journal Comparative Physiology B* 165: 571–579.

Shears, M.A. and Fletcher, G.L. 1983. Regulation of Zn^{2+} uptake from the gastrointestinal tract of a marine teleost, the winter flounder, *Pseudopleuronictes americanus*. *Canadian Journal of Fisheries and Aquatic Science* 40: 197–205.

Shephard, K.L. 1982. The influence of mucus on the diffusion of ions across the esophagus of fish. *Physiological Zoology* 55: 23–34.

Shrable, J.B., Tiemeier, O.W. and Deyoe, L.W. 1969. Effects of temperature on the rate of digestion by channel catfish. *Progressive Fish Culturist* 31: 131–138.

Shuljak, G.S. 1968. Morpho-histological studies on the alimentary canal of the gut of silver carp. *Gidrobiologicheskii Zhurnal* 4: 30–38.

Sibly, R.M. 1981. Strategies of digestion and defecation. In *Physiological Ecology, An Evolutionary Approach to Resource Use*. Townsend, C.R. and Calow, P. (eds), pp. 109–139. Sinauer Associates, Sunderland, MA.

Simon, K. and Lapis, K. 1984. Carcinogenesis studies on guppies. *National Cancer Institute Monographs* 65: 71–81.

Sims, P., Grover, P.L., Swaisland, A., Pal, K. and Hewer, A. 1974. Metabolic activation of benzo(a)pyrene proceeds by a diol-epoxide. *Nature* 252: 326–328.

Singh, G.P. 1960. The blood vascular system of some fresh water teleosts. *Indian Journal of Zoology* 1: 27–58.

Singh, R. 1966. Morpho-histological studies on the alimentary canal of *Bagarius bagarius* Ham. *Agra University Journal of Research* 16: 69–82.

Sire, M.F. and Vernier, J.M. 1992. Intestinal absorption of protein in teleost fish. *Comparative Biochemistry and Physiology* 103A: 771–781.

Sire, M.F., Lutton, C. and Vernier, J.M. 1981. New views on intestinal absorption of lipids in teleostean fishes; and ultrastructure and biochemical study in the rainbow trout. *Journal of Lipid Research* 22: 81–93.

Skou, J.C., Norby, J.G., Maunsbach, A.B. and Esmann, M. (eds). 1987. *The Na$^+$,K$^+$pump. Part A. Molecular Aspects.* Alan R. Liss, New York.

Slaga, T.J., Thompson, S., Berry, D.L., DiGiovanni, J., Juchau, M.R. and Viahe, A. 1977. The effects of benzoflavones on polycyclic hydrocarbon metabolism and skin tumor initiation. *Chemico-Biological Interactions* 17: 297–312.

Smith, L.S. 1989. Digestive functions in teleost fishes. In *Fish Nutrition*, 2nd edn. Halver, J.E. (ed.), pp. 331–421. Academic Press, San Diego.

Smith, L.S. and Bell, G.R. 1975. A practical guide to the anatomy and physiology of the Pacific salmon. *Canadian Fisheries and Marine Service Miscellaneous Publication* 27: 14pp.

Smith, T.B., Wahl, D.H. and Mackie, R.I. 1996. Volatile fatty acids and anaerobic fermentation in temperate piscivorous and omnivorous fish. *Journal of Fish Biology* 48: 829–841.

Smithson, K.W., Millar, D.B., Jacobs, L.R. and Gray, G.M. 1981. Intestinal diffusion barrier: Unstirred water layer or membrane surface mucous coat. *Science* 214: 1241–1243.

Solomon, D.J. and Brafield, A.G. 1972. The energetics of feeding, metabolism and growth of perch (*Perca fluviatilis* L.). *Journal of Animal Ecology* 41: 699–718.

Spannhof, L. and Plantikow, H. 1983. Studies on carbohydrate digestion in rainbow trout. *Aquaculture* 30: 95–108.

Spigarelli, S.A., Thommes, M.M. and Prepejchal, W. 1983. Thermal and metabolic factors affecting PCB uptake by adult brown trout. *Environmental Science and Technology* 17: 88–94.

Srivastava, P.N. 1958. Primitive features of the alimentary canal of *Gadusia chapra* (Hamilton). *Current Science* 27: 144–145.

Statham, C.N., Melancon, M.J. and Lech, J.J. 1976. Bioconcentration of xenobiotics in trout bile; A proposed monitoring aid for some waterborne chemicals. *Science* 193: 680–681.

Steffensen, J.F. and Lomholt, J.P. 1992. The secondary circulation. In *Fish Physiology: Cardiovascular Systems*, Vol. 12. Hoar, W.S., Randall, D.J. and Farrell, A.P. (eds), p. 185. Academic Press, New York.

Steffensen, J.F., Lomholt, J.P. and Vogel, W.O.P. 1986. *In vivo* observations on a specialized microvasculature, the primary and secondary vessels in fishes. *Acta Zoologica* 67: 1933–2000.

Stegeman, J.J. and Hahn, M.E. 1994. Biochemistry and molecular biology of monooxygenases: Current perspectives on forms, functions and regulation of cytochrome P450 in aquatic species. In *Aquatic Toxicology: Molecular, Biochemical and Cellular Perspectives*. Malins, D.C. and Ostrander, G.K. (eds), pp. 87–206. Lewis Publishers, Boca Raton.

Stevenson, S.V. and Groves, D.J. 1977. The extrinsic innervation of the stomach of the plaice, *Pleuronectes platessa* L-1. The vagal nerve supply. *Comparative Biochemistry and Physiology* 58C: 143–151.

Stragand, J.J. and Hagemann, R.F. 1977. Effect of lumenal contents on colonic cell replacement. *American Journal of Physiology* 233: E208–E211.

Stremmel, W., Lotz, G., Strohmeyer, G. and Berk, P.D. 1985. Identification, isolation and partial characterization of a fatty acid binding protein from rat jejunal microvillus membranes. *Journal of Clinical Investigations* 75: 1068–1076.

Stroband, H.W.J. and Debets, F.M.H. 1978. The ultrastructure and renewal of the intestinal epithelium of the juvenile grasscarp, *Ctenopharyngodon idella* (Val.). *Cell Tissue Research* 187: 181–200.

Stroband, H.W.J. and Kroon, A.G. 1981. The development of the stomach in *Clarias lazera* and the intestinal absorption of protein macromolecules. *Cell Tissue Research* 215: 397–416.

Stroband, H.W.J. and Van der Veen, F.H. 1981. Localization of protein absorption during transport of food in the intestine of the grasscarp *Ctenopharyngodon idella. Journal of Experimental Zoology* 218: 149–156.

Sugawara, I., Kataoda, I., Morishita, Y., Hamada, H., Tsuruo, T., Itlyama, S. and Mori, S. 1988. Tissue distribution of P-glycoprotein encoded by a multidrug-resistant gene as revealed by a monoclonal antibody, MRK. *Cancer Research* 48: 1926–1929.

Suyehiro, Y. 1942. A study on the digestive system and feeding habits of fish. *Japanese Journal of Zoology* 10: 1–303.

Swarup, K. 1959. The morphology and histology of the alimentary tract of *Hilsa ilisha* (Hamilton). *Proceedings of the National Academy of Sciences, India* B29: 109–126.

Takahashi, Y.I. and Underwood, B.A. 1974. Effect of long and medium chain length lipids upon aqueous solubility of α-tocopherol. *Lipids* 9: 855–859.

Takahashi, N., Miranda, C.L., Henderson, M.C., Buhler, D.R., Williams, D.E. and Bailey, G.S. 1995. Inhibition of *in vitro* aflatoxin B_1-DNA binding in rainbow trout by CYP1A inhibitors: α-naphthoflavone, β-naphthoflavone and trout CYP1A peptide antibody. *Comparative Biochemistry and Physiology* 110C: 273–280.

Takahashi, N., Harttig, U., Williams, D.E. and Bailey, G.S. 1996. The model Ah-receptor agonist β-naphthoflavone inhibits aflatoxin B_1-DNA binding *in vivo* in rainbow trout at dietary levels that do not induce CYP1A enzymes. *Carcinogenesis* 17: 79–87.

Tanabe, S., Maruyama, K. and Tatsukawa, R. 1982. Absorption efficiency and biological half-life of individual chlorobiphenyls in carp (*Cyprinus carpio*) orally exposed to Kanechlor products. *Agricultural Biology and Chemistry* 46: 891–898.

Thaker, J., Chhaya, J., Nuzhat, S., Mittal, R., Mansuri, A.P. and Kundu, R. 1996. Effects of chromium (VI) on some ion-dependent ATPases in gills, kidney and intestine of a coastal teleost *Periophthalmus dipes. Toxicology* 112: 237–244.

Thiebaut, F., Tsuruo, T., Hamada, H., Gottesman, M.M., Pastan, I. and Willinglam, M.C. 1987. Cellular localization of the multidrug-resistance gene product p-glycoprotein in normal human tissues. *Proceedings of the National Academy of Sciences of the United States of America* 84: 7735–7738.

Thomann, R.V. 1989. Bioaccumulation model of organic chemical distribution in aquatic food chains. *Environmental Science and Technology* 23: 699–707.

Thomann, R.V. and Connolly, J.P. 1984. Model of PCB in the Lake Michigan lake trout food chain. *Environmental Science and Technology* 18: 65–71.

Thomann, R.V., Connolly, J.P. and Nelson, A.T. 1986. The Great Lakes ecosystem – modelling the fate of PCBs. In *PCBs and the Environment*, Vol. III, pp. 153–180. CRC Press, Boca Raton.

Thomson, A.B.R. and Dietschy, J.M. 1981. Intestinal lipid absorption: major extracellular and intracellular events. In *Physiology of the Gastrointestinal Tract*. Johnson, L.R. (ed.), pp. 1147–1220. Raven Press, New York.

Thomson, A.B.R., Olatunbosun, D. and Valberg, L.S. 1971. Interrelation of intestinal transport system for manganese and iron. *Journal of Laboratory and Clinical Medicine* 78: 642–655.

Thorarensen, H., McLean, E., Donaldson, E.M. and Farrell, A.P. 1991. The blood vasculature of the gastrointestinal tract in chinook, *Oncorhynchus tshawytscha* (Walbaum), and coho, *O. kisutch* (Walbaum), salmon. *Journal of Fish Biology* 38: 525–531.

Thorndyke, M. and Holmgren, S. 1990. Bombesin potentiates the effect of acetycholine on isolated strips of fish stomach. *Regulatory Peptides* 30: 125–135.

Thyagarajan, K., Sand, D.M. Brockman, H.L. and Schlenk, H. 1979. Oxidation of fatty alcohols to acids in the caecum of a gourami (*Trichogaster cosby*). *Biochimica et Biophysica Acta* 575: 318–326.

Tocher, D.R. and Sargent, J.R. 1984. Studies on triacylglycerol, wax ester and sterol ester hydrolases in intestinal caeca of rainbow trout (*Salmo gairdneri*) fed diets rich in triacylglycerols and wax esters. *Comparative Biochemistry and Physiology* 77B: 561–571.

Tong, Z. 1996. Xenobiotic-metabolizing enzymes and the intestinal bioavailability and biotransformation of benzo(a)pyrene phenols and conjugates in the channel catfish, *Ictalurus punctatus*. Ph.D. Thesis. University of Florida, Florida.

Tong, Z., Morris, K.M., James, M.O., Venugopalan, C.S. and Kleinow, K.M. 1995. Bioavailability and biotransformation of [ring-G-3H]-3-hydroxybenzo(a)pyrene (3-OHBAP) in an *in situ* intestinal preparation in uninduced and β-naphthoflavone-induced channel catfish. *The Toxicologist* 15: 38.

Trevisan, P. 1979. Histomorphological and histochemical researches on the digestive tract of the freshwater grass carp, *Ctenopharyngodon idella* (Cypriniformes). *Anatomy Anzeiger* 145: 237–248.

Turner, C. 1936. The absorptive processes in the embryos of *Parabrotula dentiens*, a viviparous deep-sea brotulid fish. *Journal of Morphology* 59: 313–325.

Turner, S.G. and Barrowman, J.A. 1978. Enhanced intestinal lymph formation during fat absorption. The importance of triglyceride hydrolysis. *Quarterly Journal of Experimental Physiology* 63: 255–264.

Tyler, A.V. 1970. Rates of gastric emptying in young cod. *Journal Research Board of Canada* 27: 1177–1189.

Uematsu, K. 1985. Effects of drugs on the responses of the ovary to field and nerve stimulation in a tilapia *Sarotherodon niloticus*. *Bulletin of the Japanese Society of Scientific Fisheries* 51: 47–53.

Uematsu, K. 1986. The autonomic innervation of the ovary of the dab. *Limanda yuokohamae*. *Japanese Journal of Ichthyology* 33: 293–303.

Uematsu, K., Holmgren, S. and Nilsson, S. 1989. Autonomic innervation of the ovary in the Atlantic cod *Gadus Morhua*. *Fish Physiology and Biochemistry* 6: 213–219.

Vahouny, G.V., Roy, T., Gallo, L.L., Story, J.A., Kritchevsky, D. and Cassidy, M.M. 1980. Dietary fibers. III. Effect of chronic intake on cholesterol absorption and metabolism in the rat. *American Journal of Clinical Nutrition* 33: 2182–2191.

Van der Blick, A.M., Kooiman, P., Schneider, C. and Borst, P. 1988. Sequence of MDR 3 cDNA encoding a human P-glycoprotein. *Gene* 71: 410–411.

Van Veld, P.A. 1990. Absorption and metabolism of dietary xenobiotics by the intestine of fish. *Review of Aquatic Science* 2: 185–203.

Van Veld, P.A., Vetter, R.D., Lee, R.F. and Patton, J.S. 1987. Dietary fat inhibits the intestinal metabolism of the carcinogen benzo(a)pyrene in fish. *Journal of Lipid Research* 28: 810–817.

Van Veld, P.A., Stegeman, J.J., Woodin, B.R., Patton, J.S. and Lee, R.F. 1988a. Induction of monooxygenase activity in the intestine of spot (*Leiostomus xanthurus*), a marine teleost, by dietary polycyclic aromatic hydrocarbons. *Drug Metabolic Disposition* 16: 659–665.

Van Veld, P.A., Patton, J.S. and Lee, R.F. 1988b. Effect of preexposure to dietary benzo(a)pyrene (BP) on the first-pass metabolism of BP by the intestine of toadfish (*Opsanus tau*): *in vivo* studies using portal vein-catheterized fish. *Toxicology and Applied Pharmacology* 92: 255–265.

360 Kevin M. Kleinow and Margaret O. James

Van Veld, P.A., Ko, U., Vogelbein, W.K. and Westbrook, D.J. 1991. Glutathione S-transferase in intestine, liver and hepatic lesions of mummichog (*Fundulus heteroclitus*) from a creosote-contaminated environment. *Fish Physiology and Biochemistry* 9: 369–376.

Van Veld, P.A., Vogelbein, W.K., Cochran, M.K., Goksoyr, A. and Stegeman, J.J. 1997. Route-specific cellular expression of cytochrome P450A(CYP1A) in fish (*Fundulus heteroclitus*) following exposure to aqueous and dietary benzo(a)pyrene. *Toxicology and Applied Pharmacology* 142: 348–359.

Venugopalan, C.S., Holmes, E.P., Jarboe, H.H. and Kleinow, K.M. 1994. Non-adrenergic non-cholinergic (NANC) excitatory response of the channel catfish intestine. *Journal of Autonomic Pharmacology* 14: 229–238.

Venugopalan, C.S., Holmes, E.P. and Kleinow, K.M. 1995. Evidence for serotonin involvement in the NANC excitatory neurotransmission in the catfish intestine. *Journal of Autonomic Pharmacology* 15: 37–48.

Verighina, I.A. 1969. Ecologo-morphological peculiarities of the alimentary tract of some Cypriniformes. In *Itogi Nauki, Zoology*, 1968, pp. 79–109. Institut Nauchnoi informatsii Akademii, Nauk SSSR.

Vernier, J.M. and Sire, M.F. 1986. Is the Golgi apparatus the obligatory step for lipoprotein secretion by intestinal cells? *Tissue Cell* 18: 447–460.

Vetter, R.D., Carey, M.C. and Patton, J.S. 1985. Coassimilation of dietary fat and benzo(a)pyrene in the small intestine: an absorption model using the killifish. *Journal of Lipid Research* 26: 428–434.

Vickers, T. 1962. A study of the intestinal epithelium of the goldfish *Carassius arratus*: its normal structure, the dynamics of cell replacement, and the changes induced by salts of cobalt and manganese. *Quarterly Journal of Mircoscopical Science* 103: 93–110.

Vogel, W.O.P. 1985. Systemic vascular anastomoses, primary and secondary vessels in fish, and the phylogeny of lymphatics. In *Cardiovascular Shunts*. Johansen, K. and Bruggen, W.W. (eds), pp. 143–151. Munksgaard, Copenhagen.

Vogel, W.O.P. and Claviez, M. 1981. Vascular specialization in fish, but no evidence for lymphatics. *Zeitschrift Naturforschung* 36C: 490–492.

Vukovic, T. 1966. Duzina crevog trakta sapaca (*Barbus meridionalis petenii* (Heck.) I krukusa (*Gobio gobio* L.) iz razlietiha lokaliteta. *Ribarstvo Yugoslamia* 21: 132–133.

Walker, C.H. 1985. Bioaccumulation in marine food chains – A kinetic approach. *Marine Environmental Research* 17: 297–300.

Walsh, C.T. 1997. Toxicokinetics: Oral exposure and absorption of toxicants. In *Comprehensive Toxicology*. Vol. I. *General Principles*. Sipes, G., McQueen, C.A. and Gandolfi, A.J. (eds), pp. 51–61. Elsevier Science, New York.

Wardle, C.S. 1971. New observations on the lymph system of the plaice *Pleuronectes platessa* and other teleosts. *Journal of the Morphology and Biology Association of the UK* 51: 977–990.

Watanabe, Y. 1981. Ingestion of horseradish peroxidase by the intestinal cells in larvae or juveniles of some teleosts. *Bulletin of the Japanese Society of Scientific Fisheries* 47: 1299–1307.

Watanabe, Y. 1982a. Ultrastructure of epithelial cells of the anteromedian intestine and the rectum in larval and juvenile teleosts. *Bulletin of the Faculty of Fisheries, Hokaido University* 33: 217–228.

Watanabe, Y. 1982b. Intracellular digestion of horseradish peroxidase by the intestinal cells of teleost larval and juveniles. *Bulletin of the Japanese Society of Scientific Fisheries* 48: 37–42.

Watanabe, Y. 1984a. An ultrastructural study of intracellular digestion of horseradish peroxidase by the rectal epithelium cells in larvae of a freshwater cottid fish, *Cottus nozawae*. *Bulletin of the Japanese Society of Scientific Fisheries* 50: 409–416.

Watanabe, Y. 1984b. Morphological and functional changes in rectal epithelium of masu salmon (*Oncorhynchus masou*). *Bulletin of the Japanese Society of Scientific Fisheries* 50: 805–814.

Watanabe, Y. 1984c. Postembryonic development of intestinal epithelium of masu salmon (*Oncorhynchus masou*). *Bulletin of the Tohoku Regional Fish Research* 46: 1–14.

Watkins, J.B. and Klaassen, C.D. 1997. Absorption, enterohepatic circulation and fecal excretion of toxicants. In *Comprehensive Toxicology*. Vol. 9. *Hepatic and Gastrointestinal Toxicology*. McCuskey, R.S. and Earnest, D.L. (eds), pp. 625–637. Elsevier Science, New York.

Weinreb, E.L. and Bilstad, N.M. 1955. Histology of the digestive tract and adjacent structures of the rainbow trout *Salmo gairdneri irideus*. *Copeia* 3: 194–204.

Western J.R.H. 1969. Studies on the diet, feeding mechanism and alimentary tract in two closely related teleosts, the freshwater *Cottus gobio* L. and the marine *Parenophrys bubalis* Euphrasen. *Acta Zoologica* 50: 185–205.

Western J.R.H. 1971. Feeding and digestion in two cottid fishes, the freshwater *Cottus gobio* L. and the marine *Enophrys bubalis* (Euphrasen). *Journal of Fish Biology* 3: 225–246.

White, F.C., Kelly, R., Kemper, S., Schumacker, P.T., Gallagher, K.R. and Laurs, R.M. 1988. Organ blood flow dynamics and metabolism of the albacore tuna *Thunnus alalunga* (Bonnaterre). *Experimental Biology* 47: 161–169.

Whitlock, J.P. 1993. Mechanistic aspects of dioxin action. *Chemistry Research and Toxicology* 6: 754–763.

Williams, D.E. and Buhler, D.R. 1983. Purified form of cytochrome P-450 from rainbow trout with high activity toward conversion of aflatoxin B_1 to aflatoxin B_1–2,3-epoxide. *Cancer Research* 43: 4752–4756.

Windell, J.T., Hubbard, J.D. and Horak, D.L. 1972. Rate of gastric evacuation in rainbow trout fed three pelleted diets. *Progressive Fish Culturist* 34: 156–159.

Windell, J.T., Kitchell, J.F., Norris, D.O., Norris, J.S. and Foltz, J.W. 1976. Temperature and rate of gastric evacuation by rainbow trout (*Salmo gairdneri*). *Transactions of the American Fisheries Society* 6: 712–717.

Windell, J.T., Norris, D.O., Kitchell, J.F. and Norris, J.S. 1969. Digestive response of rainbow trout (*Salmo gairdneri*) to pellet diets. *Journal of the Fisheries Research Board of Canada* 26: 1801–1812.

Winter, M.E. 1988. *Basic Clinical Pharmacokinetics Applied Therapeutics*. Spokane, Washington.

Woodward, D.F., Farag, A.M., Bergman, H.L., DeLonay, A.J., Little, E.E., Smith, C.E. and Barrows, F.T. 1995. Metals-contaminated benthic invertebrates in the Clark Fork River, Montana: effects on age-0 brown trout and rainbow trout. *Canadian Journal of Fisheries and Aquatic Science* 52: 1994–2004.

Wourms, J.P. and Lombardi, J. 1979. Cell ultrastructure and protein absorption in the trophotaenial epithelium, a placental analogue of viviparous fish embryos. *Journal of Cellular Biology* 83: 399.

Wurth, M.A. and Musacchia, X.J. 1964. Renewal of intestinal epithelium in freshwater turtle, *Chrysemis picta*. *Anatomical Record* 148: 427–439.

Yasutake, W.T. and Wales, J.H. 1983. *Microscopic Anatomy of Salmonids: an Atlas*. Research Publication 150. US Fish and Wildlife Service, Washington, DC.

Yeh, G.C., Lopaczynska, J., Poore, C.M. and Phang, J.M. 1992. A new functional role for p-glycoprotein: Efflux pump for benzo(a)pyrene in human breast cancer MCF-7 cells. *Cancer Research* 52: 6692–6695.

Yoshitomi, H., Nishihata, T., Frederick, G., Dillsaver, M. and Higuchi, T. 1987. Effect of triglyceride on the small intestinal absorption of cefoxitin in rats. *Journal of Pharmacy and Pharmacology* 39: 887–891.

Young, J.Z. 1931. On the autonomic nervous system of the teleostean fish *Uranoscopus scaber*. *Quarterly Journal of Microscopical Science* 74: 491–535.

Zambriborsch, F.S. 1953. On the histomorphology of the alimentary canal in mullet (*Mugil*). *Sbornik Biologicheskogo Fakulteta Odesskogo Universiteta* 6: 107–118.

Zimniak, P. und Awasthi, Y.C. 1993. ATP-dependent transport systems for organic anions. *Hepatology* 17: 330–339.

Zitko, V. 1974. Uptake of chlorinated paraffins and PCB from suspended solids and food by juvenile Atlantic salmon. *Bulletin of Environmental Contamination and Toxicology* 12: 406–412.

Index

absorption: carbohydrates, gastrointestinal system 289; cutaneous 168–82; environmental factors, percutaneous 180; intestinal mucosal fold surface 296; investigative focus on percutaneous 193–6; kinetics of percutaneous 168–71; lipid 287–8; mathematical modeling, percutaneous 173–8; morphometric factors, percutaneous 179; nutrients in gastrointestinal system 336–7; physicochemical factors, percutaneous 181–2; physiologic factors, percutaneous 179–80; protein 286–7; rate-limiting factors, percutaneous 179–82; reabsorption, Na$^+$ and Cl$^-$ by proximal tubule 107; reabsorption, organic solutes 104–6; techniques for quantification of percutaneous 171–8; xenobiotics by skin 166–82

accessory cells of the branchial epithelium 11

acetaminophen (AP) 236–7

acetylcholine (ACh) 134–5

acetylcholinesterase-inhibiting insecticides 134–6

acid pH, skin toxicity and 184, 200–1

acid–base regulation of the gills 18–22

adenomas of the liver 252–3

adenosine triphosphatases (ATPases) 136, 334–5

aggregate culture studies of liver 250

aglomerular: fluid secretion 110–15; nephrons 94; species 91–2, 94, 104, 110–15

AHH [benzo(a)pyrene-3-hydroxylase] 305–7, 314

aldrin 130

alkaline: exposure 35–7; phosphatase 337

allyl formate (AF) 236, 237–9

aluminum, toxicology of 37–40

amino acid: homeostasis 234; reabsorption 104–5

amino sugars 159

aminoglycosides 137–8

ammonia: intestinal histopathology, toxicant induced 327; toxic responses of skin 203; toxicology of 58–60

ammonium sulfate 185, 202

anatomy: external of gills 1–3; hepatic, rodent/fish comparison 226; internal of gills 3–5; of liver 225–31; mammalian/fish distinctions 225–6; microanatomy of renal corpuscle 94–6; structure of skin 153–8

anchovies 272

angiotensin-converting enzyme (ACE) 120

anglerfish 93, 94, 107–8, 123

antibiotics, toxic effects of 137–8

apoptosis 127

arginine vasotocin (AVT) 117–20

aromatic hydrocarbon pollution 195

arterial system of kidneys 97

arterial vasculature, gastrointestinal system 279–81

arteriovenous anastomoses (AVAs) 4

atrazine 136–7

atrial natriuretic peptide (ANP) 121–3

autonomic innervation of gastrointestinal system 278

barb 132, 133, 135

barrier function of gastrointestinal system 269

basolateral uptake of liver 243–4

basophilic characteristics of liver 252–3

benzo(a)pyrene 188, 205, 249, 293, 302–3, 323

benzo(a)pyrene/cadmium 323

beta-naphthoflavone (BNF) 293–4, 310, 312–5
bile: biliary epithelial cells 229–30, 253; biliary passageways 246; canalicular organization 244; excretion 224; formation 224, 243; pigments 244–5; stasis 245–6
binding models, gills 30–2
biomagnification 291
biotransformation: liver 235–6; research needed on skin 193–6; skin 163–6; xenobiotics in gastrointestinal system 303–15
bitterling 205
black goby 166
blennies, comtooth 272
blood: gastrointestinal flow 282–3; lamellar flow 4–5; supply to kidneys 97–8
Bowman's capsule 96, 104
branchial epithelium 1, 5–13
branchial toxicity of ammonia 60
bullhead, black 190, 205
bullhead, brown: acid pH stress, toxic responses of the skin 200; biotransformation of benzo(a)pyrene in 249; detection of liver neoplasms in 251; epizootic neoplasms in 188, 190
bullhead, brown, epizootic skin neoplasms in 188; PAH and toxic responses of the skin 204; tumor incidence in 191, 195
bullhead, yellow 165

cadmium: gill exposure to 48–51; intestinal histopathology, toxicant induced 325; toxic effects on kidneys 129–32; toxic responses of skin to 197
cadmium chloride 318, 320, 325
calcium secretion in kidneys 109
carbohydrases, toxicant effect on 333
carbon tetrachloride (CCl₄) 236, 239–41
carcinogenesis studies of the liver 251–4
cardiac stomach 272
carp: cadmium, toxic response of skin to 197; lead, toxic response of the skin to 198; lipid solubility in 180; morphometric factors in 179; organic fertilizer, toxic response of skin to 204; renal alkaline phosphatase activity in 137; stomach of 272, 274; subepidermal capillary network of (amur carp) 158; tubular necrosis in 132; tubular secretion in 110

catecholamines 5, 28
catfish: ammonium sulfate, toxic response of skin to 202; BaP accumulation in 302–3; BaP uptake in 188; biotransformation and skin excretion in the 165–6; blood flow rate to skin in 179; cadmium distribution in 129; chloride cells in collecting tubule of 94; chromium, toxic response of skin to 198; copper, toxic response of skin to 197; cutaneous absorption of organic chemicals in 171; esophagus of 271; fundic stomach of 274; gastric evacuation in 282; intestinal cytosol of 310; lipid content of skin of 180; malachite green, toxic response of skin to 203; mercury exposure in 133, 199; muscularis mucosa lacking in 277; naphthoflavonine exposure in 329; pyloric intestine of 276; salinity stress in 187; skin disposition in 153, 155; thymidine cell labeling in 285–6
ceca, pyloric 274–5, 318–19
celiacomesenteric artery 280–1
cellular biology, lack of fish-specific 254
cellular damage compensation in the kidney 127–8
cellular injury of the kidney 127
cellular systems of the liver 248–51
char 159
chemical mixtures, skin toxicity and 185, 205–6
Chimaeridae 272
chloramine-T 203
chlordecone 130
chloride cells of the gills 9–11, 23–4, 26; *see also* branchial epithelium
chlorinated biphenyl 325
chlorinated paraffins (CPs) 292
chlorinated sewage 205
chlorine, toxicology of 62–4
chromatoblastoma, dermal 190
chromium 198
chronic toxicity of the liver 251–4
Cirrhina mrigala 201
Cl⁻ and Na⁺: reabsorption by proximal tubule 107; secretion in kidneys 110
clear cell characteristics of liver 252–3
co-culture studies of liver 250
cobalt, toxicology of 51–2
cod: Atlantic 254; Pacific 188
collecting tubule of the kidney 94, 115–17
copper: toxic responses of skin to 197; toxicology of 41–6

cortisol in kidneys 123
croaker 194, 203, 205
cutaneous: absorption/exchange 168–82;
  respiration 162–3, 164
cysteine conjugates in the gastrointestinal
  system 313
cytochrome P4501A (CYP1A) 235–6,
  237, 249–50, 310, 312, 317
cytochromes P450 (various) 304–7
cytopathological studies of the liver 241–
  3

DDT (dichlorodiphenyl-trichloroethane)
  130, 136, 291
dermis 156–8
detergents: skin toxicity and 184, 201–2;
  toxicology of 56–8
diazinon 322
dieldrin 130
dietary lipids 287–8
diethylnitrosamine (DEN) 252, 254, 331
digestive enzymes, toxicant effect on
  331–5
Dimecron 319–20, 322
dimethoate 322
dinitrosopiperazine (DNP) 331
distal nephron 93
distal tubule, functions of 115–17
digestive processes 286–9
DNA (deoxyribonucleic acid) 251–2, 285,
  312
dodecylbenzene sodium sulfonate 201
dose–response paradigm 191–2
duct (collecting) of the kidney 94
dynamic environment of gastrointestinal
  system 269–70

eel, freshwater: American 115; European
  58, 160
electrophysiology of the gastrointestinal
  system 335–6
elimination, xenobiotics from the
  gastrointestinal system 315–17; *see
  also* excretion
Elsan 323
endocytosis of proteins 277
endothelial cell layers of the renal
  corpuscle 95
endothelial cell neoplasms of the liver
  253–4
endosulfan 130
endrin 130
enteric neurotransmitters 281
environment: acidity in the 32–5;

carcinogenesis of the liver and the
  251–4; dynamic, of gastrointestinal
  system 269–70; percutaneous
  absorption and the 180–2
enzymes: angiotensin-converting 120;
  histochemistry of the liver 252;
  hydroxysteroid dehydrogenase 165;
  hydrolytic 307–8; ion transports 334–
  5; oxidative 304–7; serum analysis
  254–5; UDP-glucose dehydrogenase
  165
eosinophilic characteristics of the liver
  252–3
epidermal papillomas (EP) 188, 190, 192
epidermis 153–5
epithelial cell proliferation 285
epithelial cell renewal 284–6
epithelial rupture 23
epizootic skin neoplasms 188–91
epoxide hydrolase 307–8
EROD (ethoxyresorufin-*O*-deethylase)
  249, 305–7
esophagus 270–1, 318–19
euryhaline species 11, 90–1, 109, 114–15
excretion: ammonia at the gills 16–18;
  bile 224; carbon dioxide at the gills
  15–16; gastrointestinal system and
  269; nitrogenous waste 13–18; skin
  and 163–6; *see also* elimination
exogenous glucose load of the liver 232–3

fatty acid synthesis 234
Fick's law 168–9, 180
flatfish, Pacific rim 188
flounder: *Pleuronectes americanus* 111,
  114, 133, 244; kidney configuration in
  93; seawater-adapted 109; southern
  115, 290; starry (*Platichthys stellatus*)
  124
fluid secretion in the kidneys 114–15
food deprivation 233
freshwater drum 190
fugacity concept of dietary chemical
  absorption 299
function: control in kidneys 117–24; of
  gastrointestinal system 269; of kidneys
  (freshwater) 98; of kidneys (seawater)
  98–100; of skin 158–66

gadiforms 278
gastric evacuation rates 294
gastrointestinal system: alkaline
  phosphatase 337; arterial vasculature
  279–81; autonomic innervation 278;

barrier function of 269; biotransformation of xenobiotics 303–15; blood flow 282–3; carbohydrases, toxicant effect on 333; carbohydrate absorption 289; cysteine conjugates 313

gastrointestinal system degenerative change through toxicant exposure 328–31; digestive enzymes, toxicant effect on 331–5; digestive processes 286–9; dynamic environment of 269–70; electrophysiology 335–6; elimination of xenobiotics 315–17; enteric neurotransmitters 281; epithelial cell renewal 284–6; epoxide hydrolase 307–8; esophagus 270–1, 318–19; function of 269; glucuronides 308; glutathione-S-transferase (GST) 309–10; gut motility 281–2; histopathology 317–31; hydrolytic enzymes 307–8; innervation 278; interface setting of 269–70; intestinal mucosa, reactions in 304–13; intestine 275–7, 289–90, 322–7; ion transport enzymes, toxicant effect on 334–5; lipases, toxicant effect on 333–4; lipid absorption 287–8; metabolism, integrated 313–15; mucus 284; neoplasia 331; nuclear change through toxicant exposure 330–1; nutrient absorption/transport 336–7; observations *in vivo* 338–9; operative multiplicity of toxicants 339; osmoregulation 289–90; oxidative enzymes 304–7; P-glycoprotein transporter 316–17; pH characteristics 283–4; proteases, toxicant effect on 332–3; protein absorption 286–7; pyloric ceca 274–5, 318–19; radiation toxicity 337–8; rectum 277, 318–19; shortcomings in understanding of 339–40; stomach 272–4, 320–1; sulfatase 308; sulfotransferase (ST) 312–13; toxicant effects on 270; tracts, representative diagrams of 273; trophotaenial placenta 277–8; UDP-glucuronosyltransferase (UGT) 311–12; ultrastructural change through toxicant exposure 330; vasculature 278–81; venous vasculature 279–81; xenobiotic absorption 294–301; xenobiotic bioavailability 290–4; xenobiotic distribution 301–3

general skin stress response syndrome (GSSRS) 186–7

gentamicin 138

gills: accessory cells of the branchial epithelium 11; acid–base regulation 18–22; aluminum and the 37–40; ammonia and the 58–60; binding models 30–2; blood-to-water diffusion 14–15; branchial epithelium 5–13; cadmium and the 48–51; chloride cells 9–11, 23–4, 26; chlorine and the 62–4; cobalt and the 51–2; copper and the 41–6; detergents and the 56–8; epithelial rupture 23; external anatomy 1–3; internal anatomy 3–5; ionic exchange 18–22; lamellae, aneurisms of the 26; lamellae, blood flow and the 4–5; lamellae, blood sinus constriction/dilation of the 26; lamellae, disposition of the 3–4; lamellae, fusion of neighboring 23; leukocyte infiltration 24, 26; marginal channels, vascular congestion of the 26; mercury and the 52–6; mucus, description of cells 12; mucus, excess secretion of 26; mucus, proliferation of cells 26; multifunctional nature of 1; necrosis 23–4; neuroepithelial cells 13; nitrite and the 61–2; nitrogenous waste excretion 13–18; pavement cells, description of 6–8; pavement cells, hypertrophy of 25–6; pavement cells, lifting of 23; pavement cells, proliferation of 26; pH, high 35–7; pH, low 32–5; physiologic responses to toxicants 28–9; respiratory gas exchange 13–18; silver and the 40–1; structural responses to toxicants 22–8; surface active toxicity 30–2; toxic responses of the 1–65; water hardness, protective effects of 29–30; zinc and the 46–8

glomerular: basement membrane 96; filtration barriers 97; filtration rate (GFR) 100–3, 115, 118, 120–2, 124, 127; fluid secretion 110–15; teleosts 92, 110–15, 117; urine formation 100–3

glucose: amino acids in kidneys 104–5; regulation in liver 232–4

glucuronides 308

glutathione (GSH) 128, 129, 134, 238–9

glutathione-S-transferase (GST) 309–10

glycogen levels in liver 233–4

glycogenesis 234

glycoproteins 244

goblet cells *see* mucous cells

goldfish 138, 170, 205
Golgi apparatus 242
goodeid fish 278
gross structure of kidneys 91
guppy 170, 171, 331
gut motility 281–2

halogenated hydrocarbons 325
Henderson–Hasselbalch equation 18–19, 180
hepatic anatomical features, rodent/fish comparison 226
hepatic deterioration 225
hepatic lesion generation 224–5
hepatic lipid storage 234
hepatic neoplasms 251–4
hepatobiliary system 243–7
hepatocellular adaptation 241–3
hepatocellular carcinomas 253
hepatocytes: arrangement of 227–30; freshly isolated, studies with 248–9; necrosis of 237–41
hepatotoxicants, studies of 236–41
herbicides, toxic effects of 136–7
herring 272
hexachlorobiphenyl (HCB) 299
hexachlorobutadiene, toxic effects of 138–9
hexachloroethane (HCE) 172–3, 176, 177
histopathology of the gastrointestinal system 317–31
homeostasis maintenance of the liver 224
hydrocarbons (halogenated and petroleum) 324, 325
hydrocarbons, aromatic *see* polycyclic aromatic hydrocarbons (PAHs)
hydrolytic enzymes 307–8
hydroxysteroid dehydrogenase (HSD) 165

icefish 111
ictalurid fish 230–1
in vitro: liver model systems 247–51; mercury concentration experiments 133; percutaneous penetration assessment 171–3
in vivo: gastrointestinal absorption studies 293; gastrointestinal system observations 338–9; liver imaging studies 245; percutaneous penetration assessment 171–3
innervation, gastrointestinal 278
inorganics, skin toxicity and 184–5, 202–3
interface, gastrointestinal system 269–70
intermediate segment, kidneys 92–3

intestinal mucosa, reactions in 304–13
intestine 275–7, 289–90, 322–7
investigative focus: liver 254–5; percutaneous absorption 193–6; skin 191–209
ion concentrations/filtration rates 112–13
ion reabsorption mechanisms 105
ion secretory mechanisms 108
ion transport enzymes, toxicant effect on 334–5
ionic exchange in gills 18–22
ionocytes *see* chloride cells
ionoregulatory toxicants *see* mercury
isolated liver cells, toxicologic studies with 248

kidneys: acetylcholinesterase-inhibiting insecticides 134–6; aglomerular fluid secretion 110–15; aglomerular nephrons 94; antibiotics, toxic effects of 137–8; arginine vasotocin (AVT) 117–20; arterial system 97; atrial natriuretic peptide (ANP) 121–3; blood supply 97–8; cadmium, toxic effects of 129–32; calcium secretion 109; cellular injury 127; collecting tubule, functions of 115–17; collecting tubule/duct, description of 94; compensation for cellular damage 127–8; cortisol and the 123; distal nephron 93; distal tubule, functions of 115–17; fluid secretion, freshwater-adapted teleosts 115; fluid secretion, seawater-adapted teleosts 114–15; freshwater function 98; function control 117–24; function (overall) 98–100; glomerular fluid secretion 110–15; glomerular urine formation 100–3; glucose/amino acids 104–5; gross structure 91; herbicides, toxic effects of 136–7; hexachlorobutadiene, toxic effects of 138–9; intermediate segment 92–3; localization of damage 128; macromolecules 104; magnesium secretion 107–9; mercury, toxic effects of 132–4; metals, toxic effects of 128–34; Na^+ and Cl^- secretion 110; neck segment 92; nephrotoxic agents 128–39; nephrotoxic mechanisms 126–8; nomenclature of the nephron 91–4; organic anion secretion 106; organic solutes, reabsorption/secretion of 104–6; organochlorine insecticides 130, 136; pesticides, toxic effects of 134–6; phosphate secretion 110; physiological

differences, freshwater/marine species
124–6; prolactin and the 123–4;
proximal tubule, description of 92;
proximal tubule, reabsorption of Na⁺
and Cl⁻ by the 107; proximal tubule,
role of 104; proximal tubule, secretory
mechanisms in the 107–10; proximal
tubule, transepithelial fluxes of
inorganic ions in the 106–7; renal
corpuscle, description of 92; renal
corpuscle, microanatomy of 94–6;
renal portal system 97–8; renin–
angiotensin system (RAS) 120–1;
repair and regeneration 127–8;
seawater function 98–100; sulfate
secretion 109; tubular fluid secretion,
mechanism for 111–14; ultrafiltrate
modification along nephron 104–10;
urinary bladder 94, 115–17; urine
formation 100–17
killifish 111, 114, 115
Kupffer cells 230–1

lamellae: aneurisms of the 26; blood flow
and the 4–5; blood sinus constriction/
dilation of the 26; disposition of 3, 4;
fusion of neighboring 23
Langmuir absorption isotherm analysis 31
lead nitrate 321, 326
lead, toxic response of skin to 198
lemon sole 236
leukocyte infiltration of the gills 24, 26
linear alkylate sulfonate (LAS) 184, 201,
202
lipases, toxicant effect on 333–4
lipid absorption 287–8
lipid solubility 181
lipid storage 234
lipophilic contaminant movement 298
lipoprotein *see* VLDL
liver: acclimatization to toxic stress 225;
acetaminophen (AP) 236–7; adenomas
252–3; aggregate cultures/co-culture
systems, studies with 250; allyl
formate (AF) 236, 237–9; amino acid
homeostasis 234; anatomy 225–31;
aspects of toxicity 224–5; basolateral
uptake 243–4; basophilic
characteristics 252–3; bile canalicular
organization 244; bile excretion 224;
bile formation 224, 243; bile pigments
244–5; bile stasis 245–6; biliary
epithelial cells 229–30, 253; biliary
passageways, model of 246;
biotransformation 235–6; carbon

tetrachloride (CCl₄) 236, 239–41;
carcinogenesis studies 251–4; cellular
model systems 248–51; chronic
toxicity 251–4; clear cell
characteristics 252–3; comparison,
rodent/fish hepatic anatomies 226;
cytochrome P4501A (CYP1A) 235–6,
237, 249–50; cytopathological studies
241–3; endothelial cell neoplasms
253–4; environmental carcinogenesis
251–4; enzyme characteristics 252;
eosinophilic characteristics 252–3;
exogenous glucose load 232–3; fatty
acid synthesis 234; fish-specific
cellular biology, lack of 254; freshly
isolated hepatocytes, studies with 248–
9; future investigative focus 254–5;
glucose regulation 232–4; glycogen
levels 233–4; hepatic deterioration
225; hepatic lesion generation 224–5;
hepatic neoplasms 251–4; hepatic
tubules, architecture of within 231;
hepatobiliary system 243–7;
hepatocellular adaptation 241–3;
hepatocellular carcinomas 253;
hepatocyte arrangement 227–30;
hepatotoxicants, studies of 236–41;
hierarchy of organization of 247;
homeostasis maintenance function
224; *in vitro* model systems 247–51; *in
vivo* imaging studies 245; integrated
investigative approaches, need for
254–5; isolated liver cells, toxicologic
studies with 248; Kupffer cells 230–1;
lipid storage 234; liver cell lines,
studies with 250–1; liver slices, studies
with 250; mammalian, anatomically
distinct, and 225–6; metabolic role of
231–5; metabolic zonation 226–7;
microvasculature of piscine livers
227–30; monolayer cultures, studies
with 249–50; neoplastic process 252–
4; parenchyma, volume densities in
227; perfusion of the liver 248;
perisinusoidal cells 253–4; phenotype
demonstration 252–3; physiology 231–
5; plasma protein secretion 231–2;
proliferating cell nuclear antigen
(PCNA) 252; serum enzyme analysis
254–5; sinusoids of 230–1; somatic
index (LSI) 232; stereologic analysis
242–3; toxicant exposure 224, 225;
transporters, ATP-dependent 244;
xenobiotic metabolism 224, 235–6
loach 160

localization of kidney damage 128
longjaw mudsucker 123
lumen, intestinal 277
luminal intestinal membrane 297
luminal pH 300
lungfish 272
lysosomes 243

macromolecules of the kidney 104
macrophages 230–1, 243
magnesium secretion in the kidney 107–9
malachite green 203
marginal channels, vascular congestion of the 26
meal size and gastric efficiency 282
medaka, Japanese 170, 171, 176, 253
melanoma, dermal 190
mercuric chloride 318, 321, 326
mercury: toxic effects on kidneys 132–4;
    toxic response of skin to 199;
    toxicology of 53–6
mesangium layer, renal corpuscle 95–6
metabolic integration of the
    gastrointestinal system 313–15
metabolic role of liver 231–5
metabolic zonation of liver 226–7
metallothionein (MT) 131
metals: skin toxicity and 183–4, 197–9;
    toxic effects on kidney 128–34
methyl parathion 323
methylcholanthrene 205
micelle solubility of TCB 295
Michaelis–Menten equation 19–20, 47
microvasculature of piscine livers 227–30
minnow: European 199; fathead 169, 171,
    175, 176, 180; stomach of 272
mirex 130
mitochondrial-rich cells *see* chloride cells
mitochondrial diversity 242
MNNG (methyl-*N'*-nitro-*N*-
    nitrosoguanidine) 331
monolayer cultures, studies with 249–50
monosaccharides 159–60
morphometric factors in percutaneous
    absorption 179
mucosa: esophageal 270–1; gastric 272–3;
    intestinal 275–7, 304–13
mucosal intestinal membrane 297–8, 300
mucus: change through toxicant exposure
    in the GIT 329; excess secretion in
    gills 26; gastrointestinal system and
    284; macromolecular components 161;
    production and the skin 159–62;
    significance of epidermal 161–2

mucous cells in gills 12, 26; *see also*
    branchial epithelium
mullet 133, 135, 237, 272, 284
mummichog 236, 310
muscularis mucosa 277

Na$^+$ and Cl$^-$: reabsorption by the proximal
    tubule 107; secretion in the kidneys
    110
naphthoflavone (β) 324, 329
neck segment of the kidney 92
necrosis in gills 23–4
neoplasia: in feral fish 189; of the
    gastrointestinal system 331; of the
    liver 252–4
nephrogenesis 128
nephron: aglomerular 94; distal 93;
    illustrative diagram 95; ion transport
    in the 125; nomenclature of the 91–4;
    ultrafiltrate modification along 104–10
nephrotoxic agents 128–39
nephrotoxic mechanisms 126–8
neuroepithelial cells (NECs) 13; *see also*
    branchial epithelium
nifurpirinol (NP) 194, 205
nitrite, toxicology of 61–2
nitrogenous waste excretion 13–18
nitrosodimethylamine (DMN) 331
nitrosomorpholine (NM) 331

oral papillomas (OP) 188, 190
Oreochromis mossambicus 180
organic anion secretion 106
organic fertilizer 204
organic solute secretion 104–6
organics, skin toxicity and 185, 203–5
organochlorine insecticides 130, 136
organophosphates (OPs) 134–5
osmoregulation: cortisol and 123;
    differences, freshwater/seawater
    acclimated fish 99; intestine and 289–
    90
oxidative enzymes 304–7
oxolinic acid bioavailablity 293
oxygen tension 239

paddlefish 208
parenchyma, volume densities in 227
parrotfish 272
pavement cells in gills: description of 6–
    8; hypertrophy of 25–6; lifting of 23;
    proliferation of 26; *see also* branchial
    epithelium
pentachloroethane (PCE) 172–3, 176, 177

peptidases, intestinal 332–3
perch 205
percutaneous absorption: environmental factors in 180; investigative focus on 193–6; kinetics of 168–71; mathematical modeling of 173–8; morphometric factors in 179; physicochemical factors in 181–2; physiologic factors in 179–80; rate-limiting factors 179–82
perfusion of the liver 248
pericentral/portal necrosis in liver 239
perisinusoidal cells 253–4
perivenous necrosis 237, 239
peroxisomes 242–3
pesticides, toxic effects of 134–6
petroleum hydrocarbons 324
P-glycoprotein transporter 316–17
pH characteristics: gastrointestinal system 283–4; gills 32–7; skin and environmental levels 180–1, 184
phenotype demonstration in the liver 252–3
phosphate secretion in the kidney 110
physiology: freshwater/marine species, differences in kidneys 124–6; gills, physiologic responses to toxicants 28–9; of the liver 231–5; mercury action and 55; percutaneous absorption, physiologic factors in 179–80
physostome fish 271
pike 274
pinocytosis 301
plaice 133, 159, 185, 206, 207
plasma osmolality 90
plasma protein secretion 231–2
polychlorinated biphenyls (PCBs) 291, 299, 325
polycyclic aromatic hydrocarbons (PAHs): dermal papilloma link 185; glutathione-S-transferase and 309; micelle association 295; neoplasia link studies 188–91; sulfotransferase and 313; toxic responses of skin to 204; metabolites of 292–3
prolactin 123–4
proliferating cell nuclear antigen (PCNA) 252
proteases, toxicant effect on 332–3
protein absorption 286–7
proximal tubule: description of 92; reabsorption of Na^+ and Cl^- by the 107; role of 104; secretory mechanisms in the 107–10;

transepithelial fluxes of inorganic ions in the 106–7
pseudobranchial tumors (PB) 188
pulp mill waste 205
pyloric ceca 274–5, 318–19
pyloric intestine 275–7
pyloric stomach 272

radiation toxicity 185–6, 206–8, 337–8
ray 274
reactive oxygen species (ROS) 129, 134
rectum 277, 318–19
renal corpuscle: description of 92; endothelial cell layer 95; glomerular basement membrane 96; mesangium layer 95–6; microanatomy of 94–6; visceral layer 96
renal portal system 97–8
renin–angiotensin system (RAS) 120–1
repair/regeneration of kidneys 127–8
respiration and skin 162–3
respiratory cells *see* pavement cells
respiratory gas exchange 13–18
Rhine river water 206
Rita rita 201
Rivulus marmoratus 254
roach 129
rock whiting 272

salmon: Atlantic 159, 200; chinook 163; coho 185
salmonids 11, 234
scales, nature of 156
Schmidt–Nielson equation 179
sculpin, long-horn 115
seawater function of kidneys 98–100
secretion: aglomerular fluid 110–15; calcium in kidneys 109; fluid in kidneys 114–15; glomerular fluid 110–15; magnesium in kidneys 107–9; organic anion 106; organic solute 104–6; phosphate in kidneys 110; plasma protein 231–2; proximal tubule and 107–10; sulfate in kidneys 109; tubular fluid 111–14
sensory receptors in skin 166
serotonin 5, 13
serum enzyme analysis 254–5
shark 274
short-chain fatty acids (SCFAs) 289
silver, toxicology of 40–1
skate (*Raja erinacea*) 245, 246
skin: absorption, research needs on 193–6; acid pH, skin toxicity and 184, 200–

1; biotransformation and excretion 163–6; biotransformation, research needs on 193–6; categories of toxicity 182–6; chemical mixtures, skin toxicity and 185, 205–6; color 157; compositional diversity of 151; dermis 156–8; detergents, skin toxicity and 184, 201–2; dose–response paradigm 191–2; environmental factors in percutaneous absorption 180–2; epidermis 153–5; epizootic skin neoplasms 188–91; excretion and biotransformation 163–6; excretion, research needs on 196; function of 158–66; general skin stress response syndrome (GSSRS) 186–7; *in vitro/ vivo* assessment of percutaneous penetration 171–3; inorganics, skin toxicity and 184–5, 202–3; mathematical modeling of percutaneous absorption 173–8; metabolizing enzymes in 163–5; metals, skin toxicity and 183–4, 197–9; morphometric factors in percutaneous absorption 179; mucus production 159–62; organics, skin toxicity and 185, 203–5; percutaneous absorption, kinetics of 168–71; percutaneous absorption, rate-limiting factors 179–82; physiologic factors in percutaneous absorption 179–80; quantification of skin exchange, techniques for 171–8; research needs 191–209; respiration 162–3; sensory receptors 166; structure of 153–8; temperature, skin toxicity and 186, 209; toxic responses of the 182–91; toxic responses, contribution to 152; toxicity tests (lack of specific) 151–2; tumor epizootics, causative agents 192–3; ultraviolet-B radiation, skin toxicity and 185–6, 206–8; xenobiotics, absorption of 166–82
snakeheads 272
SNGFR (single nephron glomerular filtration rate) 100–1, 121
sole, English 249, 251
somatic index of the liver (LSI) 232
spongiosis hepatis (SH) 254
squamous epithelial cells *see* pavement cells
stenohaline species 92
stereologic analysis of the liver 242–3
steroidogenesis 132

stickleback 132, 205
stomach 272–4, 320–1
stone loach 129
sucker: stomach of 272, 274; white 188, 190, 251
sulfadimethoxine (SDM) 301
sulfate of aluminum 39–40
sulfate secretion 109
sulfatase 308
sulfhydryl 128
sulfobromophthalein (BSP) 245
sulfotransferase (ST) 312–13
sumithion 322
superoxide dismutase (SOD) 129, 134
surface active toxicity and gills 30–2
systemic toxicity of ammonia 59

temperature: gastric evacuation rates and 282; percutaneous absorption and 180; skin toxicity and 186, 209; stress through 209
tetrachloro-dibenzodioxin (TCDD) 181
tetrachlorobiphenyl (TCB) 175, 295, 298–9
tetrachloroethane (TCE) 172–3, 176
tetrachloroethylene (PCE) 139
tilapia 138, 284
toadfish 111, 114, 138
toxaphene 130
toxic effects: acetylcholinesterase-inhibiting insecticides 134–6; acid pH and skin 184, 200–1; aluminum 37–40; ammonia 58–60; antibiotics 137–8; cadmium and gills 48–51; cadmium on kidneys 129–32; on carbohydrases 333; chemical mixtures and skin 185, 205–6; chlorine 62–4; cobalt 51–2; copper 41–6; detergents and skin 184, 201–2; detergents and gills 56–8; herbicides 136–7; hexachlorobutadiene 138–9; inorganics and skin 184–5, 202–3; ion transport enzymes 334–5; on lipases 333–4; mercury and gills 52–6; mercury on kidneys 132–4; metals on kidneys 128–34; metals on skin 83–4, 197–9; nephrotoxicity 126–40; nitrites 61–2; operative multiplicity in the gastrointestinal system 339; organics on skin 185, 203–5; organochlorine insecticides 130, 136; pesticides on liver 134–6; physiologic responses in the gills 28–9; radiation 185–6, 206–8, 337–8; silver 40–1; structural

responses in gills 22–8; surface active toxicity and gills 30–2; temperature and skin 186, 209; toxicant exposure of the liver 224, 225; toxicants on the gastrointestinal system 270; ultraviolet-B radiation on skin 185–6, 206–8; zinc 46–8

toxicological studies: aggregate culture studies of liver 250; alkaline exposure 35–7; aluminum 37–40; ammonia 58–60; cadmium 48–51; chlorine 62–4; co-culture studies of liver 250; cobalt 51–2; copper 41–6; cytopathological studies of the liver 241–3; detergents 56–8; environmental acidity 32–5; gill responses to toxicants 22–8; on hepatocytes, freshly isolated 248–9; hepatotoxicants 236–41; isolated liver cells 248; liver cell lines 250–1; liver imaging *in vivo* 245; liver slices 250; liver systems *in vitro* 247–51; mercury 52–6; monolayer cultures 249–50; nitrite 61–2; percutaneous penetration 171–3; silver 40–1; zinc 46–8

transitional fundic stomach 272, 274

transports: ATP-dependent 244; electrogenic of Na$^+$ and Cl$^-$ 21–2; ion enzymes 334–5; ionic/acid–base at the gills 17, 18–22; ions in nephrons 125; nutrients in the gastrointestinal system 336–7; P-glycoprotein (PGP) 316–17

triphenyltin 199

trophotaeniae 278

trophotaenial placenta 277–8

trout, brook 200, 201

trout, brown 123, 160, 231

trout, lake 180, 291

trout, rainbow: absorption of halogenated organic chemicals 292; aluminum precipitation in gill microenvironment 39; arterial blood disturbances/water calcium concentration in 34; arterial blood plasma exposed to high pH 36; blood flow to skin in 179; branchial flux response/silver concentration exposure 41; cadmium (waterborne) exposure 50; cadmium distribution in 129; calcium/zinc, influence of waterborne 45; chloramine-T, toxic responses of skin to 203; chlorine, cyclical exposure to 64; copper exposure (sublethal) 43; cutaneous absorption of organic chemicals 171; gill binding model for cobalt 32;

hepatic tubules (cross-section) 230; hepatotoxicant studies in 236; intestinal histopathology, cadmium induced 325; kidney configuration in 93; lipid content of skin 180; mercury (methyl/organic) exposure 52; oxolinic acid bioavailability 293; Rhine river water, toxic responses of skin to 206; skin disposition in 153–4, 156; temperature stress, toxic responses of skin to 209; UV-B radiation, toxic responses of skin to 207–8; water chloride concentration/plasma nitrite levels 61

tubular fluid secretion, mechanism for 111–14

tumor epizootics, causative agents 192–3

UDP-glucuronosyltransferase (UGT) 311–12

UDP-glucose dehydrogenase 165

ultrafiltrate modification along nephron 104–10

ultraviolet-B (UV-B) radiation, skin toxicity and 185–6, 206–8

urinary bladder: description of 94; functions of 115–17

urine flow rates (UFR) 100–3, 115, 120, 122, 124

urine formation 100–17

vascular congestion, marginal channels 26

vasculature, gastrointestinal system 278–81

venous vasculature, gastrointestinal tract 279–81

visceral layer of renal corpuscle 96

VLDL (very low-density lipoprotein) 242

water hardness, protective effects of 29–30

wax esters 287–8

wrasses 272

xenobiotics: absorption by gastrointestinal system 294–301; absorption by skin 166–82; bioavailability in gastrointestinal system 290–4; distribution in gastrointestinal system 301–3; metabolism in liver 224, 235–6

zebra fish 166, 331

zinc: toxic response of skin to 198; toxicology of 46–8

9 780367 455309